Gallium Nitride (GaN) I

SEMICONDUCTORS
AND SEMIMETALS
Volume 50

Semiconductors and Semimetals

A Treatise

Edited by R. K. Willardson
CONSULTING PHYSICIST
SPOKANE, WASHINGTON

Eicke R. Weber
DEPARTMENT OF MATERIALS SCIENCE
AND MINERAL ENGINEERING
UNIVERSITY OF CALIFORNIA AT
BERKELEY

Gallium Nitride (GaN) I

SEMICONDUCTORS
AND SEMIMETALS

Volume 50

Volume Editors

JACQUES I. PANKOVE

ASTRALUX INC., BOULDER, COLORADO AND
DEPARTMENT OF ELECTRICAL AND COMPUTER ENGINEERING
UNIVERSITY OF COLORADO
BOULDER, COLORADO

THEODORE D. MOUSTAKAS

DEPARTMENT OF ELECTRICAL ENGINEERING
BOSTON UNIVERSITY
BOSTON, MASSACHUSETTS

ACADEMIC PRESS
San Diego   London   Boston
New York   Sydney   Tokyo   Toronto

This book is printed on acid-free paper.

COPYRIGHT © 1998 BY ACADEMIC PRESS

All rights reserved.
NO PART OF THIS PUBLICATION MAY BE REPRODUCED OR TRANSMITTED IN ANY FORM OR BY ANY MEANS, ELECTRONIC OR MECHANICAL, INCLUDING PHOTOCOPY, RECORDING, OR ANY INFORMATION STORAGE AND RETRIEVAL SYSTEM, WITHOUT PERMISSION IN WRITING FROM THE PUBLISHER.

The appearance of the code at the bottom of the first page of a chapter in this book indicates the Publisher's consent that copies of the chapter may be made for personal or internal use of specific clients. This consent is given on the condition, however, that the copier pay the stated per-copy fee through the Copyright Clearance Center, Inc. (222 Rosewood Drive, Danvers, Massachusetts 01923), for copying beyond that permitted by Sections 107 or 108 of the U.S. Copyright Law. This consent does not extend to other kinds of copying, such as copying for general distribution, for advertising or promotional purposes, for creating new collective works, or for resale. Copy fees for pre-1998 chapters are as shown on the title pages; if no fee code appears on the title page, the copy fee is the same as for current chapters. 0080-8784/98 $25.00

ACADEMIC PRESS
525 B Street, Suite 1900, San Diego, CA 92101-4495, USA
1300 Boylston Street, Chestnut Hill, MA 02167, USA
http://www.apnet.com

ACADEMIC PRESS LIMITED
24–28 Oval Road, London NW1 7DX, UK
http://www.hbuk.co.uk/ap/

International Standard Serial Number: 0080-8784
International Standard Book Number: 0-12-752158-5

PRINTED IN THE UNITED STATES OF AMERICA
97 98 99 00 01 IC 9 8 7 6 5 4 3 2 1

Contents

Preface xi
List of Contributors xiii

Chapter 1 Introduction
J. I. Pankove and T. D. Moustakas

    References 9

Chapter 2 Metalorganic Chemical Vapor Deposition (MOCVD) of Group III Nitrides
S. P. DenBaars and S. Keller

    I. Introduction 11
    II. Background of MOCVD Technique 12
    III. Chemistry and Reactor Design 13
        1. MOCVD Reaction Chemistry 13
        2. MOCVD System and Reactor Design Issues 17
        3. MOCVD Systems for Production 20
    IV. MOCVD Growth Issues 20
        1. Substrates 20
        2. GaN Bulk and Doping 26
        3. Growth of AlGaN and AlGaN/GaN Heterostructures 29
        4. Growth of InGaN and InGaN/GaN Heterostructures 29
    V. Conclusions 35
    References 35

Chapter 3 Growth of Group III–A Nitrides by Reactive Sputtering
W. A. Bryden and T. J. Kistenmacher

    I. Introduction 39
    II. The Sputtering Process 40
        1. Inert Gas Diode Sputtering 41

2. Reactive Diode Sputtering 42
   3. Magnetron Sputtering Devices 43
III. Results and Discussion 44
   1. Sputtered AlN Thin Films 44
   2. Sputtered GaN Thin Films 46
   3. Sputtered InN Thin Films 48
   References 51

Chapter 4 Thermochemistry of III–N Semiconductors

N. Newman

   I. Introduction 55
  II. Thermodynamics 58
   1. Synopsis of Formalism 58
   2. Thermochemical Values of Binary Reactants and Products 59
   3. Free Energies of Binary Reactions 67
   4. Phase Equilibria of Ternary $Ga_{1-x}In_xN$, $Ga_{1-x}Al_xN$, and $In_{1-x}Al_xN$ and
      Their Reactions 73
 III. Thermal Stability of GaN, InN, and AlN: The Decomposition Reaction . . . 79
 IV. Thermochemical Analysis of GaN Thin Film Growth 84
   1. Molecular Beam Epitaxy 84
   2. Chemical Vapor Deposition 92
  V. Summary 98
   References 98

Chapter 5 Etching of III Nitrides

S. J. Pearton and R. J. Shul

   I. Introduction 103
  II. Wet Etching 104
 III. Dry Etching 107
 IV. Issues of Etch Selectivity and Damage 118
  V. Summary 123
   References 124

Chapter 6 Indium-based Nitride Compounds

S. M. Bedair

   I. Introduction 127
  II. Indium Surface Segregation 129
 III. Competitive Processes During the Growth of InGaN 131
   1. Indium Nitride 138
   2. Indium Gallium Nitride 141
   3. $Al_yIn_{1-y}N$ 146
   4. AlGaInN 148
 IV. Effect of Hydrogen on the Indium Incorporation in InGaN Epitaxial Films . . 150
   1. The Phase Separation Issue 153
   2. Doping of In-based Nitride Compounds 155

V. InGaN Based Heterostructures	156
VI. Conclusions	164
References	165

Chapter 7 Crystal Structure of Group III Nitrides
A. Trampert, O. Brandt, and K. H. Ploog

I. Introduction	167
II. Crystal Structures	168
1. Polarity of the Structures	172
III. Lattice Constants and Mechanical and Thermal Properties	172
1. Aluminum Nitride	173
2. Gallium Nitride	174
3. Indium Nitride	175
4. Nitrides with Zinc-blende Structure	176
IV. Phase Stability, Phase Transitions and Polytypism	178
V. Real Structures and Imaging	179
1. Defects in Nitrides	181
VI. Epitaxial Growth	184
1. Application to Nitride Epitaxy	185
2. Outlook	190
References	190

Chapter 8 Electronic and Optical Properties of III–V Nitride based Quantum Wells and Superlattices
H. Morkoc, F. Hamdani, and A. Salvador

I. Introduction	193
II. Optical Transitions in Bulk GaN	195
III. Calculation of Confined States	199
IV. Experimental Results	214
1. Band Discontinuity Determination	215
2. Optical Properties of Quantum Well Structures	221
V. AlGaN/GaN Quasi Triangular Quantum Wells	230
1. Theoretical Method	231
2. 2DEG Concentration as a Function of ΔE_{Fi} and Calculation of Δd	234
3. Band Diagrams for Normally on and Quasi Normally off Modulation Doped Field Effect Transistors	237
4. Modulation Doped Field Effect Transistors Utilizing Quasi Triangular Wells	239
VI. InGaN/GaN Quantum Wells	241
1. Light Emitting Diodes	246
VII. InGaN/InGaN Quantum Wells	248
1. Use of $In_xGa_{1-x}N/In_yGa_{1-y}N$ Quantum Wells in Laser Diodes	250
VIII. Conclusions	253
References	254

Chapter 9 Doping in the III-Nitrides
K. Doverspike and J. I. Pankove

I. Introduction	259
II. Undoped GaN	260

III. Common Dopants for III-Nitrides 262
    1. Donors 262
    2. Acceptors 265
IV. Doping of the Alloys 269
    1. InGaN 269
    2. AlGaN 271
V. Doping Techniques 271
    1. Doping During Growth 271
    2. Post-Growth Doping (Ion Implantation) 272
VI. Future Directions 274
    References 275

Chapter 10 High Pressure Studies of Defects and Impurities in Gallium Nitride

T. Suski and P. Perlin

    I. Introduction 279
    II. Pressure Dependence of the Electronic States of Defects 281
    III. Native Versus Impurity Related Donor in Undoped n-GaN 283
    IV. Oxygen and Silicon Impurities in GaN 289
    V. Yellow Luminescence 291
    VI. Mechanism of the Luminescence in GaN/InGaN/AlGaN Quantum Wells . . . 295
    VII. Summary 299
    References 300

Chapter 11 Optical Properties of GaN

B. Monemar

    I. Introduction 305
    II. Fundamental Optical Properties 306
        1. Optical Properties Above the Bandgap Energy 306
        2. The Near Bandgap Region, Exciton Effects 311
        3. Exciton Recombination of Dynamics 334
        4. Near Bandgap Optical Properties at High Carrier (Exciton) Densities . . . 339
    III. Below Bandgap Optical Properties, Refractive Index 342
    IV. Optical Properties of Cubic GaN 345
    V. Defect-Related Optical Properties 348
        1. Bound Excitons in GaN 349
        2. Other Donor- or Acceptor-Related Optical Spectra 355
        3. Optical Spectra Related to Transition Metal Centers in GaN 361
    References 363

Chapter 12 Band Structure of the Group III Nitrides

W. R. L. Lambrecht

    I. Introduction 369
    II. Overview of Calculations 370
        1. Early Semi-Empirical Studies 370
        2. Local Density Functional Calculations 372

      3. Beyond LDA 378
  III. Relations Between Brillouin Zones of Wurtzite and Zinc-blende 379
  IV. Trends in Band Structure 385
  V. Experimental Probes 391
      1. Photoemission 391
      2. UV Optics 394
      3. X-Ray Absorption 397
      4. Other Nitrides 398
  VI. Details Near the Band Edges 399
  VII. Outlook for Future Work 404
      References 405

Chapter 13 Phonons and Phase Transitions in GaN

N. E. Christensen and P. Perlin

  I. Introduction 409
  II. Lattice Stability of GaN 410
      1. Stability of the Wurtzite Phase 410
      2. Determination of the Compressibility of GaN 413
  III. Phonons in GaN 415
      1. Zone-Center Modes at Zero Pressure 415
      2. Pressure-Dependence of the Phonon Frequencies 420
      3. Temperature Dependence of the Phonon Frequencies 421
      4. Phonons as Probe of Internal Stress in the Crystal 423
      5. Two-Phonon Raman Spectra 424
      6. Local Vibrational Modes 425
  IV. Summary and Conclusions 426
      References 427

Chapter 14 Applications of LEDs and LDs

S. Nakamura

  I. Introduction 431
  II. InGaN/AlGaN Double-Heterostructure (DH) LEDs 434
  III. InGaN Single-Quantum-Well (SQW) Structure LEDs 439
  IV. Emission Mechanism of Single-Quantum-Well LEDs 445
  V. InGaN Multi-Quantum-Well (MQW) Structure LDs 448
  VI. Summary 456
      References 456

Chapter 15 Lasers

I. Akasaki and H. Amano

  I. Introduction 459
  II. Basic Structure 462
  III. Critical Layer Thickness 463
  IV. Control of Conductivity 464
  V. Threshold Current Density 466
      1. Carrier Confinement and Optical Confinement 466

    2. Transparency Carrier Density 469
    3. Low Threshold Structure 470
 VI. Summary 470
    References 471

Chapter 16 Nonvolatile Random Access Memories in Wide Bandgap Semiconductors

J. A. Cooper, Jr.

   I. Introduction 473
  II. Dynamic Memories in Gallium Arsenide 475
    1. Basic Storage Capacitor Designs 475
    2. FET-Accessed Memory Cells 477
    3. Bipolar-Accessed Memory Cells 478
 III. Nonvolatile Memories in Silicon Carbide 481
    1. Generation Mechanisms 481
    2. Memory Cell Design 484
    3. Monolithic NVRAM Demonstration Chips 486
 IV. Potential for Nonvolatile Memories in the AlGaN System 486
  V. Conclusions 489
    References 490

INDEX   493
CONTENTS OF VOLUMES IN THIS SERIES   503

Preface

The family of refractory nitrides (InN, GaN, and AlN) is one of the most promising classes of optoelectronic materials. All three binaries and their alloys are direct bandgap semiconductors and their energy gaps cover the spectral region from the red to deep ultraviolet. The recent controllable n- and p-type doping and the fabrication of homojunction and heterojunction devices out of this class of materials led to rapid commercialization of blue-green LEDs. Furthermore, laboratory prototypes of blue lasers, UV-detectors, and high temperature transistors have been reported.

These recent breakthroughs in optical, electronic, and electromechanical devices are anticipated to have a major influence in technologies such as full color displays, true color copying, optical storage, high temperature, power and microwave electronics and potentially electromechanical sensors.

The field attracted many researchers addressing both fundamental aspects as well as device applications. The amount of information in the current literature is growing exponentially.

This book consists of two volumes, addressing issues related to crystal growth and structure, doping, physical properties, and optoelectronic applications. An attempt has been made to address the various topics in a tutorial fashion, so that the book will have a lasting quality and will be useful to both newcomers and experts. All chapters present first a historical review, describe the state of the art, anticipate future development, and present a complete list of references.

We hope that this tutorial approach to the field by seasoned experts will have a positive influence in the training of a new generation of experts.

JACQUES I. PANKOVE
THEODORE D. MOUSTAKAS

List of Contributors

Numbers in parenthesis indicate the pages on which the authors' contribution begins.

ISAMU AKASAKI (459), *Department of Electrical and Electronic Engineering, Meijo University, 1-501 Shiogamaguchi, Tempaku-ku, Nagoya 468, Japan*

HIROSHI AMANO (459), *Department of Electrical and Electronic Engineering, Meijo University, 1-501 Shiogamaguchi, Tempaku-ku, Nagoya 468, Japan*

SALAH M. BEDAIR (127), *Electrical and Computer Engineering, Department EE 232 Daniels Box 7911, North Carolina State University, Raleigh, NC 27695*

OLIVER BRANDT (167), *Paul-Drude-Institut fur Festkorperelektronik, Hausvogteiplatz 5-7, 10117 Berlin, Germany*

W. A. BRYDEN (39), *Applied Physics Laboratory, The Johns Hopkins University, Laurel, MD 20723-6099*

N. E. CHRISTENSEN (409), *Institute of Physics and Astronomy, University of Aarhus, DK-8000 Aarhus, Denmark*

JAMES A. COOPER, JR. (473), *School of Electrical and Computer Engineering, Purdue University, West Lafayette, IN 47907*

STEVEN P. DENBAARS (11), *Materials Engineering Department, University of California at Santa Barbara, Santa Barbara, CA 93106*

K. DOVERSPIKE (259), *CREE Research, 2810 Meridian Parkway, Suite 176, Durham, NC 27713*

FAYCAL HAMDANI (193), *University of Illinois at Urbana-Champaign, Materials Research Laboratory and Coordinated Sciences Laboratory, 104 South Goodwin Avenue, Urbana, IL 61801*

STACIA KELLER (11), *University of California at Santa Barbara, Department of Electrical and Computer Engineering, Santa Barbara, CA 93106*

T. J. KISTENMACHER (39), *Applied Physics Laboratory, The Johns Hopkins University, Laurel MD 20723-6099*

WALTER R. L. LAMBRECHT (369), *Department of Physics, Case Western Reserve University, Cleveland, OH 44106-7079*

B. MONEMAR (305), *Department of Physics and Measurement Technology, Materials Science Division, Linköping University, S-581 83 Linköping, Sweden*

HADIS MORKOC (193), *University of Illinois at Urbana-Champaign, Materials Research Laboratory and Coordinated Sciences Laboratory, 104 South Goodwin Avenue, Urbana, IL 61801*

T. D. MOUSTAKAS (1), *Photonics Center, Department of Electrical Engineering, Boston University, Boston, MA 02215*

SHUJI NAKAMURA (431), *R & D Department, Nichia Chemical Industries, Ltd., 491 Oka, Kaminaka, Anan, Tokushima 774, Japan*

NATE NEWMAN (55), *Center for Quantum Devices, Department of Electrical and Computer Engineering, Northwestern University, Evanston, IL 60208*

JACQUES PANKOVE (1, 259), *Astralux Inc., 2500 Central Avenue, Boulder, CO 80301-2845 and Department of Electrical and Computer Engineering, University of Colorado, Boulder, CO 80309-0425*

STEPHEN J. PEARTON (103), *Department of Materials Science and Engineering, MS&E Rhines Hall, University of Florida, Gainesville, FL 32611-6400*

P. PERLIN (279, 409), *Center for High Technology Materials, University of New Mexico, Albuquerque, NM 87131 and UNIPRESS, High Pressure Research Center, Polish Academy of Sciences, ul. Sokolowska 29, 01-142 Warszawa, Poland*

KLAUS H. PLOOG (167), *Paul-Drude-Institut fur Festkorperelektronik, Hausvogteiplatz 5-7, 10117 Berlin, Germany*

ARNEL SALVADOR (193), *University of Illinois at Urbana-Champaign, Materials Research Laboratory and Coordinated Sciences Laboratory, 104 South Goodwin Avenue, Urbana, IL 61801*

RANDY J. SHUL (103), *Sandia National Laboratories, Albuquerque, NM 87185*

T. SUSKI (279), *UNIPRESS, High Pressure Research Center, Polish Academy of Sciences, ul. Sokolowska 29, 01-142 Warszawa, Poland and Lawrence Berkeley Laboratory, Building 66 MS 200, 1 Cyclotron Road, Berkeley, CA 94720*

ACHIM TRAMPERT (167), *Paul-Drude-Institut fur Festkorperelektronik, Hausvogteiplatz 5-7, 10117 Berlin, Germany*

Gallium Nitride (GaN) I

SEMICONDUCTORS
AND SEMIMETALS
Volume 50

CHAPTER 1

Introduction: A Historical Survey of Research on Gallium Nitride

J. I. Pankove

ASTRALUX, INC
BOULDER, CO
AND UNIVERSITY OF COLORADO AT BOULDER
BOULDER, CO

T. D. Moustakas

PHOTONICS CENTER
DEPARTMENT OF ELECTRICAL & SYSTEMS ENGINEERING
BOSTON UNIVERSITY
BOSTON, MA

This chapter will start with a brief history of research on gallium nitride. There will be a short summary of each chapter, sometimes with additional relevant information.

Not all chapters were received in time to fit the publisher's schedule, so rather than hold up publication, it was decided that late chapters would be inserted into a second volume.

Gallium nitride (GaN) was synthesized by Juza and Hahn (1940) who passed ammonia over hot gallium. This method produced small needles and platelets. Their purpose was to study the crystal structure and lattice constant of GaN as part of a systematic study of many compounds. Grimmeiss and Koelmans (1959) used the same technique to produce small crystals of GaN for the purpose of measuring their photoluminescence (PL) spectra. Maruska and Tietjen (1969) used a chemical vapor deposition technique to make a large area layer of GaN on sapphire. All GaN made at that time was very conducting n-type even when not deliberately doped. Donors were believed to be nitrogen vacancies. Later this model was questioned and oxygen was proposed as the donor (Seifert *et al.*, 1983). Oxygen with its six valence electrons on a N site (N has five valence electrons) would be a single donor.

The accomplishment of Maruska and Tietjen led to a flurry of activity in many laboratories especially when Zn doping produced the first blue light

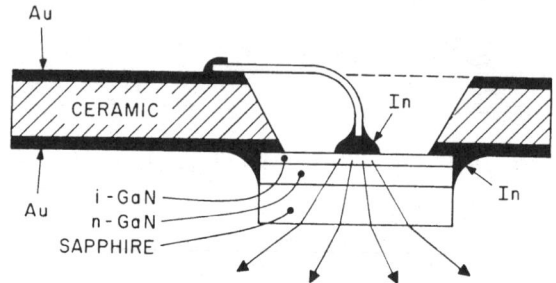

Fig. 1. Structure of GaN M-i-n LED.

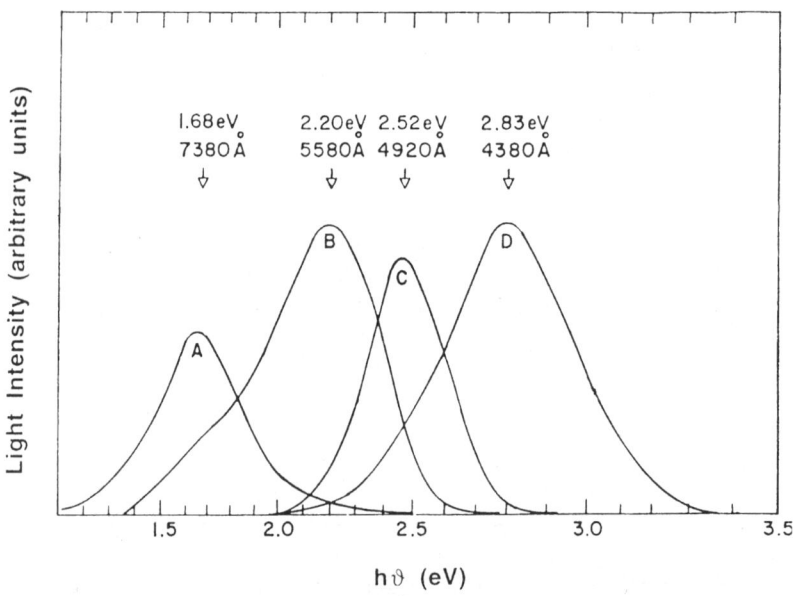

Fig. 2. Emission spectra from GaN M-i-n LEDs using Zn doped i-layers.

emitting device (LED) (Pankove, Miller, and Berkeyheiser, 1972a). This was an M-i-n type of device (M:metal) (Fig. 1) that could emit blue, green, yellow, or red light depending on the Zn concentration in the light emitting region (Fig. 2) (Pankove, 1973). Note that light is emitted only from the cathode. If the Zn concentration is different at the two edges of the Zn compensated region, reversing the polarity of the bias (making the opposite interface of the i-layer the cathode) could cause a change in color, that is, the device could switch from blue to green or to yellow. Maruska *et al.*

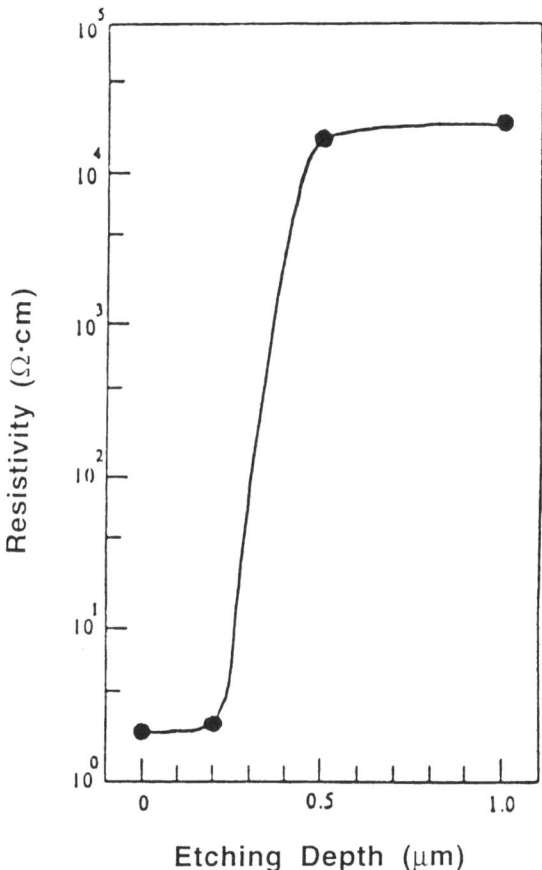

Fig. 3. Resistivity change of a Mg doped GaN film as a function of etching depth from the surface. (From Amano et al. (1990.)

(1973) was also the first to use Mg as a luminescent center in a M-i-n diode emitting violet light.

Other discoveries made with the new single crystal were: antistokes LEDs (2.8 eV photons emitted with only 1.5 V applied) (Pankove, 1975), negative electron affinity (Pankove and Schade, 1974), surface acoustic wave generation (Duffy et al., 1973), and solar-blind UV photovoltaic detectors (Pankove and McIntyre, 1971). But conducting p-type GaN was still too elusive to launch a massive effort on devices. It was the perseverance of Dr. Isamu Akasaki that eventually paid off in the pursuit of conducting p-type GaN. Actually, this was an accidental discovery: Drs. Akasaki and Amano were observing cathodoluminescence of GaN:Mg in a scanning electron microscope (SEM) and noticed that brightness increased with further raster

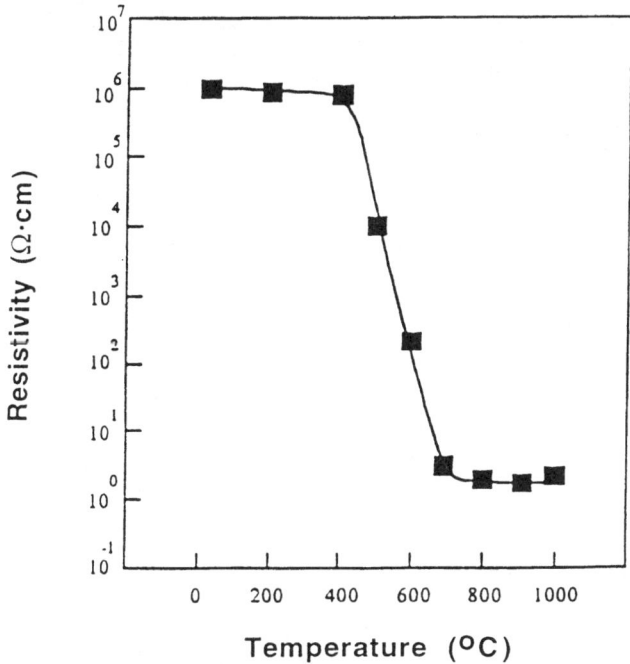

Fig. 4. Resistivity of p-type GaN:Mg after annealing at indicated temperatures. (From Nakamura et al., 1992.)

scanning. A PL study of the sample before and after the low energy electron beam irradiation (LEEBI) treatment showed that by the time luminescence was saturated, luminescence efficiency had increased by two orders of magnitude (Akasaki et al., 1988). A Hall effect measurement showed that the layer had become p-type and conducting. As shown in Fig. 3, etching away layers of LEEBI treated GaN:Mg revealed that high conductivity extended only 0.3 μm deep (i.e., the penetration depth of the electron beam). This surprising phenomenon of beam-induced type conversion was explained by van Vechten, Zook, and Horning (1992) who proposed that the shallow acceptor level of Mg was compensated by a hydrogen atom complexing with the Mg acceptor (just as H complexes with acceptors in Si (Pankove, Zanzucchi, and Magee, 1985)). The energy of the electron beam releases hydrogen atom from this complex that becomes a shallow acceptor approximately 0.16 eV above the valence band (Akasaki et al., 1991). Nakamura et al. (1992) found that annealing GaN:Mg above 750°C in N_2 or vacuum also converted the material to conducting p-type (Fig. 4). However, annealing in NH_3 reintroduced atomic hydrogen and made GaN:Mg insulating again (Nakamura et al., 1992). All this recent work led to the brightest visible

LEDs available today, epecially in the blue part of the spectrum (Nakamura et al., 1995).

A burning question in everyone's mind has been "when shall we see a UV injection laser?". Such a device is the much hoped for solution to the high information density compact disc vision since areal packing density is inversely proportional to the square of wavelength. A 360 nm laser would allow a factor of five increase in information storage in compact discs. Although optically pumped stimulated emission in GaN had been demonstrated by Dingle et al. (1971), the electrically pumped version has remained elusive. Making a p-n junction is necessary but not sufficient. The material must be of exceptionally high quality. Nichia bright LEDs have an enormous concentration (10^{11} cm^{-2}) of defects (Lester et al., 1995). Furthermore these LEDs have an extremely large concentration of impurities: the p-type region has 100 times more Mg than holes, the luminescent region is an alloy of InN and GaN with undoubtedly locally varying composition. The active region is loaded with Zn that is the luminescent center, and the donor in undoped (and doped) regions is N vacancies or O atoms. Evidence for high defect concentration appears in electron microscopy studies of Lester et al. (1995) and in photoconductivity spectra of Qiu et al. (1995a). The photo-

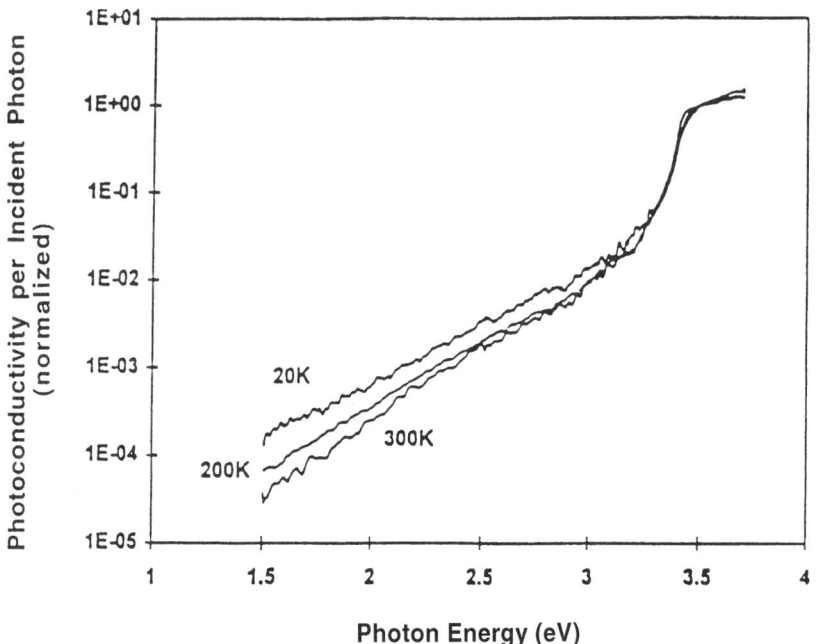

FIG. 5. Photoconductivity spectra of undoped GaN at various temperatures.

conductivity spectrum reveals the presence of a high density of states in the bandgap of GaN. Unlike GaAs that has an abrupt Urbach edge, GaN exhibits an extensive absorption tail (Fig. 5). In order to obtain a low threshold injection laser with GaN, one must eliminate absorption losses due to absorption at the lasing wavelength. Another factor affecting threshold current is the width of the emitted spectrum because threshold current is proportional to spectral width. The narrowest emission spectrum is that due to exciton recombination, hence the most efficient lasing should be due to stimulation of exciton recombination. However, excitons are destroyed by local fields in heavily perturbed semiconductors. Quantum wells (QWs) of InN and GaN barriers are suitable for a wide range of lasing wavelengths tunable by the width of the wells. Thin strained lattice QWs are desirable to overcome the lattice mismatch between InN and GaN. Doping the wells is undesirable because doping leads to level broadening and increased subbandgap absorption. A structure with multi-quantum wells (MQWs) spaced half a wavelength apart, $\lambda/2n$ (λ = wavelength in free space, n = refractive index of GaN at λ) had been proposed earlier (Pankove, 1992). Such a structure (Fig. 6) forms a surface emitting laser and provides coherence by distributed feedback rather than resonance in a Fabry–Perrot cavity.

A working GaN injection laser was revealed by Nakamura et al. (1995). It was a p-i-n junction with MQWs of GaInN in the i-region. On either side of the i-region a GaAlN barrier blocks the escape of injected carriers and provides a step in the refractive index to confine the radiation inside the elongated i-region that forms a waveguide for this edge-emitting Fabry–Perot structure. This early success is most encouraging. However, due to

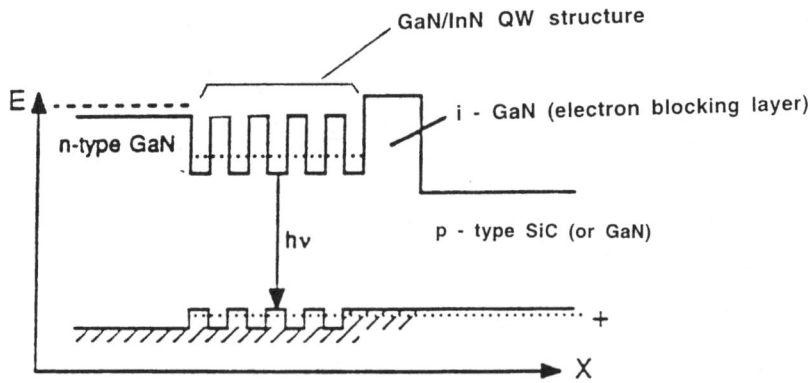

FIG. 6. Short wavelength injection laser with forward bias applied.

absorption losses already mentioned, the threshold current density is high (4 KA/cm^2) and internal resistance in the i-region causes heat dissipation, so that the laser must be driven by low duty cycle microsecond pulses. This is the first result from a device that will gradually improve.

Another point should be made about impurities in GaN. Column II acceptors can substitute for Ga to form single acceptors or substitute for N to form deeper triple acceptors. Since a priori, the acceptor can occupy both sites, a column II element could be a quadruple acceptor with one shallow and three deep levels. Thus, Zn forms levels 0.57 eV, 0.88 eV, 1.2 eV, and 1.72 eV above the valence band (Pankove *et al.*, 1972b). Mg forms levels 0.16 eV, 0.25 eV, 36 eV, and 0.49 eV above the valence band (Amano *et al.*, 1990). Note that although the 0.16 eV "shallow" level is many kTs above the valence band, thus contributing few thermally activated carriers, conductance of GaN:Mg is measurable because fewer carriers are thermally generated across the bandgap of GaN.

Another recent impact of GaN has been in a nonoptoelectronic application. GaN has been used as a heterojunction emitter for a high temperature transistor (to be discussed in Vol. II, Chapter 11 (Pankove *et al.*, 1994; Chang *et al.*, 1995; Pankove *et al.*, 1997). This transistor has operated at 535°C with a current gain of 100. A current density of 1800 A cm^{-2} has been obtained and a power density of 30 kW cm^{-2} has been sustained. A device capable of operating at elevated temperatures is suitable for high power operation. The usual limitation of a power transistor is heating due to various internal losses. This is the reason why Si power transistors are enhanced with air-cooling fins, water cooling or thermoelectric cooling, all expensive additions. A GaN/SiC heterobipolar transistor (HBT) can operate at elevated temperatures due to internal dissipation without cooling means. Hence, this new HBT is a good candidate for high power applications. The outstanding advantage of this HBT is the large difference between bandgaps of GaN and SiC, 0.43 eV is the energy that holes in the base must acquire to escape to the emitter.

Since the three primary colors can be generated in GaN, it is probable that a white light source with adjustable mood coloring will become available commercially. Also GaN is a good candidate for a full color flat panel EL display. Finally, GaN doped with erbium and oxygen is also a candidate for an electrically pumped 1.54 μm laser for optical fiber communication. Er-doped wide bandgap semiconductors are more efficient than narrow gap semiconductors (Favennec, *et al.*, 1989). In the case of GaN:Er, O the hot-electron induced luminescence is as efficient at room temperature as it is at 8°K (Qiu *et al.*, 1995b; Torvik *et al.*, 1996).

The most impressive developments in GaN related devices have been achieved by using metal organic chemical vapor deposition (MOCVD) techniques. Chapter 2 discusses the fundamental aspects of MOCVD and points out some important considerations for obtaining good quality materials: pretreatment with NH_3 to generate many GaN nucleation centers, attention to gas flow profile and boundary layers, and thickness of buffer layer. A special section is devoted to the growth of InGaN that mentions the important problem of inhomogeneous composition that may become useful if spontaneous quantum boxes are formed in what is intended to be a one-dimensional QW.

In Chapter 3 we shall discuss the sputtering technique for depositing layers of AlN, GaN, InN, and their alloys. An inert gas such as Ar is used to sputter the III metal from a target. In reactive sputtering a reactive gas (presumably atomic N from dissociated N_2 or NH_3) is used. Since this technique is not a thermal equilibrium process, it can lead to successful results.

A tutorial presentation of the thermochemistry of III–N semiconductors discussed in Chapter 4 shows the subtleties of growing III nitrides. An important factor is the large kinetic barrier that opposes III–N bond breaking.

The large kinetic barrier to the decomposition of III nitrides hinders the etching of these materials, yet etching is needed to fabricate devices. Although limited success has been obtained by wet etching we shall see in Chapter 5 that dry etching using various plasmas is very successful. Of special interest is Low Energy Electron Enhanced Etching (LE4) developed by Gillis, *et al.* (1996), because it allows anisotropic etching without damaging the remaining crystal.

Chapter 6 is devoted to the synthesis of III–N alloys. Because AlN, GaN, and InN have nonoverlapping N_2 equilibrium vapor pressures, the simultaneous formation of alloys of these compounds presents a formidable challenge.

Chapter 7 examines the crystal structures of all III nitrides and point out how lattice mismatch to the substrate induces many defects. Lack of inversion symmetry and strong ionic component in interatomic bonding cause III nitrides to be piezoelectric. Although the wurzite structure is the most stable, it is possible to produce a zinc-blende structure. Under high pressure, some III nitrides have a phase transition to the rock salt structure.

Quantum wells of InGaN are used in LEDs, laser diodes, and modulation-doped field effect transistors (MODFET). The properties of QWs and superlattices based on III nitrides are discussed in Chapter 8, where it is shown that lattice mismatched materials can be grown as heterostructures pseudomorphically if the elastic strain in the thinnest layer accommodates to the mismatch, thus preventing the formation of dislocations. Other effects such as the spin–orbit interaction and band discontinuities at the interface of two materials have important consequences such as carrier localization that are useful in many applications.

Chapter 9 addresses issues related to doping of the In-Ga-Al-N system. Donor and other defects in GaN using the powerful method of absorption spectroscopy as a function of hydrostatic pressure are discussed in Chapter 10, where it is found that the IR absorption due to free carriers in n-type GaN disappears at high pressures. This is interpreted as a conversion of shallow donors into deep localized states, while the conduction band edge moves to higher energies. Yellow luminescence was tentatively attributed to radiative transitions from the conduction band edge or donors to a deep state (perhaps an acceptor). Here again, pressure experiments showed a blue shift that agrees with the model that the initial state is either the conduction band edge or a shallow donor. As to the final state, a Ga vacancy is proposed. Comparision between intensity of yellow luminescence and Ga vacancy concentration (determined by positron annhilation) confirms the involvement of Ga vacancies. The nature of PL and electroluminescence (EL) in InGaN QWs was tested in a pressure cell. Low pressure coefficients were found indicating that localized states are involved rather than excitonic transitions. The optical properties of GaN are discussed in Chapter 11. It presents experimental and theoretical results on the fundamental absorption edge, exciton recombination dynamics and defect related transitions. The band structure of group III Nitrides is presented in Chapter 12. It presents a review of the semi-empirical and local density functional calculations, compares the Brillouin zones of Wurtzite and zinc-blende structures and discusses the experimental probes. Phonons and phase transitions phenomena are discussed in Chapter 13.

Recent progress in LEDs and laser diodes is discussed in Chapters 14 and 15. Memory devices are based on charge storage. Long term storage leads to nonvolatile or archival memories. In all semiconductors, the charge carrier lifetime depends on thermal activation of the charge out of its trapping level or the thermal activation of the opposite charge with which the stored charge will recombine. Hence, the wider the bandgap of the semiconductor, the less probable the loss of stored charge by recombination. This is why GaN is a prime candidate for application to memory devices. Chapter 16 describes structure and operation of nonvolatile random access memories and compares the figure of merit for various semiconductors. GaN is the most promising material for memory devices that can store data for more than a century at elevated temperatures. With the addition of GaN bipolar transistors, these memory devices could feature the highest read-out efficiency.

For another historical survey over the two decades preceeding 1992, see Strite and Morkoc (1992).

References

Akasaki, I., Kozowa, T., Hiramatsu, K., Sawak, N., Ikeda, K., and Ishii, Y. (1988). *J. Lumin.* **40–41**, 121.
Akasaki, I., Amano, H., Kito, M., and Hiramatsu, K. (1991). *J. Lumin.* **48–49**, 666.
Amano, H., Kitoh, M., Hiramatsu, H., and Akasaki, I. (1990). *J. Electrochem. Soc.* **137**, 1639.
Dingle, R., Shaklee, K. L., Leheny, R. F., and Zetterstrojm, R. B. (1971). *Appl. Phys. Lett.* **19**, 5.
Duffy, M. T., Wang, C. C., O'Clock, G. D., McFarlane, S. H. III, and Zanzucchi, P. J. (1973). *J. Elec. Mat.* **2**, 359.
Favennec, P. N., L'Haridon, H., Salvi, M., Moutonnet, D., Le Guillon, Y. (1989). *Electron. Lett.* **25**, 718.
Gillis, H. P., Choutov, D. A., and Martin, K. P. (1996). *J. Mat.* **48**, 50.
Grimmeiss, H., and H-Koelmans, Z. (1959). *Naturf.* **14a**, 264.
Juza, R., and Hahn, H. (1938). *Anorg. Allgem. Chem.* **234**, 282; (1940) **244**, 133.
Lester, S. D., Ponce, F. A., Craford, M. G., and Steigerwald, D. A. (1995). *Appl. Phys. Lett.* **66**, 1249.
Maruska, H. P., and Tietjen, J. J. (1969). *Appl. Phys. Lett.* **15**, 367.
Maruska, H. P., Stevenson, D. A., and Pankove, J. I. (1973). *Appl. Phys. Lett.* **22**, 303.
Nakamura, S., Iwasa, N., Senoh, M., and Mukai, T. (1992). *Jpn. J. Appl. Phys.* **31**, 1258.
Nakamura, S., Senoh, M., Iwasa, N., and Nagahama, S. (1995). *Jpn. J. Appl. Phys.* **34**, L797.
Nakamura, S., Senoh, M., Nagahama, S., Iwasa, N., Yamada, T., Matsushita, T., Kyoku, H., Sugimoto, Y. (1995). *Jpn. J. Appl. Phys.* **35**, L74.
Pankove, J. I., and McIntyre, R. (1971). Unpublished results.
Pankove, J. I., Miller, E. A., and Berkeyheiser, J. E. (1972a). *J. Lumin.* **5**, 84.
Pankove, J. I., Miller, E. A., and Berkeyheiser, J. E. (1972b). In *Luminescence of Crystals, Molecules, and Solutions* (ed. Ferd Williams), p. 426. Plenum, New York.
Pankove, J. I. (1973). *J. Lumin.* **7**, 114.
Pankove, J. I., and Schade, H. E. P. (1974). *Appl. Phys. Lett.* **25**, 53.
Pankove, J. I. (1975). *Phys. Rev. Lett.* **34**, 809; (1975) *IEEE Trans. Educ.* **22**, 721.
Pankove, J. I., Zanzucchi, P. J., and Magee, C. W. (1985). *Appl. Phys. Lett.* **46**, 421.
Pankove, J. I. (1992). "Compact Blue Green Lasers 1992 Technical Digest", Vol. 6, p. 84, *Opt. Soc. Am.* Sante Fe, New Mexico.
Pankove, J. I., Chang, S. S., Lee, H. C., Molnar, R., Moustakas, T. D., and Van Zeghbroeck, B. (1994). *Proc. IEDM* 389; Chang, S. S., Pankove, J. I., Leksono, M. W., and Van Zeghbroeck, B. (1995). 53rd Annual Device Research Conference, Charlottesville, VA; Pankove, J. I., Leksono, M., Chang, S. S., Walker, C., and Van Zeghbroeck (1997). *MRS-Internet-NSR* **1**, 39.
Qiu, C. H., Hoggart, C., Melton, W., Leksono, M. W., and Pankove, J. I. (1995a). *Appl. Phys. Lett.* **66**, 2712.
Qiu, C. H., Leksono, M. W., Pankove, J. I., Torvik, J. T., Feuerstein, R. T., and Namavar, F. (1995b). *Appl. Phys. Lett.* **66**, 562; Torvik, J. T., Feuerstein, R. J., Pankove, J. I., Qiu, C. H., and Namavar, F. (1996). *Appl. Phys. Lett.* **69**, 2098.
Seifert, W., Franzheld, R., Buttler, E., Sobotta, H., and Riede, V. (1983). *Crystal Res. and Technol.* **18**, 383.
Strite, S. and Morkoc, H. (1992) *J. Vac. Sci. Technolog.* **B10**, 1237.
van Vechten, J. A., Zook, J. D., and Horning, R. D. (1992). *Jpn. J. Appl. Phys.* **31**, 3662.

CHAPTER 2

Metalorganic Chemical Vapor Deposition (MOCVD) of Group III Nitrides

S. P. DenBaars and S. Keller

MATERIALS AND ECE DEPARTMENTS
UNIVERSITY OF CALIFORNIA
SANTA BARBARA, CA

I. INTRODUCTION	11
II. BACKGROUND OF MOCVD TECHNIQUE	12
III. CHEMISTRY AND REACTOR DESIGN	13
1. *MOCVD Reaction Chemistry*	13
2. *MOCVD System and Reactor Design Issues*	17
3. *MOCVD Systems for Production*	20
IV. MOCVD GROWTH ISSUES	20
1. *Substrates*	20
2. *GaN Bulk and Doping*	26
3. *Growth of AlGaN and AlGaN/GaN Heterostructures*	29
4. *Growth of InGaN and InGaN/GaN Heterostructures*	29
V. CONCLUSIONS	35
References	35

I. Introduction

In the past few years, metalorganic chemical vapor deposition (MOCVD) has evolved into a leading technique for production of III–V compound semiconductor optoelectronic devices and electronic devices. For commercial GaN device applications MOCVD has emerged as the leading candidate because of the achievement of super-bright blue light emitting diodes (LEDs) (Nakamura, Mukai, and Senoh, 1994) and the large scale manufacturing potential of the MOCVD technique. The majority of all GaN based p-n junction LEDs typically employ impurity related transition for blue and green emission (Pankove, 1972; Nakamura, Mukai, and Senoh, 1994; Akasaki *et al.*, 1993; Kahn *et al.*, 1995). Recently, direct bandgap emission in the blue-green spectral region has been obtained using high In content in single quantum well (SQW) LEDs and lasers using the two-flow MOCVD technique (Nakamura et al., 1996). Full-color LED displays can now be made entirely with the MOCVD technique when combining the blue and green GaN LEDs with the very high brightness yellow and red emitting LEDs which were demonstrated in the AlGaInP materials system in Kou et al. (1990); Sugawara, Ishikawa, and Hatakoshi (1991). Understanding growth of AlInGaN/GaN based materials by MOCVD is of extreme importance in improving the properties of these optoelectronic devices. In this chapter we will give a descriptive overview of the MOCVD process, a detailed discussion of reactor design issues and key issues in epitaxial growth of GaN by its alloys with Al and In MOCVD.

II. Background of MOCVD Technique

MOCVD is a nonequilibrium growth technique which relies on vapor transport of the precursors and subsequent reactions of group III alkyls and group V hydrides in a heated zone. The MOCVD technique originated from the early research of Manasevit (1968) who demonstrated that triethylgallium (TEGa) and arsine deposited single crystal GaAs pyrolytically in an open tube cold-water reactor. Manasevit and Simpson (1969); Manasevit and Hess (1979) subsequently expanded the use of this technique for growth of $GaAs_{1-y}Py$, $GaAs_{1-y}Sb_y$, and Al containing compounds. Figure 1 illustrates a basic horizontal flow reactor design for the growth of group III nitrides. Composition and growth rate are controlled by precisely controlling mass flow rate and dilution of various components of the gas stream. Organometallic group III sources are either liquids, such as trimethylgallium (TMGa) and trimethylaluminum (TMAl), or solids such as trimethylindium (TMIn). The organometallic sources are stored in bubblers through which a carrier gas (typically hydrogen) flows. The bubbler temperature is to precisely control the vapor pressure over source material. Carrier gas will saturate with vapor from the source and transport vapor to the heated substrate. Group V sources are most commonly gaseous hydrides, for example, ammonia NH_3 for nitride growth. Dopant materials can be metal organic precursors (diethylzinc (DEZn), cyclopenta dienyl magnesium (Cp_2Mg)) or hydrides (silane or disilane Si_2H_6). The substrate usually rests on a block of graphite called a susceptor that can be heated by a radio frequency (RF) coil, resistance heated, or radiantly by a strip heater. An important feature of the MOCVD process is that walls are kept substan-

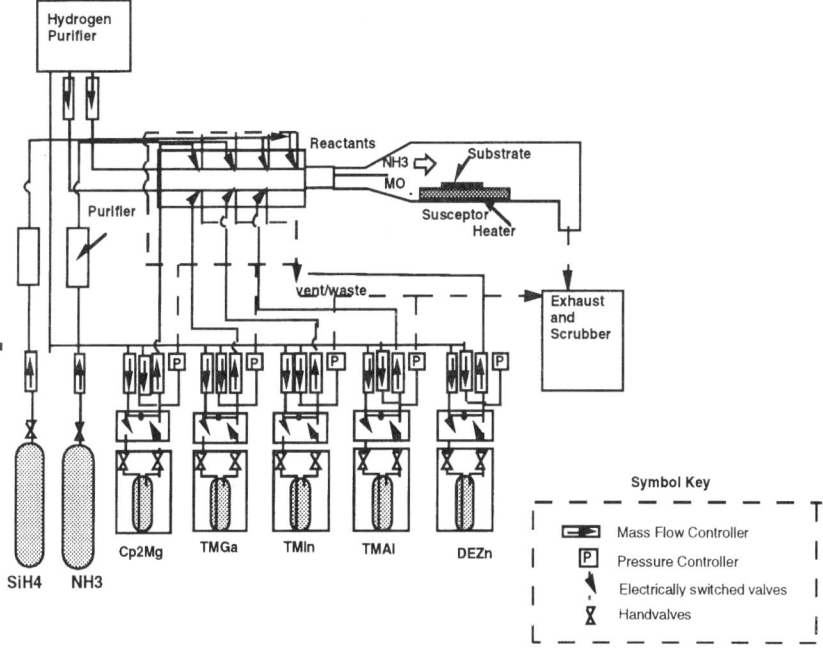

FIG. 1. Schematic of horizontal type GaN MOCVD system.

tially colder than the heated interior substrate which reduces reactant depletion effects that hot walls cause.

III. Chemistry and Reactor Design

1. MOCVD REACTION CHEMISTRY

The basic MOCVD reaction describing the GaN deposition process can be written:

$$Ga(CH_3)3(v) + NH_3(v) \rightarrow GaN(s) + 3CH_4(v)$$

where v = vapor and s = solid.

This balanced expression ignores that the specific reaction path and reactive species are largely unknown. Details of the reaction are not well known and intermediate reactions are thought to be complex. A more likely reaction pathway leading to growth of GaAs epitaxial layers involves homogeneous decomposition of TMGa as reported by Nishizawa, Abe, and Kurabayashi (1985); DenBaars et al. (1986).

$$Ga(CH_3)3(v) \rightarrow Ga(CH_3)2(v) + CH_3(v)$$
$$Ga(CH_3)2(v) \rightarrow GaCH_3(v) + CH_3(v)$$
$$Ga(CH_3)(v) \rightarrow Ga(v) + CH_3(v)$$

The group V hydride source is thought to decompose heterogeneously on the GaN surface or reactor walls to yield atomic nitrogen, or a nitrogen containing radical at high growth temperatures. Abstraction of the first hydrogen bond is thought to be the rate limiting step in decomposition of ammonia.

$$NH_3(s/v) \rightarrow NH(3-x)(s/v) + xH(s/v)$$

Therefore one possible growth mechanism of GaN that might occur at the solid vapor interface could be expressed as follows:

$$GaCH_3(s/v) + NH(s/v) \rightarrow GaN(s) + 1/2H_2$$

However, the level of understanding of growth process is inadequate at best. The most difficult topic, and certainly the least developed, is the area of kinetics of process and growth mechanisms occurring at solid/vapor interface during MOCVD growth. Figure 2 illustrates several possible reaction pathways that may occur in an MOCVD reactor. Pyrolysis and diffusion of group III source through boundary layer is the main pathway controlling growth rate. However, parasitic side reactions such as solid adduct formation between TMAl and ammonia will decrease growth rate by limiting flux of group III sources to the growing interface.

Optimization of MOCVD growth is typically done by empirical studies of external parameters such as growth temperature, V/III ratio, substrate tilt, and mass flow rates. These studies have identified three regions of growth: mass transport limited, desorption, and surface kinetically limited regimes. Conventional GaN MOCVD is usually performed in mass transport limited regime that occurs over a wide temperature range (600–1100°C). In this temperature region growth is limited by mass transport of the column III reactant to the growing interface. Because the diffusion

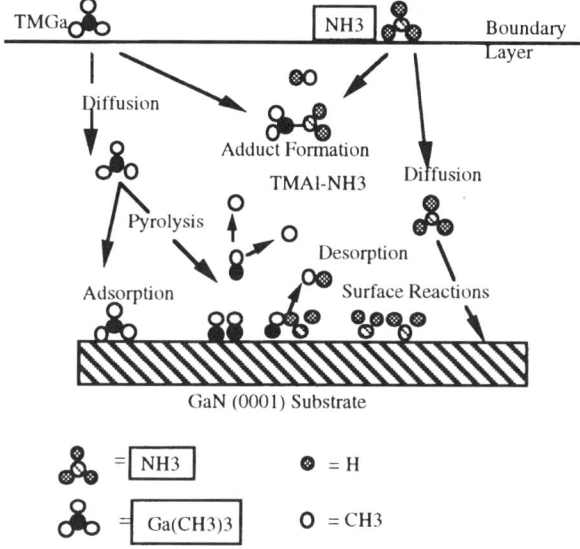

FIG. 2. Possible chemical reactions and rate limiting steps in MOCVD growth of GaN using ammonia and TMGa.

process is slightly temperature dependent, there exists a slight increase in growth rate in this temperature range. From Fick's first law of diffusion we can describe the flux of column III elements toward the substrate and can be written in the mass transport limited regime as:

$$J = D\left(\frac{\partial C}{\partial X}\right)$$

Where D is the growth species' diffusion coefficient and C is concentration in the gas phase.

For flow over a horizontal plate as shown in Fig. 3 the concentration gradient in gallium is established across a boundary layer of thickness $dX = \delta b$. Solving Fick's first law of diffusion and rewritting in terms of input partial pressures yields:

$$J_{Ga} = \frac{D_{Ga}(P^o_{Ga} - P_{Ga})}{\delta_b RT}$$

P_0 is the input gas stream partial pressure and P is partial pressure at the gas–solid interface. For temperatures above 600°C, metal alkyl pyrolysis

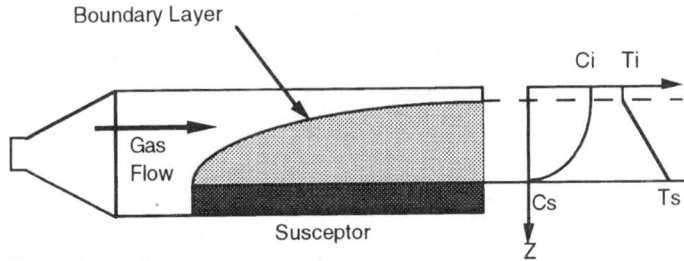

FIG. 3. Schematic illustration of boundary layer in a horizontal MOCVD reactor. Note temperature and concentration profiles in the boundary layer.

efficiency is unity, and one can assume $P^o_{Ga} \approx P_{TMGa}$. At the interface the Ga partial pressure is small in comparison to input concentration of TMGa ($P_{TMGa} \gg P_{Ga}$), so that the column III flux at the interface can be reduced to:

$$J_{Ga} = \frac{D_{Ga} P_{TMGa}}{\delta_b RT}$$

Under common growth conditions for hydride MOCVD the ratio of group V to group III partial pressures is $\gg 1$, because high thermal stability of NH_3 and high active nitrogen are pressure required to stabilize the GaN surface at high growth temperatures. The growth rate can then be calculated by:

$$\text{growth rate} = \frac{J_{Ga} MW_{Ga}}{r(\text{Ga in GaN})}$$

Where MW_{Ga} is the molar weight of Ga and ρ is the GaN density. Under these conditions growth rate will be controlled by diffusion of Ga species toward the surface. From the above equation we can see that control of growth rate is obtained by controlling partial pressure of metal alkys injected into the reactor. Precise control of growth rate is then achieved by use of accurate electronic mass flow controllers that regulate the flow through the metal alkyl bubbler. In addition, a uniform boundary layer δ_b ensures uniform deposition.

For growth of AlGaN alloys the solid composition is found to be proportional to flux ratio of Al and Ga at the growing interface. Since diffusion coefficients of Al and Ga are approximately equal, composition control of $Al_xGa_{1-x}N$ is achieved by linear adjustment of the molar flows of the column III sources. This simple relation of composition in the solid phase to mole fraction in the gas phase is described by the following equation:

$$X_{Al} = \frac{MF_{Al}}{MF_{Al} + MF_{Ga}}$$

This equation illustrates one of the greatest assets of MOCVD growth process which is that the distribution coefficient is essentially unity for the group III element. This makes the growth of (Al, In, Ga)N compounds feasible, which is in contrast to liquid phase epitaxy (LPE) where growth of Al-In compounds is unachievable because of the large liquid distribution coefficient (Yuan et al., 1985). For growth of AlGaN, we can then write a balanced MOCVD expression as:

$$(x)Al(CH_3)_3 + (1-x)Ga(CH_3)_3 + NH_3 \rightarrow Al_xGa_{1-x}N + 3CH_4$$

However, at near atmospheric pressure AlGaN growth, prereactions may occur in the gas phase, which are much more likely than for GaN growth (Thon and Kuech, 1995). This is caused by the stronger tendency for TMAl to prereact with ammonia and form an involatile adduct (Coates et al., 1968). For growth of InGaN the In incorporation into InGaN films follows the above scheme at growth temperatures equal or lower than 500°C (Matsuoka et al., 1992). However, InGaN films deposited at these low temperatures are not of pure quality, and InGaN growth is usually performed at temperatures between 700 and 850°C. In this temperature range, incorporation of In atoms into layers is much less efficient than gallium atoms and is instead determined by thermodynamic effects.

2. MOCVD System and Reactor Design Issues

Because of the high growth temperature (>1000°C) and reactivity of sources involved in GaN epitaxy, MOCVD reactor design needs careful consideration. The most common reactor chamber designs are vertical and horizontal configurations. In vertical design, reactants are injected through the top. Typically, the substrate is held flat on a rotating silicon (SiC) carbide coated graphite susceptor that is perpendicular to the gas flow direction. Heating is accomplished by RF induction, resistance, or infrared lamp heating, and temperature monitoring is accomplished by an infrared pyrometer or a thermocouple. Low gas velocities employed in conventional MOCVD reactor (2–4 cm/sec.) promote laminar flow conditions. The addition of a quartz disc baffle at the gas inlet promotes uniform gas distribution. Reactor walls are substantially colder than heated substrates to minimize predeposition on walls. Cooling of walls can be done by passive air-cooling or actively by a water cooling jacket.

Horizontal reactor designs utilize a susceptor that is situated approximately parallel to the glas flow direction. Minimal convection currents and highly laminar glas flow conditions exist in horizontal reactors (Giling, 1982). Uniform growth can be achieved by rotation or precise tilting of the susceptor to eliminate reactant depletion along the flow direction. Heating in this design can be accomplished by RF induction, resistance, or infrared heating from quartz-halogen lamps. Multiple wafer design typically requires wafer rotation for uniform deposition.

For horizontal flow reactor the thickness of the boundary layer can be described by:

$$\delta = 5\sqrt{\frac{vL}{Vx}}$$

Where v is the kinematic viscosity of the carrier gas, L is the distance along the susceptor in the X-direction, and Vx is the average velocity of the gas stream. This boundary layer profile results in the leading edge of the susceptor having high deposition rates and force the crystal grower to set-back the wafer to avoid getting a highly nonuniform deposition. Angling of the top wall of the reactor and/or rotating the susceptor help improve uniformity to better than $\pm 1.5\%$ across a 2 in. wafer in the horizontal configuration. In planetary rotating horizontal reactors this uniformity has been achieved across as many as 11 wafers in a single run.

Alternatively, vertical high speed rotating reactor boundary layers are controlled by rotation speed of the disc. In so-called stagnation flow vertical geometry, boundary layer can decrease substantially by increasing rotation rate of the susceptor disc to speeds greater than 200 rpm. In these reactors radial flow speed increases as inverse square root of rotation speed decreases and can be written as the following equation

$$\delta = 4\sqrt{\frac{v}{\omega(\text{rpm})}}\,(\text{cm})$$

where v is the kinematic viscosity of the gas stream and ω is the rotation rate in rpm, the constant 4 yield boundary layer thickness in centimeters.

Both atmospheric and low pressure MOCVD reactors are employed by various research and industrial groups in growth of GaN. Atmospheric

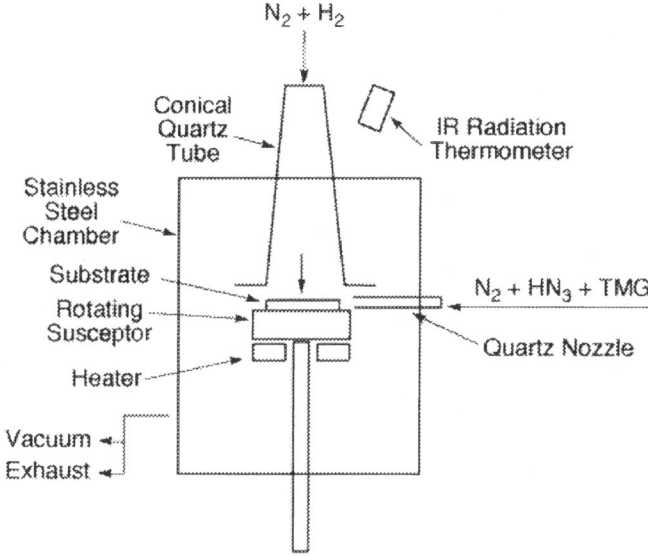

FIG. 4. Two-flow MOCVD approach. (After Nakamura et al., 1993.)

pressure reactors are favored because a high partial pressure of ammonia or nitrogen containing precursor is achievable. The majority of research groups in Japan have utilized atmospheric pressure for this reason. The breakthrough in bright blue LEDs was achieved by Nakamura, Mukai, and Senoh (1994) using a modified MOCVD system. Nichia Chemistries Inc. have employed a novel two-flow approach that yielded excellent film quality. As shown in Fig. 4, the reactor sources are supplied from a horizontal inlet, and a vertical subflow then drives the reactants to the growing film surface. Subflow from the top was found necessary to improve crystal quality and increase growth rate.

MOCVD reactor design for GaN growth must overcome problems presented by high growth temperatures, prereactions, flow and film nonuniformity. Typically in GaN growth very high temperatures are required because of high bond-strength of N–H bond in ammonia precursors. Compounding this fact is the thermodynamic tendency of ammonia to prereact with group III metalorganic compounds to form nonvolatile adducts. These factors contribute to difficulties currently facing researchers in design and scale-up of III–V nitride deposition systems. Further research and development is needed in the scale-up and understanding of the mechanism of GaN growth by MOCVD.

3. MOCVD Systems for Production

Currently, several types of MOCVD reactor geometries are being developed for mass production of GaN based materials and devices. Both atmospheric- and low-pressure systems are being produced by major MOCVD equipment manufacturers (Aixtron GmbH, Emcore Corp., Nippon Sanso, and Thomas Swan Ltd.). Three types of geometries are illustrated in Fig. 5, a high speed rotating disc reactor (RDR) for low pressure MOCVD, a closed space RDR for atmospheric pressure growth, and a two-flow horizontal flow pancake reactor. All three reactor designs are producing high quality GaN materials and it is not the intent of the authors to compare one against the other. The benefits of each approach will be specific to the ultimate device and materials being grown. For high speed vertical reactors from Emcore Corp. the main advantage is the large reactor size available (>6 wafers) and the thin uniform boundary layer caused by high rotation speed (~1500 rpm). Closed space RDR has the benefit of atmospheric pressure operation because low free height eliminates free convection. The two-flow horizontal planetary rotation® reactor from Aixtron can also be operated at near atmospheric pressure and can accommodate large wafer volumes (>7 wafers). Selection of any reactor has to be carefully considered against factors such as material quality, high throughput, reproducibility, maintenance, and source usage.

IV. MOCVD Growth Issues

Several major breakthroughs in MOCVD growth of high quality GaN and InGaN enabled the fabrication of device quality group III nitride materials. In this section we discuss several key growth issues such as (a) substrates and implementation of low temperature nucleation layers to provide a homogeneous nucleation on sapphire substrates (Hiramatsu et al., 1991), (b) achievement of p-type doping in GaN by overcoming hydrogen passivation of acceptors (Amano et al., 1989; Nakamura et al., 1992), (c) growth of GaN, AlGaN, and InGaN with high optical and electrical quality, and (d) growth of GaN/InGaN heterostructures and quantum wells (QWs).

1. Substrates

Several problems in the epitaxial growth of nitrides originate from nonavailability of single crystalline GaN substrates or other high quality

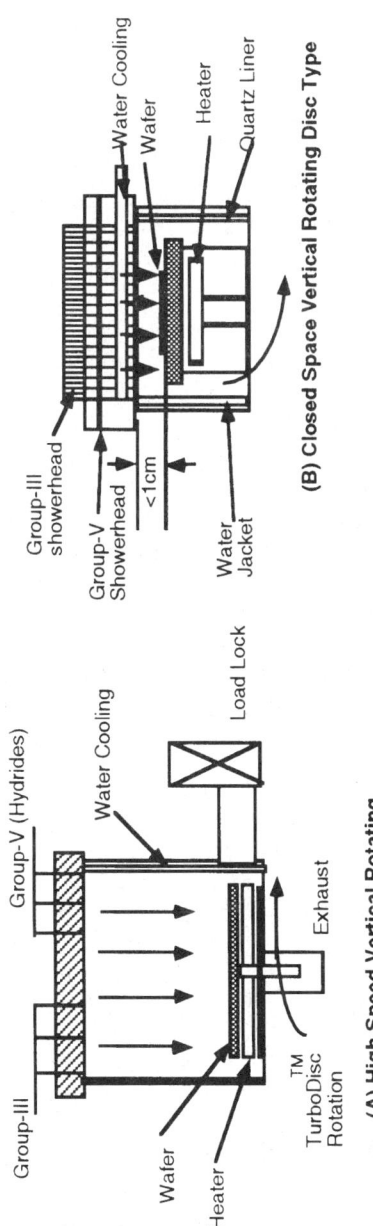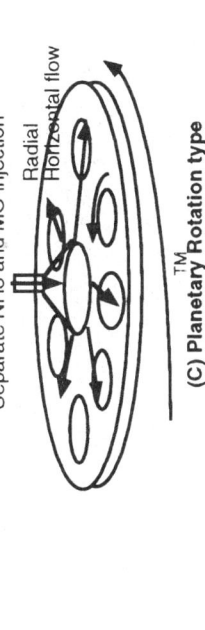

FIG. 5. Three types of MOCVD production reactors for GaN. (a) High speed vertical rotating style, (b) closed space rotating disc type, and (c) planetary rotation with radial horizontal flow.

single crystalline substrates with the same lattice parameters as GaN. For this reason most of the epitaxial growth of nitrides has been performed on sapphire or SiC substrates. In both cases, problems due to lattice mismatch between nitride epi-layer and substrate (16% sapphire and 3.5% SiC) have to be overcome. One of the major breakthroughs in growth of device quality group III nitride material was the implementation of nucleation layers. Using sapphire substrates, thin AlN or GaN nucleation layers deposited at temperatures between 500 and 750°C showed remarkable improvement in the quality of the GaN film growth at temperatures above 1000°C (Akasaki *et al.*, 1989; Nakamura, 1991). In the case of SiC substrates, growth is usually initiated with deposition of a thin AlN nucleation layer at high temperatures (Weeks *et al.*, 1995).

By this means, GaN material of comparable quality on both types of substrates could be achieved. So far most GaN growth has been performed on c-plane sapphire substrates, in the following section just the growth on c-plane sapphire will be discussed.

Additional substrate materials are currently being examined to determine if properties of GaN thin films can be enhanced by improved structural matching. Figure 6 shows that in addition to sapphire several other substrates offer potential much better latticed and thermal matching. To this end, 6H-SiC, ZnO, and 3C-SiC, MgO are alternative substrate materials.

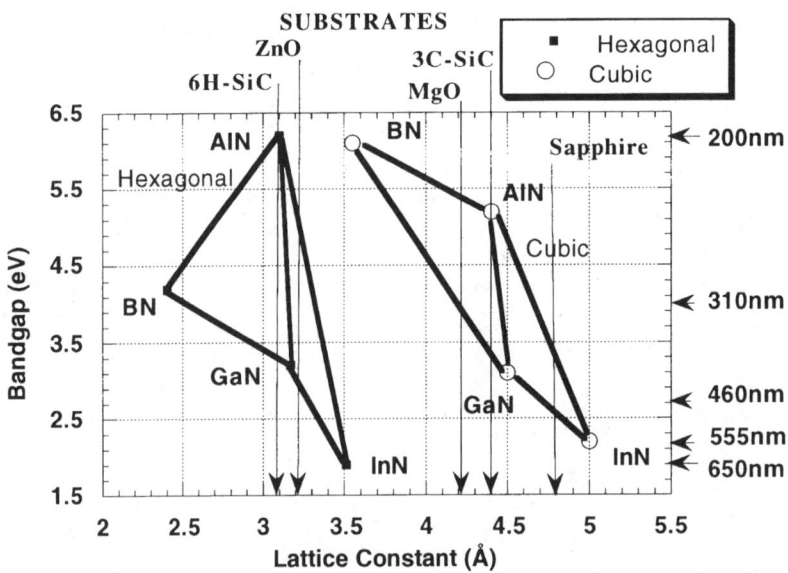

FIG. 6. Bandgap and wavelength of III–V nitrides versus lattice constant.

ZnO has a wurtzite structure with lattice constants of ($a = 3.32$ A, $c = 5.213$ A) and thus offers a better structural match to the equilibrium wurtzite nitride. 3C-SiC and MgO are both cubic zinc-blende structures having better structural and thermal match to nitides than sapphire. 3C-SiC and MgO have cubic lattice constants of $a = 4.36$ A and $a = 4.22$ A, respectively. Although nitrides are most commonly observed as wurtzite (2H) polytype, they can also crystallize in a metastable zinc-blende structure ($a = 4.52$ A) when using nonequilibrium based growth techniques. Identification of a suitable substrate material that is lattice matched and thermally compatible with GaN wurtzite structure ($a = 3.19$ A, $c = 5.185$ A) will alleviate many of the difficulties associated with deposition of device quality material.

a. Growth of GaN on c-plane Sapphire

Previous workers have revealed that thin low temperature nucleation layers composed of AlN or GaN deposited at temperatures between 450 and 600°C will drastically improve the quality of GaN film grown at temperatures above 1000°C (Akasaki et al., 1989; Nakamura, 1991; Molnar, Singh, and Moustakas, 1995). Nucleation layers, comprised from three-dimensional nuclei wetting the sapphire substrate, partially relieve strain arising from different lattice parameters of GaN and sapphire. Nucleation layer growth parameters like growth rate and temperature cause variations in size and density of nuclei and overall nucleation layer roughness. Consequently, electron mobility and photoluminescence (PL) of GaN films are very sensitive to variations of nucleation layer growth conditions (Kuznia et al., 1993; Rowland et al., 1994). These effects are strongly related to the nature and density of lattice defects present in GaN film.

In this section we discuss how the properties of thick GaN films are found to be significantly influenced by duration of exposing sapphire substrate to ammonia prior to the initiation of GaN growth. In particular, different nitridation schemes of sapphire affect dislocation structure of GaN films resulting in a decrease of dislocation density by almost two orders of magnitude for shorter NH_3 preflow times (Heying et al., 1996). The electrical, optical, and structural properties of epitaxial GaN on sapphire substrates are strongly influenced by growth conditions of low temperature GaN or AlN nucleation layer. Nucleation layers, comprised of three dimensional nuclei, may be grown so that the substrate is fully covered. Nucleation layer growth parameters such as temperature, growth rate and thickness cause variations in size and density of nuclei and all nucleation layer roughness (Wu et al., 1996). Consequently, electron mobility and PL of GaN are very sensitive to variations of nucleation layer growth conditions (Kuznia et al., 1993; Rowland et al., 1994; Keller et al., 1996a). These effects are strongly related to the nature and lattice defect density in film. The structural quality of epitaxial GaN films on sapphire as evaluated by X-ray diffraction and transmission electron microscopy (TEM) measurements indicated dislocation densities typically on the order of 10^{10} cm^{-2} with lowest reported dislocation densities of 10^8 cm^{-2} (Kapolnek, 1995).

Besides particular growth conditions of nucleation layer, it has been observed that variations in the length of ammonia exposure of sapphire prior to growth of nucleation layer, have a tremendous effect on properties of GaN films, and GaN films with completely different microstructure and electrical and optical properties may be grown (Keller et al., 1996d). Table I summarizes the results of structural characterization of GaN films. After annealing the c-plane sapphire substrate in H_2 ambient at 1050°C, sample A was exposed to ammonia flow of 3 l/min for 60 sec., sample B for 400 sec., before a 190 Å thick GaN nucleation layer was deposited at 600°C. The temperature was increased to 1080°C for growth of 1.2 μm thick GaN layer for both samples. Epitaxial films were characterized by triple axis high resolution X-ray diffraction (HRXRD) using symmetrical (002) and asymmetrical (102) reflections of GaN.

Table I summarizes the results of structural characterization of films by HRXRD. From symmetric (002) rocking curves we obtained a full width at half maximum (FWHM) as low as 40 arcsec. for sample B, but 269 arcsec. for sample A. However, symmetric reflections are insensitive to pure edge dislocations that are predominantly present in GaN films (Heying et al., 1996). For asymmetric GaN (102) reflections we obtained peak widths of 740 and 413 arcsec. for samples B and A, respectively. Therefore, we conclude that although sample B shows a highly perfect mosaic it contains a much higher number of pure edge dislocations than sample A. Characteristic for sample B is columnar microstructure with dislocation densities

TABLE I

FWHM of Symmetric and Asymmetric XRD GaN Reflections and Dislocation Densities for 1.2 μm Thick Films Grown on Sapphire with Different NH_3 Pretreatments

NH_3 Preflow	Symmetric (002) FWHM/arcsec.	Asymmetric (102) FWHM/arcsec.	Dislocation Density/cm^{-2}
60 sec. (sample B)	269	413	$4 \cdot 10^8$
300 sec. (sample A)	240	740	$1 \cdot 10^{11}$

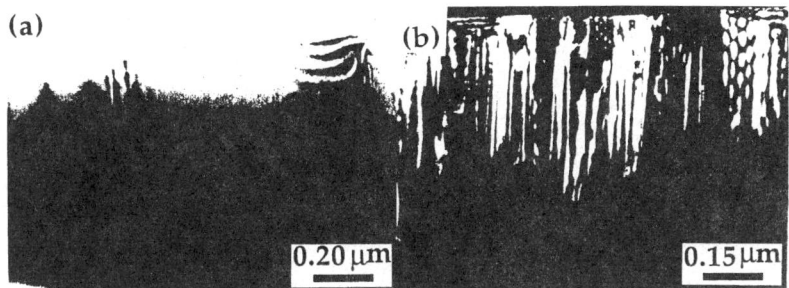

FIG. 7. TEM cross section of type A ($4 \cdot 10^8$ cm^{-2} disl) and type B material ($1 \cdot 10^{10}$ cm^{-2}).

concentrated at subgrain boundaries. From plane view and cross-sectional TEM we obtained dislocation densities of 10^{10} and 10^8 cm^{-2} for samples B and A (Wu et al., 1996), respectively, proving the superior structural quality of GaN grown on only partially nitridized sapphire. Figure 7 shows cross-sectional TEM of two different types of microstructure that illustrate large differences in defect densities.

Results of electrical characterization of both samples correspond to superior film properties for sample A (Fig. 8). At 300 K sample A shows an electron mobility of 644 cm^2 V/sec. At lower temperatures we found a significant increase in mobility with a peak of 1250 cm^2 V/sec. at 160 K. This increase is caused by reduced phonon scattering at low temperatures. The free carrier concentration is $2/1.5 \cdot 10^{17}$ cm^{-3} and $6.8 \cdot 10^{16}$ cm^{-3} at 300 and 160 K, respectively. In contrast to sample A, the 300 K electron mobility of sample B amounted to 149 cm^2 V/sec. at a free carrier concentration of $3.2 \cdot 10^{17}$ cm^{-3}. At lower temperatures mobility increased slightly to 155 cm^2 V/sec. at 230 K. This observation can be explained by a higher scattering probability at lattice defects associated with high dislocation density in sample B. For material grown under the same conditions, increase of free carriers by a factor of approximately two in sample B compared to sample A could arise from enhanced generation of electrically active defects at or near the dislocations. These results reemphasize the importance of reducing dislocation densities for obtaining superior electron transport properties in GaN films. Prenitridation of sapphire by partially replacing excess oxygen with nitrogen at the surface leads to starting conditions for nucleation layer growth that prevent an effective annihilation of misfit dislocations in early stages of growth. Thus, microstructure of GaN overlayer deteriorates causing poor electrical and degraded optical quality of films.

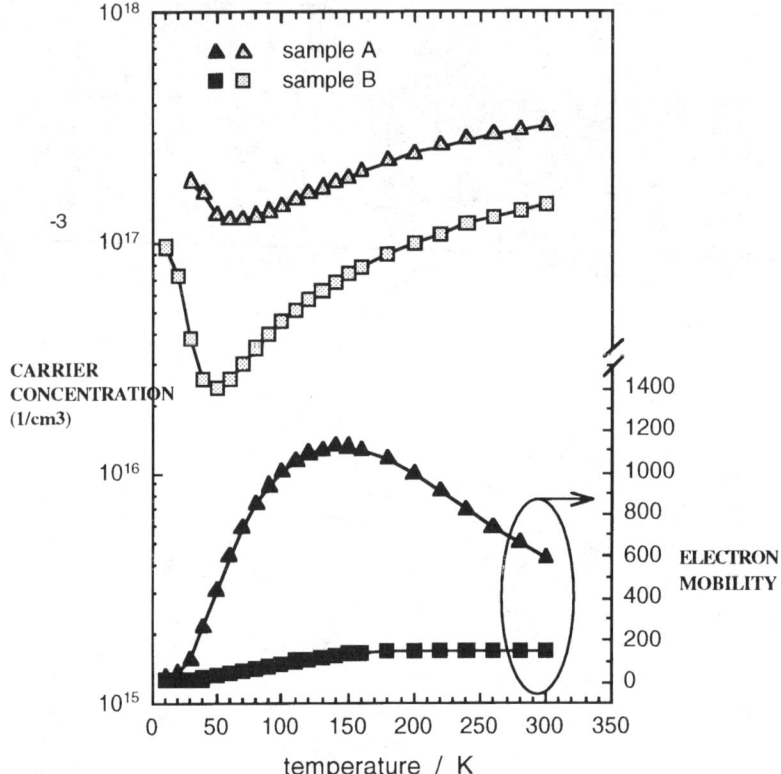

FIG. 8. Hall mobility and free cqarrier concentration as a function of temperature GaN films with 300 sec. (sample A) and 60 sec (sample B) NHa^3 nitridation of sapphire.

2. GaN Bulk and Doping

a. Growth of Bulk GaN

For growth of thicker GaN layers we found that in addition to previously mentioned defect densities, growth parameters strongly influence the resulting film properties. In particular temperature and V/III ratio strongly influence electrical mobility and optical properties of epitaxial GaN. Figure 9 illustrates the effect of growth temperature and insufficient NH_3 overpressure on Hall mobility, compensation ratio, and free carrier concentration of thicker GaN films. At temperatures less than 1050°C and V/III ratio less than 5130 (low V/III, low Tg) a sharp drop in mobility and increase in unintentionally doped carrier concentration is observed. We have included

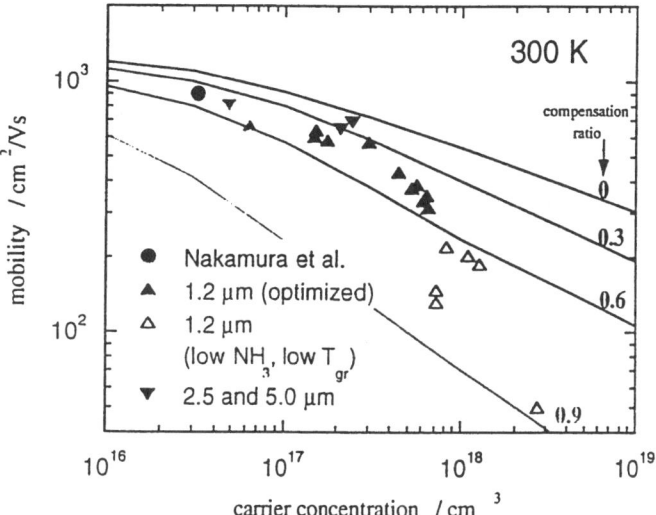

FIG. 9. Hall mobility (300 K) for bulk GaN films as a function of carrier concentration and compensation.

a theoretical estimate of different compensation ratios as a function of free carrier compensation as discussed by Chin and Transleyan (1994). As the temperature increased above 1050°C and V/III ratio increased to 5000–10,000 we reduced the background impurity level, and achieved higher Hall mobility with less compensation. We attribute this effect to more efficient cracking of ammonia and less nitrogen vacancies and defects at higher temperatures. In addition, at higher temperatures more efficient desorption of unwanted surface impurities will occur. Secondary ion mass spectroscopy (SIMS) analysis of the layers also indicated less oxygen contamination in the layers grown at 1050 and 1080°C.

The best reported mobility for bulk GaN is by Nakamura, Mukai, and Senoh (1992) in which a mobility of 900 at 300 K and a low free carrier concentration of $2 \cdot 10^{16}$ cm^{-3} were observed. Exact defect density in these films is not known. In our studies the best films were nominally undoped 4 µm thick GaN films in which we observed 300 K mobilities of 780 cm^2 V/sec. ($v300$ K = $6 \cdot 10^{16}$ cm^{-3}). On these samples dislocation densities of $4 \cdot 10^8$ cm^{-2} for GaN on sapphire substrates was observed in cross-sectional TEM measurements (Kapolnek et al., 1995). Even at excitation levels as low a 2.2 mW/cm^2, 300 K PL is dominated by near band edge emission.

b. GaN p- and n-type Doping

A key breakthrough in development of GaN technology was p doping using Mg and low energy electron beam irradiation (LEEBI) treatment. Amano *et al.*, 1989) observed that under LEEBI Mg-doped GaN exhibited much lower resistivity and PL properties drastically improved. This achievement subsequently led to development of p-n GaN diodes with good turn on characteristics. Nakamura, Senoh, and Mukai (1993) built on this fundamental breakthrough to achieve even higher p doping and uniform activation of Mg by using high temperature thermal annealing under a nitrogen ambient. Passivation requires post growth treatment for MOCVD material to activate the dopants. During growth, interstitial hydrogen is incorporated and a H-Mg acceptor complex forms that passivates the acceptor. This H-Mg bond can be broken by a high temperature annealing step under an inert environment. This work demonstrated that hydrogen compensation of Mg in MOCVD growth of GaN was the principal problem that plagued previous researchers. High room temperature p doping is further complicated by high activation energy of Mg as the most commonly used dopant (170 meV) and the passivation of acceptors with hydrogen during MOCVD growth. Binding energies of dopants are dependent on dielectric constant and effective mass of the material. The nitride system has low dielectric constant (GaN, $e_0 = 9.5$) and large effective masses (GaN, $m_e = 0.2m_0$, $mhh = 0.75m_0$) resulting in large binding energies. This is especially pronounced in p-type doping when comparing GaN to GaAs the acceptor levels are very deep because of the large hole mass. This has led to difficulty in high p-type doping and is the result of two effects: (i) high n-type background concentration compensating the p dopant and (ii) incomplete activation of dopants at room temperature. Low p-type doping (typical values are 10^{17} cm^3) leads to high contact resistance and problems with current spreading. Further work on increasing the p doping level and developing new p dopants will result in substantial payoff in producing LEDs and lasers with lower operating voltages and higher power efficiencies.

n-Type doping is rather straightforward in GaN with Si being the typical n dopant. As-grown material is typical unintentionally n-type, which is widely believed to be due to intrinsic nitrogen vacancies. The Si donor lies just below the conduction band ($E_a = 15-25$ meV). Therefore well controlled n-type doping can be easily accomplished using Si as the donor. Typical MOCVD precursor for n-type doping are silane (SiH$_4$) and disilane (Si$_2$H$_6$) which are typical diluted with hydrogen in the 200 ppm range. Doping levels between $1 \cdot 10^{17}$ and $2 \cdot 10^{19}$ cm^{-3} are easily achieved in doping GaN with silane.

3. GROWTH OF AlGaN AND AlGaN/GaN HETEROSTRUCTURES

High quality AlGaN films have been demonstrated by atmospheric pressure MOCVD as well as epitaxy performed under low pressure conditions. As mentioned in Section III.1, parasitic reaction between TMAl and NH_3 are much more likely to occur than with TMGa. Two-flow channel reactors with separate injection of groups III–V precursors (Nakamura, Mukai, and Senoh, 1994) or reactor operation under low pressure conditions have been shown to successfully prevent excessive prereactions in the gas phase.

At growth temperatures below 1100°C, mole fraction of aluminum in the AlGaN epitaxial layer was found to be almost directly proportional to mole fraction of TMAl in the gas phase. At temperatures above 1100°C, the incorporation efficiency of Ga atoms decreased. This behavior was explained by a decreased sticking probability of Ga molecules at these high temperatures (Hirosawa *et al.*, 1993). Sayyah, Chung, and Gershenzon (1986) reported a decrease in the Ga incorporation efficiency with increasing TMAl flow and explained this behavior by a surface kinetic model and competitive adsorption. The same authors also found a decrease of Ga incorporation efficiency with increasing SiH_4 dopant flow.

High quality AlGaN/GaN heterostructures are characterized by a very high mobility of two dimensional electron gas at the interface (Kahn *et al.*, 1991). Values as high as $1500\,cm^{-2}$ V/sec. at room temperature have been achieved in (Wu *et al.*, 1996). Optical properties of AlGaN/GaN QW (Kahn *et al.*, 1990) were found to be determined by both, quantum and strain related effects (Krishnankutty *et al.*, 1992a, 1992b). MOCVD AlN films showed a full width of half maximum of (002) X-ray rocking curve peak as low as 97 arcsec. (Saxler *et al.*, 1994). AlN/GaN superlattices of high structural and optical quality have also been fabricated by switched atomic layer MOCVD (Kahn *et al.*, 1993).

4. GROWTH OF InGaN AND InGaN/GaN HETEROSTRUCTURES

Growth of high quality InGaN is necessary to obtain good electrical and optical characteristics from LEDs. However, growth of high quality InGaN has proved to be more difficult than GaN. InGaN growth needs to be performed at much lower temperatures than GaN, due to low dissociation temperature of InN (Matsuoka *et al.*, 1988; Koukitu *et al.*, 1996). Furthermore, decomposition of ammonia becomes less efficient with decreasing temperature due to high kinetic barrier for breaking N–H bonds. Growth

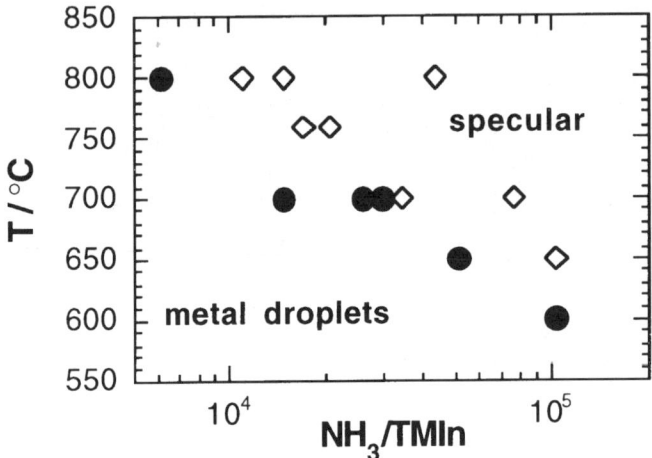

FIG. 10. Effect of growth temperature and NH$_3$/TMIn ratio on the surface morphology of InGaN films.

of InGaN has to be performed at temperatures below 850°C because of high volatility of In at common GaN growth temperatures of above 1000°C, but even on InGaN layers grown at temperatures below 800°C, In droplet formation was observed (Shimizu, Hiramatsu, and Sawaki, 1994).

Figure 10 shows the effect of growth temperature and the NH$_3$/TMIn ratio on the surface morphology of InGaN films. InGaN layers grown with a high NH$_3$/TMIn ratio were specular and showed strong band edge related luminescence (Keller et al., 1996b). At low NH$_3$/TMIn ratios, metal droplets appeared on the surface and the structural and optical quality of the InGaN layers degraded. Thereby, minimum NH$_3$/TMIn ratio necessary to prevent metal droplet formation increased with decreasing growth temperature. We believe this tendency is a result of less efficient ammonia decomposition at lower temperatures.

The In incorporation into the InGaN films strongly increases with decreasing growth temperature. Also, the relative In incorporation coefficient k_{In}, defined as

$$k_{In} = xs_{In} \times \frac{f_{TMGa}}{(1 - xs_{In}) \times f_{TMIn}} \quad (1)$$

(xs_{In-In} composition in the solid) increases with increasing growth rate at a given temperature, as shown in Fig. 11. This indicates that incorporation of

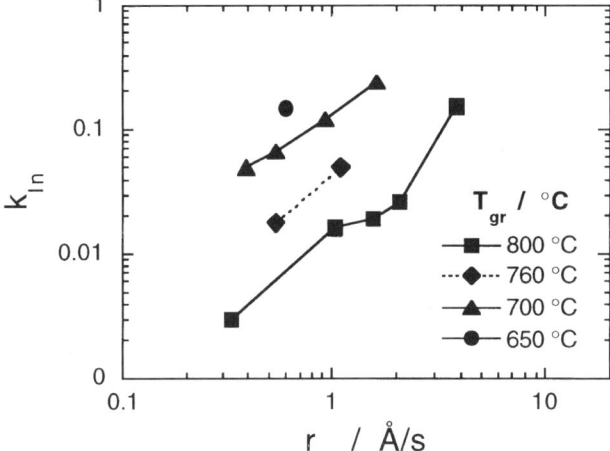

FIG. 11. Dependence of relative In segregation coefficient, k_{In}, on growth rate determined at different growth temperatures.

In is limited by evaporation of In species from the surface. The tendency for evaporation decreases at lower temperature and/or increasing growth rate, when the In species become trapped by the growing layer. But at growth temperatures below 760°C, high quality InGaN films could be obtained only by reducing growth rate to values equal or lower than 3 Å/s in the authors' reactor. These films showed intense band edge related luminescence at room temperature (Nishizawa, Abe, and Kurabayashi, 1985). The FWHM of X-ray diffraction peaks was 6.1 and 6.6 arcmin for layers containing 9 and 20% In, respectively.

Furthermore, In incorporation into the layers is affected by hydrogen flow in the reactor (Piner et al., 1997). In incorporation into the film strongly increases while decreasing the hydrogen flow (Fig. 12). But films grown at a hydrogen partial pressure below 5×10^{-3} ($T_{gr} = 810°C$) were insulating and showed pronounced deep level related luminescence. Thereby, the amount of hydrogen needed to obtain films with excellent optical properties increased with decreasing growth temperature.

Photoluminescence properties of InGaN SQWs were found to be strongly dependent on the structure of barrier layers. Thus, SQWs embedded in InGaN of graded composition showed superior properties compared to SQWs with $In_{0.04}Ga_{0.84}N$ barriers of constant composition (Keller et al., 1996c). Figure 13(a) shows the 300 K PL spectra of a 50 and 20 Å thick $In_{0.16}Ga_{0.84}N$ SQW embedded in $In_{0.04}Ga_{0.96}N$. Emission wavelength of the 1000 Å thick reference film was 407 nm, corresponding to In composi-

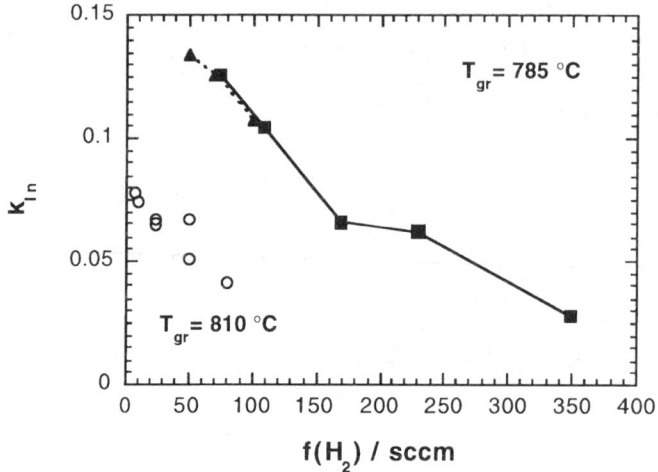

FIG. 12. Dependence of relative In segregation coefficient, k_{In}, on hydrogen flow rate.

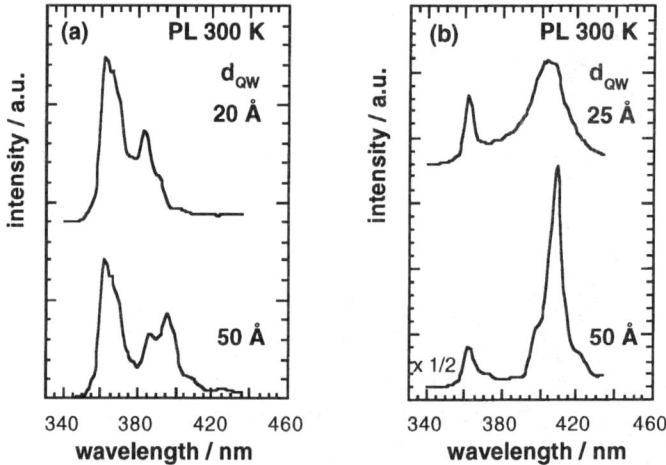

FIG. 13. The 300 K PL spectra of $In_{0.16}Ga_{0.84}N$ SQWs of (a) 20 and 50 Å thickness, embedded in $In_{0.04}Ga_{0.86}N$ (structure A); (b) 25 and 50 Å thickness, embedded in InGaN of graded composition (structure B).

tion of 15.5%. Due to quantum confinement effects, the peak emission wavelength shifted to 396 and 383 nm for the 50 and 20 Å thick SQW, respectively. In comparison, Fig. 13(b) shows the 300 K PL spectra of a 50 and 25 Å thick $In_{0.06}Ga_{0.84}N$ SQW embedded in InGaN of graded composition. Peak emission wavelength of the 1000 Å thick reference film for these two samples was 409 nm (In composition 16%; the slight differences in the composition are caused by use of different mass flow controllers in two sets of experiments). In agreement with theoretical considerations, blue shift due to quantum confinement for structures with graded barriers was much less pronounced. Thus, peak emission wavelength of 409 nm measured for the 50 Å thick SQW was the same as that for the reference film. The 20 Å thick SQW showed an emission peak at 404 nm. Besides the less pronounced blue shift, the 300 K luminescence of SQWs embedded in InGaN of graded composition was much more intense compared to SQWs embedded in $In_{0.04}Ga_{0.86}N$ barriers. For a QW thickness of 50 Å, the intensity ratio between QW and GaN band edge luminescence was 5.6 and 0.6, for structure B and structure A, respectively. Also, FWHM of QW luminescence (50 Å, 7.9 nm; 20 Å, 21.5 nm) was considerably lower in structures with graded barriers. Time resolved PL (TRPL) lifetime measurements on 50 Å thick SQWs of both types revealed a lifetime of 190 psec. in the structure with graded barriers compared to 140 psec for the SQW with constant barrier compositon, for measurements performed at 7 K with a generated carrier density in the order of $5 \times 10^{12} \, cm^{-2}$. At 300 K, the luminescence lifetime was 300 psec in the structure with graded barriers and 100 psec. for the structure with constant barrier composition (Sun *et al.*, unpublished). We believe that superior properties of SQWs buried in InGaN of graded composition originate from greater crystalline quality of InGaN layers in these structures. Additional diffusion of carriers into the QW region using graded bandgap may further enhance intensity of QW related luminescence in these structures as well.

Multi-quantum well (MQW) structures have also been grown in the blue to blue-green spectral range. These quantum structures will play a key role in high efficiency LEDs and laser diodes. We have been able to grow between 5 and 40 period $In_{0.2}Ga_{0.8}N$/GaN superlattices. Strong PL and superlattice fringes are clearly observed for even 10 periods of 2.5 nm $In_{0.2}Ga_{0.8}N$/50 nm GaN. The 300 K PL spectra of two MQW samples grown in the authors' laboratory are shown in Fig. 14. Sample (a) emits at 430 nm and is a 12 period MQW with 21 Å InGaN thick wells and 36 Å thick GaN barriers and sample (b) emits at 490 nm and consists of a 5 period MQW with 16 Å thick InGaN wells and 50 Å thick $In_{0.04}Ga_{0.86}N$ barriers. FWHM of blue MQW related luminescence is 19 nm in sample (a) and of blue-green MQW related luminescence is 26.5 nm in sample (b). Both values are among the best values reported for InGaN/GaN MQW structures (Nakamura *et al.*, 1993).

Growth of InGaN/GaN based MQW structures is particularly crucial to high efficiency blue LEDs and laser diodes. These MQW structures have been incorporated into blue LEDs which emit bright blue direct gap

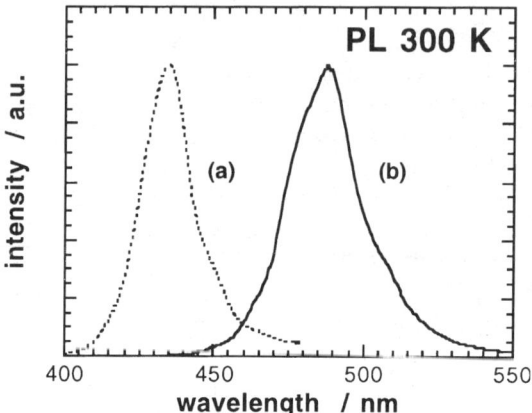

FIG. 14. InGaN MQW superlattices exhibit bright luminescence from the blue to blue-green spectrum.

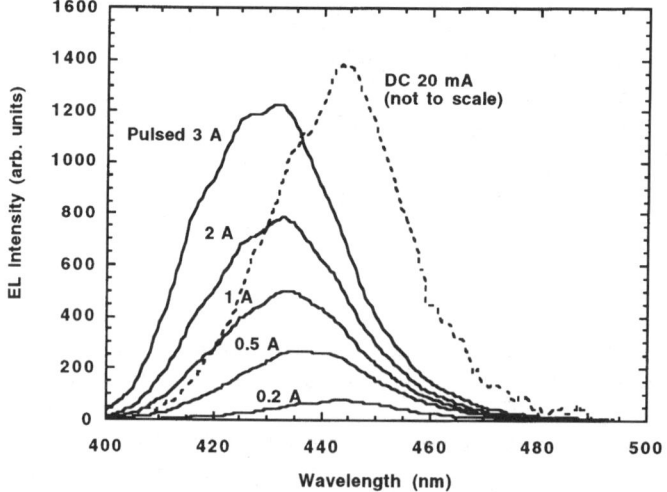

FIG. 15. Electroluminescence spectra from blue InGaN MQW LED showing blue shift at high forward currents. Wavelength shifts considerably from low current (20 mA) values due to filling of the lower bandgap InGaN (higher In composition) regions of the QW.

luminescence at a peak wavelength of 450 nm and FHWM of 28 nm. Luminescence peaked at 450 nm and external quantum efficiency was 4.5% for the MQW structure which is among the highest reported for GaN QW LEDs (Kozodoy et al., 1997). A high peak power output of 55 mW was measured under pulsed condition of 1 A for 3 msec. pulse width, which is higher than pulsed power output for SQW LED. As shown in Fig. 15 the electroluminescence (EL) peak show a significant blue shifting under forward bias. This large energy shift is caused by the filling of higher In regions of QWs which act as localized states. Chichibu et al. (1996) and Nakamura et al. (1997) reported that InGaN QWs phase separates into regions of high In content which then act as localized states in continuous wave (CW) lasers.

V. Conclusions

MOCVD has been demonstrated to be a key technology in achievement of high quality epitaxial layers of GaN and its alloys with Al and In. Near atmospheric pressure operation of MOCVD and laminar flow regime has allowed the scale-up to mass production of GaN along with several other traditional III–V semiconductors. Well controlled doping, low background carrier concentrations, and high mobilities exceeding 800 cm^2 V/sec. has been demonstrated by several research groups. Use of low temperature (500–600°C) GaN buffer layers and surface nitridation resulted in high crystalline qulity material on sapphire. Optimization of MOCVD growth of InGaN/GaN based quantum structures has enabled high efficiency blue LEDs and laser diodes to be achieved.

REFERENCES

Akasaki, I., Amano, H., Koide, Y., Hiramatsu, H., and Sawaki, N. (1989). *J. Cryst. Growth* **89**, 209.
Akasaki, I., Amano, H., Murakami, H., Sassa, M., Kato, H., and Manabe, K. (1993). *J. Cryst. Growth* **128**, 379.
Amano, H., Akasaki, I., Koide, Y., Hiramatsu, K., and Sawaki, N. (1989a). *J. Cryst. Growth* **98**, 209.
Amano, H., Kito, M., Hiramatsu, K., and Akasaki, I. (1989b). *Jpn. J. Appl. Phys.* Pt. II **28**(12) L2112.
Chichibu, S., Azuhata, T., Sota, T., and Nakamura, S. (1996). *Appl. Phys. Lett.* **69**, N27:4188–4190.
Chin, V. W. L., and Transleyan, T. L. (1994). *J. Appl. Phys.* **75**, 7365.
Coates, G. E., Green, M. L. H., Powell, P., and Wade, K. (1968). *Principles of Organometallic Chemistry*, Methuen, London.
DenBaars, S. P., Maa, B. Y., Dapkus, P. D., and Lee, H. C. (1986). *J. Cryst. Growth* **77**, 188.
Giling, L. J. (1982). *J. Phys. (Paris)*, Colloq. **C5**, 235.
Heying, B., Wu, X. H., Keller, S., Li, Y., Kapolnek, D., Keller, B. P., Mishra, U. K., DenBaars, S. P., and Speck, J. S. (1996). *Appl. Phys. Lett.* **68**, 643–5.
Hiramatsu, K., Itoh, S., Amano, H., Akasaki, I., Kuwano, N., Shiraishi, T., and Oki, K. (1991). *Cryst. Growth* **115**, 628.

Hirosawa, K., Hiramatsu, K., Sawaki, N., and Akasaki, I. (1993). *Jpn. J. Appl. Phys.* **32**, L1030.
Kahn, M. A., Skogman, R. A., van Hove, J. M., Krishnankutty, S., and Kolbas, R. M. (1990). *Appl. Phys. Lett.* **56**, 1257.
Kahn, M. A., van Hove, J. M., Kuznia, J. N., and Olson, D. T. (1991). *Appl. Phys. Lett.* **58**, 2408.
Kahn, M. A., Kuznia, J. N., Olson, D. T., George, T., and Pike, W. T. (1993). *Appl. Phys. Lett.* **63**, 3470.
Kahn, M. A., Chen, Q., Skogman, R. A., and Kuznia, J. N. (1995). *Appl. Phys. Lett.* **66**, 2046.
Kapolnek, D., Wu, X. H., Heying, B., Keller, S., Keller, B. P., Mishra, U. K., DenBaars, S. P., and Speck, J. S. (1995). *Appl. Phys. Lett.* **67**, 1541–1543.
Keller, S., Kapolnek, D., Keller, B. P., Wu, Y.-F., Heying, B., Speck, J. S., Mishra, U. K., and DenBaars, S. P. (1996a). *J. Appl. Phys.* **35**, L285.
Keller, S., Keller, B. P., Kapolnek, D. Abare, A. C., Masui, H., Coldren, L. A., Mishra, U. K., and DenBaars, S. P. (1996b). *Appl. Phys. Lett.* **68**, 3147.
Keller, S., Keller, B. P., Kapolnek, D., Mishra, U. K., DenBaars, S. P., Shmagin, I. K., Kolbas, R. M., and Krishnankutty, S. (1996c). *J. Cryst Growth.*
Keller, S., Keller, B. P., Wu, Y.-F., Heying, B., Kaplonek, D., Speck, J. S., Mishra, U. K., and DenBaars, S. P. (1996d). *Appl. Phys. Lett.* **68**, 1525.
Kou, C. P., Fletcher, R. M., Ostenowski, T. D., Lardizabal, M. C., Craford, M. G., and Robbins, V. M. (1990). *Appl. Phys. Lett.* **57**, 2937.
Koukitu, A., Tkahashi, N., Taki, T., and Seki, H. (1996). *Jpn. J. Appl. Phys.* **35**, L673.
Kozodoy, P., Abare, A., Mack, M., Sink, C., Keller, S., Coldren, L. A., Mishra, U., DenBaars, S., and Steigerwald, D. (1997). Materials Research Society. Proceedings of Spring Meeting, Vol. 468, paper D8.18. San Francisco.
Krishnankutty, S., Kolbas, R. M., Kahn, M. A., Kuznia, J. N., van Hove, J. M., and Olson, D. T. (1992a). *J. Electron. Mat.* **21**, 437.
Krishnankutty, S., Kolbas, R. M., Kahn, M. A., Kuznia, J. N., van Hove, J. M., and Olson, D. T. (1992b). *J. Electron. Mat.* **21**, 609.
Kuznia, J. N., Kahn, M. A., Olson, D. T., Kaplan, R., and Freitas, J. (1993). *Appl. Phys. Lett.* **73**, 4700.
Matsuoka, T., Tanaka, H., Sasaki, T., and Katsui, A. (1988). "GasAs and Related Compounds," Institute of Physics Conference No. 106, p. 141.
Matsuoka, T., Yoshimoto, N., Sasaki, T., and Katsui, A. (1992). *J. Electron. Mat.* **21**, 157.
Manasevit, H. M. (1968). *Appl. Phys. Lett.* **12**, 156.
Manasevit, H. M., and Simpson, W. I. (1969). *Electrochem. Soc.* **116**, 1725.
Manasevit, H. M., and Hess, K. L. (1979). *J. Electrochem. Soc.* **126**, 2031.
Molnar, R. J., Singh, R., and Moustakas, T. D. (1995). *Appl. Phys. Lett.* **66**, 2046.
Nakamura, S. (1991). *Jpn. J. Appl. Phys.* **30**, L1705.
Nakamura, S., Mukai, T., and Senoh, M. (1992). *J. Appl. Phys.* **71**, 5543.
Nakamura, S., Iwasa, N., Senoh, M., and Mukai, T. (1992). *Jpn. J. Appl. Phys.* **31**, 1258.
Nakamura, S., Mukai, T., Senoh, M., Nagahama, S., and Iwasa, N. (1993). *J. Appl. Phys.* **74**, 3911.
Nakamura, S., Senoh, M., and Mukai, T. (1993). *Jpn. J. Appl. Phys.* **32**, L8–L11.
Nakamura, S., Mukai, T., and Senoh, M. (1994). *Appl. Phys. Lett.* **64**, 1687.
Nakamura, S., Senoh, M., Nagahama, S., Iwasa, N., Yamada, T., Matsushita, T. Kiyoku, H., and Sugimoto, Y. (1996). *Jpn. J. Appl. Phys.* **35**, L74–L76.
Nakamura, S., Senoh, M., Nagahama, S., Iwasa, N., Yamada, T., Matsushita, T., Sugimoto, Y., and Kiyoku, H. (1997). *Appl. Phys. Lett.* **70**(7), 868–870.
Nishizawa, J., Abe, H., and Kurabayashi, T. (1985). *J. Electrochem. Soc.* **132**, 1197.
Pankove, J. I. (1972). *J. Electrochem. Soc.* **119**, 1118.

Piner, E. L., Behbehani, M. K., El-Masry, N. A., McIntosh, F. G., Roberts, J. C., Boutros, K. S., and Bedair, S. M. (1997). *Appl. Phys. Lett.* **70**, 461.
Rowland, L. B., Doverspike, K., Gaskill, D. K., and Freitas, Jr., J. A. (1994). *Mat. Res. Sym. Proc.*, **339**, 47.
Saxler, A., Kung, P., Sun, C. J., Bigan, E., and Razeghi, M. (1994). *Appl. Phys. Lett.* **64**, 339.
Sayyah, K., Chung, B.-C., and Gershenzon, M. (1986). *J. Cryst. Growth* **77**, 424.
Shimizu, M., Hiramatsu, K., and Sawaki, N. (1994). *J. Cryst. Growth* **145**, 209.
Sugawara, H., Ishikawa, M., and Hatakoshi, G. (1991). *Appl. Phys. Lett.* **58**, 1010.
Sun, C.-K., Keller, S., Wang, G., Minsky, M. S., Bowers, J. E., and DenBaars, S. P. Unpublished.
Thon, A., and Kuech, T. F. (1995). First International Symposium, Boston (eds. F. A. Ponce, R. D. Dupuis, S. Nakamura, and J. A. Edmond), *Mat. Res. Soc.*, p. 97, 1996.
Weeks, Jr., T. W. Bremser, M. D., Ailey, K. S., and Carlson, E., and co-workers (1995). *Appl. Phys. Lett.* **67**, 401.
Wu, X. H., Kapolnek, D., Tarsa, E. J., Heying, B., Keller, S., Keller, B. P., Mishra, U. K., DenBaars, S. P., and Speck, J. S. (1996). *Appl. Phys. Lett.* **68**, 1371.
Wu, Y.-F., Keller, B. P., Keller, S., Kapolnek, D., Kozodoy, P., DenBaars, S. P., and Mishra, U. K. (1996). *Appl. Phys. Lett.* **69**, 1438.
Yuan, J. S., Hsu, C. C., Cohen, R. M., and Stringfellow, G. B. (1985). *J. Appl. Phys.*, 1380.

CHAPTER 3

Growth of Group III–A Nitrides by Reactive Sputtering

W. A. Bryden and T. J. Kistenmacher

APPLIED PHYSICS LABORATORY
THE JOHNS HOPKINS UNIVERSITY
LAUREL, MD

I. INTRODUCTION 39
II. THE SPUTTERING PROCESS 40
   1. *Inert Gas Diode Sputtering* 41
   2. *Reactive Diode Sputering* 42
   3. *Magnetron Sputtering Devices* 43
III. RESULTS AND DISCUSSION 44
   1. *Sputtered AlN Thin Films* 44
   2. *Sputtered GaN Thin Films* 46
   3. *Sputtered InN Thin Films* 48
   References 51

I. Introduction

After a veritable hiatus of 20 years, an already substantial and rapidly expanding initiative has been evolving for the preparation of group III–A metal nitrides (AlN, GaN, InN, and their alloys) in thin film form (Strite and Morkoc, 1992; Morkoc and Mohammad, 1995). This renewed effort partially stems from the potential of these direct, wide bandgap materials for various microelectronic and optoelectronic applications (e.g., passivation barriers, dielectric layers, and substrates in integrated circuits, short wavelength emitters and lasers and solar-blind detectors, and surface acoustic wave devices), and partially from realization that an understanding and control of materials physics of these semiconductors offers singular challenges indeed.

One particular difficulty is the absence of a readily available, thermodynamically stable, single-crystal phase for homoepitaxial growth. Thus, heteroepitaxial growth is a practical necessity and the choice of substrate is critical. This problem is well recognized, and there have been a number of studies on the effect on structural, electrical, and morphological properties of thin films of these compound semiconductors. In the main, however, deposition onto sapphire has dominated early (Strite and Morkoc, 1992; Morkoc and Mohammad, 1995) and contemporary studies. Due largely to projected poor in-plane lattice mismatch (ranging from 13 to 29%) between the (00.1) oriented films of these metal nitrides and (00.1) sapphire, depositions utilizing a variety of synthetic approaches have commonly led to three dimensional island growth and films comprised of a mosaic (typical grain size on the order of a few hundred angstroms to a few microns) of weakly interacting columnar grains. As such, these films are of only limited utility for device applications.

A first step in improvement of device applicability of group III–A nitride films deposited on (00.1) sapphire would be to alter film growth, to the exclusion of columnar islands and secondary grain growth. In principal, this is no simple task for it is well recognized that as the lattice mismatch for heteroepitaxial systems increases, either the Stranski–*Krastanov* or *Volmer*–Weber (three dimensional island) growth modes are favored over Frank-van der Merwe (two dimensional layer-by-layer) growth. However, it is becoming increasingly appreciated that nucleation layers can alter growth mechanism and dependent physical properties, and such nucleation layers have become an integral factor in growth of GaN and InN by a variety of techniques (metalorganic chemical vapor deposition (MOCVD), molecular beam epitaxy (MBE), and reactive sputtering).

In this chapter, we briefly explore thin film studies of AlN, GaN, and InN prepared by reactive sputtering techniques, with an emphasis on nucleated growth and device applications.

II. The Sputtering Process

Sputtering is a vacuum based process whereby atoms or ions of a solid material target are ejected into the gas phase by momentum exchange with energetic particles (Fig. 1). Thin films can be produced, using the sputtering process, by placing a substrate in the path of gas phase atoms or ions, which tend to condense on the surface. A wide range of materials including metals, alloys, semiconductors, and insulators have been deposited using this technique. A full treatment of the sputtering process and applications to thin film formation can be found in any number of leading references (Ohring, 1992). Only a few basic concepts necessary to the discussion will be outlined here.

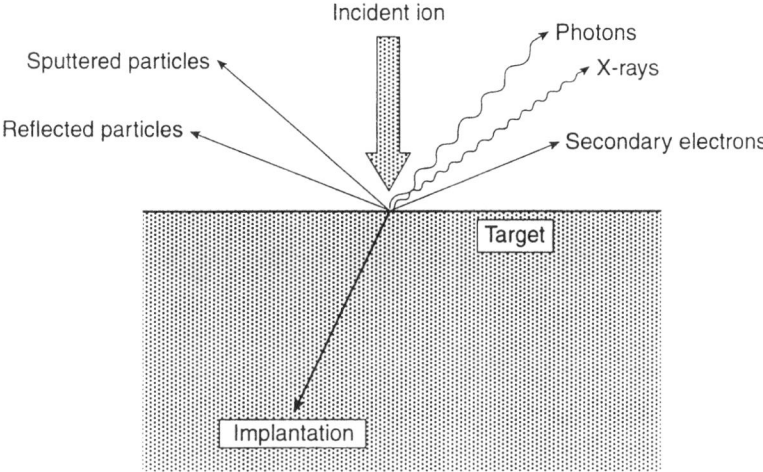

FIG. 1. Microscopic results of energetic charged particles incident on a solid surface. The incident ion may be reflected from or implanted in the target, or a combination of sputtered target particles (photons, X-rays, and secondary electrons) may be emitted from the target surface. The desired species are sputtered particles from the target surface that impinge on, and cover the substrate.

1. INERT GAS DIODE SPUTTERING

An experimental setup for the simplest technique, diode sputtering with an inert gas, is shown in Fig. 2. Typically, a vacuum chamber containing diode sputtering electrodes, target material, and the substrate is pumped down, using high vacuum pumps, to a level of 10^{-6} to 10^{-8} Torr. An inert gas such as argon is metered into the system using mass flow controllers such that a constant pressure, in the range of 10^{-2} to 10^{-3} Torr, is maintained. A high voltage (dc or rf) is applied to electrodes and an electrical discharge (i.e., glow discharge) is initiated between diode electrodes. Resulting ionized gas is used as the active element of the sputtering process. Positive ions are accelerated toward the cathode where they strike the attached target material and eject (typically) atomic species into the gas phase. The substrate, mounted onto the anode, is coated with target material (desirable) and bombarded with energetic electrons (and other species) from plasma (generally undesirable). The net result of this process is a thin film coating of target material onto the substrate. However, inert gas diode sputtering has several characteristics that make it undesirable for preparation of high quality films of compound semiconducting materials. These drawbacks fall into two general categories: (1) severe deviations from

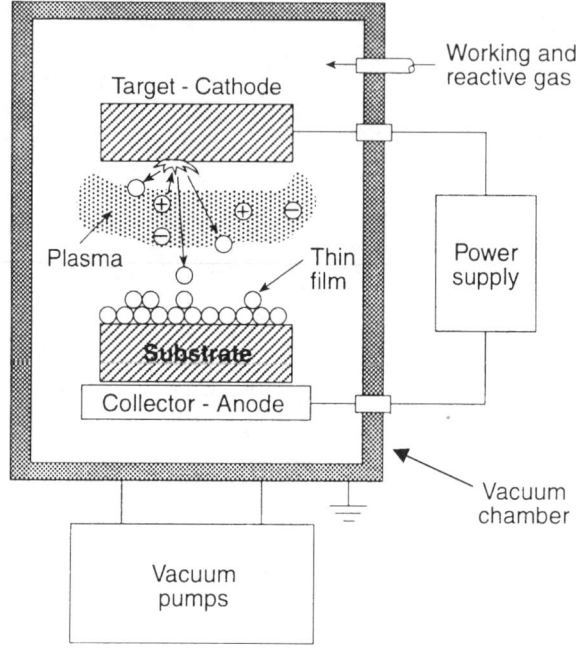

FIG. 2. Essential components of a diode sputtering system, including (a) electrode arrangements (cathode holding the sputtering target and anode holding the substrate), (b) vacuum system, (c) gas inlet manifold, and (d) electrical power supply.

desired film stoichiometry for sputtering from compound targets, and (2) uncontrolled effects of energetic particle irradiation on growing film. The first class of difficulties can be remedied by employing techniques such as reactive sputtering and other preparative methods that provide reactive species at the surface of growing film. The second class of difficulties can be addressed by employing alternative sputtering techniques such as triode devices and several types of magnetron arrangements.

2. REACTIVE DIODE SPUTTERING

The physics of reactive sputtering is quite similar to inert gas sputtering in that a plasma is maintained in an electrical discharge and positive ions are accelerated to the target, dislodging material. However, the chemistry of the reactive process is quite different from inert gas sputtering. In reactive sputtering, the plasma is formed from a chemically reactive gas, or a mixture of inert and reactive gas, so when atoms or ions are sputtered from the target, a chemical compound tends to form at the substrate surface. Chemical compounds can then be formed from elemental (or compound) targets and stoichiometry can be altered by variations in sputtering parameters such as partial pressures of sputtering gases, substrate bias, and geometrical factors. In a typical reactive sputtering arrangement (Fig. 2), a mixture of inert and reactive gases are utilized to produce compound materials. While reactive diode sputtering has been used successfully to produce many compound materials, the process is difficult to optimize and control. A more controllable deposition process involves use of alternative ion sources such as electron cyclotron resonance (ECR) to produce a high density of reactive species at the substrate. Reactive diode sputtering has, however, been used to produce nitride semiconductor films as described later.

3. Magnetron Sputtering Devices

A major drawback to diode sputtering configuration is the fact that the substrate is bombarded by a relatively high density of energetic electrons during the sputtering process. This electron flux is generally deleterious to growth of high quality epitaxial films that are desirable for semiconductor materials. Alternative sputtering geometries including triode and magnetron configurations have been developed to address this problem. A typical planar magnetron device configured for reactive sputtering is shown in Fig. 3. As mentioned previously, this arrangement uses a mixture of inert gas (injected near the target) and reactive gas (injected near the substrate) to produce compound materials. This geometry tends to keep the target surface from reacting with reactive gas (a process that often slows the sputtering rate) and enhances concentration of reactive gas in a reaction zone at the substrate surface. A specially shaped magnetic field is developed in the target area to effect trapping of plasma electrons in this region, thus the substrate may be remotely placed away from deleterious effects of plasma electrons. A second advantage to magnetron configuration is much higher sputtering rates obtainable. This technology has also been developed with different geometries such as coaxial, cylindrical, and a large variety of magnetically and electrically unbalanced configurations. These alternatives allow tailoring of physics of deposition procession for specific types of coatings. When magnetron sources are coupled with controllable reactive sources, good regulation of thin film deposition can be achieved.

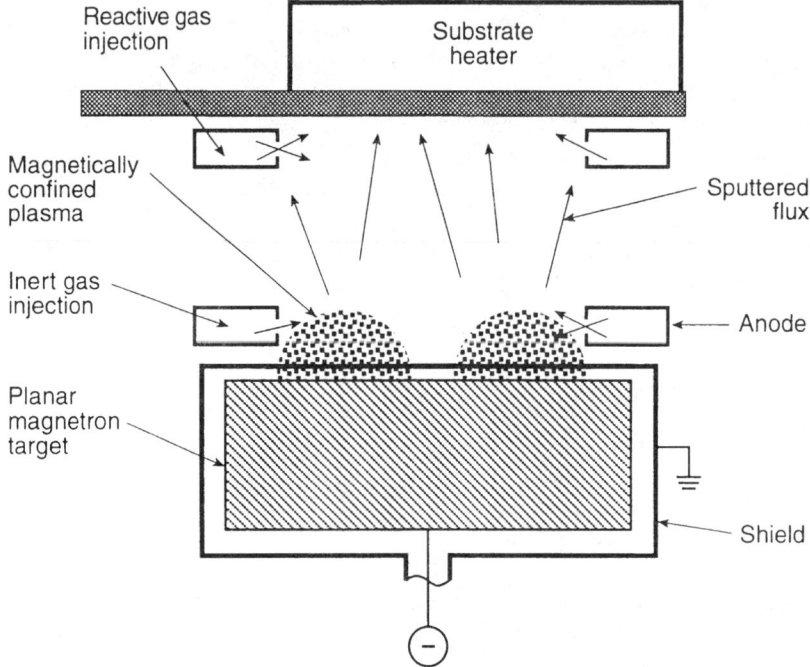

FIG. 3. Reactive magnetron sputtering configuration using a mixture of inert gas (injected near the target) and reactive gas (injected near the substrate) to produce compound materials. This geometry tends to keep the target surface from reacting with reactive gas (a process that often slows the sputtering rate) and enhances concentration of reactive gas in the reaction zone at the substrate surface.

III. Results and Discussion

1. Sputtered AlN Thin Films

Reactive sputtering has played a very important role in growth of AlN thin films (Chuskus, Reeder, and Oaradis, 1974; Shioaski *et al.*, 1980; Tsuboughi, Sugai, and Mikoshiba, 1981; Tsuboughi, Sugai, and Mikoshiba, 1982; Aita, 1982; Aita and Gawlak, 1983; Cakchard *et al.*, 1990; Meng, Heremans, and Cheng, 1991; Okano *et al.*, 1992; O'Toole *et al.*, 1992; Lin *et al.*, 1993; Okano *et al.*, 1994; Meng *et al.*, 1994; Lindquist, Lin, and Ketterson, 1994; Suter *et al.*, 1995; Ivanov *et al.*, 1995). One of the principal device applications for AlN thin films is as the active piezoelectric element surface acoustic wave (SAW) sensors, an application emphasized even in early work of Chuskus, Reeder, and Oaradis in 1974) where a piezoelectric coupling constant as large as 0.2% was reported for AlN thin films sputtered on (00.1) sapphire substrates. The surface acoustic wave velocity for AlN at approximately 6000 m/s is higher, by about a factor of two, than any other material that has been used in construction of SAW devices. Additionally, AlN displays two other highly desirable properties: (1) large electromechanical coupling constant and (2) very low dispersion at high frequencies. These factors combine to allow operation of AlN based sensors at frequencies between 200 MHz and 1 GHz—a decided advantage when developing sensor systems (Okano *et al.*, 1994; Suter *et al.*, 1995).

Historically, one significant drawback to application of SAW devices has been temperature dependence of the resonant frequency. Complicated schemes involving special crystal cuts and temperature stabilization systems have evolved in an attempt to ameliorate this problem. In continuing work (Shioaski *et al.*, 1980; Tsuboughi, Sugai, and Mikoshiba, 1981; O'Toole *et al.*, 1992; Suter *et al.*, 1995), the control of acoustic properties by precise control of piezoelectric film (AlN) thickness was exploited to develop nearly temperature independent SAW devices for frequency control applications, like resonators, delay lines and filters. The result of these studies showed that by controlling the product kH, where k is the wavevector ($2\pi/\lambda$) of SAW resonator and H is the AlN film thickness, devices could easily be produced with a nearly temperature independent frequency near 1 GHz. For AlN films deposited on the basal plane of sapphite, a value of product kH of 3.75 gave the desired temperature independence near room temperature.

In a series of innovative studies on the growth of AlN on a substrate other than sapphire, Meng *et al.* (1991, 1994) employed X-ray diffraction (XRD) and transmission electron microscopy (TEM) to study in depth the growth of epitaxial AlN on (111) silicon substrates by reactive magnetron sputtering. Among their many results, it was determined that at a growth temperature of 900°C AlN nucleated in its (111) oriented 3C cubic (zincblende) polymorph and this nucleation layer persisted for approximately 6 to 8 unit cell lengths. Subsequently, epitaxial growth of AlN continued as its (00.1) oriented hexagonal (wurzite) polymorph. The complete heteroepitaxial relationship was determined to be:

$$2H - AlN(00.1)/3C - AlN(111)/Si(111)$$

and

$$2H - AlN[1, -2.0]/3C - AlN[110]/Si[110]$$

Finally, in addition to their obvious interest as scientific curiosities, thin film multilayers of AlN and metallic, semiconducting, and magnetic nitrides have been investigated for a number of technological applications, including (a) two dimensionally enhanced superconducting critical fields, (b) Josephson tunnel junctions, (c) photothermal conversion of solar energy, and (d) high density recording media and recording heads. Specifically, several families of artificially structured materials involving insulating AlN have been prepared and include NbN/AlN (Murduck et al., 1987, 1988; Bhadra et al., 1989), ZrN/AlN (Meng, Sell, and Waldo, 1991; Meng et al., 1991; Meng and Heremans, 1992), and FeN/AlN (Sin and Wang, 1996) mixed nitride multilayers and a series of metal/AlN multilayers (Krishnan, Nigam, and Tessier, 1990; Krishnan et al., 1991; Rippert et al., 1995; Kikkawa et al., 1996). Because of its versatility and despite its inherent complexity, reactive sputtering has played an early and continuing role in growth of these multilayer materials.

2. Sputtered GaN Thin Films

While MOCVD and MBE have emerged as preferred techniques for deposition of GaN thin films, significant results (both early and late) have been obtained for GaN films prepared by reactive sputtering. In an early and seminal piece of work, Hovel and Cuomo (1972) studied electrical and optical properties of rf-sputtered GaN thin films deposited on (00.1) sapphire, (111) silicon, and metal substrates. Owing to the low melting point of Ga ($\sim 30°C$), sputter target for these rf-diode sputtering experiments was formed by Ni- or Mo-coated copper disks covered with elemental gallium. Light-yellow, highly transparent GaN thin films showed a marked (00.1) texture for all substrates, while measured resistivity (using the van der Pauw lead geometry) for GaN thin films on (00.1) sapphire substrates was greater than $10^8 \Omega$ cm and was taken to imply low impurity densities and possibly greatly reduced native defects (especially nitrogen atom vacancies and gallium atom interstitials). Finally, measured index of refraction for these rf-diode sputtered GaN thin films ranged from 2.1 to 2.4 at long wavelengths.

Carin (1990); Ross and Rubin (1991); Ross, Rubin, and Gustafson (1993); Newman, Ross, and Rubin (1993) investigated the effects of various deposition parameters on structure, composition, electrical, and optical properties of GaN thin films deposited by reactive rf-magnetron sputtering. In these studies, the sputtering target was ultrapure elemental gallium held in a stainless steel cup. In their initial studies (Carin, 1990; Ross and Rubin, 1991), it was found that unnucleated, epitaxial (11.0) oriented films of GaN were formed on (01.2) sapphire substrates at growth temperatures between 640 and 650°C. Maximum N/Ga atomic ratio was reached, however, at 625°C and this ratio decreased as growth temperature increased. A direct, optical bandgap of 3.4 eV was inferred from optical absorption data. The full width at half maximum (FWHM) for room temperature photoluminescence (PL) was approximately 11 meV. Collectively, these results suggest that these rf-magnetron sputtered GaN thin films are of comparable growth quality to unnucleated films grown by MOCVD or MBE methods.

In a follow-up study, Ross, Rubin, and Gustafson (1993) reported on conditions for growth of highly oriented GaN thin films on unnucleated (111) GaAs and epitaxial GaN thin films on AlN nucleated (111) GaAs. In the latter case, both GaN overlayer and AlN nucleation layer were deposited by reactive rf-magnetron sputtering. Optimal growth conditions depended on variables such as substrate temperature and nitrogen partial pressure in growth plasma. As with the deposition of GaN on (00.1) sapphire, unnucleated growth on (111) GaAs typically yielded films composed of large hexagonal grains (with diameters of order 200 nm) of GaN, while predeposition of a 200 nm AlN nucleation layer yielded much denser and smoother growth patterns and surface morphologies.

In a similar vein, Meng and Perry (1994) have examined strain effects in epitaxial GaN thin films deposited on (111) silicon utilizing AlN nucleation layers and reactive rf-magnetron sputtering. In this study, elemental aluminum and gallium sputtering targets were contained in Mo cups. Employing $\theta/2\theta$ diffractometry, the AlN nucleation layer and GaN overlayer were shown to grow with their (00.1) crystallographic planes parallel to the (111) plane of silicon substrate. The in-plane orientation relationship

$$[(11.0)GaN/(11.0)AlN/(1,-1,0)Si]$$

was deduced from plan view and cross-sectional TEM measurements. Of some note are the widths of the rocking curves (ω scans) for reflections normal to silicon substrate, the AlN nucleation layer, and the GaN overlayer: 0.09° (resolution limited) for silicon (111) reflection; 0.85° for (00.2) reflection from AlN nucleation layer; 0.61° for (00.2) reflection from GaN overlayer. The implication of this data is that structural coherence of GaN film parallel to growth direction is reasonably well developed, but structural coherence is considerably poorer than MOCVD or MBE films where widths of the corresponding GaN rocking curve for a comparable film thickness is routinely 0.1° or less.

Furthermore, Meng and Perry (1994) examined the strain present in these films as a function of the input sputtering power and growth temperature using XRD and Raman scattering. These authors found that while narrower rocking curve widths (corresponding to more structurally coherent films) were found at higher growth temperatures, increasing the rf power at a specific growth temperature had little effect on the width of the rocking curve. In contrast, lattice spacing of (00.2) reflection increased significantly with increasing rf power—implying a significant increase in out-of-plane strain with increasing power. Similarly, the resonant frequency of GaN E_2 Raman line scaled with strain induced in GaN film as a function of increasing rf power (the order of 4.4 cm^{-1}/GPa). The exact nature of defects created in GaN overlayers that are responsible for creation of enhanced lattice strain with increasing rf power are still under active investigation.

3. Sputtered InN Thin Films

A comprehensive study on reactive dc sputtering of elemental indium was made by Natarajan *et al.* (1980a, b). In a gas mixture composed of increasing amounts of O_2 in N_2, a critical O_2 concentration ($\sim 2.5\%$) was signaled by a sudden decrease in target sputtering rate, dramatic change in X-ray scattering profile, and equally striking variations in optical properties of reactive plasma. The antecedent of these abrupt changes is the creation of an indium oxide layer at the target surface, which is stabilized by relatively large heat of formation of In_2O_3 (~ 221 kcal/mol). In contrast, while there was a decrease in sputtering rate (owing primarily to a reduction in ion current) with increasing volume fraction of N_2 in an Ar/N_2 reactive gas mixture, no indication of complete nitridation of target surface or sputter ejection of (InN)$_n$ clusters was found experimentally. This latter result is largely understood based on lack of chemisorption of molecular N_2 by In and relatively small heat of formation of InN (~ 5 kcal/mol).

In addition, Hovel and Cuomo (1972) reported a carrier concentration of 7×10^{18} cm^{-3} and an electron mobility of 250 cm^2 V/sec. for deposition of InN thin films by reactive rf-diode sputtering on (00.1) sapphire substrates. Even more impressive film parameters (a carrier concentration of 10^{16}–10^{17} cm^{-3} and a low temperature electron mobility as high as 4000 cm^2 V/sec.) were reported by Tansley and Foley (1984, 1986) for growth of polycrystalline InN thin films on glass substrates. Although the semiconducting properties of these films were obviously very promising, extremely long presputtering and sputtering times and polycrystalline morphology of films are not compatible with fabrication of modern optoelectronic devices, including those based on heterostructures and quantum wells (QWs).

Unfortunately, InN thin films prepared by reactive sputtering in other laboratories (Sullivan *et al.*, 1988; Kistenmacher *et al.*, 1990, 1994; Kistenmacher and Bryden 1991, 1993; Bryden *et al.*, 1992, 1993, 1994; Bryden, Ecelberger, and Kistenmacher, 1994) have not met these expectations and have universally high carrier concentrations near 10^{20} cm^{-3} and concomitantly low electron mobility of $<100\,\text{cm}^2\,\text{V/sec}$. As has become readily apparent, deviations from stoichiometry (typically oxygen incorporation and nitrogen deficiency) and residual film strain (arising from poor lattice and thermal expansion matches) in thin films of InN are undoubtedly two of the principal impediments to realization of their potential for applications in electronics and photonics.

Utilization of nucleated substrates (particularly, AlN and GaN nucleation layers) has done much to reduce residual strain, and in the last few years, a novel method of increasing nitrogen content has been developed around ECR assisted molecular beam epitaxy. Following in this vein, an ECR source was added in our laboratories to an existing ultrahigh vacuum deposition chamber to yield what is believed to be a unique growth facility,

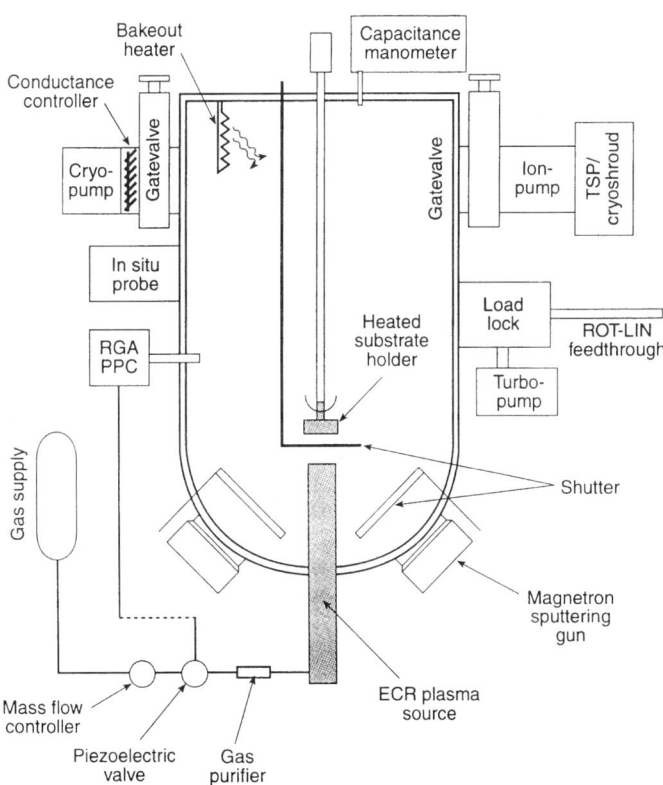

FIG. 4. Schematic of an ultrahigh vacuum, ECR assisted magnetron sputtering system for ultraclean, high reactivity sputter deposition and *in situ* monitoring of electrical transport properties. (From Bryden *et al.*, 1994.)

and results have been reported for InN grown on AlN nucleated (00.1) sapphire (Bryden, Ecelberger, and Kistenmacher (1994); Bryden et al., 1994).

A diagram of the ECR assisted magnetron sputtering system (Bryden, Ecelberger, and Kistenmacher, 1994) is shown in Fig. 4. The all stainless steel chamber is equipped with a combination of turbo-, ion-, and cryo-pumps and typically achieves a base pressure of 5×10^{-11} Torr. An ECR source is aligned parallel to rotation axis of the substrate/heater assembly and normal to the substrate surface (00.1) sapphire. Magnetron sputtering guns are at an angle of 45° to this common axis, and to facilitate uniform film thickness, the substrate/heater assembly is incrementally rotated through ±180° during deposition. Under typical sputtering conditions, substrates were heated to 850°C in vacuum and cooled to 600°C where a 40 nm nucleation layer of AlN was dc sputtered from a high-purity Al target in N_2. Subsequently, seeded substrates were cooled to growth temperature (100–400°C), the ECR source activated, and InN deposited by reactive sputtering of a high purity In target.

Structural coherence and strain normal to the film plane

$$[(00.1)_{InN}/(00.1)_{sapphire}]$$

were generally ascertained from X-ray data collected on a $\theta/2\theta$ diffractometry. Analogous in-plane $[(10.0)_{InN}/(11.0)_{sapphire}]$ structural parameters were obtained from data collected using the X-ray precession method. Room temperature electrical resistivity and Hall effect were measured using a four probe van der Pauw geometry, and optical bandgap derived from absorption spectra.

The effects of ECR plasma on magnetron sputter growth of InN films can be illustrated by comparing their electrical transport properties to those for films grown by conventional reactive magnetron sputtering. Thin film growth of InN is a crucial test as it has the lowest heat of formation of group III–A nitrides and, therefore, nitrogen deficiency is especially troublesome. Room temperature Hall mobility as a function of growth temperature was determined (Bryden, Ecelberger, and Kistenmacher, 1994) for both conventional and ECR assisted sputtered films of InN. It was apparent, except at the very lowest growth temperatures, that mobility of InN films deposited by ECR assisted sputtering are enhanced by factors as large as three in the 250–350°C growth temperature regime. It is also evident that much of the enhancement in Hall mobility is due to a reduction in carrier concentration. One probable origin of reduced carrier concentration is a more nearly stoichiometric film, leading to a reduction in nitrogen defect sites. In parallel, onset of absorption is notably sharper and optical bandgap for ECR assisted films is 10–15% larger than found for films grown by conventional reactive magnetron sputtering.

Similarly, zero level precession photographs of the combined scattering from (hh.0) reciprocal lattice planes of (00.1) sapphire substrate and (00.1) heteroepitaxial InN film revealed an improvement in signal-to-noise, largely achieved by a reduction in noise. Several processes potentially contribute to X-ray background level, including (a) fluorescence, (b) extrinsic instrumental scattering, and (c) Bremsstrahlung and diffuse scattering. While it might be expected that fluorescence and extrinsic incoherent (Compton modified) fraction of diffuse scattering are largely sample independent, the coherent scattering from various kinds of imperfections (point and line defects, stacking faults, surface roughness, and homogeneous and inhomogeneous stress) is expected to be highly sample dependent. Thus, the decreased background is likely due to reduced stress and a decrease in density of such imperfections. In that context, apparent homogeneous strain normal to growth surface for both sets of samples has been estimated from X-ray diffractometry

$$\text{strain} = \frac{d(00.2)^{\text{obs}} - d(00.2)^{\text{bulk}}}{d(00.2)^{\text{bulk}}}$$

It was apparent that homogeneous strain in ECR assisted films is significantly reduced compared to that found in films deposited by conventional reactive magnetron sputtering.

Finally, surface morphology of these ECR assisted magnetron sputtered InN thin films has been investigated by atomic force microscopy (Bryden *et al.*, 1994). On a fine scale, these films exhibit highly orientated grains with diameters on the order of a few hundred angstroms. On a larger scale, there is clear evidence of Ostwald ripening to yield well defined islands with diameters on the order of 1 μm. This is in marked contrast to conventional rf-magnetron grown films where there is no evidence of island formation, only dense packing of micrograins.

References

Aita, C. R. (1982). "Basal Orientation Aluminum Nitride Grown at Low Temperature by rf Diode Sputtering," *J. Appl. Phys.* **53**, 1807.

Aita, C. R., and Gawlak, C. J. (1983). "The Dependence of Aluminum Nitride Film Crystallography on Sputtering Plasma Composition," *J. Vac. Sci. Technol. A* **1**, 403.

Bhadra, R., Grimsditch, M., Murduck, J., and Schuller, I. K. (1989). "Elastic Constants of Metal-Insulator Superlattices," *Appl. Phys. Lett.* **54**, 1409.

Bryden, W. A., Morgan, J. S., Fainchtein, R., and Kistenmacher, T. J. (1992). "Effects of an AlN Nucleation Layer on Magnetron Sputtered Indium Nitride Films," *Thin Solid Films* **213**, 86.

Bryden, W. A., Hawley, M. E., Ecelberger, S. A., and Kistenmacher, T. J. (1993). "Growth Morphology of InN Thin Films by Scanning Tunneling and Atomic Force Microscopies and X-Ray Scattering," *Mat. Res. Soc. Symp. Proc.* **312**, 95.

Bryden, W. A., Ecelberger, S. A., and Kistenmacher, T. J. (1994). "Heteroepitaxial Growth of InN on AlN-Nucleated (00.1) Sapphire by Ultrahigh Vacuum ECR-Assisted Reactive Magnetron Sputtering," *Appl. Phys. Lett.* **64**, 2864.

Bryden, W. A., Ecelberger, S. A., Hawley, M. E., and Kistenmacher, T. J. (1994). "ECR-Assisted Reactive Magnetron Sputtering of InN," *Mat. Res. Soc. Symp. Proc.* **339**, 497.

Cachard, A., Fillit, R., Kadad, I., and Pommier, J. C. (1990). "Magnetron Sputtering Deposited AlN Waveguides: Effect of the Structure on Optical Properties," *Vacuum* **41**, 1151.

Carin, R. (1990). "Dielectric Quality of GaN Insulating Thin Films Grown by Reactive Sputtering," *Rev. Phys. Appl.* **25**, 489.

Chuskus, A. J., Reeder, T. M., and Oaradis, E. L. (1974). "rf-Sputtered Aluminum Nitride on Sapphire," *Appl. Phys. Lett.* **24**, 155.

Hovel, H. J., and Cuomo, J. J. (1972). "Electrical and Optical Properties of rf-Sputtered GaN and InN," *Appl. Phys. Lett.* **20**, 71.

Ivanov, I., Hultman, L., Jarrendahl, K., Martensson, P., Sundgren, J.-E., Hjorvarsson, B., and Greene, J. E. (1995). "Growth of Epitaxial AlN(0001) on Si(111) by Reactive Magnetron Sputter Deposition," *J. Appl. Phys.* **78**, 5721.

Kikkawa, S., Fujiki, M., Takahashi, M., Kanamaru, F., Yoshioka, H., Hinomura, T., Nasu, S., and Watanabe, I. (1996). "Interface of Iron Metal/Aluminum Nitride Multilayer Composite Film," *Appl. Phys. Lett.* **68**, 2756.

Kistenmacher, T. J., Bryden, W. A., Morgan, J. S., and Poehler, T. O. (1990). "Characterization of rf-Sputtered InN Films and AlN/InN Bilayers on (0001) Sapphire by the X-Ray Precession Method," *J. Appl. Phys.* **68**, 1541.

Kistenmacher, T. J., and Bryden, W. A. (1991). "Overgrowth of Indium Nitride Thin Films on Aluminum Nitride Nucleated (00.1) Sapphire by Reactive Magnetron Sputtering," *Appl. Phys. Lett.* **59**, 1844.

Kistenmacher, T. J., and Bryden, W. A. (1993). "Comparative Influence of Bias and Nucleation Layer Thickness on the Heteroepitaxial Growth of InN on AlN-Nucleated (111) Silicon and (00.1) Sapphire," *Appl. Phys. Lett.* **62**, 1221.

Kistenmacher, T. J., Ecelberger, S. A., Bryden, W. A., and Hawley, M. E. (1994). "Structural and Electrical Properties of Thermally Annealed InN Thin Films on Native and AlN-Nucleated (00.1) Sapphire," *Mat. Res. Soc. Symp. Proc.* **317**, 461.

Krishnan, R., Nigam, A. K., and Tessier, M. (1990). "Magnetic and Magneto-Optical Properties of Thin Single and Multilayer Films of a-Tb-Gd-Fe/AlN Multilayers," *J. Magn. Magn. Mat.* **83**, 27.

Krishnan, R., Cagan, V., Tessier, M., and Visnovsky, S. (1991). "Magnetic and Magneto-Optical Studies in Fe/AlN Multilayers," *J. Magn. Magn. Mat.* **101**, 205.

Lin, W. P., Lundquist, P. M., Wong, G. K., Rippert, E. D., and Ketterson, J. B. (1993). "Second Order Optical Nonlinearities of Radio Frequency Sputter-Deposited AlN Thin Films," *Appl. Phys. Lett.* **63**, 2875.

Lindquist, P. M., Lin, W. P., and Ketterson, J. B. (1994). "Ultraviolet Second Harmonic Generation in Radio-Frequency Sputter-Deposited Aluminum Nitride Thin Films," *Appl. Phys. Lett.* **65**, 1085.

Meng, W. J., Heremans, J., and Cheng, Y. T. (1991). "Epitaxial Growth of Aluminum Nitride on Si(111) by Reactive Sputtering," *Appl. Phys. Lett.* **59**, 2097.

Meng, W. J., Morelli, D. T., Roessler, D. M., and Heremans, J. (1991). "Electrical Transport and Optical Properties of Zirconium Nitride/Aluminum Nitride Multilayers," *J. Appl. Phys.* **69**, 846.

Meng, W. J., Sell, J. A., and Waldo, R. A. (1991). "Reactive Sputter Deposition of Zirconium Nitride/Aluminum Nitride Multilayers: Chemical Composition Effects and Structural Characterization," *J. Vac. Sci. Technol. A* **9**, 2183.
Meng, W. J., and Heremans, J. (1992). "Growth of Epitaxial Aluminum Nitride and Aluminum Nitride/Zirconium Nitride Superlattices on Si(111)," *J. Vac. Sci. Technol. A* **10**, 1610.
Meng, W. J., and Perry, T. A. (1994). "Strain Effects in Epitaxial GaN Grown on AlN-Buffered Si(111)," *J. Appl. Phys.* **76**, 7824.
Meng, W. J., Sell, J. A., Perry, T. A., Rehn, L. E., and Baldo, P. M. (1994). "Growth of Aluminum Nitride Thin Films on Si(111) and Si(001): Structural Characteristics and Development of Intrinsic Stresses," *J. Appl. Phys.* **75**, 3446.
Morkoc, H., and Mohammad, S. N. (1995). "High-Luminosity Blue and Blue-Green Gallium Nitride Light-Emitting Diodes," *Science* **267**, 51.
Murduck, J. M., Vicent, J., Schuller, I. V., and Ketterson, J. B. (1987). "Fabrication of NbN/AlN Superconducting Multilayers," *J. Appl. Phys.* **62**, 4216.
Murduck, J. M., Capone, D. W., Schuller, I. V., Foner, S., and Ketterson, J. B. (1988). "Critical Current Enhancement in NbN/AlN Multilayers," *Appl. Phys. Lett.* **52**, 504.
Natarajan, B. R., Eltoukhy, A. H., Greene, J. E., and Barr, T. L. (1980a). "Mechanisms of Reactive Sputtering of Indium I: Growth of InN in Mixed Ar-N_2 Discharges," *Thin Solid Films* **69**, 201.
Natarajan, B. R., Eltoukhy, A. H., Greene, J. E., and Barr, T. L. (1980b). "Mechanisms of Reactive Sputtering of Indium III: A General Phenomenological Model for Reactive Sputtering," *Thin Solid Films* **69**, 229.
Newman, N., Ross, J., and Rubin, M. (1993). "Thermodynamic and Kinetic Processes Involved in the Growth of Epitaxial GaN Thin Films," *Appl. Phys. Lett.* **62**, 1242.
Ohring, M. (1992). *The Materials Science of Thin Films.* Academic, San Diego.
Okano, H., Takahashi, Y., Tanaka, T., Shibata, K., and Nakano, S. (1992). "Preparation of c-Axis Oriented AlN Thin Films by Low-Temperature Reactive Sputtering," *Jpn. J. Appl. Phys.* **31**, 3446.
Okano, H., Tanaka, N., Takahashi, Y., Tanaka, T., Shibata, K., and Nakano, S. (1994). "Preparation of Aluminum Nitride Thin Films by Reactive Sputtering and Their Applications to GHz-Band Surface Acoustic Wave Devices," *Appl. Phys. Lett.* **64**, 166.
O'Toole, R. P., Burns, S. G., Bastiaans, G. J., and Porter, M. D. (1992). "Thin Aluminum Nitride Film Resonators: Miniaturized High Sensitivity Mass Sensors," *Anal. Chem.* **64**, 1289.
Rippert, E. D., Song, S. N., Thomas, C. Lomatch, S., Maglic, S. R., Ulmer, M., and Kettersen, J. B. (1995). "Intrinsically Damped Multilayered (Stacked) Nb/Al-AlN$_x$/Nb Superconducting Tunnel Junctions," *Appl. Supercond.* **3**, 567.
Ross, J., and Rubin, M. (1991). "High Quality GaN Grown by Reactive Sputtering," *Mater. Lett.* **12**, 215.
Ross, J., Rubin, M., and Gustafson, T. K. (1993). "Single Crystal Wurtzite GaN on (111)GaAs with AlN Buffer Layers Grown by Reactive Magnetron Sputtering," *J. Mater. Res.* **8**, 2613.
Shioaski, T., Yamamoto, T., Oda, T., Harada, K., and Kawabata, A. (1980). "Low-Temperature Growth of Piezoelectric AlN Films for Surface and Bulk Wave Transducers by rf Reactive Planar Magnetron Sputtering, *Proc. Ultrasonics Symp.*, pp. 451–454.
Sin, K., and Wang, S. H. (1996). "FeN/AlN Multilayer Films for High Moment Thin Film Recording Heads" In *IEEE International Magnetics Conference Proceedings*, AQ-08.
Strite, S., and Morkoc, H. (1992). "GaN, AlN, InA: A Review," *J. Vac. Sci. Technol.* **10**, 1231.
Sullivan, B. T., Parsons, R. R., Westra, K. L., and Brett, M. J. (1988). "Optical Properties and Microstructure of Reactivity Sputtered Indium Nitride Thin Films," *J. Appl. Phys.* **64**, 4144.

Suter, J. J., Bryden, W. A., Kistenmacher, T. J., and Porga, R. D. (1995). "Aluminum Nitride on Sapphire Films for Surface Acoustic Wave Chemical Sensors," *JHU/APL Technical Digest* **16**, 288.

Tansley, T. L., and Foley, C. P. (1984). "Electron Mobility in Indium Nitride," *Electr. Lett.* **20**, 1066.

Tansley, T. L., and Foley, C. P. (1986). "Optical Band Gap of Indium Nitride," *J. Appl. Phys.* **59**, 3241.

Tsuboughi, K., Sugai, K., and Mikoshiba, N. (1981). "AlN Material Constants Evaluation and SAW Properties on AlN/Al_2O_3 and AlN/Si," *Proc. Ultrasonics Symp.*, pp. 375–380.

Tsuboughi, K., Sugai, K., and Mikoshiba, N. (1982). "Zero Temperature Coefficient Surface-Acoustic-Wave Devices Using Epitaxial AlN Films," *Proc. Ultrasonics Symp.*, pp. 340–345.

CHAPTER 4

Thermochemistry of III–N Semiconductors

N. Newman

CENTER FOR QUANTUM DEVICES
DEPARTMENT OF ELECTRICAL AND COMPUTER ENGINEERING
NORTHWESTERN UNIVERSITY
EVANSTON, IL

I.	INTRODUCTION	55
II.	THERMODYNAMICS	58
	1. Synopsis of Formalism	58
	2. Thermochemical Values of Binary Reactants and Products	59
	3. Free Energies of Binary Reactions	67
	4. Phase Equilibria of Ternary $Ga_{1-x}In_xN$, $Ga_{1-x}Al_xN$ and $In_{1-x}Al_xN$ and Their Reactions	73
III.	THERMAL STABILITY OF GaN, InN, and AlN: THE DECOMPOSITION REACTION	79
IV.	THERMOCHEMICAL ANALYSIS OF GaN THIN FILM GROWTH	84
	1. Molecular Beam Epitaxy	84
	2. Chemical Vapor Deposition	92
V.	SUMMARY	98
	References	98

I. Introduction

Unusually large kinetic and thermodynamic barriers are present during the epitaxial growth of III–N systems. For this reason, many aspects of the synthesis methods and the resulting material quality differ from those typically encountered with other III–V compounds. In order to fully understand the critical issues of growth and processing, it is important that the energetic barriers of these material systems be well characterized. In this work, the principles of thermochemistry are used to determine the thermodynamic and kinetic barriers encountered during III–N synthesis and processing.

This chapter is organized in the following way. In Section I, a brief introduction to thermochemistry is presented. In Section II, the principles of thermodynamics are reviewed. Section II also provides a compilation of thermodynamic state functions (e.g. G, H, S) of chemicals used in III–N technology and the net change in these parameters which occur in commonly encountered III–N reactions.

In Section III, a quantitative analysis of the role that thermodynamic and kinetic barriers play in III–N decomposition reactions is presented. Since both GaN and InN are thermodynamically unstable in vacuum at moderate temperatures (~ 500 K), the rate of the decomposition reaction plays an important role in determining the conditions which optimize material growth and device fabrication processes. In this section, we show that the characteristics of III–N decomposition are highly-unusual—the reaction is dominated by an extremely large kinetic barrier resulting from breaking III–N bonds at the solid surface.

In Section IV, an analysis of the thermodynamic and kinetic barriers which occur during the synthesis of III–N thin films is presented. This section illustrates the unique chemistry of III–N semiconductors which occurs as a result of unusually large kinetic barriers. We demonstrate that the successful growth of III–N thin films actually depends on the presence of kinetic barriers. In the first example, an analysis of the plasma-assisted Molecular Beam Epitaxy (MBE) process indicates that the growth of GaN films occurs under meta-stable conditions with a non-equilibrium kinetically-limited reaction. The growth process is controlled by a competition between the forward reaction, which depends on the arrival of activated nitrogen species at the growth surface, and the reverse reaction, whose rate is limited by the unusually large kinetic barrier of GaN decomposition. In the second example, an analysis of the Chemical Vapor Deposition (CVD) process demonstrates that the reaction occurs under thermodynamically favorable conditions. The synthesis reaction is, however, found to be possible only under conditions in which the competing kinetically-limited $NH_3(g)$ decomposition reaction is sluggish (i.e. catalysts for the reaction are not present within the reactor).

There are two fundamental types of energetic barriers, *thermodynamic* and *kinetic*. Thermodynamic barriers result from differences in free energy, G ($G = H - T \cdot S$) between two phases of matter (Fig. 1). As pointed out by Gibbs (1873a, 1873b, 1876, 1878), reactions that decrease the net free energy occur spontaneously. The driving force for reactions can therefore be attributed to the reduction of the system free energy. The condition of minimal free energy can be used to uniquely predict the stable phases for a given material system and chemical environment.

Although thermodynamics determines if a reaction occurs spontaneously, it does not predict the rate of reaction. Kinetic barriers play an important

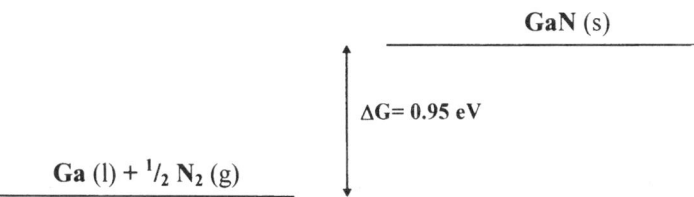

FIG. 1. A schematic diagram indicating Gibbs free energy between reactant and product for the reaction Ga(l) + 1/2N$_2$(g) = GaN(s) at 10^{-6} Torr N$_2$(g) and 800°C.

part in determining the rate of reaction (Fig. 2) (King, 1964). In contrast to thermodynamic barriers, kinetic barriers depend on the reaction path and presence of catalysts. For reactions involving nitrogen-containing molecules and compounds, the kinetic barrier of breaking the unusually strong nitrogen bond is often the rate limiting reaction step (Searcy, Ragone, and Colombo, 1970). For this reason, these systems have been used extensively as a model system for investigating the kinetics of solid state chemical reactions (Searcy, Ragone, and Colombo, 1970; Schoonmaker, 1965; Mar and Searcy, 1968; Munir, and Searcy, 1965).

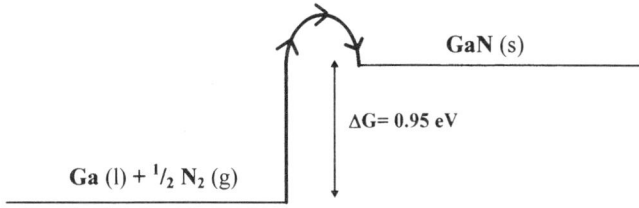

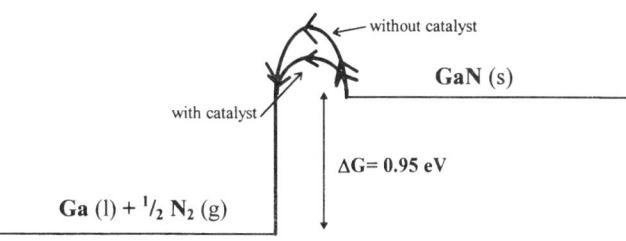

FIG. 2. A schematic diagram representing the reaction in Fig. 1 which also includes a kinetic barrier, (a) to form GaN(s) and (b) for GaN(s) to decompose.

II. Thermodynamics

1. Synopsis of Formalism

This section outlines the formalism adopted by thermochemists. It is written as a tutorial and presents essential background and commonly encountered methodology of thermochemistry. Readers familiar with this field can advance to Section III and use this section only as a reference source for tabulated thermodynamic values.

Because changes in temperature and pressure are accompanied by an exchange of energy between matter and surroundings, Gibbs introduced the definition of internal energy to differentiate energy associated with matter from actual mechanical energy (Gibbs, 1873a, 1873b, 1876, 1878). Using this concept, the free energy of a chemical system can be uniquely defined by the state functions, temperature (T), pressure (P), volume (V), entropy (S), and internal energy (U). Of these, only the internal energy cannot be determined absolutely, as it can only be defined to within an indeterminate constant.

To establish a convenient formalism that removes the ambiguity associated with this indeterminate constant, the enthalpy is typically defined with respect to pure elements in their standard states at a temperature of 298.15 K and a pressure of 1 bar (i.e., $H^{1\ \text{bar}, 298.15} = H_o = 0$) (Chase et al., 1985; Gaskell, 1981; Van Wylen and Sonntag, 1985; Knacke, Kubaschewski, and Hesselmann, 1991). The subscript o is used to indicate that the state parameters should be evaluated at standard pressure–temperature conditions. Parameters at 1 bar and an arbitrary temperature T are denoted by the following notation, H_o^T, S_o^T, and G_o^T. Similarly, parameters at 298.15 K and an arbitrary pressure P are denoted by H_o^P, S_o^P, and G_o^P.

From the knowledge of free energy of reactants and products, phase diagrams can be derived. For binary reactions 1 and 2, the free energy difference as a function of temperature and pressure can be determined from Eqs. (3) and (4):

$$X(l) + 1/2 N_2(g) = XN(s) \tag{1}$$

$$X(l) + NH_3(g) = XN(s) + 3/2 H_2(g) \tag{2}$$

$$\Delta G^{P,T} = G^{P,T}_{XN(s)} - G^{P,T}_{X(l)} - 1/2 G^{P,T}_{N_2(g)} \tag{3}$$

$$\Delta G^{P,T} = G^{P,T}_{XN(s)} + 3/2 G^{P,T}_{H_2(g)} - G^{P,T}_{X(l)} - G^{P,T}_{NH_3(g)} \tag{4}$$

where X represents the group III atom (i.e., Al, Ga, In).

In equilibrium under constant pressure and temperature conditions, the free energy change is zero (i.e., $\Delta G^{P,T} = 0$) (Gibbs, 1873a, 1873b, 1876, 1878).

This criterion enables the determination of the thermodynamically stable phases and their concentration.

Phase diagrams are used to graphically illustrate thermodynamically stable products and their relative concentration as a function of independent variables, typically temperature, pressure, and composition.

The phase rule determines the number of degrees of freedom (F) present in reaction equilibria (Gaskell, 1981). For reaction 1, the number of phases (P) is three (i.e., XN(s), X(l), and N_2(g)). The number of components (C) is two[1] (as indicated by two elements, the group III element and nitrogen). The phase rule ($F = C + 2 - P = 2 + 2 - 3 = 1$) indicates equilibria can be characterized with only one degree of freedom. For reaction 2, the phase rule with three components (nitrogen, hydrogen and group III element) and four phases (XN(s), X(l), NH_3(g), H_2(g)) indicates that one degree of freedom characterizes reaction equilibria (i.e., $F = C + 2 - P = 3 + 2 - 4 = 1$).

The fundamental basis for this conclusion can be readily understood by noting that the equation for each system (Eqs. (3) and (4) with $\Delta G^{P,T} = 0$) has two independent variables (P and T). Three deductions follow:

1. At equilibrium, the determination of one independent variable (P or T) uniquely determines its counterpart.
2. For a given temperature T_o, there exists a specific pressure ($p_{N_2eq}^{T_o}$) such that all greater pressures will result in spontaneous formation of XN(s) and all lesser pressures result in spontaneous decomposition of XN(s) into the group III liquid X(l) and N_2(g).
3. Similarly, for a given pressure P_o, there exists a specific temperature ($T_{N_2eq}^{P_o}$) below which XN(s) spontaneously forms and above which XN(s) spontaneously decomposes into X(l) and N_2(g).

The phase diagrams shown in Fig. 3 illustrate the above deductions for several nitride semiconductor systems.

2. THERMOCHEMICAL VALUES OF BINARY REACTANTS AND PRODUCTS

Free energy of solids, liquids, and gases depends on external environmental conditions. In this section, we outline the formalism adopted by thermochemists to define thermodynamic parameters (e.g., H, S, G) (Chase

---

[1] Typically component C is the number of elements present in the reaction. However, if two or more atoms occur together in the same proportion in each phase, the group represents a single component.

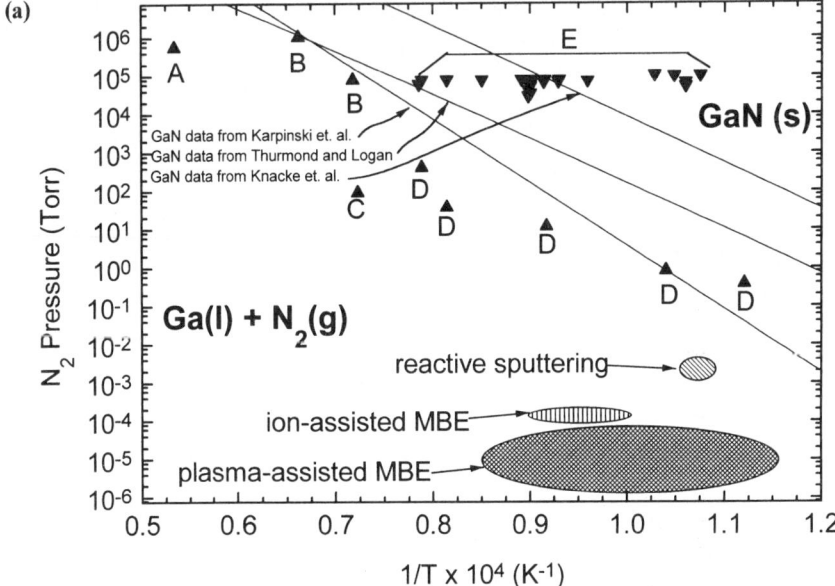

FIG. 3. (a) Phase diagram for reaction $Ga(l) + N_2(g) \to GaN(s)$. Triangles indicate lower (▲) and upper (▼) bounds of critical stability line for phase equilibrium. Points A, B, C, D, and E correspond to MacChesney, Bridenbaugh, and O'Connor (1970); Karpinski, Jun, and Porowski (1984); Juza and Hahn (1940); Lorenz and Binkowski (1962); Gilleson, Schuller, and Struck (1977), respectively. The theoretical critical stability line, as calculated using parameters from Thurmond and Logan (1972); Karpinski (1984); Knacke, Kubaschewski, and Hesselmann (1991) are also included. Note that the fit from Thurmond and Logan is consistent with experiment. Also shown are typical gas pressure and substrate temperature used for successful production of GaN thin films (Newman, Ross, and Rubin, 1993; Fu et al., 1995; Gassmann et al., 1996; Lei et al., 1992; Newman et al., 1994; Powell et al., 1990; Rubin et al., 1994). (From Newman, Ross, and Rubin, 1993.) (b) Phase diagram for reaction $Ga(l) + NH_3(g) \to GaN(s) + 3/2H_2(g)$. Arrows indicate lower (↑) and upper (↓) bounds of critical stability line of phase equilibrium. Theoretical critical stability line, as calculated by Thurmond and Logan (1972) is also included. Also shown are typical gas pressures and substrate temperatures used for the successful production of GaN thin films (Sato et al., 1990; Powell, Lee, and Greene, 1992; Yang, Li, and Wang, 1995; Liu and Stevenson, 1978; Nakamura, 1991). (From Newman, Ross, and Rubin, 1993.) (c) Phase diagram for reaction $In(l) + NH_3(g) \to InN(s) + 3/2H_2(g)$. Critical stability lines, as determined in Grzegory et al. (1995) (solid line) and MacChesney, Bridenbaugh, and O'Connor (1970) (dashed line), are also included. Also shown are typical gas pressures and substrate temperatures used for successful synthesis of InN thin films. Points A, B, C, and D correspond to Matsuoka et al. (1989); Kistenmacher, Ecelberger, and Bryden (1994); Bryden et al. (1994); Abernathy et al. (1995), respectively. Data points which are boxed indicate growth conditions which results in the presence of metallic indium within films.

(b)

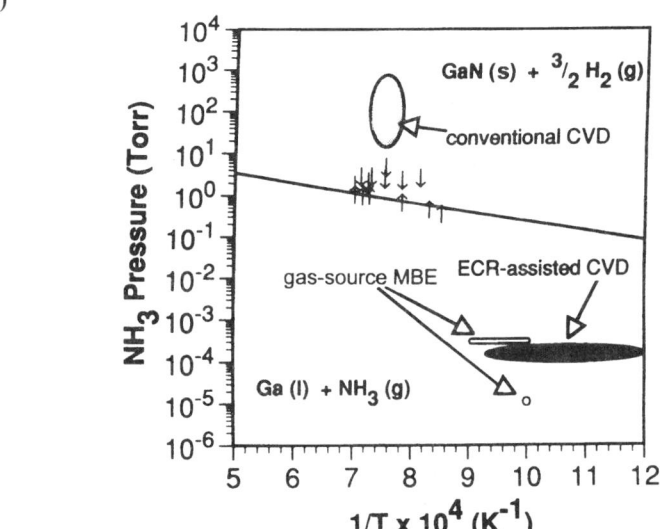

(c)

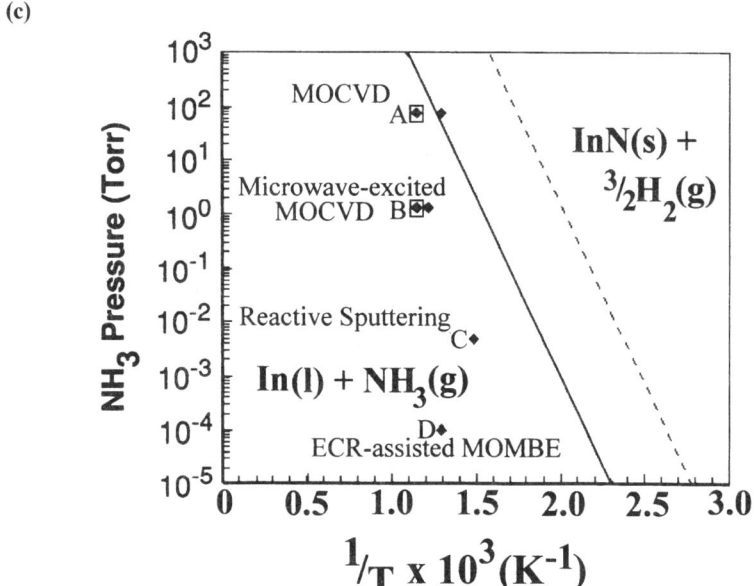

Fig. 3. (*Continued*).

et al., 1985; Gaskell, 1981; Van Wylen and Sonntag, 1985; Knacke, Kubaschewski, and Hesselmann, 1991). Also included is a compilation of thermodynamic data of chemicals used in synthesis and processing of binary and ternary III–N semiconductors.

For gases, the thermodynamic parameters used in this work are derived from models employing ideal gas laws and spectroscopically derived energies of rotational, vibrational, and electronic states.[2] For liquids, elemental solids and binary solids, the thermochemical parameters used in this work are derived from experimental data (see footnote 2). Only for ternary III–N alloy systems are experimentally determined thermochemical parameters not available. For these alloy systems, the thermodynamic parameters are inferred from regular solution model in conjunction with theoretically determined mixing enthalpies (van Schilfgaarde, Sher, and Chen, 1997) and the experimentally determined binary parameters.

a. Heat Capacity, Enthalpy, Entropy, and Free Energy of Condensed Phases

The heat capacity of a solid c, is defined to be the temperature derivative of internal energy (i.e., dU/dT). This quantity is important in that it uniquely determines the temperature dependence of the state variables, including enthalpy (H), thermal entropy (S_{th}), and free energy (G). c_p and c_v are used to denote heat capacity at constant pressure and volume conditions, respectively.

Above ~ 10–20 K (where free carrier electronic excitation can dominate), the main contribution to heat capacity of solids emanates from phonons (Ashcroft and Mermin, 1976). Acoustic modes give the largest contribution to the heat capacity when the thermal energy is small compared to the optical phonon energies. Under these conditions, the Debye theory gives excellent agreement with experiment, with $c_v = 9R(T/\Theta_D)^3 \int_0^{\Theta_D/T} x^4 e^{-x}/(1 - e^{-x})^2 \, dx$, where $\Theta_D = h\nu_D/k$, $x = h\nu/kT$ and ν_D is the phonon frequency. At temperatures in which the thermal energy is on the order of the optical phonon energies, the dominant contributor to the heat capacity is the excitation of optical phonons and the Einsten theory accurately models experimental results, with $c_v = 3R(h\nu/kT)^2 e^{\Theta_E/T}/(e^{\Theta_E/T} - 1)^2$, where $\Theta_E = h\nu/R$.

The complex equations from the theories of Debye and Einstein do not permit the derivation of simple analytic expressions for other state variables. To accomplish this, thermochemists have traditionally fit heat capacity to the following power–law equation over the temperature ranges of interest

---

[2]In this work, thermodynamic parameters are derived from the work of Knacke et al. (1991), except when specifically referenced otherwise.

(Gaskell, 1981; Van Wylen and Sonntag, 1985; Knacke, Kubaschewski, and Hesselmann, 1991).

$$c_p = a + bT + cT^{-2} \tag{5}$$

From this, the temperature dependence of the state variables can be calculated.

$$H_o^T = H_o^{298} + \int_{298}^T c_p dT = H_o^{298} + aT + bT^2/2 - cT^{-1} + C_1$$

$$= H^+ + aT + bT^2/2 - cT^{-1} \tag{6}$$

$$S_o^T = S_o^{298} + \int_{298}^T c_p T^{-1} dT = S_o^{298} + a \ln T + bT - 1/2cT^{-2} + C_2$$

$$= S^+ + a \ln T + bT - 1/2cT^{-2} \tag{7}$$

$$G_o^T = H_o^T - TS_o^T = (H^+ + aT + bT^2/2 - cT^{-1})$$
$$- T(S^+ + a \ln T + bT - 1/2cT^{-2})$$
$$= (a - S^+)T - aT \ln T - bT^2/2 - 1/2cT^{-1} + H^+ \tag{8}$$

Since $H^+(=H_o^{298} + C_1)$ and $S^+(=S_o^{298} + C_1)$ include an integration constant, no physical significance is directly associated with these quantities.

Thermodynamic parameters for solids and liquids of interest are listed in Table I. For solids and liquids, the heat capacity, enthalpy, entropy and free energy are virtually independent of pressure (typically of order $\Delta G \delta P$ (in bar) $\simeq 0.01 \Delta G \delta T$ (in K)) (King, 1964). For the experimental conditions of interest to this work, the influence of pressure on thermodynamic quantities of solids and liquids is insignificant and will not be included in our analysis.

b. *Heat Capacity, Enthalpy, Entropy, and Free Energy of Gases*

For an ideal gas, the heat capacity is a function of temperature alone (Van Wylen and Sonntag, 1985; Knacke, Kubaschewski, and Hesselmann, 1991; Kennard, 1938; Jeans, 1948; Knudsen, 1950; Knudsen, 1915; Langmuir, 1913; Hertz, 1882). The heat capacity of *atomic* gases arise from thermally induced excitation of kinetic energy and electronic states. For *polyatomic* gases, the heat capacity also includes contributions from excitation of rotational and vibrational states (Van Wylen and Sonntag, 1985; Knacke, Kubaschewski, and Hesselmann, 1991; Kennard, 1938; Jeans, 1948; Knudsen, 1950; Knudsen, 1915; Langmuir, 1913; Hertz, 1882).

The classical statistical equipartition theorem predicts that each translational, vibrational, and rotational degree of freedom contributes $RT/2$ per

TABLE I

Thermodynamic Parameters of Relevant Chemical Phases[a]

Phase	a J K^{-1} mol^{-1}	b J K^{-2} mol^{-1}	c J K^2 mol^{-1}	d J K^{-3} mol^{-1}	H$^+$ J mol^{-1}	S$^+$ J K^{-1} mol^{-1}
AlN(s)	47.823	1.849×10^{-3}	-1.674×10^6		-338.356×10^3	-262.292
GaN(s)	38.074	8.996×10^{-3}			-121.373×10^3	-189.908
GaN(s)[b]	45.3631	6.500×10^{-3}			-125.10262×10^3	-218.91686
GaN(s)[c]	39.593	3.565×10^{-3}	0.191		-163.520×10^3	-206.225 ($T < 700$)
GaN(s)[c]	41.777	1.272×10^{-3}	-0.119		-164.892×10^3	-219.244 ($T > 700$)
InN(s)	38.074	12.134×10^{-3}			-149.963×10^3	-177.037
Al(l)	31.752				-0.926×10^3	-145.913
Al(g)	20.778	0.004×10^{-3}	0.054×10^6		323.549×10^3	46.476
Ga(l)	24.384	2.293×10^{-3}	0.310×10^6		-0.755×10^3	-78.636 ($T < 700$ K)
Ga(l)	26.568				-2.164×10^3	-91.655 ($T > 700$ K)
Ga(g)	24.866		0.251×10^6		264.077×10^3	29.079
In(l)	29.878	-1.381×10^{-3}			-6.501×10^3	-107.124 ($T < 900$ K)
In(l)	29.079	-0.891×10^{-3}			-6.142×10^3	-102.490 ($T > 900$ K)
In(g)	22.686	2.364×10^{-3}	-0.238×10^6		238.706×10^3	42.476
N$_2$(g)	30.418	2.544×10^{-3}	-0.238×10^6		-9.982×10^3	16.203
NH$_3$(g)	37.321	18.661×10^{-3}	-0.649×10^6		-60.244×10^3	-29.402
N(g)	20.878	0.146×10^{-3}	0.038×10^6	0.054×10^{-6}	466.587×10^3	34.599
H$_2$(g)	26.882	3.586×10^{-3}	0.105×10^6		-7.823×10^3	-22.966
N$_2$H$_4$(g)	56.275	42.509×10^{-3}	-1.665×10^6		71.159×10^3	-103.94
GaCl(g)	37.217	0.661×10^{-3}	-0.151×10^6		-92.457×10^3	27.126
HCl(g)	26.527	4.602×10^{-3}	0.109×10^6		-100.056×10^3	35.010
Cl$_2$(g)	36.610	1.079×10^{-3}	-0.272×10^6		-11.875×10^3	12.638

[a] From Knacke, Kubaschewski, and Hesselmann, 1991, except when noted.
[b] Calculated using thermochemical parameters in this table (Knacke, Kubaschewski, and Hesselmann, 1991), combined with change in thermodynamic parameters for reactions 1 and 2 from Thurmond and Logan (1972).
[c] Calculated using thermochemical parameters in this table (Knacke, Kubaschewski, and Hesselmann, 1991), combined with change in thermodynamic parameters for reaction 1 from Karpinski, Jun, and Porowski (1984).

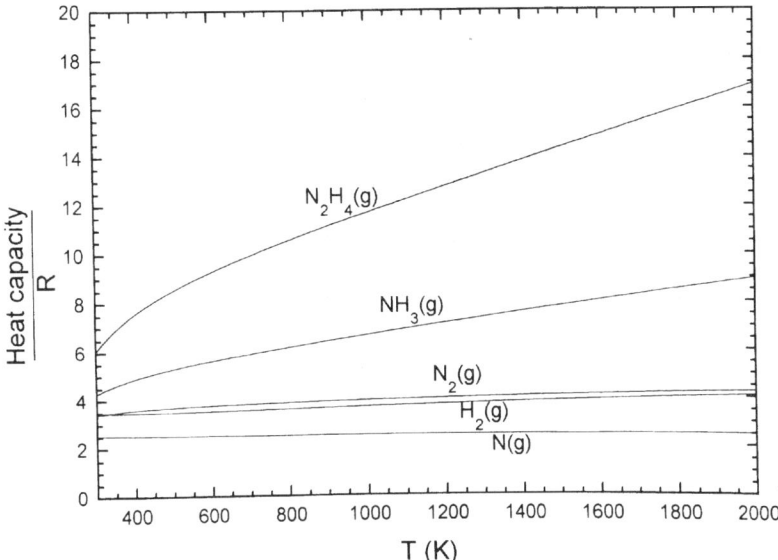

FIG. 4. Heat capacity, c_p, of several gasses as a function of temperature.

mole to the heat capacity. However, because of the quantized nature of vibrational, rotational, and electronic energies, their contribution is suppressed at low temperatures (Van Wylen and Sonntag, 1985; Knacke, Kubaschewski, and Hesselmann, 1991; Kennard, 1938; Jeans, 1948; Knudsen, 1950; Knudsen, 1915; Langmuir, 1913).

Figure 4 illustrates the temperature dependence of the heat capacity for gasses of nitrogen atoms, nitrogen molecules, hydrogen molecules, hydrazine and ammonia. In this paragraph, the physical mechanism responsible for the observed heat capacity is briefly described. For a monatomic gas, such as atomic nitrogen (N), the three degrees of translational freedom result in an internal energy (U_t) of $3/2RT$ and therefore a heat capacity of $c_v = 3/2R$ and $c_p = c_v + R = 5/2R$. In contrast to the heat capacity of monatomic gasses, polyatomic gasses are found to have a strong temperature dependence. For N_2, a homogeneous diatomic gaseous molecule, use of the classical model with approximation of two atoms rigidly bound,

$$U = U_t + U_r = 5/2RT \quad \text{and therefore} \quad c_v = 5/2R \quad \text{and} \quad c_p = 7/2R$$

gives a reasonable estimate of experimentally determined room temperature values, $c_v = 2.448R$ and $c_p = 3.493R$ (Kenard, 1938; Jeans, 1948; Knudsen, 1950). At elevated temperatures, the contribution by vibrational degrees of freedom also needs to be included, $U = U_t + U_r + U_v = 7/2RT$, that is,

$c_v = 7/2R$ and $c_p = 9/2R$. Because of the large energy required to excite electronic states of nitrogen molecules, their contribution is negligible except at temperatures greater than ~ 4800 K (Chase et al., 1985). For NH_3 at room temperature, the experimental value of $3.42R$ for c_v and $4.48R$ for c_p (Van Wylen and Sonntag, 1985) is indicative of the three center-of-mass translational degrees of freedom, two rotational degrees of freedom, and an additional vibrational contribution from bond bending distortions. The classical prediction for heat capacity is approached at high temperature (i.e., $c_v = 9R$, $c_p = 10R$).

To calculate the contribution of a given quantized rotational or vibration mode with frequency v to the heat capacity, the harmonic oscillator model can be used (Knacke, Kubaschewski, and Hesselmann, 1991). This model predicts $c_v = R[2/((\Theta_v/T) \sinh(\Theta_v/T))]^{-2}$, where $\Theta_v = (hv/k)$ is the characteristic temperature.

As was encountered in Section II.2, a simpler expression is needed to derive analytic expressions, and c_p is typically fitted to a power–law equation over the temperature ranges of interest (Gaskell, 1981; Knacke, Kubaschewski, and Hesselmann, 1991; Weast, 1973).

$$c_p = a + bT + cT^{-2} + dT^2 \qquad (9)$$

In contrast to the equation used for solids and liquids, a fourth term proportional to T^2 is included. It was initially introduced to account for excitation of electronic states (Knacke, Kubaschewski, and Hesselmann, 1991).

Using the approach outlined above, the following equations are used to describe thermodynamic parameters of gasses as a function of temperature.

$$H_o^T = H^+ + aT + bT^2/2 - cT^{-1} + \frac{d}{3}T^3 \qquad (10)$$

$$S_o^T = S^+ + a \ln T + bT - 1/2cT^{-2} + \frac{d}{2}T^2 \qquad (11)$$

$$G_o^T = (a - S^+)T - aT \ln T - bT^2/2 - \frac{d}{6}T^3 - 1/2cT^{-1} + H^+ \qquad (12)$$

To determine the change in free energy as a function of gas pressure, the ideal gas law can be utilized (Van Wylen and Sonntag, 1985; Knacke, Kubaschewski, and Hesselmann, 1991; Kennard, 1938; Jeans, 1948; Knudsen, 1950; Knudsen, 1915; Langmuir, 1913; Hertz, 1882).

$$G^{T,P} = G_o^T + \int_{P_o}^{P} V\,dP = G_o^T + RT \int_{P_o}^{P} \frac{RT}{P}\,dP = G_o^T + RT \ln \frac{P}{P_o}$$
$$= G_o^T + RT \ln P \qquad (13)$$

where P is in bars and P_o is defined at standard pressure conditions (i.e., 1 bar).

Intermolecular interactions in gases and solutions (e.g., mutual attractive force between molecules and finite volume of gaseous species) cause deviations from predictions of ideal stochastic models (Van Wylen and Sonntag, 1985; Knacke, Kubaschewski, and Hesselmann, 1991). Under these conditions, phases do not exhibit properties of ideal gasses, pure solids, or ideal solutions. Deviations from ideal gas law are particularly important at high pressures used in bulk growth of GaN (Karpinski, Jun, and Porowski, 1984).

Under these conditions, the fugacity f, is used to represent the activity of a species. Fugacity, f, (Searcy, Ragone and Colombo, 1970; Van Wylen and Sonntag, 1985) is defined as $dG = RT\, d\ln f$ where dG is at constant temperature. The activity a (Searcy, Rigone and Colombo, 1970) is defined as $a = f/f^0$ where f^0 is the fugacity of the component in the standard state.

Free energy of gasses, including those that do not exhibit ideal behavior, can be represented by (Van Wylen and Sonntag, 1985):

$$G^{P,T} = G^T + RT \ln f \qquad (14)$$

3. Free Energies of Binary Reactions

In this section, we establish the effect of chemical environment on the free energy of reaction.

An equilibrium constant K, equal to $e^{-\Delta G_{T,P}/RT}$ can be derived from thermodynamic statistical partition function, and can be used to determine all thermodynamic properties, including reactant and product concentrations (Reif, 1965).

In an analysis using equilibrium constants, it is preferred that formulas be expressed as a function of activities, rather than concentrations, so that they are as simple and general as possible. This formalism is particularly valuable when investigating chemistry of phases which do not have properties of ideal gasses, pure solids or ideal solutions (Searcy, Ragone and Colombo, 1970; Van Wylen and Sonntag, 1985). Thus, for reactions 1 and 2:

$$\Delta G^{T,P} = G^{T,P}_{XN(s)} - G^{T,P}_{X(l)} - 1/2 G^{T,P}_{N_2(g)} = -RT \ln K \qquad (15)$$

where $K = a_{XN(s)}/(a_{X(l)} a^{1/2}_{N_2(g)})$,

$$\Delta G^{T,P} = G^{T,P}_{XN(s)} + 3/2 G^{T,P}_{H_2(g)} - G^{T,P}_{X(l)} - G^{T,P}_{NH_3(g)} = -RT \ln K \qquad (16)$$

where $K = a_{XN(s)} a^{3/2}_{H_2(g)}/(a_{X(l)} a_{NH_3(g)})$.

The activity of pure liquids and pure solids in the standard state is unity. Activity of ideal gasses is P/P_o where P_o is the reference pressure (by convention, the standard pressure of 1 bar) (Searcy, Ragone and Colombo, 1970; Gaskell, 1981; Van Wylen and Sonntag, 1985; Knacke, Kubaschewski, and Hesselmann, 1991; Kennard, 1938; Jeans, 1948; Knudsen, 1950; Knudsen, 1915; Langmuir, 1913; Hertz, 1882).

Neglecting the insignificant effect of pressure induced changes in the free energy of solids and liquids (discussed in Section II.1), the change in free energy at arbitrary temperature and pressure becomes:

$$\Delta G^{T,P} = G^T_{XN(s)} - G^T_{X(l)} - 1/2(G^T_{N_2(g)} + RT \ln P_{N_2(g)}) \tag{17}$$

$$\Delta G^{T,P} = G^T_{XN(s)} + 3/2(G^T_{H_2(g)} + RT \ln P_{H_2(g)}) - G^T_{X(l)} - G^T_{X(l)}$$
$$- (G^T_{NH_3(g)} + RT \ln P_{NH_3(g)}) \tag{18}$$

where P is in bar.

For reactions 1 and 2, the free energy change and equilibrium pressure are plotted as a function of temperature in Figs. 5(a)–(f).

This allows the equilibrium constant of the reaction to be written as:

$$K = p_{N_2}^{-1/2}$$
$$K = p_{H_2}^{3/2}/p_{NH_3} \tag{19}$$

The relationship between equilibrium pressure of N_2 and NH_3, and the temperature, establishes the critical stability line in phase diagrams of Figs. 3 and 5(d)–(f).

The N_2 molecule is relatively inert due to the strong triple bond. For this reason, more reactive forms of nitrogen have been utilized to synthesize III–N semiconductors (Newman, Ross, and Rubin, 1993; Anders *et al.*, 1996; Bryden *et al.*, 1994; Fu *et al.*, 1995; Gassmann *et al.*, 1996; Lei *et al.*, 1992; Paisley *et al.*, 1989; Martin *et al.*, 1991; Harris, unpublished; Newman *et al.*, 1994; Powell *et al.*, 1990; Rubin *et al.*, 1994; Sato *et al.*, 1990), including hydrazine, atomic nitrogen, excited molecular nitrogen, and ionized molecular nitrogen. Section IV.2 will, in fact, show that use of reactive species is a necessary criterion for the synthesis of GaN and InN under typical molecular beam epitaxy (MBE) vacuum conditions. Figure 5 illustrates ΔG and the equilibrium pressure of nitrogen-containing species for reactions with N_2, NH_3, and several representative reactive species (i.e., hydrazine and atomic nitrogen).

$$X(l) + 1/2 N_2H_4(g) = XN(s) + H_2(g) \tag{20}$$
$$X(l) + N(g) = XN(s) \tag{21}$$

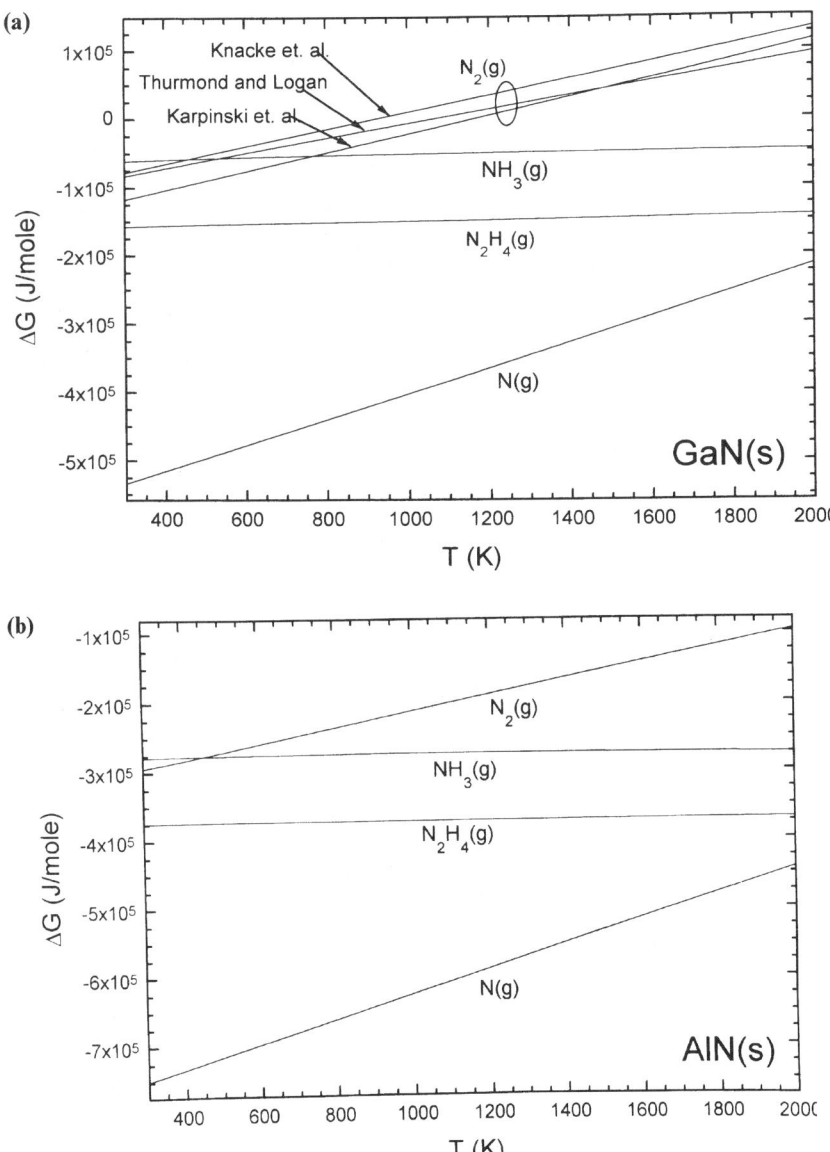

FIG. 5. Change in Gibbs free energy and equilibrium pressures for the following reactions: (1) X(l) + 1/2N$_2$(g) = XN(s), (2) X(l) + NH$_3$(g) = XN(s) + 3/2H$_2$(g), (3) X(l) + 1/2N$_2$H$_4$(g) = XN(s) + H$_2$(g), (4) X(l) + N(g) = XN(s), where X represents group III atom (i.e., Al, Ga, In).

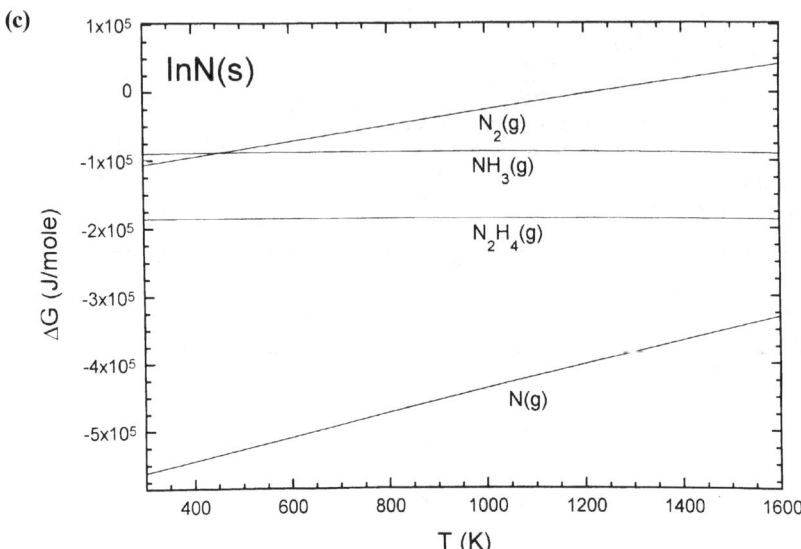

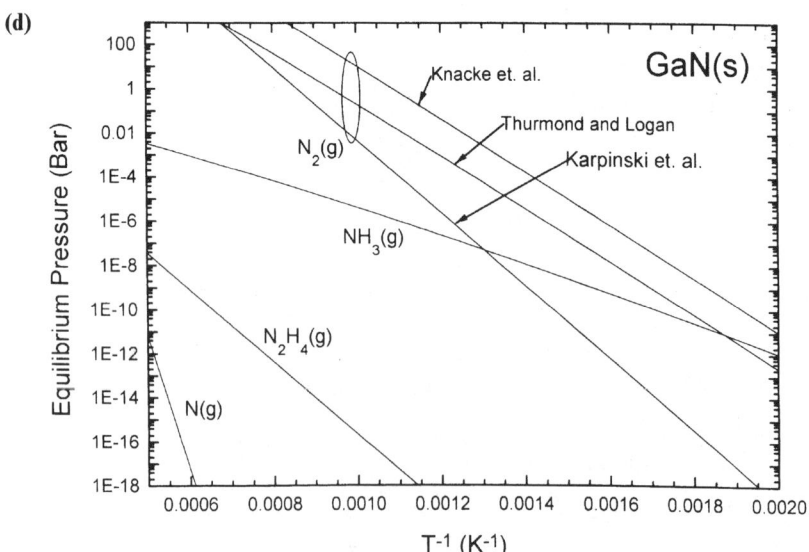

FIG. 5. (*Continued*).

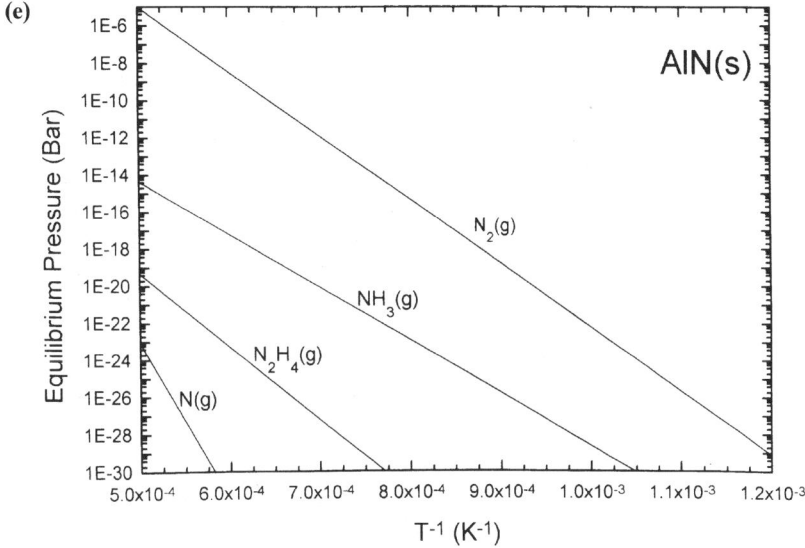

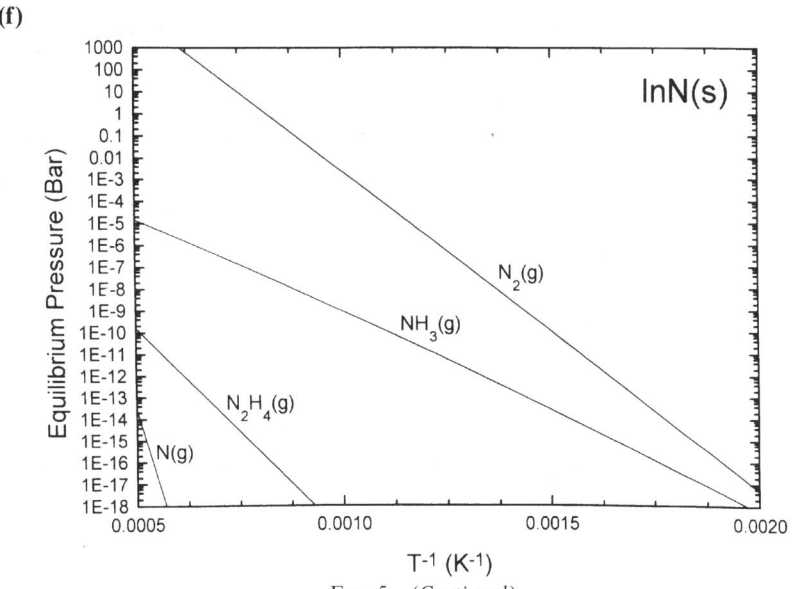

FIG. 5. (*Continued*).

The reader is referred to Section IV for a more complete discussion of reactive synthesis of GaN.

So far, we have focused on accurately determining the temperature and pressure dependence of thermodynamic parameters. Although this analysis facilitates precise predictions, additional intuition into the fundamentals of phase diagrams (Fig. 3) can be obtained from a less rigorous approach. The state variables ΔH and ΔS are not expected to vary greatly over moderate ranges in temperature since, in principle, the change in enthalpy (ΔH) arises primarily from differences in bond strength and the change in entropy (ΔS) from the difference in configurational disorder between reactants and products (Gaskell, 1981). In practice, ΔH and ΔS are often found to be relatively independent of temperature (implicit in the validity of this approximation is that $c_p \simeq 0$) (Gaskell, 1981). Under this assumption, it is readily shown (i.e., $\ln P_{N_2}^{eq} = 2\Delta G_o/RT = 2\Delta H_o/R(1/T) - 2\Delta S/R$) that Fig. 3(a) has a slope and intercept equal to $2\Delta H_o/R$ and $-2\Delta S_o/R$, respectively. Similarly, it can be shown (i.e., $\ln P_{NH_3}^{eq} = \Delta G_o/RT = \Delta H_o/R(1/T) - \Delta S_o/R$), that Fig. 3(b, c) would have a slope and an intercept equal to $\Delta H_o/R$ and $-\Delta S_o/R$, respectively. Note that although ΔS is typically relatively temperature independent, the entropy contribution to the free energy (i.e., $T\Delta S$) increases with temperature and is the driving force for formation of high entropy phases (e.g., gasses and alloys) at elevated temperatures.

Because of the large kinetic barriers to III–N formation and decomposition, experimental values used to determine thermochemical state variables have large error bounds. A lower bound for the critical stability line can be estabished when a III–N compound decomposes under well defined pressure and temperature conditions. Similarly, an upper bound for the critical stability line can be established when a III–N compound is synthesized from elemental group III metal and nitrogen-containing reactant. As can be seen in the phase diagrams in Fig. 3, the experimental data contains large differences between experimentally determined upper and lower bounds for the critical stability line, particularly in the lower range of temperatures. When establishing the phase diagram, it is extremely important that experiments actually observe a phase change rather than just recording phases present under well defined pressure and temperature conditions. This criterion ensures that the observed phases are thermodynamically stable and not an artifact of a very slow reaction resulting from large kinetic barriers. Karpinski, Jun, and Porowski (1984) reported that bulk GaN samples decompose to 50% of their volume in ~ 1 day at 1000°C and ~ 1.5 h at 1200°C.

Measurements at higher temperature are expected to be less prone to this error. In the case of GaN and InN, the extremely high nitrogen pressures required to achieve equilibrium conditions at elevated temperatures results in nonideal gas properties, complicating the determination of thermochemical parameters. Work to improve the accuracy of the experimental measurements for GaN and InN in the low temperature regions and to experimentally establish ternary and quaternary III–N alloy system phase diagrams is needed.

4. Phase Equilibria of Ternary $Ga_{1-x}In_xN$, $Ga_{1-x}Al_xN$ and $In_{1-x}Al_xN$ and Their Reactions

Because the state variables, enthalpy, entropy, and free energy have not been experimentally determined for the ternary nitride systems, knowledge of the thermochemistry for these important compounds is still in the rudimentary stages. In this section, we will use the regular solution model, in combination with theoretically calculated values for heats of mixing (van Schilfgaarde, Sher, and Chen, in press), to investigate the alloy phase diagrams (Chen and Sher, 1995).

The regular solution model, which is derived from statistical theory, presumes a random configuration of atoms and a zero volume change in mixing. If interatomic forces between dissimilar components are assumed to have only short range interactions, the mixing enthalpy, H_{AB}, has the form (Gaskell, 1981; Chen and Sher, 1995)

$$H_{AB} = xH_A + (1-x)H_B + \Omega[x(1-x)] \qquad (22)$$

where Ω is the interaction energy.

For lattice mismatched systems, the quadratic term in this equation is dominated by strain, for which a valence force field (VFF) approach (Shih et al., 1985; Hwang et al., 1988), or the bond–orbital approximation (Harrison, 1980), can be used to model mixing enthalpy. Starting from a VFF approach, Shi et al. (1985) considered the relaxation around an isolated impurity in a zinc-blende lattice, allowing only nearest neighbors to relax. To illustrate this model, we use a cation impurity, labeled B and a host lattice, labeled A–N. Neglecting angular forces, the B–N bond in an alloy is predicted to relax from the A–N bond length to three-quarters of the equilibrium B–N bond length (Shi, 1985). This result is in excellent agreement with experimental X-ray absorption fine structure (XAFS) work for a large number of semiconductors (Mikkelsen and Boyce, 1982). The model also predicts that lattice relaxation results in a decrease in strain energy to one-quarter of the value of an unrelaxed system. Chen and Sher (1995) considered a similar but more sophisticated treatment that included the contribution of relaxation of more distant neighbors (lowers energy) and angular forces (increases energy). Both make small corrections of opposite

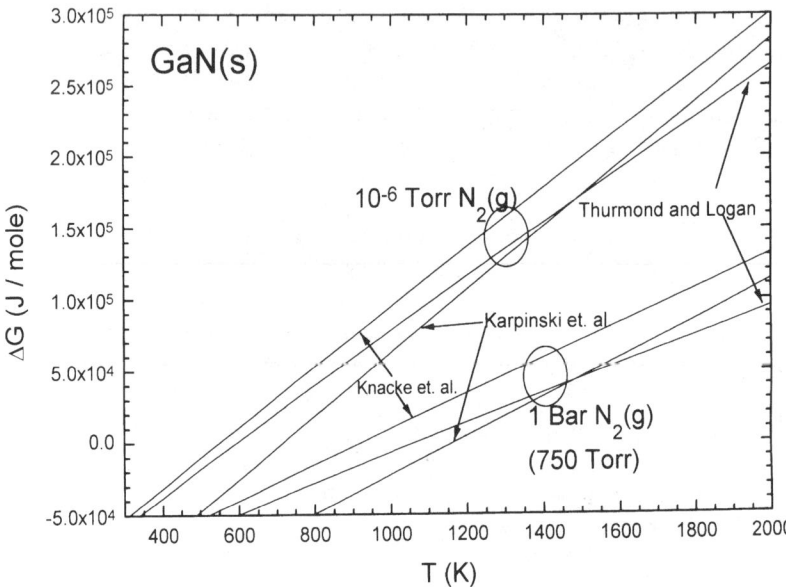

FIG. 6. Change in Gibbs free energy for reaction $X(l) + 1/2 N_2(g) = XN(s)$ at 1 bar and at 10^{-6} Torr $N_2(g)$ pressure.

signs (~15%) to the simple model of Shi et al. (1985), which very nearly cancel each other. Thus, to a reasonably good approximation, the mixing enthalpy may be calculated using Eq. (22), as derived from a local, nearest neighbor model (Chen and Sher, 1995; Shih et al., 1985; Hwang et al., 1988).

In principle, the heat of reaction $(BN)_x + (AN)_{1-x} = A_{1-x}B_xN$ is equal to $\Omega[x(1-x)]$ (Chen and Sher, 1995). Therefore, Ω can be calculated from the difference between the total energy of random alloys and separate binary compounds. We use theoretically determined values of van Shilfgaarde, Sher, and Chen (1997). In this work, a local density functional calculation was performed using full-potential linear Muffin Tin Orbital theory (Methfessel, 1988). The difference in the total energy of randomly configured 50–50% cation solutions (i.e. $A_{0.5}B_{0.5}N$) of 32-atom supercells and weighted average of separate binary solids (i.e., AN, BN) were determined (van Schilfgaarde, Sher, and Chen 1997). Results are tabulated in Table I. Note that when In is a constituent, Ω is large in comparison to kT; indicating these solutions will have a strong tendency to undergo spinodal decomposition.

In addition to the enthalpy, the entropy of random mixing must be included in the determination of free energy (Gibbs 1873a, 1873b, 1876,

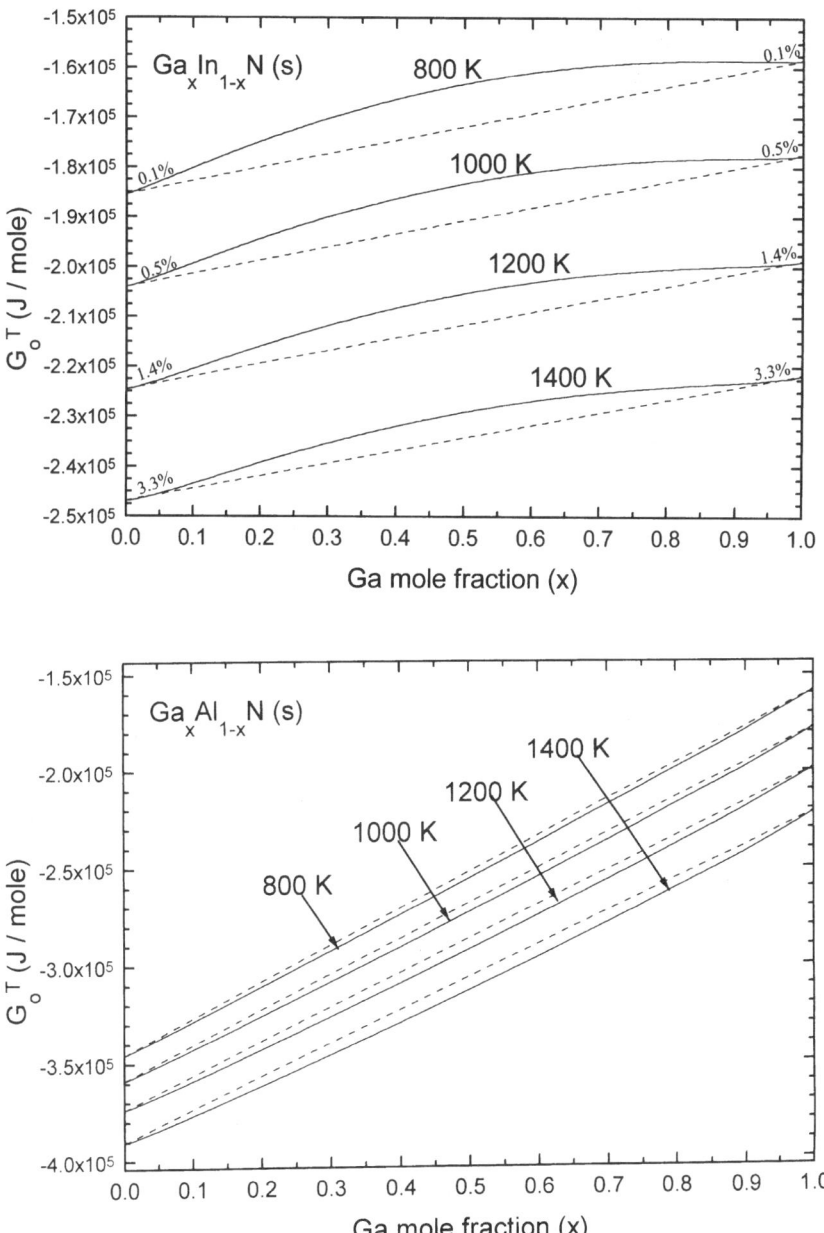

FIG. 7. Gibbs free energy of ternary III–N alloys. The ternary alloys are the thermodynamically favored phase when the alloy free energy (solid line) is below that of the binary phases (dashed line). The maximum solubility of the minority component for systems with a miscibility gap are given. Thermodynamic parameters used for GaN are from Thurmond and Logan (1972).

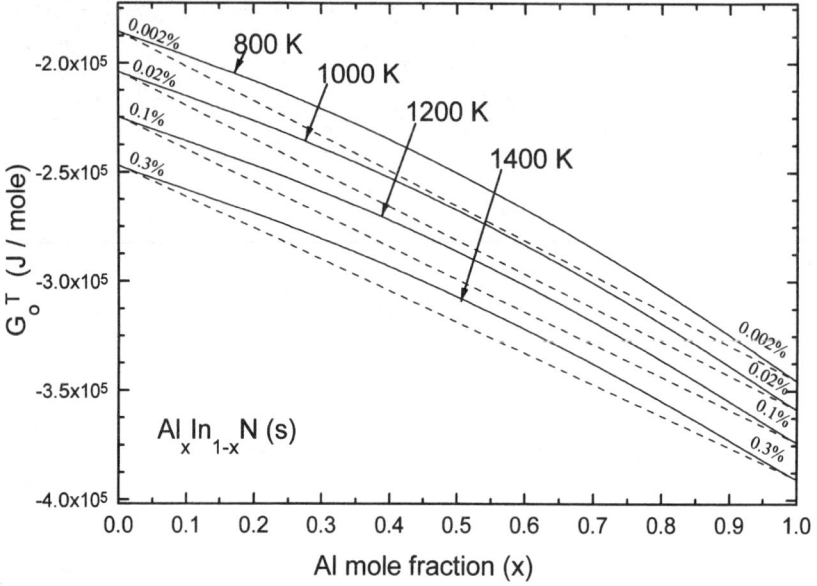

FIG. 7. (*Continued*).

1878; Gaskell, 1981; Shih et al., 1985; Hwang et al., 1988). The entropy change can be derived to be (Gibbs, 1873a, 1873b, 1876, 1878):

$$\Delta S = R \ln W = R \ln \left[\frac{(N_A + N_B)!}{N_A! N_B!} \right] \quad (23)$$

where W is the number of possible arrangements of N_A and N_B cations (Gaskell, 1981). From Stirling's approximation,

$$\Delta S = -R \left[\frac{N_A}{N_A + N_B} \ln \left(\frac{N_A}{N_A + N_B} \right) + \frac{N_B}{N_A + N_B} \ln \left(\frac{N_B}{N_A + N_B} \right) \right] \quad (24)$$

$$= -R[X_A \ln(X_A) + X_B \ln(X_B)] \quad (25)$$

where X_A and X_B are defined to be mole fraction of component A and B, respectively.

Figure 7 illustrates calculated results of the free energy of III–N ternary alloys. For comparison, the weighted average of the binary components is also graphed in order to delineate the deviation from Raoultian behavior (Gaskell, 1981; Shih et al., 1985; Hwang et al., 1988). Figure 8 illustrates the

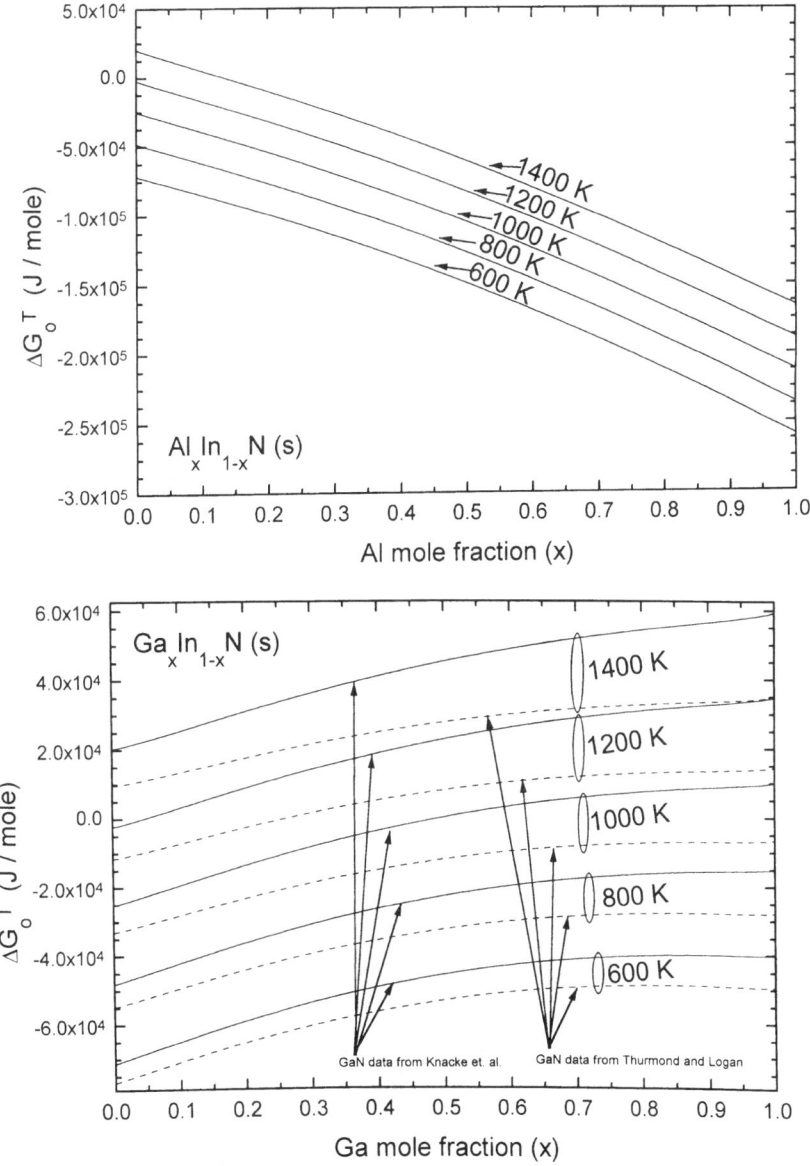

FIG. 8. Change in Gibbs free energy for ternary III–N reaction $xA(l) + (1-x)B(l) + 1/2N_2(g) = A_xB_{1-x}N(s)$ and the corresponding equilibrium pressure for the $Ga_xAl_{1-x}N$ reaction.

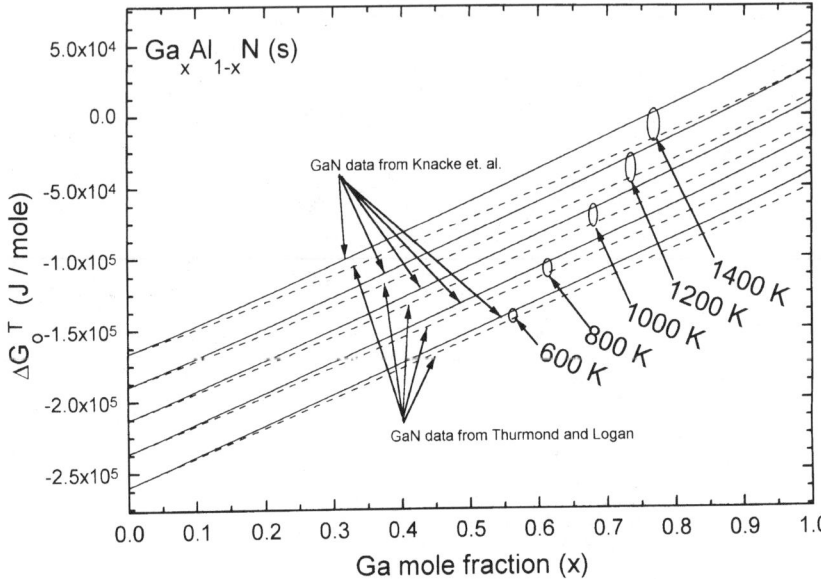

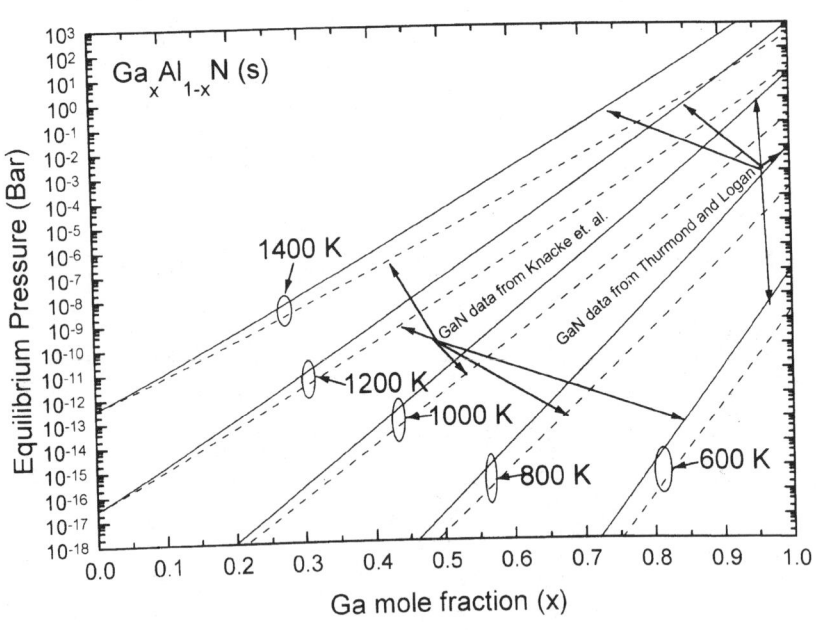

FIG. 8. (*Continued*)

TABLE II

INTERACTION ENERGIES Ω, FOR III–N SEMICONDUCTOR ALLOYS

Alloy System	Ω (meV)
AlGaN	48
AlInN	818
GaInN	549

Source: Calculated by van Schilfgaarde, Sher, and Chen (1997).

The energies were calculated with the local denity approximation, using a full potential version of the linear muffin tin orbitals method. Conversion to units used in this work, J/mol, can be accomplished by multiplying Ω by 96.490 J/meV.

resulting equilibrium pressure and free energy changes for several ternary nitride reactions.

$$X_A A(l) + X_B B(l) + 1/2 N_2(g) = A_{X_A} B_{X_B} N(s) \qquad (26)$$

where A and B represent a group III atom (i.e., Al, Ga, In) and X_A and X_B are the mole fraction of the respective group III atom, with $X_A + X_B = 1$.

The minimum temperature in which all compositions of mixed alloys are thermodynamically favored over the separate binary compound is labeled the critical temperature T_c. In the regular solution model, the critical temperature is equal to $\Omega/2R$ (Gaskell, 1981; Chen and Sher, 1995). Using the interaction energies given in Table II (van Schilfgaarde, Sher, and Chen, 1997), critical temperatures for GaInN, AlGaN, and AlInN alloys are calculated to be 3980, 280, and 4740 K, respectively.

III. Thermal Stability of GaN, InN, and AlN: The Decomposition Reaction

The characteristics of III–N evaporation are in strong contrast to those of almost all other known material systems. In most cases, bulk thermodynamic data can be used to accurately predict the rate of evaporation of each chemical species, indicative that kinetic barriers to the reverse reaction are absent (Searcy, Ragone and Colombo, 1970; Somorjai, 1968; Somorjai and Lester, 1967). The thermodynamic barrier establishes the theoretical upper limit to the reaction rate since it is bounded from above by a process which is able to sustain conditions present in thermodynamic equilibria (Searcy, Ragone and Colombo, 1970). The explicit relationship between these quantities is given by the Hertz–Langmuir equation (Searcy, Ragone and Colombo, 1970; Knudsen, 1915; Langmuir, 1913; Hertz, 1882).

$$J_{eq} = (2\pi MRT)^{-1/2} P_{eq} \qquad (27)$$

where J_{eq} is equilibrium flux, P_{eq} is equilibrium pressure, and M is the mass of the gaseous species.

Metal surfaces are found to vaporize as atoms at a rate that falls within experimental error of the thermodynamically predicted value (Searcy, Ragone and Colombo, 1970). Many other solid surfaces including, for example, Al_2O_3, Ga_2O_3, In_2O_3, and LiCl, decompose at a rate ~ 1 to 3 times slower than the maximum predicted value due to the presence of a kinetic barrier (Knudsen, 1915; Langmuir, 1913; Hertz, 1882; Somorjai, 1968; Somorjai and Lester, 1967). The ratio of observed rate to thermodynamically predicted rate is labeled the evaporation coefficient α (Knudsen, 1915; Langmuir, 1913; Hertz, 1882).

$$J = \alpha J_{eq} = \alpha (2\pi MRT)^{-1/2} P_{eq} \qquad (28)$$

For strongly bound nitrogen-containing compounds, a rate limiting step associated with surface disassociation occurs. In the case of AlN (Hildenbrand and Hall, 1964) BN (Hildenbrand and Hall, 1964), Mg_3N_2 (Somjorjai, 1968; Somjorjai and Lester, 1967), and GaN (Munir and Searcy, 1965; Groh et al., 1974), the process of breaking strong surface metal–N bonds results in evaporation rate being kinetically limited with $\alpha < 6 \times 10^{-3}$. The presence of a strong chemical bond is the most important factor for having a small evaporation coefficient. Surprisingly, the occurrence of bond breaking and subsequent bond formation is not a sufficient condition to result in a small evaporation coefficient, since it has been established that several materials with near unity evaporation coefficients undergo this process prior to evaporation (i.e., BaF_2, LaF_2, and PrF_2) (Searcy, Ragone and Colombo, 1970).

The measurements of Munir and Searcy (1965) indicate that the evaporation coefficient, α, can be as small as 10^{-8} Torr under vacuum conditions in the absence of catalysts. The presence of the kinetic barrier to GaN decomposition causes the Ga(v) evaporation rate from a free surface of GaN(s) (Munir and Searcy, 1965) to be slower than that from Ga(l) (Dushman, 1949) at comparable temperatures. The consequence of this anomaly is that GaN undergoes congruent sublimation in vacuum once steady state conditions are achieved (Munir and Searcy, 1965), that is,

$$\text{GaN(s)} \rightarrow \text{Ga(g)} + 1/2\text{N}_2(\text{g}) \tag{29}$$

In the work by Zuhair, Munir, and Searcy (1965) and Mar and Searcy (1968) mass spectrometry confirmed the presence of only Ga and N_2 species at thermal energies in the vapor. In contrast to this observation, Groh et al. (1974) reported that a mass spectrometer detected only nitrogen during GaN vacuum decomposition. The presence of metallic Ga was, however, detected on cooler parts of the vacuum annealing furnace, indicating that a Ga containing species evolved from the GaN surface *undetected by the mass spectrometer*. Since Ga(l) has a significant vapor pressure under conditions present during GaN decomposition, a significant fraction of the evaporant is expected to be Ga(v), consistent with the observations of Munir and Searcy.

$$\text{GaN(s)} \rightarrow \text{Ga(l)} + 1/2\text{N}_2(\text{g}) \tag{30}$$

In contrast to the observations of Munir and Searcy, Groh et al. (1974) also reported the presence of Ga droplets on the GaN surface during decomposition in vacuum (reaction 30). More accurate experimental work is required to understand the discrepancy in the observation of Ga(l) found on decomposition of GaN in work by Groh et al. and its absence in work by Zuhair, Munir, and Searcy. All of these early studies were performed with low quality polycrystalline GaN samples with impurity levels on the order of a few percent. Use of current state-of-the-art GaN material in analogous studies is expected to be important in clarifying these issues.

Using an Arrhenius analysis, Groh et al. (1974) reported an activation energy of 74.72 kcal/mol for the decomposition reaction. This is within experimental value found when a similar analysis is performed on data from the earlier study by Munir and Searcy (1965) ($\sim$73 kcal/mol).[3] The energy of activation is significantly below the cohesive energy of surface adatoms, indicating that the decomposition process involves interaction with other species on the surface. The rate limiting step is commonly associated with bond breaking at a surface defect, such as a kink or step site on the surface (Somorjai, 1968; Somorjai and Lester, 1967).

---

[3]Groh et al. and Munir and Searcy used a different formalism for their data analysis. For the reaction $2\text{Ga(g)} + \text{N}_2(\text{g}) \rightarrow 2\text{GaN(s)}$, the enthalpy of sublimation was defined as $-R\partial/\partial(1/T)$ $(\ln(p_{Ga}^2 p_{N_2}))$ by Munir et al. which would be three times the value given by $-R\partial/\partial(1/T)$ $(\ln(p_{N_2}))$, as used by Groh, for congruent sublimation. In order to directly compare the results, the method of analysis of Groh et al. was used.

The observed decomposition of GaN(s) into elemental phases is in contrast to the earlier speculation that stoichiometric molecular species (i.e., $(GaN)_x$) evolve during GaN decomposition (Sime and Margrave, 1956; Margrave and Sthapitanonda, 1955; Sime, unpublished). Evaporation by predominantly stoichiometric molecular species is most commonly encountered in strongly bound ionic compounds (e.g., mono-molecular units of CsCl, KCl, KI, BaO, B_2O_3, TiO, ZrO_2, HfO_2, ThO_2, VO, and as polymeric units of $(LiCl)_2$ for LiCl, $(LiBr)_2$ for LiBr, $(MoO_3)_3$, $(MoO_3)_4$, $(MoO_3)_5$ for MoO_3, and $(WO_3)_3$, $(WO_3)_4$, $(WO_3)_5$ for WO_3) (Searcy, Ragone and Colombo, 1970; Chandrasekharaiah, 1967; Bauer and Porter, 1964).

Ga(l) and In(l) are known to catalytically enhance the decomposition rate of GaN(s) (Schoonmaker, Buhl, and Lemley; 1965). The catalytic action presumably entails the formation of In–N or Ga–N bonds across the liquid–solid interface which results in a reduced cohesive energy of the top atomic layer of solid. The subsequent dissolution of GaN within the liquid and evaporation of N_2 from liquid surface completes the reaction process. Since the process of breaking the Ga–N bond at solid surface is the rate limiting step for this reaction, this model explains the large catalytic enhancement of decomposition reaction by In(l) and Ga(l).

Similar to the case of vacuum annealing, GaN is thermodynamically unstable in a pure hydrogen environment and is expected to decompose, consistent with the experimental observations of Morimoto (1974) and Sun et al. (1994). In the work by Sun et al. (1994) GaN thin films deposited on different substrates were found to exhibit a wide range of thermal stability. After annealing in hydrogen at 1000°C for 1 h, GaN films grown on (0001) sapphire completely decomposed, leaving condensed Ga(l) on the substrate surface. For similar annealing conditions, GaN grown on Si terminated 6H:SiC did not show evidence of significant decomposition. For 1 h 900°C anneals in hydrogen, Sun et al. also reported that $(11\bar{2}0)$ GaN films deposited on $(01\bar{1}2)$ sapphire have a higher decomposition rate than (0001) GaN films deposited on (0001) sapphire. The authors attributed the differences in decomposition rate to the varying stability of the GaN surfaces in the hydrogen environment, with the N terminated (0001) GaN surface (grown on (0001) Si terminated 6H:SiC) being most stable, followed by $(11\bar{2}0)$ GaN surface (grown on $(01\bar{1}2)$ sapphire) and Ga terminated (0001) GaN surface (grown on (0001) sapphire) being least stable (Sun et al., 1994). It should be noted that some caution should be exercised when comparing experimental results of dissimilar materials since the density and type of defects present in GaN thin films depend on the substrate and its orientation. For other solid systems, it has also been firmly established that the decomposition rate strongly depends on the crystal perfection (Somorjai, 1968; Somorjai and Lester, 1967) and topography. In classic work, Somorjai demonstrated the dependence of decomposition rate on both the density of dislocations and bulk impurity dopants (Somorjai, 1968; Somorjai and Lester, 1967).

The role hydrogen plays in GaN surface reactions has not yet been established. It is not clear *a priori* that H_2 will influence the rate of GaN decomposition. Work by Munir and Searcy (1965) indicates that $\sim 4\,\mu m$ of polycrystalline GaN would be expected to decompose *in vacuum* for the same annealing treatment (i.e., 1000°C for 1 h) used by Sun *et al.* (1994). The much greater thermal stability of GaN films on (0001) Si terminated 6H:SiC, as reported by Sun *et al.* (1994), suggests that H_2 may be responsible for lowering the surface free energy, resulting in a decrease in sublimation rates in hydrogen environments compared to those found for vacuum annealing. Unfortunately, a final conclusion cannot be made because of the differences in crystal perfection for the GaN used in the work of Zuhair, Munir, and Searcy and Sun *et al.*

Morimoto (1974) and Sun *et al.* (1994) report that rate of thermal decomposition is insignificant when GaN(s) is exposed to a temperature of 1000°C and a pressure of 1 atm of $N_2(g)$. The significantly slower rate of decomposition found in these experiments, as compared to those encountered during vacuum annealing, may be attributed to enhanced thermal stability of GaN(s) in atmospheric pressures of $N_2(g)$ (Searcy, Ragone and Colombo, 1970). This relationship is a direct result of law of mass action due to the enhanced activity of reactant N_2 gas with increasing nitrogen pressure. In order to better understand this, we will first explore the influence of the chemical environment on the thermodynamically determined maximum reaction rate. The Hertz–Langmuir equation and the equation, $P_{eq,Ga}(P_{eq,N_2})^{1/2} = e^{-\Delta G/RT}$, (Searcy, Ragone and Colombo, 1970) (pressure in bar) can be used to show that maximum rate of evaporation (Eq. 27) depends exponentially on the free energy of reaction ΔG:

$$J_{Ga,max} = 2J_{N_2} = 2^{-1/6}(\pi RT)^{-1/2} M_{Ga}^{-1/3} M_{N_2}^{-1/6} e^{-2\Delta G/3RT}$$

where M_{Ga} is the atomic weight of Ga and M_{N_2} is the molecular weight of N_2. In order to also include the influence of kinetic barriers in our analysis, we assume that the same pathway is taken in both forward and reverse reaction. Under this assumption, the kinetic energy barrier is the same for both reaction directions since they share the same activated complex (Fig. 2) (King, 1964). For conditions in which GaN(s) is the thermodynamically favored phase (i.e., $\Delta G < 0$ for synthesis reaction), the activation barrier to decomposition increases with decreasing ΔG. Therefore, we conclude that the rate which GaN(s) decomposes in atmospheric $N_2(g)$ pressures ($\Delta G > 0$) is expected to be significantly slower than in vacuum conditions ($\Delta G < 0$), as is experimentally observed. In addition, evaporation rate at near atmospheric pressures are further reduced due to significant probability that the evaporant is redirected to the solid by gas scattering, effectively increasing the near surface Ga(g) vapor pressure (Dushman, 1949). High gas pressures may also alter surface chemistry, significantly influencing the kinetic barriers of decomposition.

These examples give important information on thermal stability of GaN under a wide range of chemical environments. Clearly, more work towards understanding the influence of chemical environment and material quality on evaporation rates in these kinetically limited systems is needed.

IV. Thermodynamic Analysis of GaN Thin Film Growth

1. MOLECULAR BEAM EPITAXY[4]

Growth of III–V semiconductor thin films by MBE is a well established technology for producing device-quality single crystal thin films (Cho, 1995; Tsao, 1993). It was initially developed by Cho and co-workers at Bell Laboratories in the 1970s. Within a few years of its conception, MBE was found to offer several advantages over other thin film growth techniques. It is able to precisely tailor thickness and alloy composition of each vertical layer due to the unprecedented control of atomic layer by layer growth (Cho, 1995). In addition, techniques to fabricate complex device structures in new materials systems can be rapidly developed because the quality of MBE grown material is relatively insensitive to growth conditions (Tsao, 1993).

In this section, the understanding of growth chemistry of conventional III–V semiconductors is extended to wide bandgap III–N semiconductors (Newman, Ross, and Rubin, 1993). We will demonstrate that MBE growth of GaN occurs with a unique metastable growth process. In fact, it is shown that crystal growth of III–N systems is not only more strongly influenced by kinetics, successful MBE synthesis of several nitride systems actually *requires* the presence of kinetic barriers (Newman, Ross, and Rubin, 1993). For this reason, GaN is a model system for understanding such phenomena due to the dominance of large well defined energetic barriers involved in both the forward and reverse reactions.

In the following discussion, we will outline the most important aspects of MBE growth of conventional III–V systems which are relevant to under-

---

[4]This section is an extension of an earlier analysis, as reported in (N. Newman, *J. Crystal Growth*, 178, 102 (1997).

standing GaN synthesis. A recent chemical analysis of MBE growth by Tsao has shown that bulk thermodynamics accurately predicts conditions that result in the single phase synthesis of conventional III–V semiconductors (Tsao, 1993). This work demonstrates that a necessary criterion for the fabrication of high quality films is that growth occurs when only the III–V compound and vapor are thermodynamically stable.

Despite the utility of Tsao's approach, the analysis does not directly address the influence of kinetic processes, including surface diffusion, nucleation and adatom incorporation into the lattice. Although the level of sophistication required to accurately model microscopic kinetics for the MBE process has not yet been achieved, a number of empirical observations have been used to guide experimental work in overcoming the kinetic barriers of epitaxy. For example, the optimal substrate temperature used to grow almost all high quality single crystal thin films is ∼50% to 75% of melt temperature (Cho, 1995; Tsao, 1993; Vossen and Kern, 1978, 1991). In the case of GaAs, MBE is performed at 600°C (Tsao, 1993) which is 58% of melt temperature. This "rule of thumb" has also been found to be applicable to other methods of thin film growth (Vossen and Kern, 1978, 1991), including metalorganic chemical vapor deposition (MOCVD), sputtering and pulsed laser deposition.

To determine if the general principles developed for MBE growth of conventional III–V semiconductors can be extended to wide bandgap III–V nitrides, Fig. 3a shows the phase diagram, as well as conditions used for successful growth of GaN thin films (Newman, Ross, and Rubin, 1993; Newman *et al.*, 1994). In all cases, synthesis occurs under conditions in which GaN(s) is *not* the thermodynamically stable phase. Also, note that the MBE growth temperature (typically <1123 K) is significantly below 50% of the theoretically predicted melt temperature ($T_m \simeq 2793$ K) (Van Vechton, 1973). Clearly, the concepts which have guided MBE synthesis of conventional III–V semiconductors cannot be directly applied to the growth of GaN.

In the following discussion, we will describe an alternative approach that can be used to understand the chemistry of plasma assisted MBE growth of GaN and related compounds. We will show synthesis involves a metastable growth process that is controlled by a competition between the forward reaction, which depends on the arrival of activated nitrogen species at growing surface, and the reverse reaction whose rate is limited by unusually large kinetic barrier to decomposition (Newman, Ross, and Rubin, 1993; Newman *et al.*, 1994). We also include in our discussion the influence of interaction of activated species on the rate of these processes.

a. Driving the Forward Reaction

As can be seen in Fig. 3, typical MBE growth conditions for GaN fall within an area of the phase diagram in which Ga(l) and N_2(g) have the lowest free energy. In order to provide sufficient energy to form GaN, the use of more reactive species is therefore required. Figure 9 compares the energy of excited nitrogen to difference in Gibbs free energy of GaN reaction for typical MBE conditions. The excited neutral molecular nitrogen ($A^3\Sigma_u^+$, $B^3\Pi_g$, $a^1\Pi_g$, $C^3\Pi_u$), ionized molecular nitrogen ($^2\Sigma_g^+$), and atomic nitrogen (^4S, ^2P, ^2D) all provide sufficient energy to form GaN. For example,

$$Ga(l) + 1/2N_2^*(g) \to GaN(s) \qquad (31)$$

where $N_2^* = A^3\Sigma_u^+, B^3\Pi_g, a^1\Pi_g, C^3\Pi_u$,

$$Ga(l) + 1/2N_2^+(g) + e^- \to GaN(s) \qquad (32)$$

where $N_2^+ = {}^2\Sigma_g^+$,

$$Ga(l) + N(g) \to GaN(s) \qquad (33)$$

where $N = {}^4S$,

$$Ga(l) + N^*(g) \to GaN(s) \qquad (34)$$

where $N = {}^2P, {}^2D$.

Note that the combination of the free energy difference ($\Delta G \sim 1.9\,eV$) and the kinetic barrier to synthesis ($\Delta G < 1.3\,eV$) (King, 1964; Munir and Searcy, 1965; Groh et al., 1974) is significantly smaller than the potential energy of *all* activated nitrogen species. The value of the kinetic barrier to synthesis was inferred assuming that the forward and reverse reaction follow a common reaction path (King, 1964; Karpinski, Jun, and Porowski, 1984). Although the accuracy of the magnitude of the kinetic barrier is questionable due to the implicit assumptions, it is expected to be of the correct order. For a more complete description of approximations in this approach, see Karpinski, Jun, and Porowski, 1984. This analysis and a similar analysis of high pressure GaN solution growth by Karpinski (1984) demonstrate that GaN is a model system for studying kinetic phenomena in solids.

For GaN MBE growth, the most common source of the group III element is Ga evaporation from a Knudsen cell (Fu et al., 1995; Lei et al., 1992; Paisley et al., 1989; Martin et al., 1991; Harris, unpublished; Newman et al., 1994; Powell, Lee, and Greene, 1992; Yang, Li, and Wang, 1995). The energy supplied by reactive gallium vapor (i.e., Ga(v)) could also potentially drive the forward reaction. However, because of the rapid thermalization of the

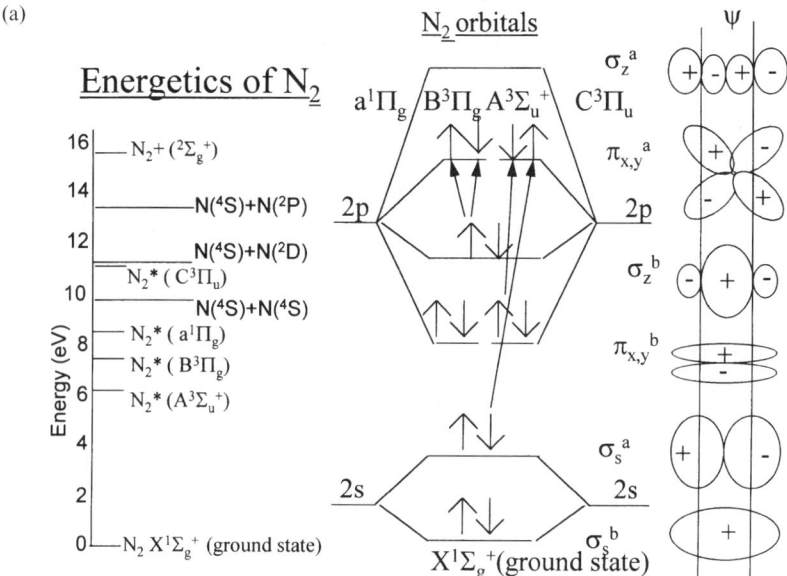

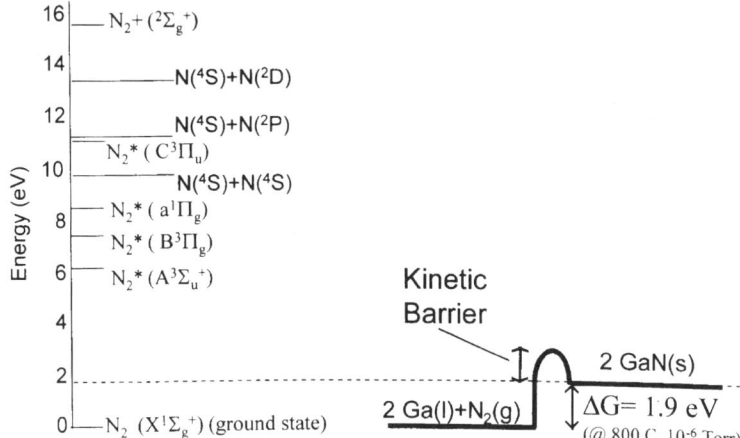

FIG. 9. (a) Potential energy, electronic states and wavefunctions of nitrogen species (von Engel, 1993; Wright and Winkler, 1968). (b) Comparison of potential energy of activated nitrogen (von Engel, 1993; Wright and Winkler, 1968) with Gibbs free energy for reaction $2\text{Ga}(l) + N_2(g) \rightarrow 2\text{GaN}(s)$ at typical growth conditions (a temperature of 800°C and an effective pressure of 10^{-6} Torr nitrogen). Note that the reaction energy represents formation of two molecular units in order that it can be directly compared to $N_2(g)$ and its activated forms. An eV is equal to $\sim 96,490$ Joules, the unit used predominantly in this work.

heat of Ga condensation on the GaN surface, as well as the very unreactive nature of the groundstate nitrogen molecule, the reaction to form GaN is *not* expected. Consistent with these arguments, it has been experimentally realized that GaN *cannot* be formed under MBE vacuum conditions using only groundstate molecular nitrogen (i.e., without activation) (Madar *et al.*, 1975).

In contrast, when the more reactive NH_3 gas is used as the nitrogen source in MBE synthesis of GaN (Powell, Lee, and Greene, 1992; Yang, Li, and Wang, 1995), the additional energy of Ga(v) is found to be important in driving the forward reaction. In this case, GaN is able to form in the unstable region of the GaN(s) and Ga(l)/NH_3(g) phase diagram (Fig. 3(*b*)), even without plasma assist. Equation (15), with Ga(g) as the reactant, is relevant. Figure 10 illustrates the phase diagram for Eq. (15), as well as typical thin film growth conditions.

$$Ga(g) + NH_3(g) \rightarrow GaN(s) + 3/2 H_2 \qquad (35)$$

Similar conclusions can be derived for MBE growth of InN with NH_3, as illustrated in Fig. 3(*c*).

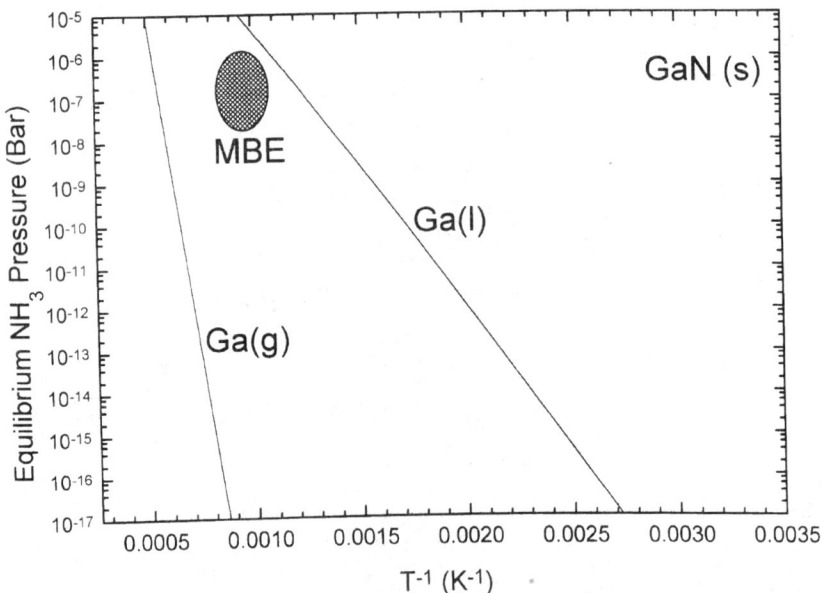

FIG. 10. Change in Gibbs free energy and equilibrium pressure for the following reactions: (1) Ga(l) + 1/2N_2(g) = GaN(s), (2) Ga(g) + 1/2N_2(g) = GaN(s).

The decomposition of metastable $NH_3(g)$ at elevated substrate temperature is also a contributing factor to a small energy of activation for NH_3 MBE reaction. A small activation barrier for reactions with $NH_3(g)$ has been well documented for the metastable reaction between α-Fe and $NH_3(g)$ at 600°C. This reaction results in supersaturation of soluble nitrogen ($>2\%$) in α-Fe(s) and is driven thermodynamically by the high chemical potential of metastable $NH_3(g)$. The rapid rate of the reaction is found to be a direct consequence of $NH_3(g)$ decomposition activated at the solid surface (Darken and Gurry, 1953).

b. The Reverse Reaction

Since GaN(s) is thermodynamically unstable under MBE conditions, the binary compound will undergo decomposition during growth (Eqs. 29 and 30). Because of the unusually large kinetic barrier to decomposition (Zuhair, Munir, and Searcy, 1965), the reaction rate is sufficiently slow, allowing successful synthesis at the growth conditions summarized in Fig. 3. With this process, metastable growth is feasible for all compounds with an evaporation coefficient (α) significantly smaller than one.

Because Ga(l) is known to catalytically enhance the decomposition rate (Schoonmaker, Buhl, and Lemley, 1965), growth under Ga rich conditions and high temperatures is typically avoided in order to inhibit the reverse reaction. Experimental evidence for the presence of this reaction at high growth temperatures ($\gg 850°C$) comes from the observation of a rapid decrease in growth rate and formation of macroscopic Ga droplets within GaN films under these conditions.

c. Criterion for Metastable Growth

For materials in which the kinetic barrier to decomposition is large, growth of high quality thin films can be performed under metastable conditions if the forward reaction is driven with activated species at a sufficient rate. Successful growth results when the rate of arrival of activated species (i.e., maximum rate of forward reaction) is greater than the thermal decomposition reaction rate (i.e., minimum rate of reverse reaction). For MBE growth of GaN using an ion source (Newman et al., 1994; Powell et al., 1990; Rubin et al., 1994), Fig. 11 compares the rate of thermal decomposition with the rate at which activated nitrogen ($^2\Sigma_g^+$) arrives at the growing GaN surface (Newman, Ross, and Rubin, 1993). Since the rate of the decomposition reaction is orders of magnitude slower than predicted from thermodynamic values, driving the forward reaction at a sufficient rate facilitates growth at a higher temperature and/or lower pressure than are thermodynamically stable for GaN(s) (Fig. 3).

Because of the practical limitations in producing a high flux of activated nitrogen, the rate of decomposition does, however, limit maximum temperature of GaN MBE growth to ~40% (900°C) of the melt temperature ($T_m \simeq 2700°C$). Another detrimental effect attributed to the presence of the decomposition reaction is the rapid increase in the background free carrier concentration of film for growth at elevated temperature (Newman, Ross, and Rubin, 1993; Newman et al., 1994). This experimental observation suggests that the decomposition reaction may be responsible for facilitating the formation of auto-doping donor levels, presumably the nonstoichiometric nitrogen vacancy (V_N).

Abernathy (private communication) reported successful growth at temperatures as high as 950°C. According to data shown in Fig. 11, the rate of decomposition would be expected to limit growth rate and crystal quality under these conditions. However, since the rate of decomposition in Fig. 11 (Munir and Searcy, 1965) was determined using polycrystalline GaN samples of questionable quality, additional studies using recently developed low defect density crystals are needed before a final conclusion can be reached.

d. The role of the Kinetic Energy of Plasma Species

The impingement of high kinetic energy species from the plasma (>40 eV) on the film surface can cause an enhancement in the decomposition rate (i.e., plasma enhanced decomposition or sputtering) (Winters and Kay, 1967). For the case of GaN films produced using a high energy process such as sputtering, the rate of plasma enhanced decomposition was reported to be significant at temperatures as low as 680°C (Newman, Ross, and Rubin, 1993). This is clearly illustrated in Fig. 11 where the sputtering growth rate drops rapidly as the substrate temperature is increased above 680°C. With sputtering, the enhancement of the reverse reaction by high kinetic energy plasma species prevents growth of GaN at temperatures sufficient to promote high quality epitaxy.

The impact of high energy species on the GaN surface is also expected to result in subsurface damage with the subsequent formation of defects in the growing film (Brice, Tsao, and Picraux, 1989). In order to experimentally determine the influence of the energy of plasma species on film quality, growth of GaN was performed as a function of the kinetic energy of molecular nitrogen ions (Newman et al., 1994; Rubin et al., 1994). Strong band edge photoluminescence (PL) and cathodoluminescence (CL) are found for growth conditions using 10 eV ion kinetic energies, while lumines-

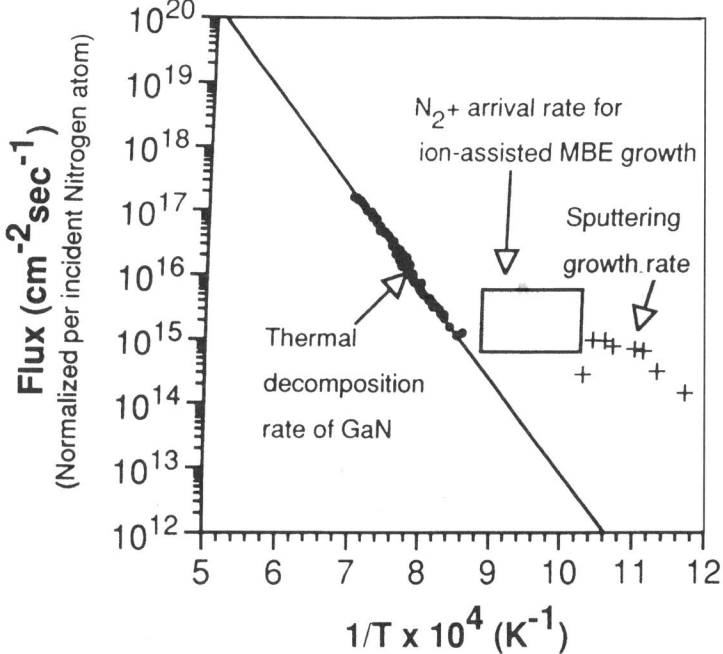

FIG. 11. A comparison of flux of activated nitrogen (N_2^+, $^2\Sigma_g^+$) impinging on film surface during growth (i.e., maximum forward reaction rate) (Newman et al., 1994; Powell et al., 1990; Rubin et al., 1994) with the experimentally determined sublimation rate of GaN in vacuum (Zuhair, Munir, and Searcy, 1965) due to the kinetic barrier of decomposition (minimum reverse reaction rate). Also shown is the flux of nitrogen incorporated into films during sputter deposition of GaN (Newman, Ross, and Rubin, 1993). Note that in the case of the highly energetic process of sputtering, there is evidence for significant decomposition at temperatures as low as 680°C. (From Newman, Ross, and Rubin, 1993.)

cence was not detectable when ion kinetic energy exceeded 18 eV (Fu et al., 1995). This study clearly demonstrated that the quality of GaN films grown by plasma-assisted MBE is limited by damage from the impingement of high energy species (> 18 eV) on the growth surface (Fu et al., 1995).

From a recent study using an ultra-low energy plasma source called the Hollow anode source, it was clearly demonstrated that excited neutral molecular nitrogen, even with kinetic energies of ~ 1 eV, can efficiently drive the synthesis reaction (Anders et al., 1996; Gassmann et al., 1996; Phatak et al., private communication). As was discussed earlier, this is not surprising since the lowest energy excited neutral nitrogen molecule has over 6 eV potential energy, which is considerably more than the energy required to drive the forward reaction.

e. Are Ultra-low Kinetic Energies Optimal for Growth?

Although experiments with the Hollow anode clearly demonstrate that low energy activated nitrogen molecules can drive the reaction, it is not clear whether use of low energy species can optimize GaN thin film quality. A low-energy activated nitogen beam is desired to minimize damage and decomposition. However, the presence of moderate energy species (5–10 eV) may produce a substantial improvement in surface diffusion rate, resulting in higher quality GaN epitaxy (Fu et al., 1995; Gassmann et al., 1996; Liliental-Weber et al., 1995; Liliental-Weber, 1996). Since growth is limited to $\sim 33\%$ of the melt temperature by the decomposition reaction, the use of precisely controlled low energy beams could be a possible path to substantially improve the quality of GaN thin films using MBE.

Brice, Tsao, and Picraux (1989) have performed theoretical calculations of surface and bulk displacements due to the impingement of ion beams on several solid surfaces. This work reported the occurrence of an "energy window" in which only surface displacements result. This "window" is bounded from above by energies that produce bulk displacements and subsequent creation of defects. From below, it is bounded by energies not sufficient to produce surface motion. Brice's analysis was performed only for C, Si, and Ge, although the authors noted that the upper threshold energy occurs at approximately three times the cohesive energy of the solid. For GaN, this energy is approximately 27 eV (Weast, 1973). Ideally, the impingement of a neutral or ion beam would be tuned to satisfy the resonant energy and momentum conditions which enhance the surface diffusion rate without significantly enhancing desorption or decomposition.

In summary, the MBE growth process is controlled by a competition between the forward reaction, which depends on arrival of activated nitrogen species at growing surface, and the reverse reaction whose rate is limited by the unusually large kinetic barrier to decomposition. One very important aspect is the role that kinetic energy of plasma species plays in the quality of deposited thin films. Impingement of high energy species on the growth surface results in subsurface damage to the growing film, as well as an enhancement of the reverse decomposition reaction rate. An interesting area which needs further exploration is whether the precise control of energies of ions, atoms, and molecules can be used to improve film quality by enhancing the surface diffusion rate.

2. CHEMICAL VAPOR DEPOSITION

Chemical vapor deposition (CVD) has become the industry standard for producing thin film structures used in commercial III–V devices. Uniformity, high yield, and high throughput make the method well suited for large scale production. Despite these advantages, wide spread use did not occur overnight. The methods currently used in the state-of-the-art reactor evolved over the last 30 years. Chemical vapor deposition was first reported for growth of GaAsP alloys by Tietjen and Amick (1966). The method, now called chlorine transport CVD, was applied to growth of GaN by Maruska and Tietjen (1969).

For most new material systems, CVD technique has typically taken considerably longer to develop than other thin film growth techniques such as MBE or liquid phase epitaxy (LPE) due to the need to develop exotic precursors, to minimize the rate of competing reactions, and to precisely control transients involved in multilayer growth (Vossen and Kern, 1978, 1991). However, since the earliest work on nitride based semiconductors, CVD has consistently been the technique which has achieved the highest quality thin film material. In this section, we show that this anomaly can be attributed to the thermodynamic stability of GaN under high pressures (0.1–1.0 bar) and high temperature (1000–1100°C) conditions typically used in CVD, in contrast to the metastable growth required for vacuum deposition techniques (see Section IV.1).

The chemistry of the CVD process relies on the introduction of an input gas with higher reactant concentrations than is stable at the elevated substrate temperature (Stringfellow, 1983, 1984a, 1984b). When other competing reactions are suppressed, many aspects of CVD growth can be successfully modeled by assuming a kinetic process for transport of reactants through a heated boundary layer to a vapor/solid surface in thermodynamic equilibrium (Stringfellow, 1983, 1984a, 1984b). The inclusion of kinetically limited transport of reactants through the boundary layer is necessary to account for the observation that the growth rate typically falls an order of magnitude lower than predicted by thermodynamics alone (Stringfellow, 1984b).

For conventional III–V semiconductors, the assumption that equilibrium is achieved at vapor/solid interface is supported by experimental observation that thermodynamics accurately predicts the composition of solid products which are synthesized (Stringfellow, 1983, 1984b), indicative of rapid surface reaction rates in comparison to gas phase diffusion. The presence of metastable alloy formation for compounds with a miscibility gap does, however, indicate a deviation from the assumption that thermodynamic equilibrium is achieved, due to presence of finite atomic diffusion rates (Stringfellow, 1984a).

For the synthesis of III–N semiconductors, we will show that an analogous analysis can be used to successfully model III–N growth when the competing NH_3 decomposition reaction is inhibited. The thermochemistry of GaN growth using CVD was addressed in important papers by Ban (1972) and Liu and Stevenson (1978). The authors modeled both thermodynamic and kinetic factors for CVD growth using the chlorine-transport CVD (CTCVD). Our analysis follows the earlier work, using more recent thermochemical data. When this is done, some important conclusions of earlier researchers need to be revised. Our analysis does not directly address the modern MOCVD technique. Nevertheless, principles introduced in this analysis can be readily extended to this more modern variant. For a discussion of MOCVD technique, the reader is referred to Chapter 2

The CTCVD technique uses a two step reaction process to produce GaN(s). GaCl gas is synthesized when HCl gas is flowed over liquid Ga in the source zone (Ban, 1972).

$$\text{Ga(l)} + \text{HCl(g)} = \text{GaCl(g)} + \text{H}_2\text{(g)} \tag{36}$$

Under typical processing conditions, this reaction is highly efficient. Using mass spectroscopy, Ban (1972) found that GaCl(g) is the only gallium-containing reaction product generated. Using a chemical acid-base titration technique, Liu and Stevenson (1978) demonstrated that conversion efficiency into GaCl(g) at 1223 K is approximately 97 to 99% under typical growth conditions. Using mass spectrometry, Ban (1973) found a significantly smaller conversion efficiency ($<83\%$) at lower temperatures (973–1073 K).

When GaCl(g) reactant gas is mixed with NH_3 in the deposition zone, GaN(s) is synthesized (Ban, 1972; Liu and Stevenson, 1978):

$$\text{GaCl(g)} + \text{NH}_3\text{(g)} = \text{GaN(s)} + \text{HCl(g)} + \text{H}_2\text{(g)} \tag{37}$$

Ban (1972) demonstrated experimentally that 80% of the GaCl(g) introduced reacts when He is the carrier gas and 50% when H_2 is the carrier gas. This difference was attributed to the effect of the law of mass action in Eq. (37).

In the next section, we use thermochemistry to gain additional insight into the limiting factors involved in growth.

The change in free energy of reaction (37) is

$$\Delta G = \Delta G_o + RT \ln\left(\frac{a_{\text{GaN}} a_{\text{HCl}} a_{\text{H}_2}}{a_{\text{GaCl}} a_{\text{NH}_3}}\right) \tag{38}$$

Assuming GaN is a pure solid phase and gasses are ideal:

$a_{\text{GaN}} = 1$, $a_{\text{HCl}} = \sim 0$, $a_{\text{GaCl}} = \sim p_{\text{HCl}}$, $a_{\text{H}_2} = \sim p_{\text{H}_2} + \frac{3\alpha}{2} p_{\text{NH}_3} + \frac{1}{2} p_{\text{HCl}}$

$a_{\text{NH}_3} = (1-\alpha) p_{\text{NH}_3}$, $p_{\text{TOT}} = p_{\text{H}_2} + p_{\text{NH}_3}$

$$\Delta G = G^o_{\text{GaN}} + G^o_{\text{HCl}} + G^o_{\text{H}_2} - G^o_{\text{GaCl}} - G^o_{\text{NH}_3} + RT\left[\frac{\ln(1+(3\alpha/2-1)p_{\text{NH}_3})}{(1-\alpha)p_{\text{NH}_3}}\right] \tag{39}$$

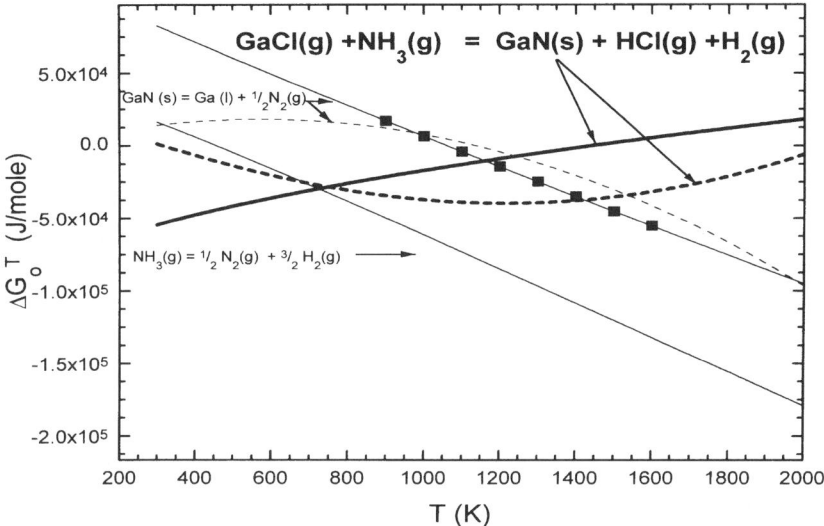

FIG. 12. Gibbs free energy for chlorine transport reaction (bold). The solid bold line indicates that Gibbs free energy is positive for temperatures exceeding ∼1400 K, approximately the upper bound of range of optimized growth (1373 K) (Liu and Stevenson, 1978). Also shown are dashed lines which come from analysis of Liu and Stevenson. In their study, the poor fit to the Thurmond and Logan (1972) data (boxes) gave them a fortuitous fit of a minimum in chlorine transport reaction at ∼1350 K, close to the optimized growth temperature. Also shown are the Gibbs free energy of two competing GaN(s) and NH$_3$(s) decomposition reactions which become favorable at elevated temperatures.

In Fig. 12, we compare results using thermodynamic value from Section II, with analysis of Liu and Stevenson (1978). Figure 12 shows that the fit by Liu and Stevenson (bold dashed line) does not accurately model data which the authors' reference (Thurmond and Logan, 1972), nor does it fit other more recent thermochemical data (Knacke, Kubaschewski, and Hesselmann, 1991; Karpinski, Jun, and Porowski, 1984). This error results in the erroneous conclusion that a minimum in free energy exists near 1050°C. Instead, it is apparent that the Gibbs free energy using Thurmond and Logan data, combined with other parameters from Section I, increases with temperature and the reaction becomes thermodynamically unfavorable ($\Delta G > 0$) at temperatures exceeding 1100°C. This value is equal to the upper bound at which CVD growth process is optimized (1100°C) (Liu and Stevenson, 1978). Also included in Fig. 12 are the free energy change of two competing reactions. The GaN decomposition reaction into Ga(l) and 1/2N$_2$(g) may also play a role in limiting the growth temperature; although the large kinetic barrier to the reaction is expected to result in a relatively slow decomposition process (see Section III). Another competing reaction

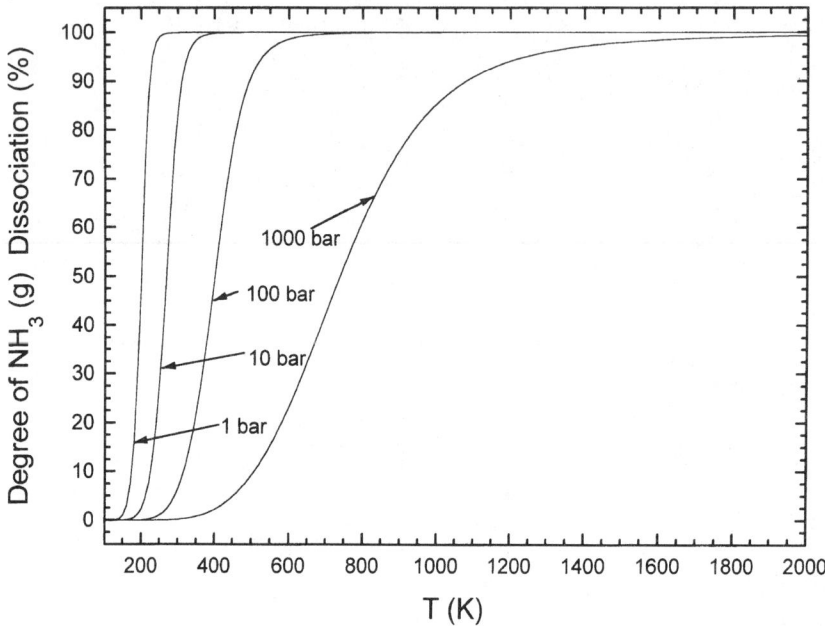

FIG. 13. Equilibrium dissociation of NH_3.

involves the decomposition of NH_3. This turns out to be an important factor and is discussed in the next paragraph.

Despite the apparent simplicity of the CVD synthesis reaction, the chemistry of the CVD process is nontrivial due to one critical factor; ammonia is thermodynamically unstable at the elevated temperature used in CVD growth (see Fig. 13). The ammonia decomposition reaction results in the production of $N_2(g)$; a chemical species which does not react with GaCl(g) in the CVD reactor environment (Ban, 1972; Liu and Stevenson, 1978). For this reason, the ammonia decomposition reaction must be inhibited to facilitate GaN synthesis.

$$NH_3(g) \rightarrow \tfrac{1}{2}N_2(g) + 3/2H_2(g) \tag{40}$$

Figure 13 illustrates the extent of diassociation of ammonia, as predicted by thermodynamics for a wide range of temperature and pressure conditions. This figure shows that ammonia is thermodynamically unstable (i.e., >99.9925% dissociated) for the conditions used in CVD growth (i.e., ~1000–1100°C, ~0.1–1.0 bar). The CVD reaction relies on an extremely slow disassociation reaction rate resulting from the kinetic barrier to ammonia decomposition.

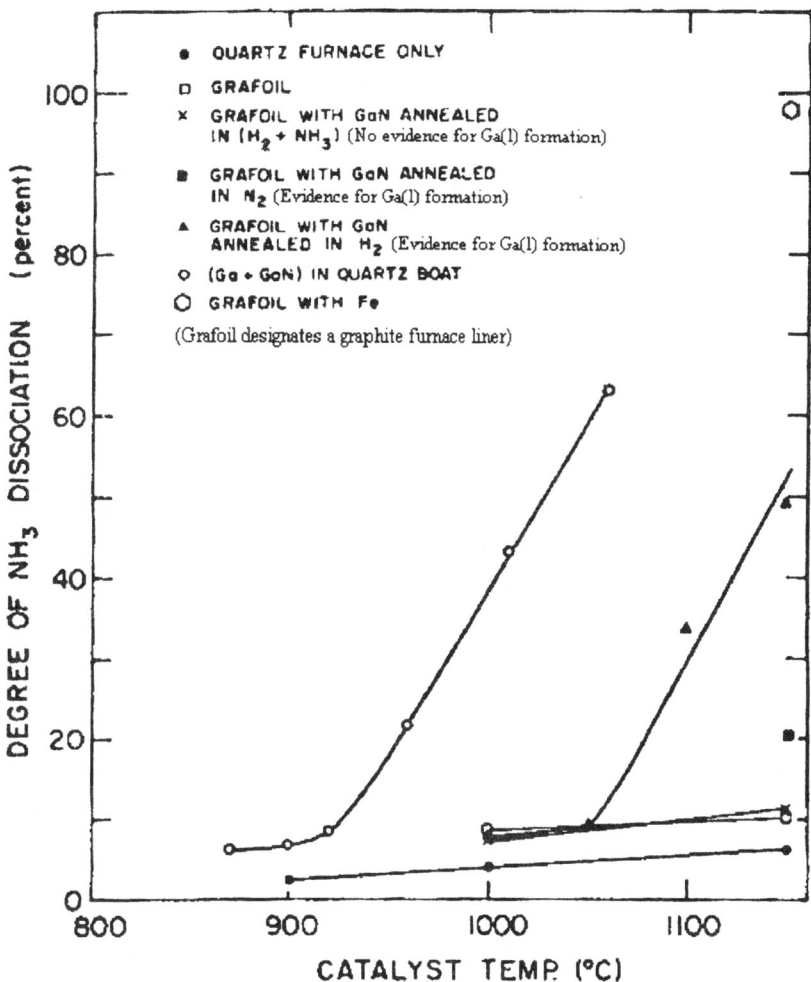

FIG. 14. Kinetically limited dissociation of NH_3 in several environments, including ones which act catalytically. (From Liu and Stevenson, 1978.)

The rate of disassociation reaction is sluggish due to the kinetic barrier of breaking the N–H bond. Experimentally, Ban (1972); Rocasecca, Saul, and Lorimor (1974); Liu and Stevenson (1978) also found that the decomposition rate increases with temperature, although with varying efficiency, depending on the presence of catalysts (Fig. 14). Tungsten, platinum and iron are commonly used for catalyzing the decomposition of ammonia (Hinshelwood, 1940). Graphite has also been reported to act as a catalyst, a material commonly used in the removable liner to shield reactor walls (Liu and Stevenson, 1978; Rocasecca, Saul, and Lorimor, 1974). (Grafoil refers to the brand name of the graphite liner.) The combination of Ga(l) and GaN(s) is found to be surprisingly effective as a catalyst, although Ga(l) or GaN(s) alone were not found to increase the decomposition rate. This work suggests that the product deposition on walls of the reactor will be detrimental to growth under conditions in which GaN(s) and Ga(l) are present. Under these conditions, a rapid drop in the growth rates has been reported, presumably due to the reduction of NH_3 concentration (Liu and Stevenson, 1978).

V. SUMMARY

In this work, a thermochemical approach was used to investigate and analyze chemistry of III–N synthesis and processing. This chapter was specifically written to review the formalism of thermochemistry and demonstrate the utility of this approach to researchers in III–N technology. Although by no means complete, the important concepts used in understanding the role that kinetics and thermodynamics play in synthesis and processing were illustrated by several examples of the rich chemistry of III–N semiconductors.

ACKNOWLEDGMENTS

This work was supported by the Office of Naval Research under contract no. N00014-96-1-1002. The author would like to thank Colin Wood, Alan Searcy, Arden Sher, and Mark van Schilfgaarde for critically reviewing the manuscript and generously suggesting many changes and additions which improved the manuscript.

REFERENCES

Abernathy, C., private communication.
Abernathy, C. R., MacKenzie, J. D., Bharatan, S. R., Jones, K. S., and Pearton, S. J. (1995). *J. Vac. Sci. Technol. A* **13**, 716.
Anders, A., Newman, N., Rubin, M., Dickinson, M., Thomson, A., Jones, E., Phatak, P., and Gassmann, A. (1996). *Rev. Sci. Instr.* **67**, 905.
Ashcroft, N. W., and Mermin, N. D. (1976). *Solid State Physics*, Holt, Rinehart, and Winston, New York.
Ban, V. S. (1972). *J. Electrochem. Soc.* **119**, 761.

Bauer, S. H., and Porter, R. F. (1964). *Molten Salt Chemistry* (ed. M. Blander), Interscience, New York, p. 607.
Brice, D. K., Tsao, J. Y., and Picraux, S. T. (1989). *Nucl. Instr. Meth.* B **44**, 68.
Bryden, W. A., Ecelberger, S. A., Hawley, M. E., and Kistenmacher, T. J. (1994). *Mat. Res. Soc. Symp. Proc.* **339**, 497.
Chandrasekharaiah, M. S. (1967) In *The Characterization of High Temperature Vapors*, Appendix 3 (ed. J. L. Margrave), Wiley, New York, p. 395.
Chase, Jr., M. W., Davies, C. A., Downey, Jr., J. R., Frurip, D. J., McDonald, R. A., and Syverud, A. N. (1985). *JANAF Thermochemical Tables*, J. Phys. Chem. Ref. Data. American Chemical Society and American Institute of Physics, Vol. 14, Supp. 1, p. 1.
Chen, A.-B., and Sher, A. (1995). *Semiconductor Alloys: Physics and Materials Engineering*, Plenum, New York.
Cho, A. Y. (1995). MRS Bulletin, p. 21.
Darken, L. S., and Gurry, R. W. (1953). *Physical Chemistry of Metals*, McGraw Hill, New York, pp. 372–377.
Dushman, S. (1949). *Scientific Foundations of Vacuum Technique*, New York, pp. 743–781.
Fu, T. C., Newman, N., Jones, E., Chan, J. S., Liu, X., Rubin, M. D., Cheung, N. W., and Weber, E. R. (1995). *J. Electon. Mat.* **24**, 249.
Gaskell, D. (1981). *Introduction to Metallurgical Thermodynamics*, Hemisphere Publishing Corp., New York.
Gassmann, A., Suski, T., Lilental-Weber, Z., Newman, N., Kisielowski, C., Jones, E., and Weber, E. R. (1996). *J. Appl. Phys.* **80**, 2195.
Gibbs, J. R. (1873a). *Trans. Conn. Acad.* **2**, 309.
Gibbs, J. R. (1873b). *Trans. Conn. Acad.* **2**, 382.
Gibbs, J. R. (1876). *Trans. Conn. Acad.* **3**, 108.
Gibbs, J. R. (1878). *Trans. Conn. Acad.* **3**, 343.
Gilleson, K., Schuller, K.-H., and Struck, B. (1977). *Mater. Res. Bull.* **12**, 955.
Grzegory, I., Jun, J., Bockowski, M., St. Krukowski, Wroblewski, M., Lucznik, B., and Porowski, S. (1995). *J. Phys. Chem. Solids* **56**, 639.
Groh, R., Gerey, G., Bartha, L., and Pankove, J. I. (1974). *Phys. Stat. Sol.* **26**, 353.
Harris, J. Unpublished.
Harrison, W. A. (1980). *Electronic Structure and the Properties of Solids*, Freeman, San Francisco.
Hertz, H. (1882). *Ann. Phys. (Leipzeg)* **17**, 177.
Hildenbrand, D. L., and Hall, W. F. (1964). *Condensation and Evaporation of Solids* (eds. E. Rutner, P. Goldfinger, and J. P. Hirth), Gordon and Breach, New York, p. 399.
Hinshelwood, C. N. (1940). *The Kinetics of Chemical Change*, Clarendon Press, Oxford.
Hwang, J., Pianetta, P., Pao, Y.-C., Shih, C. K., Shen, Z.-X. Lindberg, P. A. P., and Chow, R. (1988). *Phys. Rev. Lett.* **61**, 877.
Jeans, Sir James (1948). *Kinetic Theory of Gases*, Cambridge University Press, Cambridge.
Juza, R., and Hahn, H. (1940). *Z. Anorg. Z. Chem.* **244**, 111–133.
Karpinski, J., Jun, J., and Porowski, S. (1984). *J. Cryst. Growth* **66**, 1.
Karpinski, J. and Porowski, S. (1984). *J. Cryst. Growth* **66**, 11.
Kennard, E. H. (1938). *Kinetic Theory of Gases*, McGraw-Hill, New York.
Khan, M. A., Kuznia, J. N., Van Hove, J. M., Olson, D. T., Krishnankutty, S., and Kolbas, R. M. (1991). *Appl. Phys. Lett.* **58**, 526.
Khan, A. M., Kuznia, J. N., Van Hove, J. M., Olson, D. T., Krishnankutty, S., and Kolbas, R. M. (1991). *Appl. Phys. Lett.* **58**, 526.
King, E. L. (1964). *How Chemical Reactions Occur*, W. A. Benjamin, Inc., Menlo Park, CA, pp. 41–72.

Kistenmacher, T. J., and Bryden, W. A. (1993). *Appl. Phys. Lett.* **62**, 1221.
Kistenmacher, T. J., Ecelberger, S. A., and Bryden, W. A. (1994). *Mat. Res. Soc. Symp. Proc.* **339**, 509.
Knacke, O., Kubaschewski, O., and Hesselmann, K. (eds.) (1991). *Thermochemical Properties of Inorganic Substances*, Springer Verlag, Berlin.
Kennard, E. H. (1938). *Kinetic Theory of Gases*, McGraw-Hill, New York.
Knudsen, M. (1915). *Ann. Phys. (Leipzeg)* **47**, 697.
Knudsen, M. (1950). *The Kinetic Theory of Gasses*, Methuen, London.
Kung, P., Saxler, A., Zhang, X., Walker, D., Wang, T. C. Ferguson, I., and Razeghi, M. (1995). *Appl. Phys. Lett.* **66**, 2958.
Langmuir, I. (1913). *J. Am. Chem. Soc.* **35**, 931.
Lei, T., Mouststake, T. D., Graham, R. J., He, Y., and Berkowitz, S. J. (1992). *J. Appl. Phys.* **71**, 4933.
Liliental-Weber, Z., Sohn, Hyunchul, Newman, N., and Washburn, J. (1995). *J. Vac. Sci. Technol. B* **13**, 1578.
Liliental-Weber, Z., Ruvimov, S., Kisielowski, C., Chen, Y., Swider, W., Washburn, J., Newman, N., Gassman, A., Liu, X., Schloss, L. Weber, E. R., Grzegory, I., Bockowski, M., Jun, J., Suski, T., Pakula, K., Baranowski, J. Porowski, S., Amano, H., and Akasaki, I. (1996). *Mater. Sci. Forum.* **395**, 351.
Liu, S. S., and Stevenson, D. A. (1978). *J. Electrochem. Soc.* **125**, 1161.
Lorenz, M. R., and Binkowski, B. B. (1962). *J. Electrochem. Soc.* **109**, 24.
MacChesney, J. B., Bridenbaugh, P. M., and O'Connor, P. B. (1970). *Mat. Res. Bull.* **5**, 783.
Madar, R., Jacob, G., Hallais, J., and Fruchart, R. (1975). *J. Cryst. Growth* **31**, 197.
Mar, R. W., and Searcy, A. W. (1968). *J. Chem. Phys.* **49**, 182.
Margrave, J. L., and Sthapitanonda, P. (1955). *J. Phys. Chem.* **59**, 31.
Martin, G., Strite, S., Thornton, J., and Markoc, H. (1991). *Appl. Phys. Lett.* **58**, 2375.
Maruska, H. P., and Tietjen, J. J. (1969). *Appl. Phys. Lett.* **15**, 327.
Matsuoka, T., Tanaka, H., Sasaki, T., and Katsui, A. (1989). *Inst. Phys. Conf. Ser.: GaAs and Related Compounds* **106**, 141.
Methfessel, M. (1988). *Phys. Rev. B* **38**, 1537.
Mikkelsen, Jr., J. C., and Boyce, J. B. (1982). *Phys. Rev. Lett.* **49**, 1412.
Morimoto, Y. (1974). *J. Electrochem. Soc.* **121**, 1383.
Munir, A., and Searcy, A. W. (1965). *J. Chem. Phys.* **42**, 4223.
Nakamura, S. (1991). *Jpn. J. Appl. Phys.* **30**, L1705.
Newman, N., Ross, J., and Rubin, M. (1993). *Appl. Phys. Lett.* **62**, 1242; *ibid.* (1993), *Appl. Phys. Lett.* **63**, 424.
Newman, N., Fu, T. C., Liu, X., Liliental-Weber, Z., Rubin, M., Chan, J. S., Jones, E., Ross, J. T., Tidswell, I., Yu, K. M., Cheung, N., and Weber, E. R. (1994). *Mater. Sci. Forum* **339**, 483.
Paisley, M. J., Sitar, Z., Posthill, J. B., and Davis, R. F. (1989). *J. Vac. Technol. A* **7**, 701.
Phatak, P., Anders, A., Chan, J., and Newman, N., private communication.
Powell, R. C., Tomasch, G. A., Kim, Y.-W., Thornton, J. A., and Greene, J. E. (1990). *Mater. Res. Soc. Symp. Proc.* **162**, 525.
Powell, R. C., Lee, N.-E., and Greene, J. E. (1992). *Appl. Phys. Lett.* **60**, 2505.
Reif, F. (1965). *Fundamentals of Statistical and Thermal Physics*, McGraw Hill, New York.
Rubin, M., Newman, N., Chan, J. S., Fu, T. C., and Ross, J. T. (1994). *Appl. Phys. Lett.* **64**, 64.
Rocasecca, D. D., Saul, R. H., and Lorimor, O. G. (1974). *J. Electrochem. Soc.* **121**, 962.
Sato, H., Sasaki, T., Matsuoka, T., and Katsui, A. (1990). *Jpn. J. Appl. Phys.* **29**, 1654.
Schoonmaker, R. C., Buhl, A., and Lemley, J. (1965). *J. Phys. Chem.* **69**, 3455.

Searcy, A. W., Ragone, D. W., and Colombo, U. (eds.) (1970). *Chemical and Mechanical Behavior of Inorganic Materials*, Wiley, New York.
Shih, C. K., Spicer, W. E., Harrison, W. A., and Sher, A. (1985). *Phys. Rev. B* **31**, 1139.
Sime, R. J., and Margrave, J. L. (1956). *J. Phys. Chem.* **60**, 810.
Sime, R. J., B.S. thesis, unpublished.
Somorjai, G. A., and Lester, J. E. (1967). *Progress in Solid State Chemistry*, Vol. 4 (ed. H. Reiss), Pergamon, New York.
Somorjai, A. (1968). *Science* **162**, 755.
Sun, C. J., Kung, P., Saxler, A., Ohsato, H., Bigan, E., Razeghi, M., and Gaskill, D. K. (1994). *J. Appl. Phys.* **76**, 236.
Stringfellow, G. B. (1983). *J. Cryst. Growth* **62**, 225.
Stringfellow, G. B. (1984a). *J. Cryst. Growth* **68**, 111.
Stringfellow, G. B. (1984b). *J. Cryst. Growth* **70**, 133.
Thurmond, C. D., and Logan, R. A. (1972). *Solid State Sci. Technol.* **119**, 622.
Tietjen, J. J., and Amick, J. A. (1966). *J. Electrochem. Soc.* **113**, 724.
Tsao, J. Y. (1993). *Materials Fundamentals of Molecular Beam Epitaxy*, Academic, New York.
van Schilfgaarde, M., Sher, A., and Chen, A.-B. (1997). *J. Crystal Growth* **178**, 8.
van Vechton, J. A. (1973). *Phys. Rev. B* **4V**, 1479.
van Wylen, G. J., and Sonntag, R. E. (1985). *Fundamentals of Classical Thermodynamics*, Wiley, New York, pp. 40–44, 380–400.
von Engel, A. (1993). *Ionized Gases*, American Institute of Physics, New York.
Vossen, J. L., and Kern, W. (1978). *Thin Film Processes*, Academic, New York.
Vossen, J. L., and Kern, W. (1991). *Thin Film Processes II*, Academic, New York.
Weast, R. C. (1973). *CRC Handbook of Chemistry and Physics*, Chemical Rubber Company Press, Cleveland, OH, 54th edn.
Winters, H. F., and Kay, E. (1967). *J. Appl. Phys.* **38**, 3928.
Wright, A. N., and Winkler, C. A. (1968). *Active Nitrogen*, Academic, New York.
Yang, Z., Li, L. K., and Wang, W. I. (1995). *Appl. Phys. Lett.* **67**, 1686.

CHAPTER 5

Etching of III Nitrides

S. J. Pearton

DEPARTMENT OF MATERIALS SCIENCE AND ENGINEERING
UNIVERSITY OF FLORIDA
GAINESVILLE, FL

R. J. Shul

SANDIA NATIONAL LABORATORIES
ALBUQUERQUE, NM

I. INTRODUCTION	. . . .	103
II. WET ETCHING	. . . .	104
III. DRY ETCHING	. . . .	107
IV. ISSUES OF ETCH SELECTIVITY AND DAMAGE	. . . .	118
V. SUMMARY	. . . .	123
References	. . . .	124

I. Introduction

Group III nitrides have high bond energies compared to other semiconductors and are generally quite inert chemically. GaN has a bond energy of 8.92 eV/atom, AlN 11.52 eV/atom, and InN 7.72 eV/atom, in comparison with GaAs, 6.52 eV/atom (Harrison, 1980). In material prepared by reactive sputtering it is possible to use a variety of acid and base solutions for etch removal, but heteroepitaxial material is much more difficult to remove. Essentially all of the device patterning to date has been done using dry etching techniques. For example, commercially available blue and green light emitting diodes have the mesa etched by Cl_2 based reactive ion etching to expose the n layer in the heterostructure (Nakamura, Mukai, and Senoh, 1994), and the first GaN based laser diode employed a dry etched facet (Nakamura *et al.*, 1996). In this chapter we will briefly summarize the status of wet etching of group III nitrides and then describe relative effectiveness of different dry etching techniques, as well as plasma chemistries employed.

We will also mention some problem areas such as ion induced damage, creation of nonstoichiometric surfaces and sidewall roughness resulting from mask degradation.

II. Wet Etching

Relatively little success has been obtained in developing wet etch solutions for III–V nitrides (Pearton et al., 1993a; Walker and Tarn, 1991). For AlN, a number of different solutions have been reported for amorphous or polycrystalline material. For example, hot ($\leqslant 85°C$) H_3PO_4 has been found to etch AlN deposited on Si by plasma enhanced chemical vapor deposition at low rates ($\leqslant 500$ Å/min^{-1}) (Sheng, Yu, and Collins, 1988; Pauleau, 1982). A variety of other solutions, including hot ($\sim 100°C$) HF/H_2O (Taylor and Lenie, 1960; Long and Foster, 1959; Barrett et al., 1985), HF/HNO_3 (Aita and Gawlak, 1983), or NaOH (Kline and Lakin, 1983) can etch sputtered or reactively evaporated amorphous AlN. For GaN, there were several early reports of wet etching in NaOH that progressed by formation of an insoluble gallium hydroxide (GaOH) coating (Chu, 1971; Pankove, 1972).

TABLE I

BINARY NITRIDE ETCHING RESULTS IN ACID AND BASE SOLUTIONS. (Samples were tested at room temperature (25°C) unless otherwise noted.)

Etch Solution	GaN Etch Rate (Å/min)	InN Etch Rate (Å/min)	AlN Etch Rate (Å/min)
Citric acid	0 (75°C)	0	0 (75°C)
Succinic acid	0 (75°C)	0	0
Oxalic acids	0 (75°C)	0 (74°C)	0
Nitric acid	0 (85°C)	0	0 (85°C)
Phosphoric acid	0 (82°C)	0 (82°C)	0 (82°C)
Hydrochloric acid	0 (80°C)	0	0 (80°C)
Hydrofloric acid	0	0	0
Hydroiodic acid	0	0	0
Sulfuric acid	0 (82°C)	Lifts off layer (83°C)	0
Hydrogen peroxide	0	0	0
Potassium iodide	0	0	0
2% Bromine/methanol	0	0	0
n-methyl-2-pyrrolidonone	0	0	0
Sodium hydroxide	0	Lifts off layer	500 (75°C)
Potassium hydroxide	0	Lifts off layer	22,650
AZ400K photoresist developer	0	Lifts off layer	~ 60–10,000

This film had to be removed by continual jet action. Others have reported that H_3PO_4 will remove GaN at a very slow rate. For InN, aqueous KOH and NaOH solutions were found to produce etch rates of a few hundred angstrom/min at 60°C (Guo, Kato, and Yoshida, 1992). There has been particular difficulty in finding reliable wet etchants for single crystal nitrides.

Table I shows a compilation of data obtained for wet etching of binary nitrides (Mileham et al., 1995). We did not find any etchant for GaN or InN at temperatures below ~80°C. However, strong base solutions (KOH, NaOH or photoresist developer, in which active ingredient is KOH) were found to etch single crystal AlN at controllable rates whose magnitude was strongly dependent on material quality.

Figure 1 shows an Arrhenius plot of etch rates in AZ400K photoresist developer of three different AlN samples.

1. Data designated by circles is from polycrystalline AlN grown on GaAs. Etch rates for this material are much faster than for two single crystal samples grown on Al_2O_3 (MacKenzie et al., 1995; Abernathy, 1995).
2. Data designated by triangles is from a 1 μm thick layer with a double crystal X-ray diffraction peak width of 400 arcsec.
3. Data designated by squares is from material with a peak width ~200 arcsec.

The etching is thermally activated with an activation energy of ~15.5 kCal/mol in each case. This is consistent with reaction limited etching, and the etch depth was also found to be a linear function of time with an absence

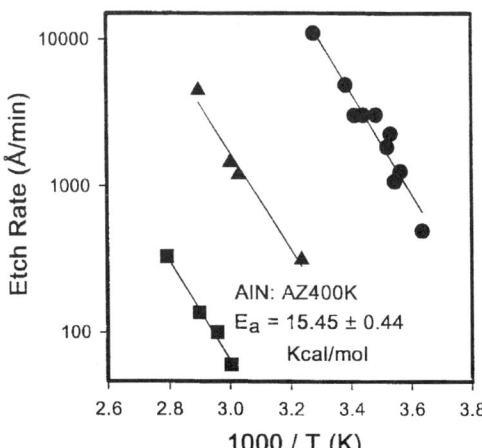

FIG. 1. Arrhenius plot of etch rate of AlN samples in AZ400K developer solution.

of dependence on agitation. If the etching was diffusion limited we would expect an activation energy below ~ 6 kCal/mol, a $\sqrt{t}$ dependence of etch on time, and a strong dependence of etch rate on degree of solution agitation. Higher rates for lower crystalline quality materials is expected on the basis of greater number of dangling or defective bonds that can be attacked by the OH^- ions in solution. Therefore the successful attempt frequency is higher under these conditions and etch rate R is higher. The process is well described by the relation

$$R = R_o \exp\left(-\frac{E_a}{kT}\right)$$

where R_o is the successful attempt frequency for breaking of an Al–N bond and formation of a soluble etch product, E_a is the activation energy (15.5 kCal/mol), k is Boltzmann's constant and T is absolute temperature of etch solution.

We have observed a strong effect of annealing on subsequent wet etch rate of sputtered AlN films in KOH solutions, with over an order of magnitude decrease in rate after annealing at 1100°C (Vartuli et al., 1996a). Similarly the etch rate for $In_{0.2}Al_{0.8}N$ grown on Si was approximately three times higher in KOH based solutions than for material grown on GaAs, which is consistent with the superior crystalline quality of the latter. Etching of $In_xAl_{1-x}N$ was also examined as a function of In composition, with etch rate initially increasing up to 36% In and then decreasing to zero for InN (Vartuli et al., 1996a).

Other researchers have found that only molten salts (KOH, NaOH, P_2O_5) will etch GaN at temperatures above 300°C, making handling and masking of material impractical (Sakai, private communication).

Minsky, White, and Hu (1996) reported laser enhanced, room temperature wet etching of GaN using dilute HCl/H_2O or 45% KOH/H_2O, with rates up to a few thousand angstrom/min for HCl and a few thousand angstrom/min for K_OH. The mechanism is believed to be photoenhancement of oxidation and reduction reactions in what amounts to an electrochemical cell. Etch rates were linearly dependent on incident HeCd laser power.

Zory et al. (1996) have employed a pulsed electrochemical cell combining 40 parts ethylene glycol, 20 parts water and 1 part 85% H_3PO_4 to etch p-GaN and InGaN epitaxial layers at rates up to 1.5 μm/h. Cell voltage (220 V) was pulsed at 100 Hz (300 μm/sec. pulse width). This technique was used in fabricating a double heterostructure p-GaN/InGaN QW/n-GaN light-emitting diode using a liquid contact.

III. Dry Etching

Ion milling rates for nitrides are low because of their high bond strengths (Pearton *et al.*, 1994). Figure 2 shows Ar^+ mill rates for binary nitrides and $In_{0.5}Ga_{0.5}N$ as a function of acceleration energy. These mill rates are about a factor or two slower than for GaAs and InP under the same conditions, and since there is virtually no selectivity between mask materials and semiconductors it would be impossible to provide a thick enough mask for deep etching applications.

Leonard and Bedair (1996) obtained photoassisted dry etching of GaN at ~80 Å/min using 193 nm ArF excimer laser enhancement of HCl vapor etching at 200–400°C. There was no etching with HCl vapor alone at 650°C. Use of laser pulses to promote product formation and desorption suggests the possibility of controlled, atomic layer etching.

Various ion enhanced etch techniques have been employed for obtaining anisotropic profiles at practical rates using conventional mask materials such as photoresist, SiN_x or SiO_2. These techniques include reactive ion etching (RIE), magnetron enhanced RIE (MERIE), electron cyclotron resonance (ECR), chemically assisted ion beam etching (CAIBE), and inductively coupled plasma (ICP) (Shul, 1996). The basic principle of these methods is similar, in that etching occurs through a combination of both chemical and physical means. Reactive neutral atoms of chlorine, iodine, bromine or other elements, produce chemical etching by forming volatile

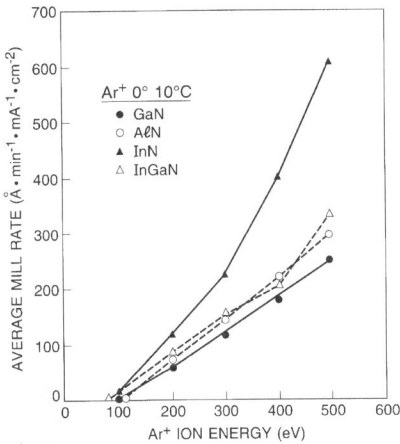

FIG. 2. Ion mill rates normalized to the beam current for nitride materials as a function of Ar^+ ion energy.

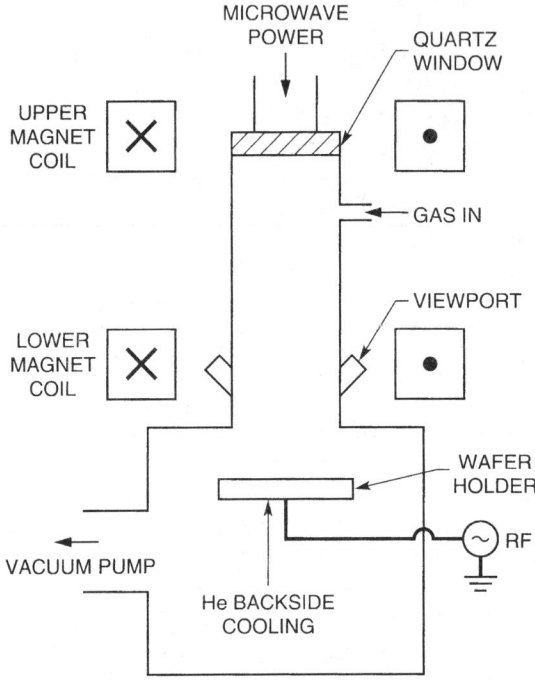

FIG. 3. Schematic of a high profile ECR etch reactor. In some configurations the source is shortened and lower magnets placed under the sample position.

products. This type of etching is generally rapid, isotropic, and damage free. Physical removal of material by sputtering is produced by impinging ions. This type of etching is slow, anisotropic, and produces a lot of lattice disorder. The combination of two mechanisms is able to produce anisotropic etching at practical rates with varying degrees of damage.

A typical configuration for a high ion density plasma etch system is shown in Fig. 3. A remote plasma is created in a source above the sample position, at a frequency of 2.45 GH_z for ECR, or 2 MH_z for ICP and other rf antenna designs. Neutral gas atoms and molecules from this plasma are incident on the sample surface, as well as ions whose energy is controlled by separate rf biasing of the electrode. This configuration obviously allows for separate control of plasma density and ion energy, which not the case in RIE.

A schematic representation of the situation around the sample during plasma etching is shown in Fig. 4. The body of plasma is neutral, containing

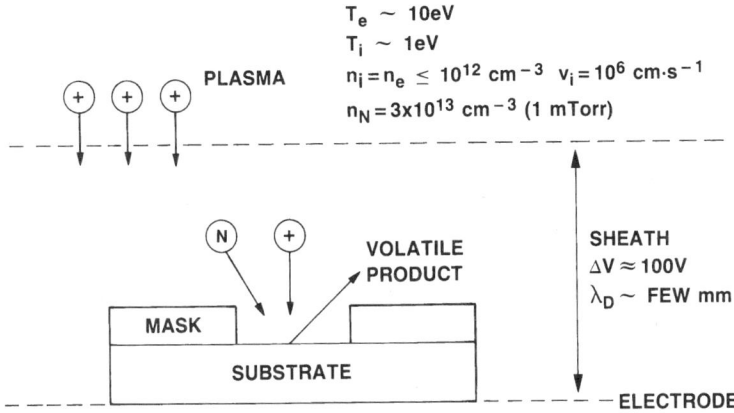

FIG. 4. Schematic of configuration of dry etching of a semiconducting sample.

equal numbers of electrons and ions (typically $\sim 10^9 \mathrm{cm}^{-3}$ for an RIE system, $\sim 5 \times 10^{11} \mathrm{cm}^{-3}$ for ECR and ICP), but is at a positive potential with respect to the sample position. Ions in plasma that stray near the sheath region (where electrons are excluded by the negative potential of sample position) are accelerated across it and strike the sample at near vertical incidence (Pearton, 1994). Neutral reactive gas atoms and molecules which constitute the most prevalent state in plasma ($\sim 3 \times 10^{13} \mathrm{cm}^{-3}$ at 1 mTorr) are also incident on the surface. Volatile etch products formed by adsorption and reaction of these species on the sample are removed by ion assisted processes at a more rapid rate than would occur in the absence of ion bombardment. Sputter assisted removal of these etch products exposes a fresh surface for the process to occur again, and in this fashion there is a synergy between chemical and physical etch mechanisms.

Table II shows a compilation of etch rates obtained with various techniques and various plasma chemistries (Pearton, Abernathy, and Ren, 1994; Shul et al., 1995b, 1995d; Pearton et al., 1993b; Lin et al., 1994; McLane et al., 1995; Adesida et al., 1993; Ping et al., 1994; Pearton, Abernathy, and Vartuli, 1994a; Adesida et al., 1994; Gillis et al., 1996c; Vartuli et al., 1996b; Vartuli et al., 1996a; Shul et al., 1996; Matsutani et al., 1997; Zhang et al., 1996; Lee, Oberman, and Harris, 1996; Ping et al., 1996; Gillis, Choutov, and Martin, 1996c. The conclusions from this data are:

(i) etch rates under high ion density conditions, ECR, ICP, or MERIE are much higher than for conventional reactive ion etching,

TABLE II

PLASMA CHEMISTRIES FOR DRY ETCHING OF GaN

Plasma	Technique	GaN Etch Rate (Å/min)	dc Bias	Refs.
Cl_2/H_2	ECR	1,100	−150	27, 28
Cl_2/SF_6	ECR	900	−150	28
$CH_4/H_2/Ar$	ECR	400	−250	27
$Cl_2/CH_4/H_2/Ar$	ECR	3,100	−125	28
$Cl_2/H_2/Ar$	ECR	2,200	−100	29
CCl_2F_2	ECR	400	−250	30
BCl_3	RIE	500	−230	31
BCl_3	MIE	3,500	−75	32
$SiCl_4(Ar, SiF_4)$	RIE	500	−400	33
$HBr(H_2/Ar)$	RIE	600	−400	34
HI/H_2	ECR	1,100	−150	35
HBr/H_2	ECR	900	−150	35
Ar	Ion milling	250	−400	23
Cl_2/Ar	CAIBE	2,000	−600	36
H_2	LEEEE	75	0	37
IBr	ECR	3,000	−150	38
ICl	ECR	13,000	−150	39
$Cl_2/H_2/Ar$	ICP	6,875	−280	40
Cl_2	RIBE	240	−1000	41
$SiCl_4/Ar$	ECR	950	−280	42
CHF_3, C_2ClF_5	RIE	450	−500	43
HCl	CAIBE	2,000	−500	44
H_2/Cl_2	LEEEE	2,500	0	45

(ii) halogen based plasma chemistries produce much higher rates than for CH_4/H_2, regardless of the etching technique,

(iii) highest etch rates are generally $\leqslant 7000$ Å/min, which is slower by a factor of 3–5 than conventional III–V's like GaAs under the same conditions.

The question of why etch rates are lower for nitrides is illuminated by considering the volatility of possible etch products, shown in Table III. All of the nitrogen etch products are extremely volatile (more so than As, P, or Sb products for other III–V materials). The rate limiting step therefore cannot be desorption of these products, but rather the initial bond breaking which must precede etch product formation. Since higher ion densities in ECR, ICP, and MERIE will enhance this bond breaking, etch rates are much higher with these techniques. An example is shown in Fig. 5 for ECR or RIE etching of GaN and AlN in Cl_2/Ar or CH_4/H_2Ar. These experiments

TABLE III

BOILING POINTS OF III–V ETCH PRODUCTS

Species	Boiling Point (°C)
$GaCl_3$	201
$GaBr_3$	279
GaI_3	Sub 345
$(CH_3)Ga$	55.7
$InCl_3$	600
$InBr_3$	>600
InI_3	210
$(CH_3)_3In$	134
$AlCl_3$	183
$AlBr_3$	263
AlI_3	191
$(CH_3)_3Al$	126
NCl_3	<71
NI_3	Explodes
NF_3	−129
NH_3	−33
N_2	−196
$(CH_3)_3N$	2.9
PCl_3	76
PBr_5	106
PH_3	−88
$AsCl_3$	130
$AsBr_3$	221
AsH_3	−55
AsF_3	−63

were performed in the same reactor chamber, with 1000 W (ECR conditions) or 0 W (RIE conditions) being applied to ECR source. The rf power and pressure were held constant at 150 W and 1.5 mTorr, respectively. What is clear from Fig. 5 is that etch rates are much higher for ECR conditions and also rates increase as ion energy increases due to more efficient bond breaking.

Figure 6 shows a comparison of GaN etch rates in $Cl_2/CH_4/H_2/Ar$ plasmas at fixed pressure (1 mTorr) and high density source power (750 W for ECR and ICP) for three different techniques performed on the same reactor. In this case, ICP source replaced ECR source for purposes of direct comparison. Rates for ICP source are slightly higher than for ECR source at rf powers above 150 W, and may be due to a higher ion and active neutral density achieved in this design. Note that once again rates for both high density techniques are substantially higher than for RIE.

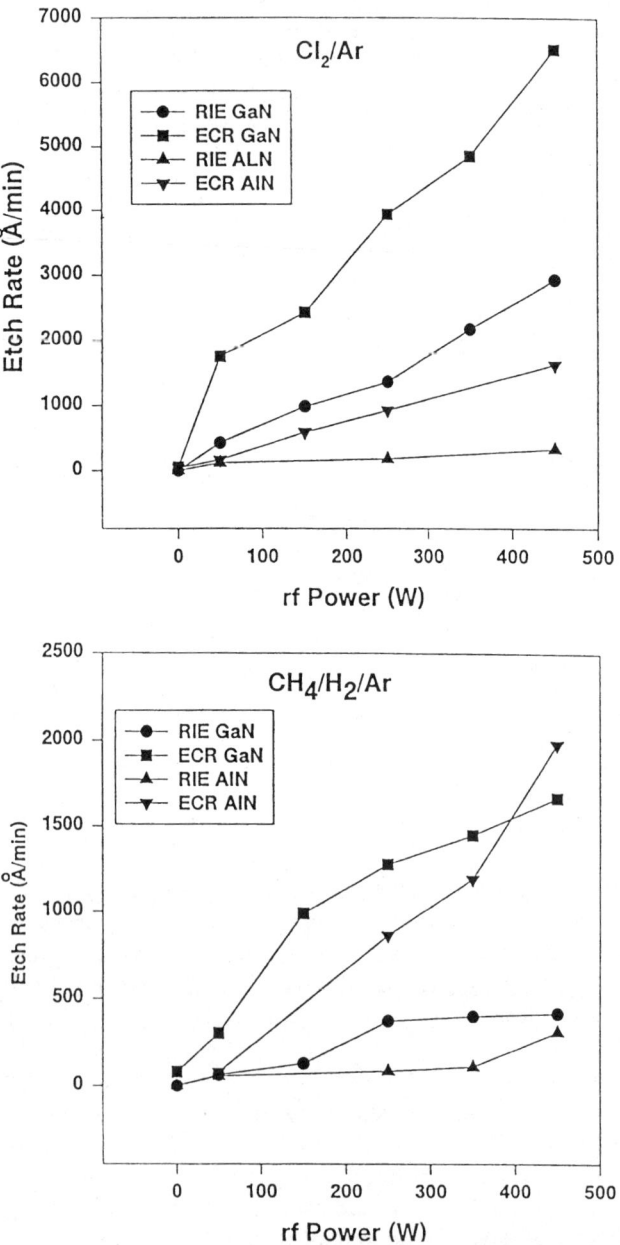

FIG. 5. Etch rates of GaN and AlN in RIE or ECR Cl_2/Ar (top) or CH_4/H_2/Ar (bottom) plasmas as a function of rf chuck power. The ECR power is 1000 W and the pressure 1.5 mTorr.

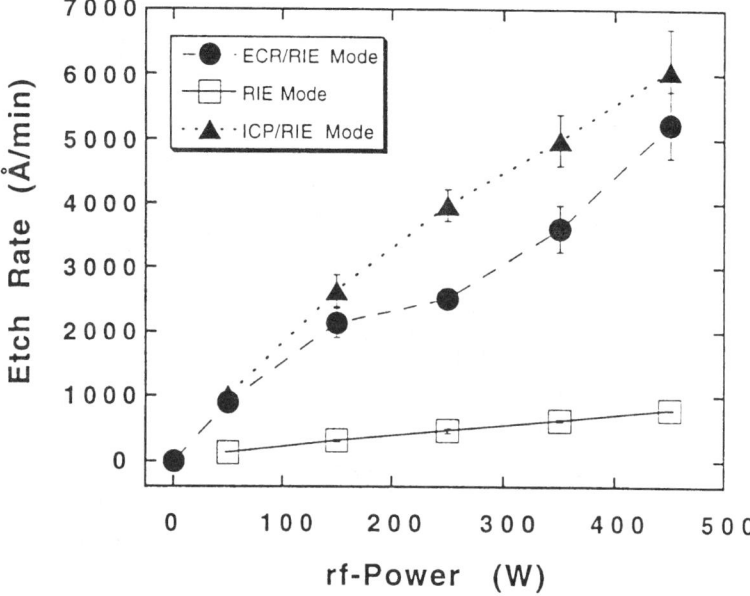

FIG. 6. Comparison of GaN etch rates in $Cl_2/CH_4/H_2/Ar$ plasmas using different dry etching methods.

A scanning electron microscopy (SEM) micrograph of a feature etched into a 1.8 μm thick GaN epilayer grown on Al_2O_3 is shown in Fig. 7. An ICP $Cl_2/H_2/Ar$ plasma at 750 W source power, 150 W rf power and 1 mTorr was used to form this simulation of a laser facet. The photoresist mask has been removed prior to taking the micrograph. There are several features to note. First, etching is highly anisotropic, showing it is possible to achieve high fidelity pattern transfer at moderate ion energies. Second, facet sidewall is quite smooth indicating an absence of mask degradation and material redeposition, and third, craters observed on the field are often seen for GaN grown on Al_2O_3 and are probably related to replication of defects in the substrate.

One problem with employing ion energies above ~100 eV is that it is difficult to avoid preferential loss of lighter N atoms from the plasma etched GaN surface (Shul et al., 1995c; McLane, Pearton, and Abernathy, 1995; Shul et al., 1995a). Figure 8 shows Auger electron spectroscopy (AES) surface scans of GaN after etching in ECR $Cl_2/CH_4/H_2/Ar$ discharges (750 W microwave power) at progressively higher rf powers. As ion energy increases under these conditions there is a Ga enrichment in the top 100 Å

FIG. 7. SEM micrograph of GaN feature sidewall etched with an ICP $Cl_2/H_2/Ar$ plasma.

of the surface. Onset of preferential nitrogen loss is accompanied by surface roughening, as shown in atomic force microscopy (AFM) images (Fig. 9) for both GaN and InN. At the highest rf powers (275 W) there are group III droplets evident on the surface, which is clearly unsuitable for further processing.

The highest etch rate reported for GaN is 1.3 μm/min for ECR etching in an ICl plasma chemistry (Vartuli *et al.*, 1996b). Iodine has some attractive features, particularly since InI_3 is much more volatile than $InCl_3$ (Pearton, Abernathy, and Vartuli, 1994b). Some characteristics of ICl/Ar dry etching of GaN and related compounds are shown in Fig. 10. Rates in general increase with ICl percentage (top), and increase rapidly for GaN, InN, and InGaN above 150 W rf chuck power in ECR ICl/Ar discharges (1000 W microwave power). An advantage to the use of ICl is that it is a very weakly bound molecule (melting point 27°C) and therefore microwave powers as low as 400 W are sufficient to basically fully dissociate them into their component iodine and chlorine atoms. We observed essentially no dependence of nitride etch rate on microwave power between 400–1000 W. The

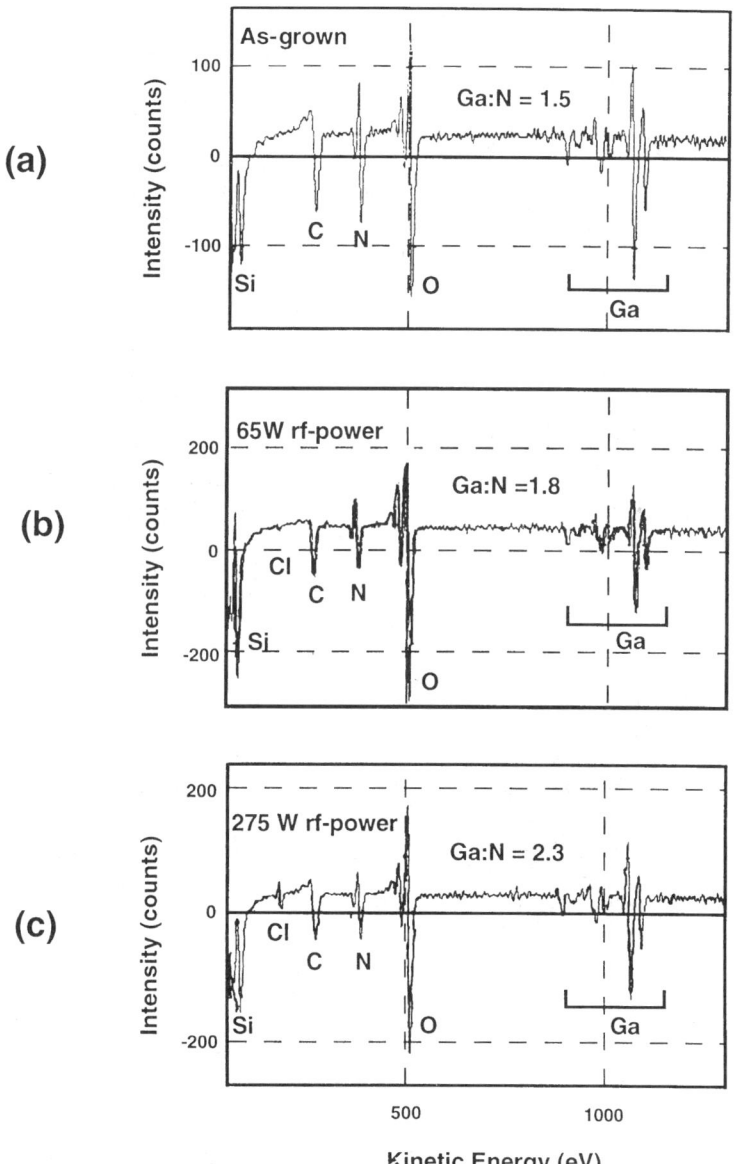

FIG. 8. AES surface scans of GaN surfaces before and after etching in $Cl_2/CH_4/H_2/Ar$ ECR discharges with different rf powers.

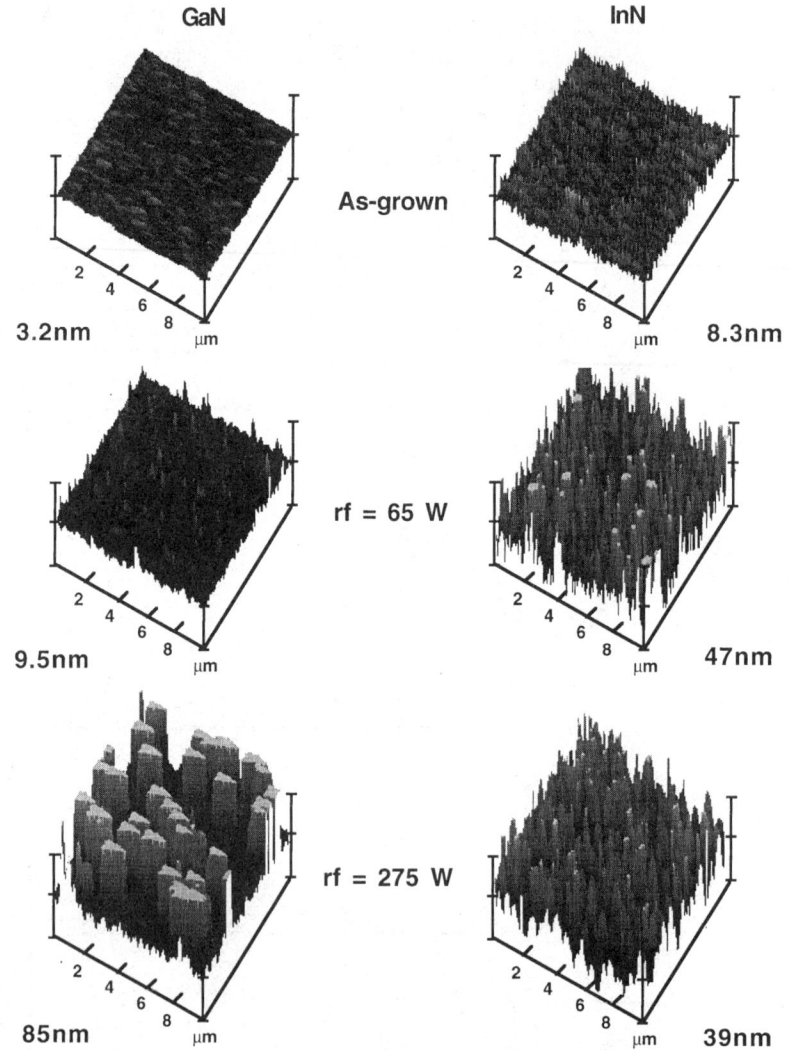

Fig. 9. AFM scans of GaN surfaces before and after etching in $Cl_2/CH_4/H_2/Ar$ ECR discharges with different rf powers.

advantage in use of low microwave powers is that simple photoresist masks can be used, rather than having to revert to the more robust dielectric or metal masks that are generally necessary with high density plasmas (Melville, Thompson, and Simmons, 1995; Lee, Crockett, and Pearton, 1996).

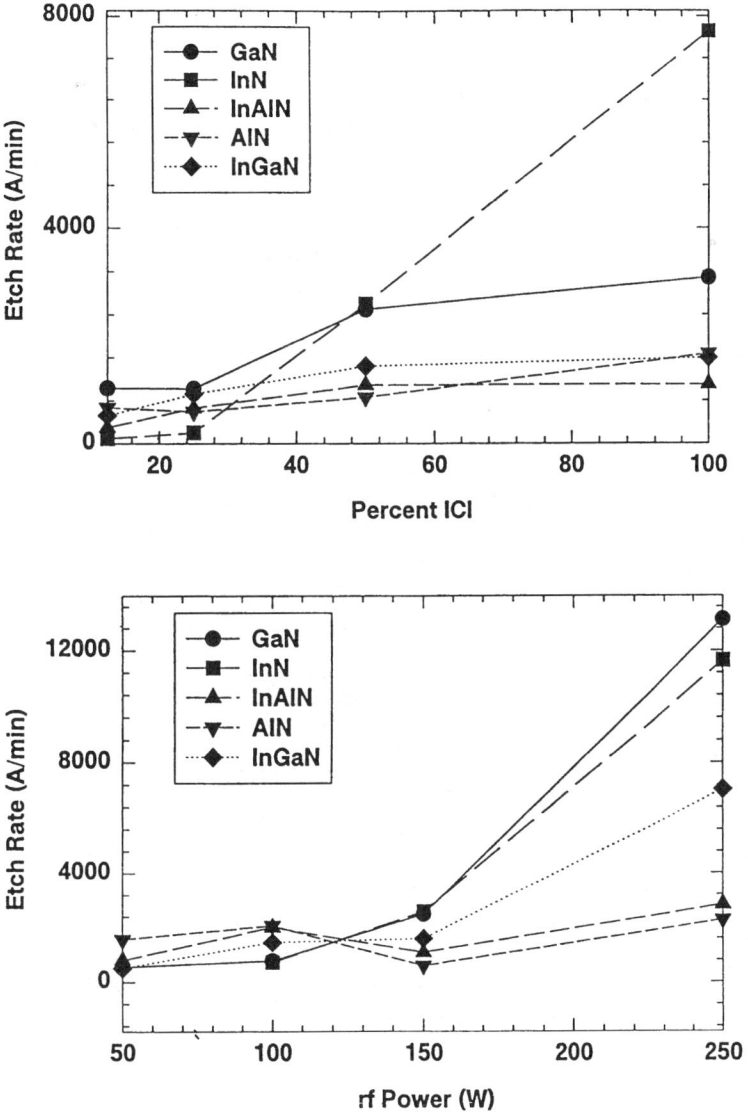

FIG. 10. Etch rates of nitride materials as a function of plasma composition or rf power in ECR ICl/Ar discharges.

IV. Issues of Etch Selectivity and Damage

RIE is an inherently high damage technique in many instances because of high ion acceleration energies. It has proved effective for robust devices such as light emitting devices (LEDs) because in these structures compound layers are comparatively thick and heavily doped, and thus are fairly resistant to damage. Moreover, the etch proceeds to an n^+GaN layer, onto which an ohmic contact is subsequently deposited. Ion damaged GaN surfaces are usually made highly conducting n-type, and thus in this particular case ohmic contact resistance can be dramatically improved. We suspect this is at least partially responsible for very low contact resistivity reported by Fang et al. (1996) for ohmic metallization in RIE etched n-GaN. However, if one needs to deposit Schottky contacts, great care is needed to avoid surface leakage currents.

ECR and ICP techniques use lower ion energies than RIE, and thus in general have lower damage levels as a result (Pearton, 1994; Shul et al., 1994). However, at high source powers ion flux incident on the surface is so high that substantial damage can be detected to depths of a few hundred angstrom. This is generally observed as a reduction in carrier concentration due to introduction of deep trap states that remove carriers from the conduction process. CAIBE typically operates with Ar ion energies >200 eV (more typically ~ 500 eV) in order to obtain sufficient ion current from Kaufman type sources employed, and thus has the inherent capability for ion induced damage.

An interesting new low damage alternative etching technique is low energy electron enhanced etching (LE4) (Gillis et al., in press; Gillis, Choutov, and Marlin, in press; Gillis et al., 1995; Gillis et al., 1996), in which electrons at energies up to 15 eV and reactive species at thermal velocities are incident on the sample. Electrons appear to enhance reactions between the etch gas (H_2, HCl, Cl_2/H_2) and semiconductor, producing anisotropic features. In pure H_2 plasmas, etch rates near room temperature are relatively low (70 Å/min) but use of pure Cl_2 produced rates of 2500 Å/min at 100°C (Gillis, Choutov, and Marlin, in press).

Since in more common etching techniques it is necessary to have a high physical component to produce bond breaking, it is difficult to achieve high selectivities for etching one material over another (Seaward and Moll, 1992). High selectivity is generally achieved by initiating an etch stop reaction on reaching a buried layer within a heterostructure. The classic system in this regard is GaAs/AlGaAs, where addition of fluorine to a chlorine based plasma chemistry creates an etch stop at AlGaAs through formation of involatile AlF_3 (Cooper, Salimian, and MacMillan, 1987; Seaward et al., 1987; Pearton et al., 1989; Pearton, 1991). However, at high ion energies

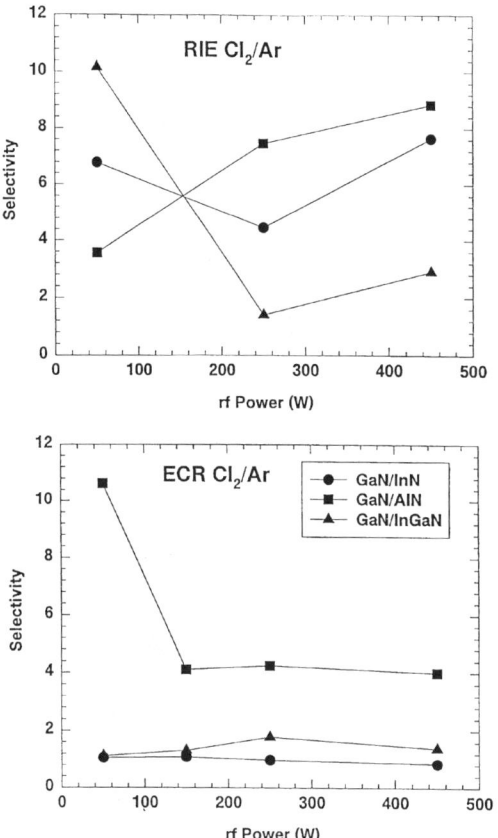

FIG. 11. Selectivities for etching GaN over InN, InGaN or AlN in RIE or ECR Cl_2/Ar plasmas, as a function of rf chuck power.

and/or fluxes it is possible to sputter away the AlF_3 and eliminate selectivity for GaAs over AlGaAs.

Figure 11 shows selectivities for GaN over InN, AlN, or $In_{0.5}Ga_{0.5}N$ in ECR and RIE Cl_2/Ar discharges as a function of applied rf chuck power. In the case of RIE, selectivities are between 4–8 for GaN over the other two binaries and reach as high as 10 for low rf power etching of GaN/InGaN. In the case of GaN/AlN selectivity increases with ion energy because of enhanced etch rate of GaN, whereas in general, selectivity decreases with bias. For ECR conditions there is essentially no selectivity for GaN/InN and GaN/InGaN, although $CH_4/H_2/Ar$ plasma chemistries do have selectivities up to ~5 for similar rf powers. Optimization of etch parameters for

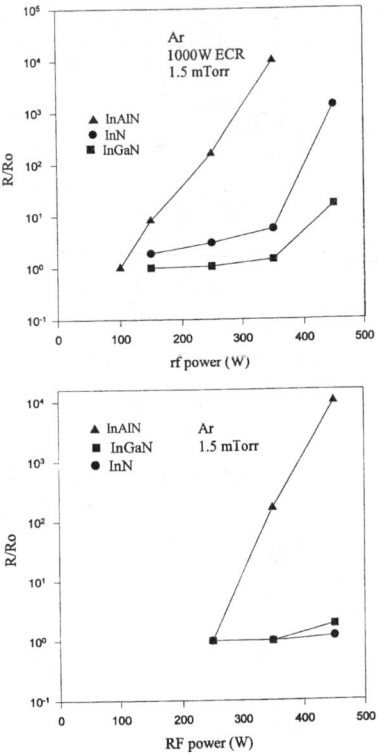

FIG. 12. Resistance ratio increase of InN, InGaN or InAlN layers exposed to ECR or RIE Ar plasmas, as a function of rf chuck power.

achieving higher selectivities will be more critical in processing of electronic devices, where layer thicknesses are in general much thinner than in photonic devices.

The issue of ion induced damage will also be more important in field effect transistors compared to light emitting diodes. We have studied a worst-case scenario for damage introduction during RIE or ECR etching of nitrides by exposing thin layers (~ 2000 Å) to pure Ar discharges under various conditions to simulate the physical ion bombardment component. In a real etch process, damage would be less severe because presence of a chemical etch component would enhance removal rate of material and reduce the amount of damage accumulation (Pearton et al., 1991; Pang, 1986; Lishan et al., 1989). Figure 12 shows the resistance ratio increase of layers of InN, $In_{0.5}Ga_{0.5}N$ or $IN_{0.5}Al_{0.5}N$ exposed to ECR or RIE Ar plasma, as a function of rf chuck power (Pearton et al., 1995). Under ECR conditions, heavily doped InN and InGaN ($\sim 10^{20}$ cm^{-3} n-type) do not show much increase in resistance until ~ 350 W, whereas more lightly doped InAlN ($\sim 10^{18}$ cm^{-3}) is already substantially damaged at 150 W. These results emphasize that damage thresholds are a strong function of layer thickness and doping level, since more damage is needed to have an effect on heavily doped thick layers than for thin, lightly doped material. In the case of RIE plasmas there is little influence of plasma damage on InN and InGaN until very high powers (450 W), while the threshold for InAlN has shifted from 150 W to ~ 250 W observed under ECR conditions. The latter is a result of much lower ion flux in RIE plasmas.

During gate mesa plasma etching of InN/InAlN field effect transistors (FETs) the apparent conductivity in the channel can be increased or decreased through three different mechanisms. If hydrogen is part of the plasma chemistry, hydrogen passivation of shallow donors in InAlN can occur (we find diffusion depths for ^{2}H of $\geqslant 0.5$ μm in 30 min at 200°C), hydrogen remains in the material until temperatures $\geqslant 700$°C. Energetic ion bombardment in SF_6/O_2 or BCl_3/Ar plasmas also compensates doping in InAlN by creation of deep acceptor states. Finally, conductivity of the immediate InAlN surface can be increased by preferential loss of N during BCl_3 plasma etching, leading to poor rectifying contact characteristics when the gate metal is deposited on this etched surface. Careful control of plasma chemistry, ion energy, and stoichiometry of the etched surface are necessary for acceptable pinch-off characteristics.

We have used ECR etching to etch through InN contact layers on InN/InAlN transistor structures in order to form the gate mesa (Ren et al., in press). On dry etch removal of the InN capping layer, a Pt/Ti/Pt/Au gate contact was deposited on exposed InAlN to complete FET processing. If pure BCl_3 was employed as plasma chemistry, we observed ohmic and not rectifying behavior for gate contact. If BCl_3/N_2 was used, there was some improvement in gate characteristics. A subsequent attempt at a wet etch clean up using H_2O_2/HCl or H_2O_2/HI produced a reverse breakdown in excess of 2 V (Fig. 13). These results suggest that InAlN surface becomes nonstoichiometric during the dry etch step, and that addition of N_2 retards some of this effect.

Figure 14 shows the drain source current (I_{DS}) values obtained as a function of dry etch time in ECR discharges of BCl_3 or BCl_3/N_2. In the former case, current does not decrease as material is etched away, suggesting that a conducting surface layer is continually being created. By contrast BCl_3/N_2 plasma chemistry does reduce the drain source current as expected, even though breakdown characteristics of gate metal deposited on this surface are much poorer than would be expected.

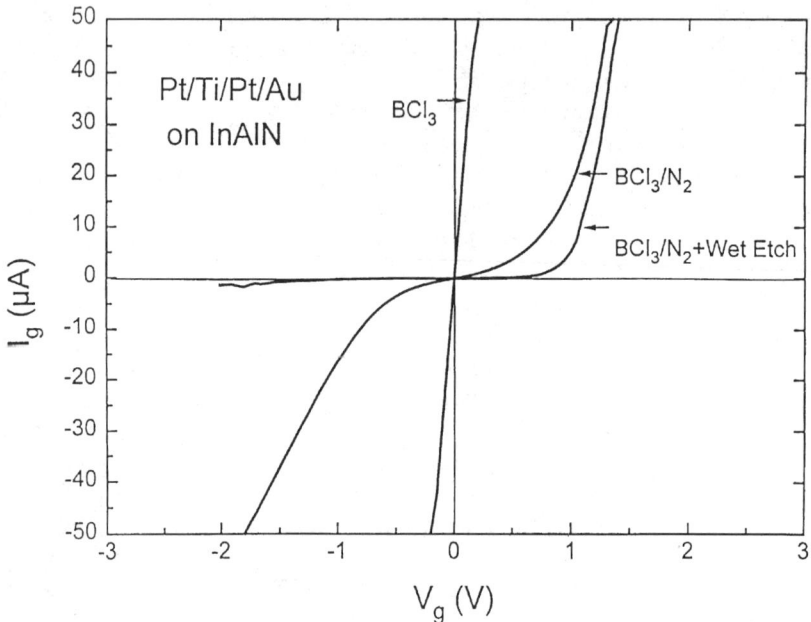

FIG. 13. I–V characteristics of Pt/Ti/Pt/Au contacts on InAlN exposed to BCl_3 or BCl_3/N_2 plasmas.

AES analysis of etched surfaces was employed to understand these results. The N/In ratio of near surface was decreased as a result of dry etching from 0.35 in the as-grown sample to 0.28 (BCl_3), 0.29 (BCl_3/N_2), and 0.31 (BCl_3/N_2), plus a wet etch. Thus it appears that nitrogen is preferentially lost from the etched surface, much as is the case with loss of P from InP in Cl_2 or CH_4/H_2 plasma chemistries. Deposition of a gate metal onto this In rich surface produces ohmic behavior. To overcome this problem in InP it is necessary to heat the sample during plasma etching to enhance desorption of the $InCl_3$ etch product (McNevin, 1986).

In summary, InAlN FET structures are sensitive to several effects during dry etching of the gate mesa. Firstly, if hydrogen is present in plasma there can be passivation of doping in the InAlN channel layer. Secondly, ion bombardment from the plasma can create deep acceptor states that compensate the material, and thirdly, even when these problems are avoided through use of H-free plasma chemistries and low ion energies and fluxes, preferential loss of N can produce poor rectifying gate characteristics for metal deposited on the etched surface.

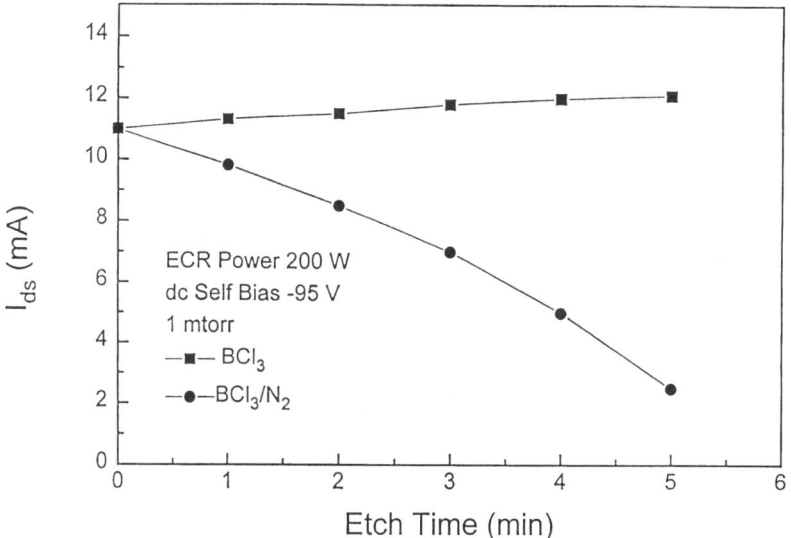

FIG. 14. I_{DS} values at 5 V bias for InAlN FETs etched for various times in BCl_3 or BCl_3/N_2.

V. Summary

The RIE, MERIE, ECR, and CAIBE etch processes for structures such as LEDs are now well established. Etch rates are much higher with high density methods, and conventional $CH_4/H_2/H_2$ or In based plasma chemistries are suitable for nitrides. The recent advent of ICP sources offers the possibility of even higher etch rates. For laser diodes where etched facets must have smooth, striation-free sidewalls to avoid loss of intensity by scattering losses (Shul *et al.*, 1995e; Constantine *et al.*, 1995), avoidance of mask erosion is critical. This is less of an issue in microdisk laser structures where there is no need for facet formation (Abernathy *et al.*, 1996).

For electronic device fabrication it is important to avoid creation of nonstoichiometric surfaces if Schottky contacts are to be deposited on the etched surface. This appears to be more of a problem with In containing nitrides because of the vast difference in atom weight between In and N, leading to preferential loss of the latter. The issue of ion induced damage will also be much more important in transistor structures because of their thinner layers and lower doping levels relative to photonic device structures. For these devices, new etching techniques such as LE4 may have an important role.

Acknowledgments

The authors acknowledge the collaboration of their co-workers, C. R. Abernathy, J. W. Lee, C. B. Vartuli, and J. D. MacKenzie at UF; S. Kilcoyne, M. Hagerott-Crawford, G. B. McClellan, and J. C. Zolper at SNL; F. Ren at Lucent Technologies, Bell Laboratories; R. F. Karlicek, Jr., and R. A. Stall at EMCORE Corp.; and C. Constantine and C. Barratt at Plasma Therm. The work at these facilities is partially supported by an DARPA grant (A. Husain) administered by AFOSR (G. L. Witt) and SNL is supported by DOE (Contract DE-AC04-94AL85000). The assistance of P. L. Glasborg at Sandia is also greatly appreciated.

References

Abernathy, C. R. (1995). *Mat. Sci. Eng. R* **14**, 203.
Abernathy, C. R., Pearton, S. J., MacKenzie, J. D., Mileham, J. R., Bharatan, S. R., Krishnamoorthy, V., Jones, K. S., Crawford, M. H., Shul, R. J., Kilcoyne, S. P., Zavada, J. M., Zhang, D., and Kolbas, R. M. (1996). *Solid State Electron.* **39**, 311.
Adesida, I., Mahajan, A., Andideh, E., Khan, M. A., Olsen, D. T., and Kuznia, J. N. (1993). *Appl. Phys. Lett.* **63**, 2777.
Adesida, I., Ping, A. T., Youtsey, C., Dow, T., Khan, M. A., Olsen, D. T., and Kuznia, J. N. (1994). *Appl. Phys. Lett.* **65**, 889.
Aita, C. R., and Gawlak, C. J. (1983). *J. Vac. Sci. Technol. A* **1**, 403.
Barrett, N. J., Grange, J. D., Sealy, B. J., and Stephens, K. G. (1985). *J. Appl. Phys.* **57**, 5470.
Chu, T. L. (1971). *J. Electrochem. Soc.* **118**, 1200.
Constantine, C., Shul, R. J., Sullivan, C. T., Snipes, M. B., McClellan, G. B., Hafich, M., Fuller, C. T., Mileham, J. R., and Pearton, S. J. (1995). *J. Vac. Sci. Technol. B* **13**, 2025.
Cooper, C. B. III, Salimian, S., and MacMillan, H. F. (1987). *Appl. Phys. Lett.* **51**, 2225.
Fang, Z., Mohammad, S. N., Kim, W., Aktas, O., Botchkarev, A. E., and Morkoc, H. (1996). *Appl. Phys. Lett.* **68**, 1672.
Gillis, H. P., Choutov, D. A., Steiner, P. A. IV, Piper, J. D., Crouch, J. H., Dove, P. M., and Martin, K. D. (1995). *Appl. Phys. Lett.* **66**, 2475.
Gillis, H. P., Choutov, D. A., Martin, K. P., and Song, L. (1996a). *Appl. Phys. Lett.* **68**, 2255.
Gillis, H. P., Choutov, D. A., and Marlin, K. P. (1996b). *J. Mater.* **48**, 50.
Gillis, H. P., Choutov, D. A., Martin, K. P., Pearton, S. J., and Abernathy, C. R. (1996c). *J. Electrochem. Soc.* **143**, L251.
Guo, Q. X., Kato, O., and Yoshida, A. (1992). *J. Electrochem. Soc.* **139**, 2008.
Harrison, W. A. (1980). *Electronic Structure and Properties of Solids*, Freeman, San Francisco.
Kline, G. R., and Lakin, K. M. (1983). *Appl. Phys. Lett.* **43**, 750.
Lee, J. W., Crockett, R. V., and Pearton, S. J. (1996). *J. Vac. Sci. Technol. B* **14**, 1752.
Lee, H., Oberman, D. B., and Harris, Jr., J. S. (1996). *J. Electron. Mater.* **25**, 835.
Leonard, R. T., and Bedair, S. M. (1996). *Appl. Phys. Lett.* **68**, 794.
Lin, M. E., Fan, Z. F., Ma, Z., Allen, H. L., and Morkoc, H. (1994). *Appl. Phys. Lett.* **64**, 887.
Lishan, D. G., Wong, H. F., Green, D. L., Hu, E. L., Mertz, J., and Kirillov, K. (1989). *J. Vac. Sci. Technol. B* **7**, 565.
Long, G., and Foster, L. M. (1959). *J. Am. Ceram. Soc.* **42**, 53.

MacKenzie, J. D., Abernathy, C. R., Pearton, S. J., Krishnamoorthy, V., Bharatan, S., Jones, K. S., and Wilson, R. G. (1995). *Appl. Phys. Lett.* **67**, 253.

Matsutani, M., Honda, T., Sakaguchi, T., Saotome, K., Koyama, F., and Iga, K. (1997). *Mat. Res. Soc. Symp. Proc.* Vol. **449**, 1029.

McLane, G. F., Pearton, S. J., and Abernathy, C. R. (1995). Proceedings Symposium: *Wide Bandgap Semiconductors and Devices*, Electrochemical Society, Pennington, Vol. 95-21, pp. 204–212.

McLane, G. F., Casas, L., Pearton, S. J., and Abernathy, C. R. (1995). *Appl. Phys. Lett.* **66**, 3328.

McNevin, S. C. (1986). *J. Vac. Sci. Technol. B* **4**, 1216.

Melville, D. L., Thompson, D. A., and Simmons, J. G. (1995). *J. Electrochem. Soc.* **142**, 2762.

Mileham, J. R., Pearton, S. J., Abernathy, C. R., MacKenzie, J. D., Shul, R. J., and Kilcoyne, S. P. (1995). *J. Vac. Sci. Technol. A* **14**, 836; ibid. (1995), *Appl. Phys. Lett.* **67**, 1119.

Minsky, M. S., White, M., and Hu, E. L. (1996). *Appl. Phys. Lett.* **68**, 1531.

Nakamura, S., Mukai, T., and Senoh, M. (1994). *Appl. Phys. Lett.* **64**, 1687.

Nakamura, S., Senoh, M., Nagahama, S., Iwasa, N., Yamada, T., Matsushito, T., Kiyoku, H., and Sugimoto, U. (1996). *Jap. J. Appl. Phys.* **35**, L74.

Pang, S. W. (1986). *J. Electrochem. Soc.* **133**, 784.

Pankove, J. I. (1972). *J. Electrochem. Soc.* **119**, 1118.

Pauleau, T. (1982). *J. Electrochem. Soc.* **129**, 1045.

Pearton, S. J., Hobson, W. S., Chakrabarti, U. K., Emerson, A. B., Lane, E., and Jones, K. S. (1989). *Appl. Phys.* **66**, 2137.

Pearton, S. J. (1991). *Mat. Sci. Eng. B* **10**, 187.

Pearton, S. J., Ren, F., Lothian, J., Fullowan, T., Kopf, R.l F., Chakrabarti, U. K., Hui, S. P., Emerson, A. B., Kostela, R. L., and Pei, S. S. (1991). *J. Vac. Sci. Technol. B* **9**, 2487.

Pearton, S. J., Abernathy, C. R., Ren, F., Lothian, J. R., Wisk, P., and Katz, A. (1993a). *J. Vac. Sci. Technol. A* **11**, 1772.

Pearton, S. J., Abernathy, C. R., Ren, F., Lothian, J. R., Wisk, P., Katz, A., and Constantine, C. (1993b). *Semicond. Sci. Technol.* **8**, 310.

Pearton, S. J. (1994). *J. Mod. Phys. B* **8**, 1781.

Pearton, S. J. (1994). In *InP HBTs: Growth, Processing and Devices* (eds. B. Jalali and S. J. Pearton), Artech House, Dedham, Ch. 3.

Pearton, S. J., Abernathy, C. R., and Ren, F. (1994). *Appl. Phys. Lett.* **64**, 2294; ibid. (1994) *Appl. Phys. Lett.* **64**, 3643.

Pearton, S. J., Abernathy, C. R., Ren, F., and Lothian, J. R. (1994). *J. Appl. Phys.* **76**, 1210.

Pearton, S. J., Abernathy, C. R., and Vartuli, C. B. (1994a). *Electron Lett.* **30**, 1985.

Pearton, S. J., Abernathy, C. R., and Vartuli, C. B. (1994b). *Electron. Lett.* **30**, 894.

Pearton, S. J., Lee, J. W., MacKenzie, J. D., Abernathy, C. R., and Shul, R. J. (1995). *Appl. Phys. Lett.* **67**, 2329.

Ping, A. T., Adesida, I., Khan, M. A., and Kuznia, J. N. (1994). *Electron. Lett.* **30**, 1895.

Ping, A. T., Schmitz, A. C., Khan, M. A., and Adesida, I. (1996). *J. Electron. Mater.* **25**, 825.

Ren, F., Lothian, J. R., MacKenzie, J. D., Abernathy, C. R., Vartuli, C. B., Pearton, S. J., and Wilson, R. G. (1996). *Solid State Electron.* **39**, 1747.

Sakai, S. (private communication).

Seaward, K. L., Moll, N. J., Coulman, D. T., and Stickle, W. F. (1987). *J. Appl. Phys.* **61**, 2358.

Seaward, K. L., and Moll, N. J. (1992). *J. Vac. Sci. Technol. B* **10**, 46.

Sheng, T. Y., Yu, Q., and Collins, G. J. (1988). *Appl. Phys. Lett.* **52**, 576.

Shul, R. J., Lovejoy, M. L., Hetherington, D. L., Rieger, D. J. Vawter, G. A., Klem, J. F., and Melloch, M. R. (1994). *J. Vac. Sci. Technol. A* **12**, 1351.

Shul, R. J., Howard, A. J., Pearton, S. J., Abernathy, C. R., and Vartuli, C. B. (1995a). Proceedings Symposium: *Wide Bandgap Semiconductors and Devices*, Electrochemical Society, Pennington, Vol. 95-21, pp. 217–227.

Shul, R. J., Howard, A. J., Pearton, S. J., Abernathy, C. R., Vartuli, C. B., Barnes, P. A., and Bozack, M. J. (1995b). *J. Vac. Sci. Technol. B* **13**, 2016.

Shul, R. J., Howard, A. J., Kilcoyne, S. P., Pearton, S. J., Abernathy, C. R., Vartuli, C. B. Barnes, P. A., and Bozack, M. J. (1995c). *Proceedings 22nd SOTAPOCS*, Electrochemical Society, Pennington, Vol. 95-6, pp. 209–218.

Shul, R. J., Kilcoyne, S. P., Crawford, M. H., Parmeter, J. E., Vartuli, C. B., Abernathy, C. R., and Pearton, S. J. (1995d). *Appl. Phys. Lett.* **66**, 1761.

Shul, R. J., Sullivan, C. T., Snipes, M. B., McClellan, G. B., Hafich, M., Fuller, C. T., Constantine, C., Lee, J. W., and Pearton, S. J. (1995e). *Solid State Electron.* **38**, 2047.

Shul, R. J. (1996). In *GaN and Related Compounds* (ed. S. J. Pearton), Gordon and Breach, New Jersey.

Shul, R. J., McClellan, G. B., Casalnuovo, S. A., Rieger, D. J., Pearton, S. J., Constantine, C., Barratt, C., Karlicek, Jr., R. F., Tran, C., and Schurmann, M. (1996). *Appl. Phys. Lett.* **69**, 1119.

Taylor, K. M., and Lenie, C. (1960). *J. Electrochem. Soc.* **107**, 308.

Vartuli, C. B. (1996). *Appl. Phys. Lett.* **69**, 1426.

Vartuli, C. B., Lee, J. W., Abernathy, C. R., MacKenzie, J. D., Pearton, S. J., and Shul, R. J. (1996). *J. Appl. Phys.* **80**, 3705.

Vartuli, C. B., Pearton, S. J., Lee, J. W., Abernathy, C. R., MacKenzie, J. D., Zolper, J. C., Shul, R. J., and Ren, F. (1996). *J. Electrochem. Soc.* **143**, 3681.

Walker, D., and Tarn, W. H. (eds.) (1991). *CRC Handbook of Metal Etchants*, Chemical Rubber, Boca Raton, pp. 118–121.

Zhang, L., Ramer, J., Brown, J., Zheng, K., Lester, L. F., and Hersee, S. D. (1996). *Appl. Phys. Lett.* **68**, 367.

Zory, P. L., Oh, J. S. and Bour, D. R. (1996). *Proc. SME*, Vol. **3002**, 117.

CHAPTER 6

Indium-based Nitride Compounds

S. M. Bedair

ELECTRICAL AND COMPUTER ENGINEERING
NORTH CAROLINA STATE UNIVERSITY
RALEIGH, NC

I. INTRODUCTION 127
II. INDIUM SURFACE SEGREGATION 129
III. COMPETITIVE PROCESSES DURING THE GROWTH OF InGaN 131
   1. *Indium Nitride* 138
   2. *Indium Gallium Nitride* 141
   3. $Al_yIn_{1-y}N$ 146
   4. *AlGaInN* 148
IV. EFFECT OF HYDROGEN ON THE INDIUM INCORPORATION IN InGaN EPITAXIAL
    FILMS 150
   1. *The Phase Separation Issue* 153
   2. *Doping of In-based Nitride Compounds* 155
V. InGaN BASED HETEROSTRUCTURES 156
VI. CONCLUSIONS 164
   References 165

I. Introduction

InGaN device quality films and their related heterostructures play a critical role in the development of nitride devices. Unfortunately, limited informative research regarding epitaxial growth of InGaN and related quantum well (QW) structures exist, and can be attributed to the following fundamental problems associated with the growth of In-based nitride compounds:

1. The high equilibrium vapor pressure of nitrogen required during growth to prevent the dissociation of the In–N bond represents a significant problem in the growth of In-based nitride compounds. Nitrogen equilibrium vapor pressures over AlN, GaN, and InN are shown in Fig. 1 and compared with those of group V elements over GaAs and InP (Matsuoka *et al.*, 1989). Nitrogen over InN has the

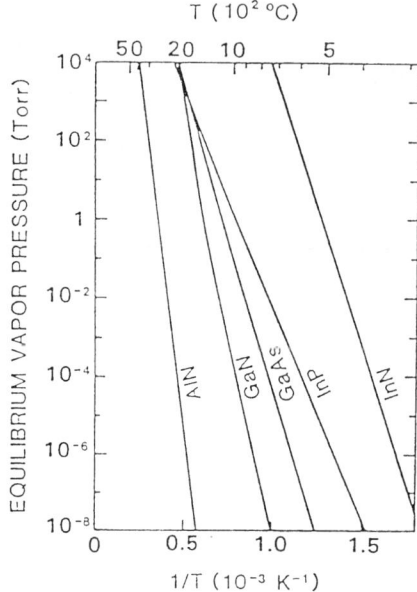

FIG. 1. Equilibrium vapor pressures of N_2 over AlN, GaN, and InN, the sum of As_2 and As_4 over GaAs, and the sum of P_2 and P_4 over InP. (From Matsuoka, 1989.)

highest equilibrium vapor pressure of all group V elements. This creates several problems during InN epitaxial growth. However, the conditions for GaN growth are quite different. For example, GaN grown by metal-organic chemical vapor deposition (MOCVD) at 1000°C is feasible due to the sufficient decomposition of NH_3 at the growing GaN surface (Doverspike et al., 1995) thus satisfying the vapor pressure requirements shown in Fig. 1. Such a high growth temperature cannot be used with In based compounds due to the weak In–N bonds which can cause In atom desorbtion from the growing surface. Some researchers have investigated new sources of nitrogen atoms that can dissociate at lower temperatures, such as hydrogen azide (HN_3) (Oberman, 1994) and hydrazide (N_2H_4) (Gaskill, Bottka, and Lin, 1986). Nitride growth by molecular beam epitaxy (MBE) (Abernathy et al., 1995) usually relies on ECR plasma sources, with fairly low growth rates.

2. AlGaN/InGaN heterostructures currently used in commercially grown device structures suffer from the fairly high lattice mismatch between these ternary alloys. For example, for green light emitting diodes

(LEDs) the thickness of the InGaN well should not exceed 30 Å to remain below the critical thickness for the generation of misfit dislocations as based on Matthew's model (Matthews and Blakeslee, 1974). Imposing such strict thickness tolerances on the InGaN active layer will make production scale-up of LEDs more difficult. This will leave AlInGaN quaternary alloys as the only viable candidate for the barrier layers, since their lattice constant and bandgap can be independently adjusted to lattice match the InGaN wells with the desired barrier at the heterointerfaces. This quaternary alloy is barely studied and its growth is also challenging due to the inherent growth temperature incompatibility between Al- and In-containing nitride compounds.
3. The inferior quality of the interfaces between the InGaN wells and the AlGaN or AlInGaN barrier layers is another source of growth related problems. Poor interfaces can result from poor nucleation (3D vs. 2D), lattice mismatch, surface reconstruction, segregation, the reaction of In atoms, and incompatible growth temperatures.
4. The potential for phase separation in this material system especially for high In concentration as was recently suggested (Moustakas, 1996).

In this chapter we review the current progress of In-based nitride compounds. We start with an overview of In surface segregation, followed by a tentative model for the reaction processes that take place during deposition, comparing the model with current experimental results. This will be followed by the growth and characterization of InN, InGaN, AlInN, and AlGaInN compounds. The effect of hydrogen on the In incorporation and the issue of phase separation will then be discussed. Finally, we will present AlGaN/InGaN and AlGaInN/InGaN double heterostructure data and their emission properties in the blue, green, and yellow regions. In-based devices will be discussed in a later chapter.

II. Indium Surface Segregation

It has been previously observed for the arsenide and phosphide compounds that there is a tendency for column III atoms to segregate at the growing surface (Massies *et al.*, 1987). The accumulation of high surface concentrations of In atoms occurs as follows: Column III atoms such as In impinge on the substrate, before being incorporated, they stay at the surface long enough to travel sizable distances (1 μm for commonly used growth rates) and seek lower energy configurations. If the driving force is provided, an exchange between surface atoms and impinging atoms are promoted, probably at defects sites. When the segregation efficiency is high, the process may repeat along the growing overlayer growth. A thin layer with different composition than the interior material floats on the growing surface. The final structure is thus composed of the desired composition, except for a segregated surface layer with thickness of one monolayer (or less) with composition that is different from this desired composition. The tendency for this segregation is larger for In than for Al and can be summarized by the relation In > Ga > Al. This In segregation results in In floating at the surface with composition x_s that is larger than the bulk composition x_b. The value of x_s can reach values close to 1 for ternary compositions with high values x_b. A simple model of local equilibrium between the surface concentration x_s and bulk composition x_b has been proposed (Moison et al., 1989). It has been argued that elastic strains play an important role in this segregation process (i.e., the largest atom comes to the surface (Parker, 1985)). This can be related to the bond length in the arsenide system, such as In–As (2.62 Å) > Ga–As (2.45 Å). Correlations with bulk bonding energies has also been suggested to explain In segregation to the surface (Nagle et al., 1993):

$$\text{In—As}\,(1.41\,\text{eV}) < \text{Ga—As}\,(1.59\,\text{eV}) < \text{Al—As}\,(1.98\,\text{eV})$$

The segregation processes were found to be very sensitive to the V/III ratio and growth temperature (Nagle et al., 1993).

The same observations and arguments are expected to be valid in the case of nitrides. Both bond length and bond strength in In based nitrides would indicate that the segregation problem will be even more severe than in both the arsenide and the phosphide compounds. Thus, it is expected that at the surface of InGaN particularly for high values of x_b, x_s can be relatively high ($x_s \gg x_b$). Moreover, if the effective V/III ratio during the nitride growth is not very high and x_b is very large, the equilibrium process between the In surface concentration (x_s) and bulk concentration (x_b) may not hold. Also, the nitrogen vapor pressure in the gas phase can be less than the equilibrium vapor pressure. Coupling this with a relatively weak In–N bond can lead to the experientially observed In metal droplet formation at the growing surface.

The presence of In segregation or In metal droplets at the growing surface will effect properties of bulk grown films, and will have more drastic effects at heterointerfaces. Experimental evidence from several sources are currently available especially for MBE growth of InGaAs and their normal and inverted structures with GaAs (Gerard, 1992). For example, in GaInAs/GaAs QW structures, large shifts induced by In segregation was observed in the photoluminescence (PL) spectrum (Nagle et al., 1993). Shifts in this arsenide system as large as 26 meV was reported. When In droplets or values of x_s approaching unity are present at the InGaN surface, more problems in heterostructures are expected. For example, with InGaN when grown by MOCVD, the size of the In droplets has been observed to increase with growth time (Shimizu, Hiramatsu, and Sawaki, 1994; Bedair et al., 1997). If a thin GaN or AlGaN film is grown on a thin InGaN film while In droplets are not generated or are still small in size, it is possible to grow a good heterointerface free to In-metal. However, for relatively thick $In_xGa_{1-x}N$ films or high values of x, In metal segregating at the surface will effect the composition of the GaN or AlGaN overgrown layer (Shimizu, Hiramatsu, and Sawaki, 1994; Bedair et al., 1997). This results, in a red shift in PL spectrum indicating that In droplets act as indium sources to form an InGaN intermediate layer near the heterointerface during the growth of GaN with higher In content than the primary layer. This structure emits a wavelength corresponding to the lower bandgap InGaN layer, resulting in a red shift as was experimentally observed in Shimizu, Hiramatsu, and Sawaki (1994). The same argument holds when AlGaN is used as the barrier layer to form a quaternary compound intermediate film. It should be mentioned that the problem of In droplet formation can be reduced with a moderately high V/III ratio, lower growth rate, and are not pronounced at all for low values of x ($x < 0.15$). With the exception of some laboratories the problem of In segregation can explain the near absence of InGaN based double heterostructures emitting in the green and yellow spectrums.

III. Competitive Processes During the Growth of InGaN

The incorporation of In to form InGaN is controlled by several simultaneous competitive processes. The salient rate process involved are (Bedair et al., 1997):

1. In incorporation in the solid ternary alloy, "F_s" (atoms/cm²/sec.),
2. In atom desorption from the growth surface, "F_d" (atoms/cm²/sec.),
3. In incorporated as In metal droplets, "F_m" (atoms/cm²/sec.).

These reaction pathways are illustrated schematically in Fig. 2. It should be noted that rate process F_d can be due to the desorption of solitary adsorbed In atoms, or the breaking of the existing In–N bond. So, for an indium incident flux, F_{in} (atoms/cm²/sec.), the following mass balance equation can

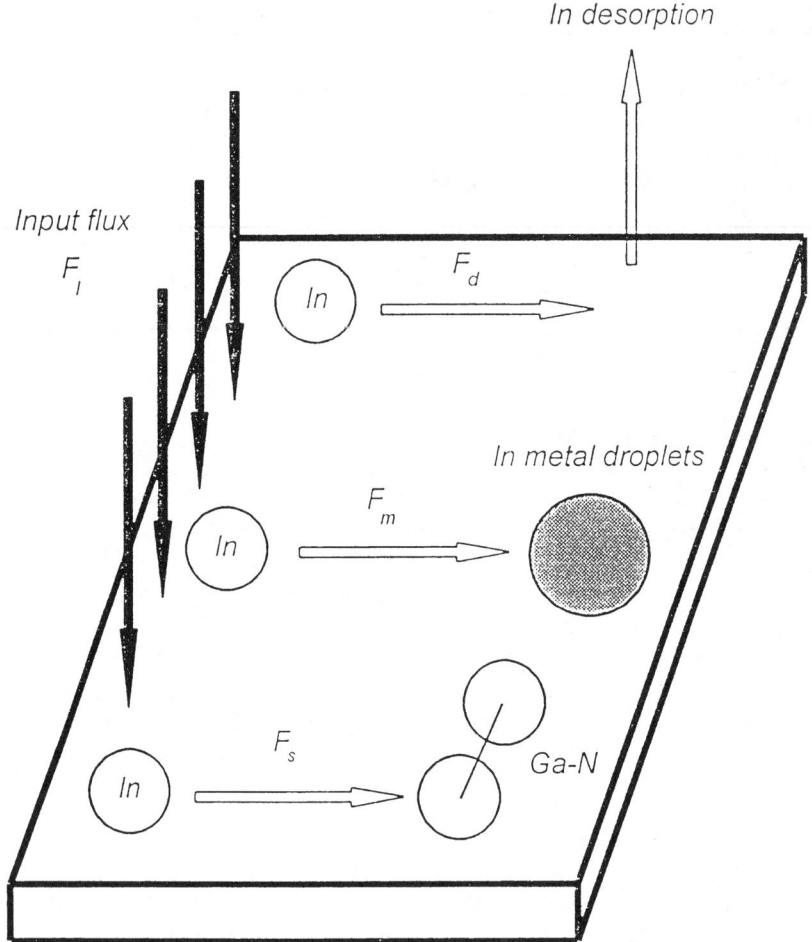

FIG. 2. Reaction pathways for the deposition of In-based nitride compounds. (From Bedair et al., 1997.)

be assumed:

$$F_{in} = F_s + F_d + F_m$$

For growth temperatures in the 600 to 800°C range it is expected that the organometallic In sources are fully decomposed to In atoms at the surface. Based on $In_xGa_{1-x}N$ growth both by MOCVD ($0 < x < 0.5$) and atomic

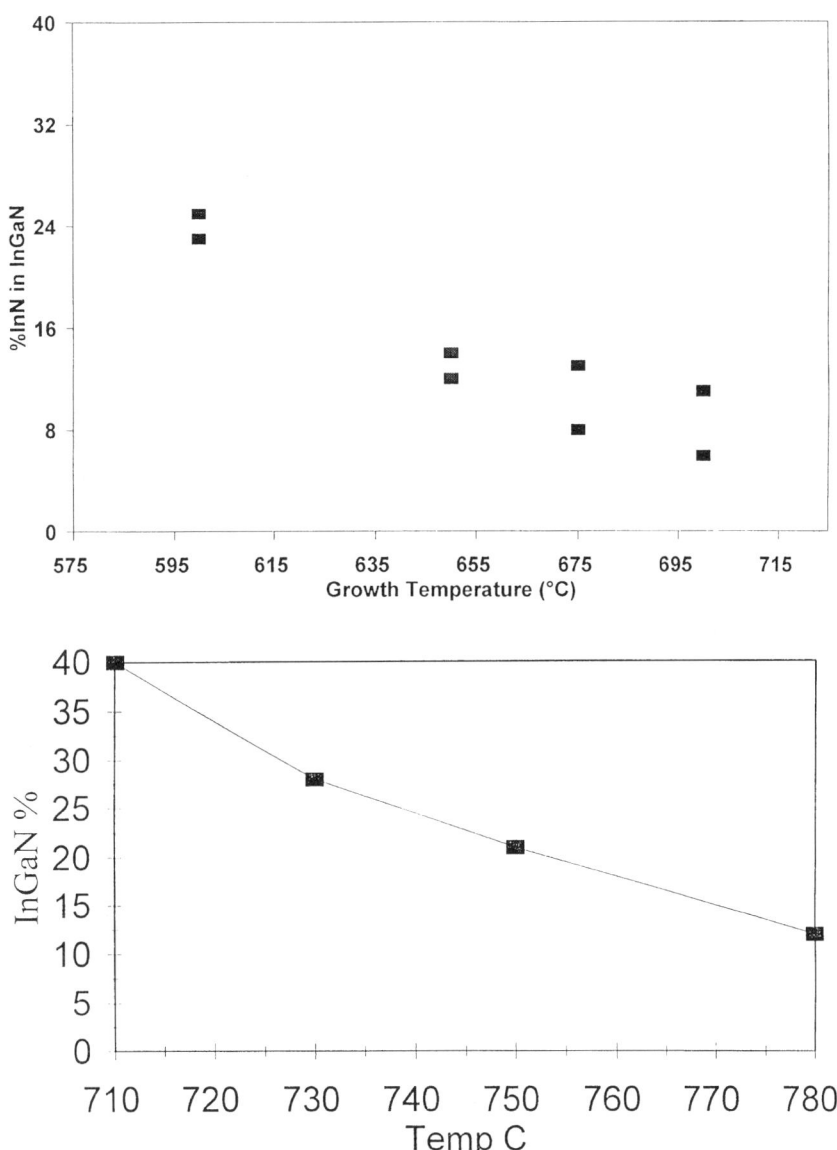

layer epitaxy (ALE) ($0 < x < 0.25$), it was found that the growth temperature plays an important role as shown in Fig. 3. In Fig. 3(a) the value of x obtained by ALE critically depends on the growth temperature, and the same trend was also observed by MOCVD (Fig. 3(b)). F_d is inversely proportional to the residence lifetime, τ, of In on the growing surface where

τ can be written as

$$\tau = \tau_0 \, e^{+E_d/kT}; \qquad \therefore F_d \propto e^{-E_d/kT}$$

where τ_0 is fixed for a given system and E_d is the activation energy for In desorption. This exponential dependence of F_d on the temperature T can be the dominant reaction pathway at high growth temperatures as observed for both ALE growth above 700°C and MOCVD growth above 800°C. The values of E_d and τ have not been experimentally determined for III nitride material system, but these parameters can be compared to those of GaAs and InGaAs. The relative stability of Ga compared to In on the nitride surface is supported by the behavior of Ga and In atoms on As terminated GaAs and InGaAs surfaces. The lifetimes of Ga and In on an As surface at 600°C are 10 and 1 sec., respectively (Arthur, 1968). The lifetime of Ga on a nitride surface may be greater than 10 sec. and that of In may be less than 1 sec. We can conclude from these assumptions that in the nitride system F_d (In atoms) $\gg F_d$ (Ga atoms) and Ga atoms at the surface will have a greater chance of being incorporated in the growing film. This explains earlier MOCVD results at 800°C (Matsuoka et al., 1992; Nakamura and Mukai, 1992) where a vapor pressure ratio of In to Ga precursors as high as 12 was used to achieve values of x of approximately 0.2 as shown in Fig. 4. Thus, the reaction pathway F_d can also be used to explain the lower operating temperature range of ALE compared to MOCVD required to achieve high In incorporation. Since ALE relies on growth interruption (Gong et al.,

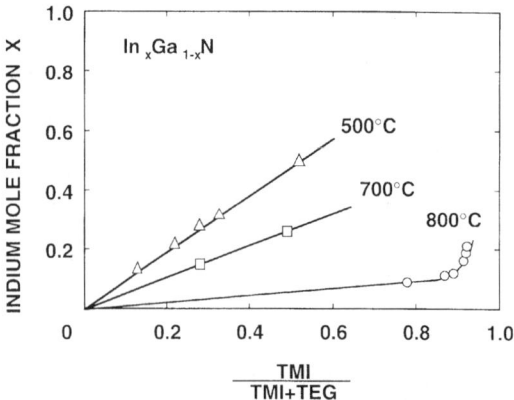

FIG. 4. Relation between indium mole fraction of InGaN and flow rate ratio of indium to the sum of group III sources. (From Matsuoka et al., 1992.)

1992) that can be much longer than the residence time τ of In atoms on the growing surfaces, values of x higher than approximately 0.25 were difficult to achieve (Boutros et al., 1995).

With reduced growth temperature, F_d will be reduced and there will be competition between F_m and F_s. In and Ga atoms on the growing surface will react with NH_3 independently or collectively to form $(InN)_x$ and $(GaN)_{1-x}$ in the ternary alloy according to the following reactions (Seki and Koukitu, 1989).

$$In + NH_3 \rightarrow (InN)_x + \tfrac{3}{2}H_2 \qquad (1)$$

$$Ga + NH_3 \rightarrow (GaN)_{1-x} + \tfrac{3}{2}H_2 \qquad (2)$$

Since the Ga–N bond is stronger than the In–N bond, the equilibrium constant for Eq. (2) is larger than Eq. (1). This allows Ga atoms to retain N more efficiently than In atoms thus forming Ga–N sites with unsaturated bonds shown in Fig. 2. Higher surface densities of unsaturated nitrogen bonds will enhance the probability of In incorporation, thus enhancing the reaction pathway F_s and resulting in higher values of x suggested by the data in Fig. 5 for ALE growth at 650°C for a fixed EDMIn/TMGa partial pressure ratio of 0.3 in the gas phase. Increasing the TMGa flux in Fig. 5 from 1 μmol/min to 6 μmol/min results in an increase in the value of x from 0.07 to 0.2, respectively. We have observed a similar trend in the MOCVD growth of InGaN at 780°C. For example, for the same flow rate of EDMIn

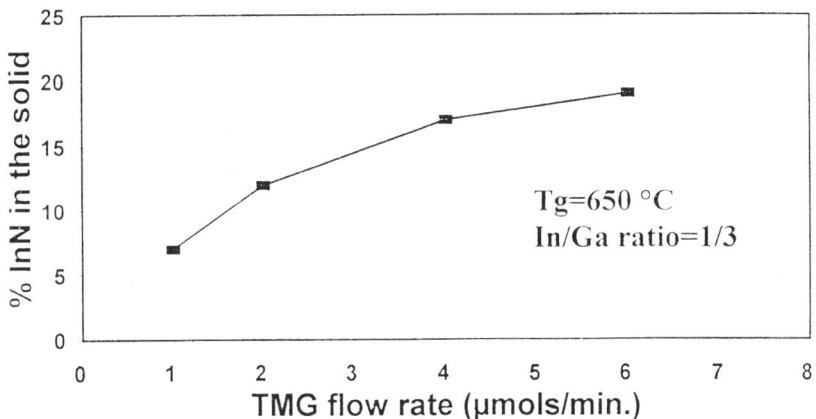

FIG. 5. Dependence of ALE grown $In_xGa_{1-x}N$ composition on TMGa flow. (From Bedair et al., 1997.)

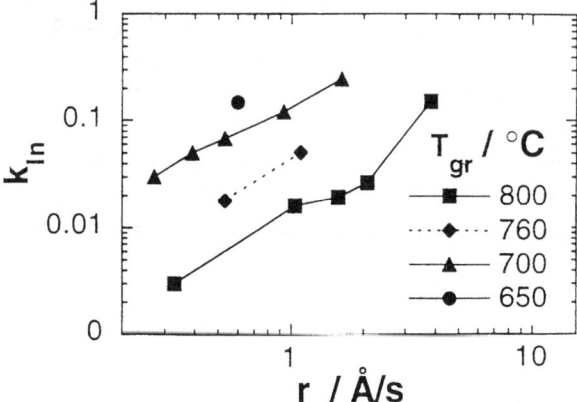

FIG. 6. Relative indium segregation coefficient k_{In}, as a function of the growth rate at different growth temperatures. (From Keller et al., 1996.)

at 2 μmole/min increasing the TMGa from 0.5 to 1 μmol/min increases the value of x from 0.07 to 0.12. It should be mentioned that, for ALE and MOCVD, increasing the TMGa flow rate also increases the growth rate and can help in trapping In atoms as shown in Fig. 6 (Keller et al., 1996). However, by increasing the growth rate $In_xGa_{1-x}N$ films with poor optical and structural quality were obtained. Increasing the density of Ga–N nucleation sites or the growth rate enhances the chance of In atoms attaching to these unsaturated bonds rather than desorbing with their short lifetime τ. Thus, increasing the TMGa flux enhances the reaction pathway F_s for In atoms and increases the value of x. These Ga–N unsaturated bonds are located either at step edges or at new nucleation sites when the growth mode is controlled by either step flow or two-dimensional (2D) nucleation, respectively. The criteria of these growth modes can be written as $\lambda p^{-1} > d$ or $\lambda p^{-1} < d$ for the step flow and 2D nucleation, respectively, where λ is the diffusion length, p is the sticking probability of adatoms to atomic steps, and d is the terrace width (distance between edges) (Morishita et al., 1995).

Intuitively, to achieve high values of x, lower growth temperatures and high values of F_{in} are needed, and that is where growth difficulties are encountered. The primary problem will be the formation of In metal droplets on the growing InGaN films that has been observed experimentally by scanning electron microscopy (SEM) (Shimizu, Hiramatsu, and Sawaki, 1994), optical microscopy, energy despersive spectroscopy (EDS) (Boutros, 1995), and X-ray diffraction (XRD) (Bedair et al., 1997). The presence of sufficient densities of In atoms diffusing across the surface, coupled with paucity in nitrogen dangling bonds due to reduced growth temperature, will

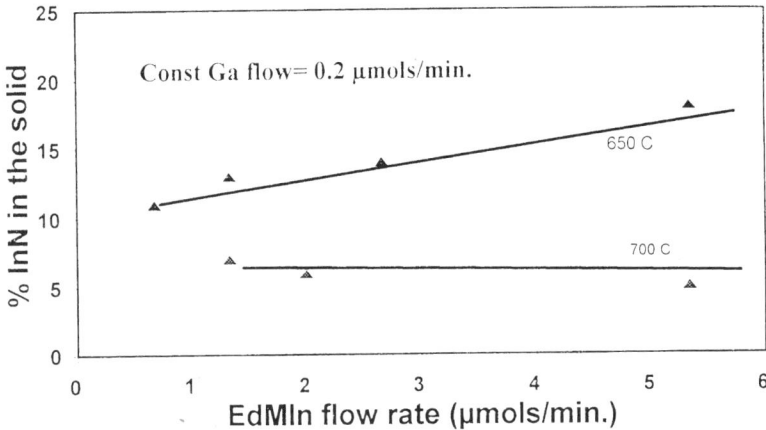

FIG. 7. Dependence of ALE grown $In_xGa_{1-x}N$ composition, x, on EdMIn flow for a fixed TMGa flow. (From Bedair et al., 1997.)

result in the formation of small In metal droplets. Once these droplets reach a critical size, they will become thermodynamically stable and continue to grow. These In droplets act as a getter for other available In atoms on the film surface, thus competing with, or even dominating, the process of In incorporation in the solid InGaN phase. Several of our experimental results can be explained by the role these In droplets play in limiting the InN% in the growing InGaN alloy. Figure 7 shows the variation of x with EDMIn flow rate for a fixed TMGa flow at $2\,\mu\text{mol/min}$, for the ALE growth of InGaN at 650 and 700°C. From this figure the value x tends to saturate for high values of EDMIn flux where In droplets were also observed as shown in Fig. 8. The same saturation trends were also observed in MOCVD growth in the temperature range 750 to 800°C and are shown in Fig. 4. In this growth regime, the value of x in the solid has very little dependence on EDMIn/(EDMIn + TMG) partial pressure gas phase ratio, a dependence commonly observed in the growth of other III–V In based compounds. This may be due to the fact that F_s is dominated by F_m or F_d. Most of the In atoms are either desorbed or lost to the In metal droplets resulting in low values of x. We have observed that increasing this ratio by reducing the TMGa flux alone did not result in increasing the value of x in the deposited films. This is either due to reduction of nucleation sites or low growth rate. We also observed that the amount of In droplets in the growing films, as indicated by the In metal peak in X-ray diffraction spectra (Fig. 9), decreases with increasing NH_3 flow. Within the accuracy of EDS, these droplets are mainly In metal with traces of carbon and no Ga signal was observed,

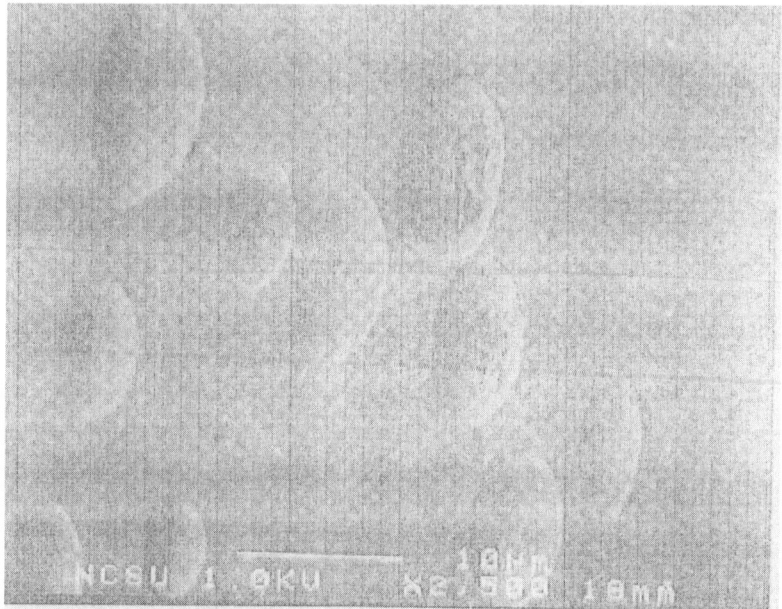

FIG. 8. SEM of ALE grown InGaN surface showing the presence of In metal droplets. (From Boutros, 1995.)

consequently they do not act as sinks for Ga atoms. It should be mentioned that the formation of these In droplets can be avoided by increasing the NH_3 flow rate or using a reactor design that makes efficient use of NH_3. From Fig. 10 the appearance of In metal droplets occured when the NH_3 flow was reduced below 1 s/m, accompanied by a reduction in the InN% in the film (Boutros, 1995; Piner et al., 1997).

1. INDIUM NITRIDE

InN is one of the least studied III–V nitride compounds. It has a direct bandgap energy of 1.9 eV. Growth of InN is difficult because its dissociation pressure is extremely high as shown in Fig. 1, thus requiring low growth temperatures. In addition to its rather poor thermal stability, other workers have cited the large disparity of atomic radii of In and N as a possible contributing factor to the difficulty in obtaining good quality InN. There have been several methods utilized in the growth of InN. Examples are

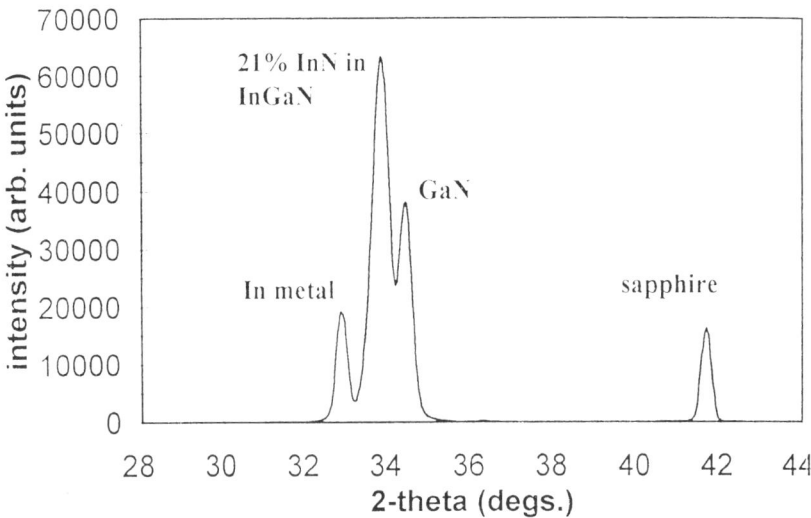

FIG. 9. XRD of InGaN grown by MOCVD showing the presence of In metal droplets. (From Bedair et al., 1997.)

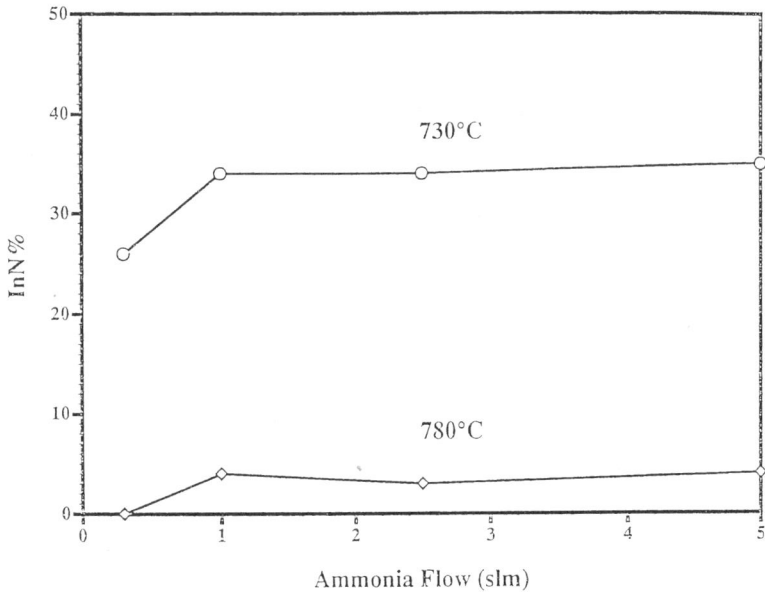

FIG. 10. InN% in InGaN as determined by XRD as a function of ammonia flow at, (a) 730°C and (b) 780°C. For an ammonia flow rate less than 1 l/min, In metal droplets appear at the surface. (From Piner et al., 1997.)

reactive evaporation (Trainer and Rose, 1974), reactive rf sputtering (Nata Rajan *et al.*, 1980; Tansley and Foley, 1986), MOCVD (Matsuaka *et al.*, 1989), Metalorganic molecular beam epitaxy (MOMBE) (Ren *et al.*, 1995), MBE (Bryden, Ecelberger, and Kistennacher, 1994), and ALE (Bedair *et al.*, 1997; Boutros, 1995; McIntosh *et al.*, 1997). Most of the reported growth of InN has exhibited the problems of unintentionally high electron concentrations and low electron mobilities where carrier concentrations in the 10^{19} to 10^{20} cm^3 range were obtained. The epitaxial growth of InN is generally accompanied by formation of In metal droplets at the surface. Matsuoka *et al.* (1989) were able to reduce the formation of In metal by using high V/III ratios, greater than 10^5. The amount of In droplets present was shown to decrease with increasing V/III ratio. For a ratio of more than 1.6×10^5, the indium metal peak observed by XRD disappeared.

ALE seems to be a suitable approach for low temperature growth of InN, since it can enhance decomposition of NH$_3$ on the In rich surface. ALE was previously used to grow device quality GaN (Karam *et al.*, 1995) at temperatures as low as 550°C. ALE of InN was carried out in our laboratory at 480°C. Lower growth temperatures resulted in polycrystalline films and with higher growth temperatures ($T > 550$°C), InN formation was not observed (Bedair *et al.*, to be published; McIntosh, *et al.*, 1997). Figure 11 shows a typical XRD of InN grown directly on an AlN buffer (mismatch

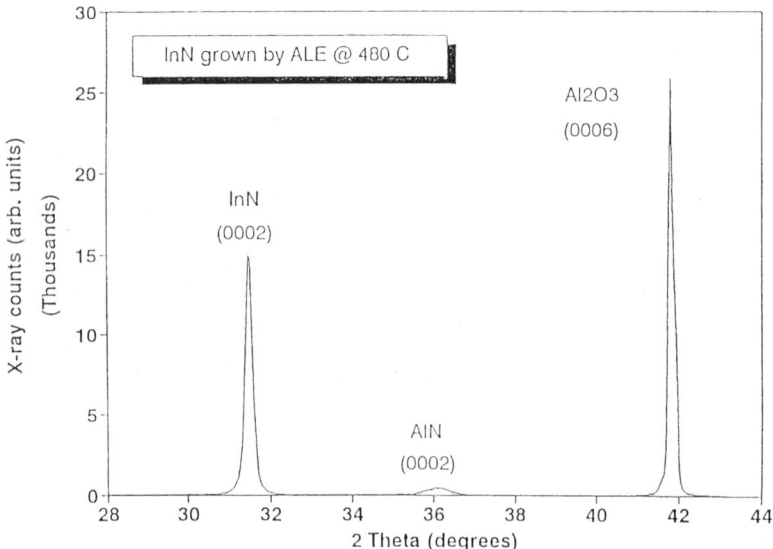

FIG. 11. XRD of InN grown by ALE at 480°C on an AlN buffer. (From Bedair *et al.*, 1997.)

~15%) after a 60 s HCL etch indicating that the film is single crystalline. The surface was also specular. Despite the apparent potential of ALE, the growth rate is extremely low, 500 Å/h. Hall measurements performed on these ALE grown films indicated n-type conduction with a carrier concentration of $\sim 10^{19}$ cm^3 and a mobility of 10 cm^2/V·sec. (Bedair et al., 1997; Boutros, 1995).

High quality single crystal InN with low carrier concentration and good PL properties have not been achieved. This can be due to the lack of suitable substrates, high concentration of nitrogen vacancies and low growth rates. Thus, the optical, electrical, and physical properties are still lagging behind other III nitride compounds.

2. INDIUM GALLIUM NITRIDE

$In_xGa_{1-x}N$ epitaxially grown films, especially with $x < 0.1$, have been well studied due to their potential applications in blue emitters. For high values of x, recent reports have begun to fill in the gaps of information concerning InGaN. Matsuoka et al. (1992) has reported growth of $In_xGa_{1-x}N$ on sapphire substrate by MOCVD. Values of x from 0 to 0.5 were obtained at 500°C, however, with poor optical properties and weak PL signal. The value of x seems to increase linearly with TMIn/TMIn+TMGa at this low growth temperature (Fig. 4). When the growth temperature was increased to 800°C lower values of x were obtained ($x < 0.23$), however, with a stronger PL signal. High TMIn/TMIn+TMG ratio was used to achieve $x \leqslant 0.23$ at 800°C and deep level emission was dominant in PL measurements both at room temperature and 77 K. Nakamura and Mukai (1992) have improved on this work and values of x as high as 0.26 were grown at 780°C. Band edge emission from PL was observed with a full width at half maximum (FWHM) of 70 and 110 meV for values of x of 0.14 and 0.23, respectively. The emission intensity decreases with increasing x and reduced growth temperature. The XRD FWHM for $x = 0.14$ and 0.24 were approximately 6 and 9 min, respectively. The partial pressure ratio TMIn/TMGa was approximately 12 to achieve values of x of approximately 0.25.

Nakamura (1994) emphasized the effect of growth rate on the quality of InGaN films. Figure 12 shows the growth rate used to achieve high quality InGaN at different growth temperatures. Figure 13 shows two PL spectra with two growth rates at 805°C for the same TMIn/TMIn+TEGa flow ratio of 0.9. Data in Fig. 13 is for silicon (Si) doped samples. It is apparent from these figures that to achieve device quality InGaN films, especially at low growth temperatures, low growth rates are recommended. This can be

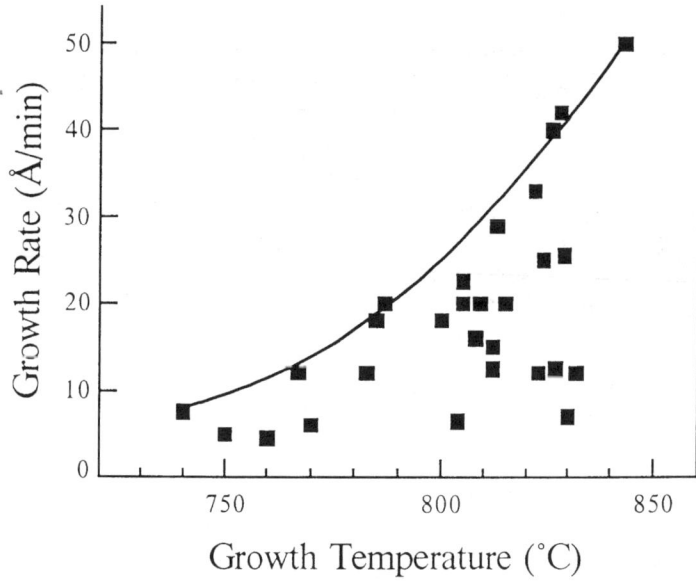

FIG. 12. The growth rate of high quality $In_xGa_{1-x}N$ films which were grown under the same [TMI/(TMI + TEG)] flow rate ratio of 0.9 as a function of the growth temperature. The curve shows the maximum growth rate at which high quality InGaN films can be grown at each temperature. (From Nakamura, 1994.)

related to insufficient cracking of NH_3 at these low growth temperatures. Carbon originating from organometallic radicals, such as $GaCH_3$ or $InCH_3$ can be scavaged when enough H atoms are available. The primary source of H active radicals is the cracked NH_3. At low growth temperature, the concentration of these hydrogen species is low, leading to the possibility of high carbon concentration in these epitaxial films.

InGaN epitaxial layers have been grown by MOCVD with values of InN up to 40% (Fig. 14(a)) (Roberts et al., 1995). These films were achieved with a In/Ga gas phase ratio of 1:2, a relatively low ratio compared to results by Matsuoka et al. (1992); Nakamura and Mukai (1992). The FWHM of the double crystal X-ray diffraction data is broad for high values of x, but is comparable to GaN grown at higher temperatures for low values of InN% in the ternary. For example, double crystal X-ray diffraction (DCXRD) data for an $In_{0.06}Ga_{0.94}N$ alloy grown at 780°C on GaN, has a FWHM of ~250 arcsec. (Fig. 14(b)).

The optical properties of InGaN films grown in the hybrid ALE/MOCVD reactor show intense or near band edge emission that is some-

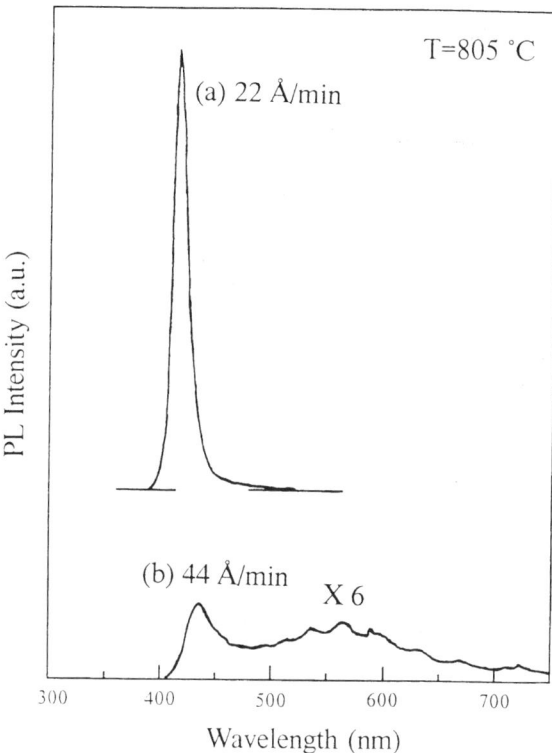

FIG. 13. Room temperature PL spectra of the InGaN films grown at 805°C with different growth rates and the same [TMI/(TMI + TEG)] flow rate ratio of 0.9. The growth rates were (a) 22 Å/min and (b) 44 Å/min. (From Nakamura, 1994.)

times accompanied by emission from deep levels. The optical properties are very sensitive to the way reactants are introduced to the substrate. Figure 15 shows room temperature PL data of several $In_xGa_{1-x}N$ films and Figs. 15(a) and (b) show room temperature PL data of an $In_{0.15}Ga_{0.85}N$ and $In_{0.35}Ga_{0.65}N$ alloy that were grown at 750°C (Roberts et al., 1995; Piner et al., 1996). Intense band edge or near band edge emission approximately 410 nm, having a FWHM of ~190 Å, was observed in the film with 15% InN. Photoluminescence spectrum of 35% InN is broad and the transition can be defect related. Samples with high InN% usually appear yellow tinted, probably due to carbon contamination from the organometallic (OM) sources. Carbon background can be reduced when ethyl rather than methyl OM sources are used. The problem with the characterization of these In based ternary alloys with high InN% is the poor uniformity in the In

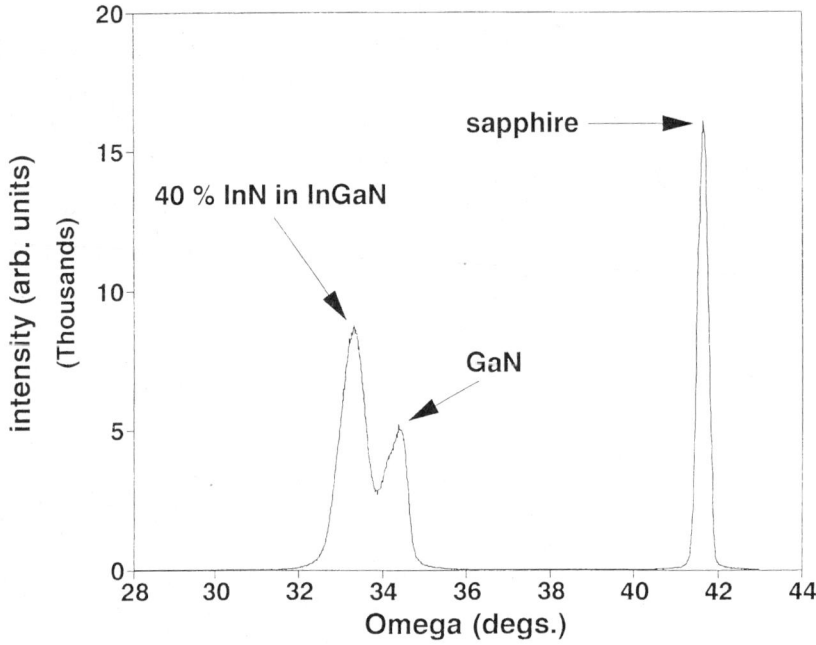
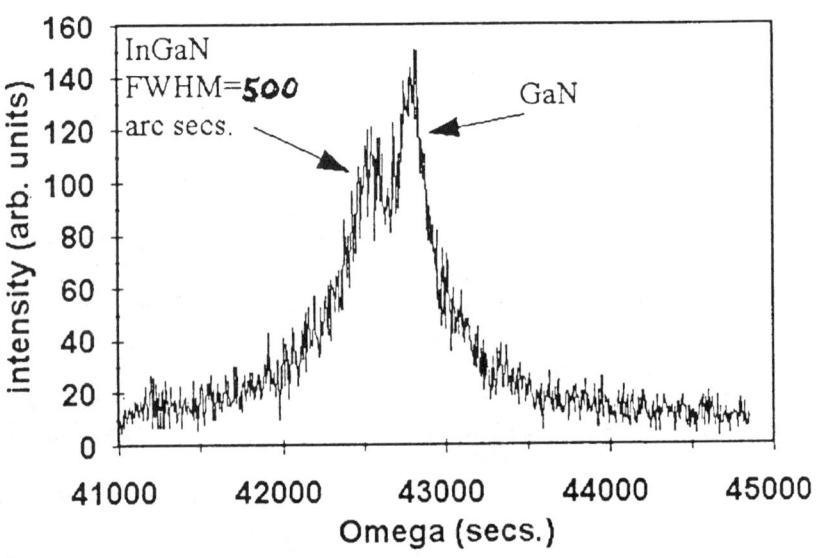

FIG. 14. (a) XRD of InGaN with 40% InN grown by MOCVD. (b) DXCRD of $In_{0.06}Ga_{09.4}N$ with FWHM of ~ 500 arcsec. (From Roberts et al., 1995.)

composition across the wafer. Photoluminescence is a localized characterization process, whereas XRD gives the average composition over a large area. This problem can create doubts about PL peak position and the value of x. The lack of composition uniformity can be due to several reasons that can affect the In incorporation process. Examples are nonuniformity in growth rate and TMIn and NH_3 fluxes. We have also discovered that hydrogen and NH_3 play a critical role in the In incorporation during the MOCVD and ALE growth of InGaN (Piner *et al.*, 1997; Piner *et al.*, to be submitted). It should be mentioned that ALE grown films (Boutros *et al.*, 1995; McIntosh *et al.*, 1997) show better composition uniformity, as determined by XRD and PL, when compared to MOCVD films. During the ALE process (Boutros *et al.*, 1995), even outside the growth conditions that lead to self limited growth (Tichler and Bedair, 1986; Suntola, 1989), the growth temperature is the dominant parameter controlling the ternary composition. It is possible that the In desorption rate (Fig. 2), is more critical in ALE relative to MOCVD due to the growth interruptions during the ALE process (Suntola, 1989).

InGaN growth was also reported by MOMBE using TEG and TMI (Abernathy *et al.*, 1995). These films have a high background n-type carrier concentration, and the optical properties were not reported.

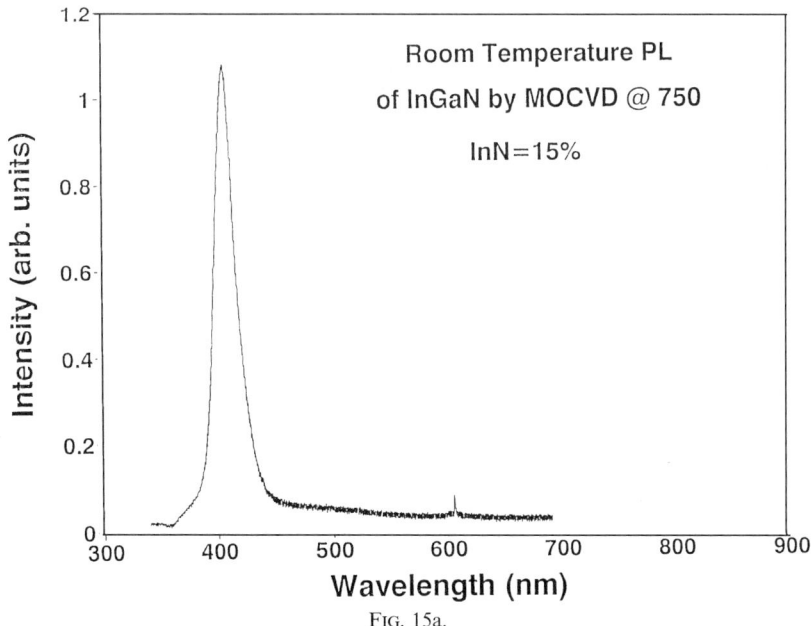

FIG. 15a.

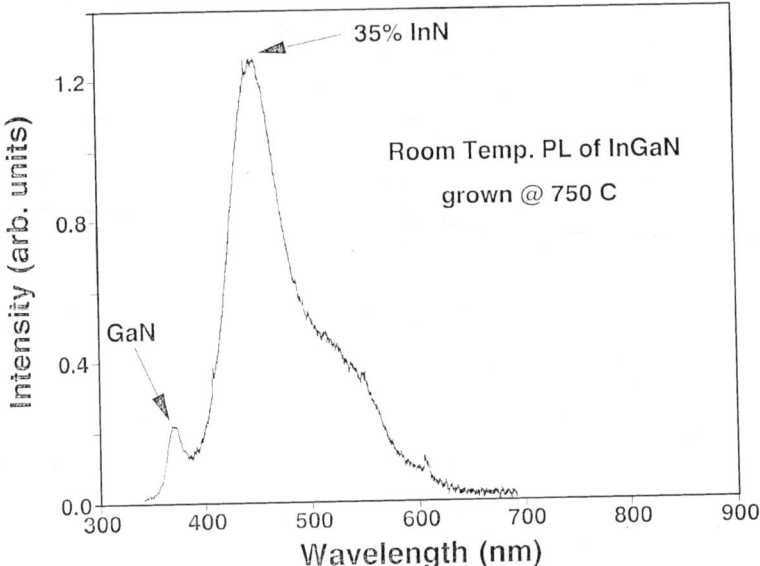

FIG. 15. Room temperature PL of MOCVD InGaN. (a) InN 15% and (b) InN 35%. (From Roberts *et al.*, 1995.)

The variation of the bandgap E_g with composition x of $In_xGa_{1-x}N$ is shown in Fig. 16. Data is collected from Bedair *et al.* (1997); Nakamura and Mukai (1992); Nakamura (1994); Roberts *et al.* (1995).

3. $Al_yIn_{1-y}N$

With AlN having an Eg of 6.2 eV and InN of 1.9 eV, the $Al_yIn_{1-y}N$ ternary alloy can cover a broad spectrum. Also, the AlInN alloy has the potential to offer lattice matched insulating barriers and heterostructures for GaN. However, very little research effort has been devoted to this ternary alloy. Starosta (1981) fabricated $Al_yIn_{1-y}N$ films using reactive multitarget sputtering and showed the existence of an alloy system throughout the range from AlN to InN. Kubota, Kobayashi, and Fujimoto (1990) prepared AlInN by sputtering, however, single crystal films were not reported. The alloy composition was estimated by applying Vegard's law to measured lattice constants and the bandgap was measured by an optical absorption

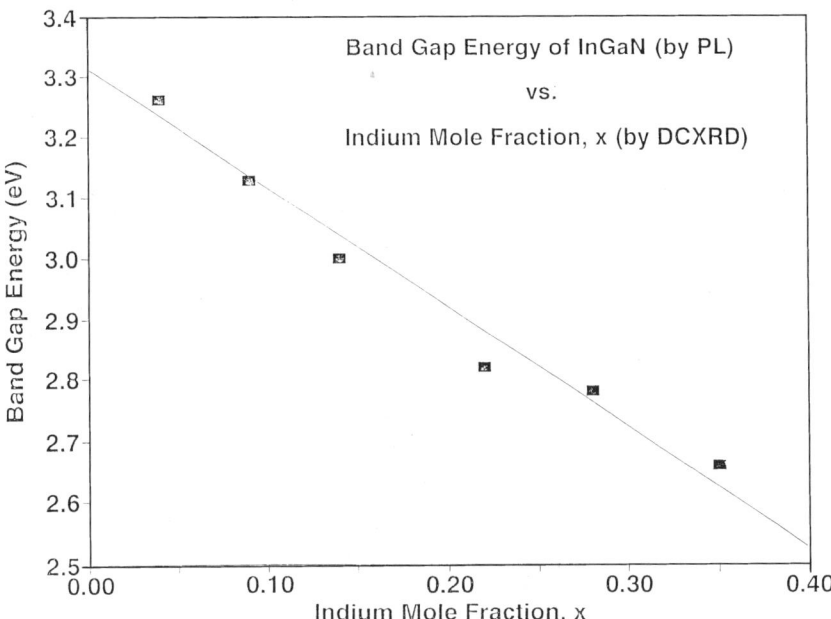

FIG. 16. Variation of bandgap E_g with composition x of $In_xGa_{1-x}N$ data are deduced from Bedair *et al.*, 1997; Nakamura and Mukai, 1992; Karam *et al.*, 1995; Roberts *et al.*, 1995.

technique. A highly nonlinear dependence of the bandgap on alloy composition was observed. Films with high In mole fraction were conductive while those with high Al mole fraction were resistive. They estimated that $Al_{0.83}In_{0.19}N$ is lattice matched to GaN with a bandgap of 3.34 eV. Guo, Ogawa, and Yoshida (1995) reported the epitaxial growth of $Al_yIn_{1-y}N$ ($0 < y < 0.14$) on sapphire substrates by microwave-excited metalorganic vapor-phase epitaxy (MOVPE) at 600°C. The composition of this ternary alloy was controlled by the ratio of the flow rates of group III MO sources in the vapor phase. In this composition range n-type carrier concentrations in the 10^{20} cm^3 range was reported. From XPS measurements, AlInN films did not show the presence of oxygen which may have been expected due to the strong Al–O bonds and the low growth temperature. They suspected that nitrogen vacancies are the main native defects in these films (Guo, Ogawa, and Yoshida, 1995). The issue of phase separation in this ternary alloy has not been addressed.

4. AlGaInN

The AlGaInN quaternary alloy offers more flexibility than AlInN, in achieving lattice matched heterostructures with GaN and InGaN while independently varying the bandgap. The potential application of this quaternary alloy for both microwave and optical devices will rely on composition that will be rich in Ga with relatively low In or Al mole fractions. This will allow more flexibility in the growth parameters such as growth temperature to allow for the reduction of In segregation and deep levels due to high Al mole fractions. This quaternary alloy avoids current problems in achieving $In_xGa_{1-x}N$ based heterostructures that rely on the AlGaN ternary alloy as the high bandgap barrier or confinement layer. The large lattice mismatch between these two ternary alloys will limit InN% in InGaN or thickness of InGaN well. For example, critical layer thicknesses for the $Al_{0.1}Ga_{0.9}N$/InGaN system with 40% and 60% InN are approximately 20 and 15 Å, respectively (Matthews and Blakeslee, 1974). This will limit the application of InGaN LEDs in the green and yellow regions. Such limitations can be avoided if AlGaN is replaced by a more versatile $Al_xGa_{1-x-y}In_yN$ quaternary compound. We can see in Fig. 17 that by varying the value of x and y, bandgap and lattice constants can each be

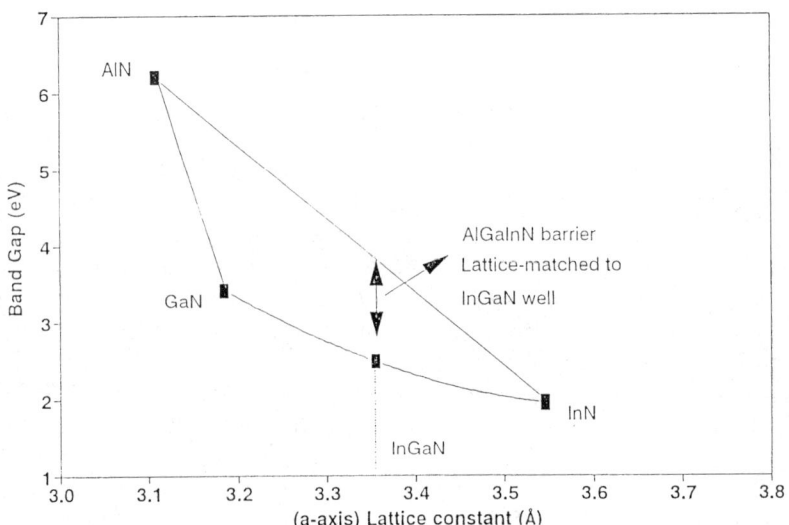

FIG. 17. Bandgap versus lattice constant for AlGaInN quaternary alloys.

independently adjusted to achieve AlGaInN/InGaN lattice matched structures (McIntosh et al., 1995).

One major obstacle in the development of quaternary AlGaInN is the determination of the optimal growth temperature. Aluminum based compounds generally require higher growth temperatures. Any residual background oxygen impurities in the deposition system or source gases will result in incorporation of oxygen in the growing films. Therefore, higher temperatures are required in order to desorb these oxides and prevent their incorporation into the epitaxial film. Lower temperatures, however, are required for indium based compounds. Indium compounds have relatively high vapor pressures (Matsuoka et al., 1989) and growth temperature must be lowered in order to increase indium incorporation and to reduce dissociation of the In–N bond (Bedair et al., 1997). The growth temperature will then govern the limits to which both In and Al can be incorporated into the AlGaInN quaternary alloy.

McIntosch et al. (1996) reported MOCVD epitaxial growth at 750°C of $Al_y In_x Ga_{1-x-y} N$ in the $(0 < y < 0.2)$ and $(0 < x < 0.15)$ composition range. Chemical compositions of AlGaInN quaternary films were obtained by EDS. First EDS was calibrated for In, Al, and Ga using standards made from InGaN and AlGaN ternary films of known compositions determined from DCXRD and using Vegard's law. The lattice constants of these quaternary alloys can also be predicted from the following equation (Williams et al., 1978):

$$a_{[Al_x Ga_{1-x-y} In_y N]} = x a_{AlN} + (1 - x - y) a_{GaN} + y a_{InN}$$

This equation is based on the assumption that a solid solution of binary constituents is present in the quaternary alloy. There is a reasonable agreement between lattice constants measured by XRD and the value from the above relation using EDS. Thus, at least in the composition range studied, a solid solution does exist between binary constituents. Figure 18 shows the XRD and PL spectrum of these quaternary alloys. AlGaInN films were reported in Matsuoka et al. (1992) with x and y of 0.022 and 0.743, respectively, however, PL and structural characterization data were not reported. The problem of phase separation in this quaternary alloy has not been addressed.

This quaternary alloy needs further development to achieve device quality films, doped both n and p-type. It will eventually have an impact on the nitride material system in the same fashion as quaternaries in other III–V compounds.

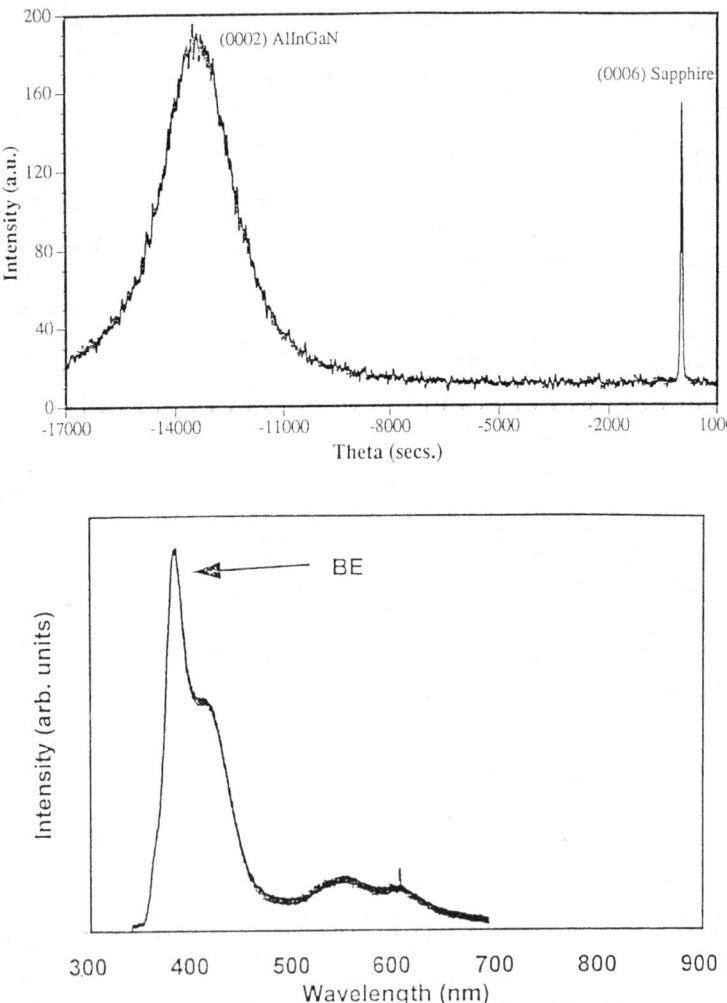

FIG. 18. Properties of MOCVD grown $Al_{0.1}Ga_{0.76}In_{0.14}N$. (a) XRD and (b) room temperature PL. (From McIntosh et al., 1996.)

IV. Effect of Hydrogen on the Indium Incorporation in InGaN Epitaxial Films

The InN% in MOCVD and ALE grown InGaN films was found to be significantly influenced by the amount of hydrogen flowing into the reactor (Piner et al., 1997). Temperature ranges for this study were 710–780°C for

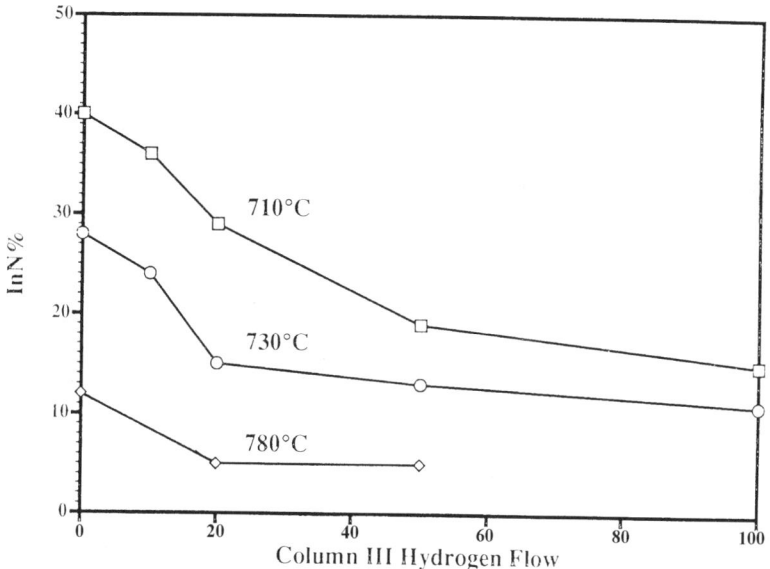

FIG. 19. InN% in InGaN as determined by θ–2θ XRD as a function of the hydrogen flow at growth temperature (a) 710°C, (b) 730°C, (c) 750°C, and (d) 780°C. (From Piner et al., 1997.)

MOCVD and 650–700°C for ALE. For a given set of growth conditions, an increase of approximately 25% InN in InGaN, as determined by XRD, can be achieved by reducing the hydrogen flow from 100 to 0 sccm.

Figure 19 plots the InN%, as determined by (θ–2θ XRD), in InGaN as a function of hydrogen flow injected with nitrogen carrier gas. All samples were grown by MOCVD using 1 sccm of TMGa (-10°C), and 90 sccm of EDMIn ($+10$°C) in 5 slm of nitrogen carrier gas and 5 slm of ammonia at the temperatures and hydrogen flows indicated. Figure 19 indicates that the InN% in the films drops significantly as the hydrogen flow increases. This trend occurs at all four temperatures investigated with good consistency. The general trend of Fig. 19 shows a rapid decrease in the InN% as the hydrogen flow is increased from 0 to 20 sccm, followed by a gradual decrease on further increased hydrogen flow from 20 to 100 sccm. All films are single crystalline, good quality, and without indium metal as-grown.

The effect of hydrogen on In incorporation was also observed for InGaN grown by ALE. The main characteristic of this growth technique is reactant gas separation, thereby eliminating the possibility for gas phase reactions as described in Tichler and Bedair (1986). In the growth temperature range of 650–700°C, a consistent reduction in the InN% is observed with increasing amounts of hydrogen (Piner et al., 1997). For example, at 650°C with

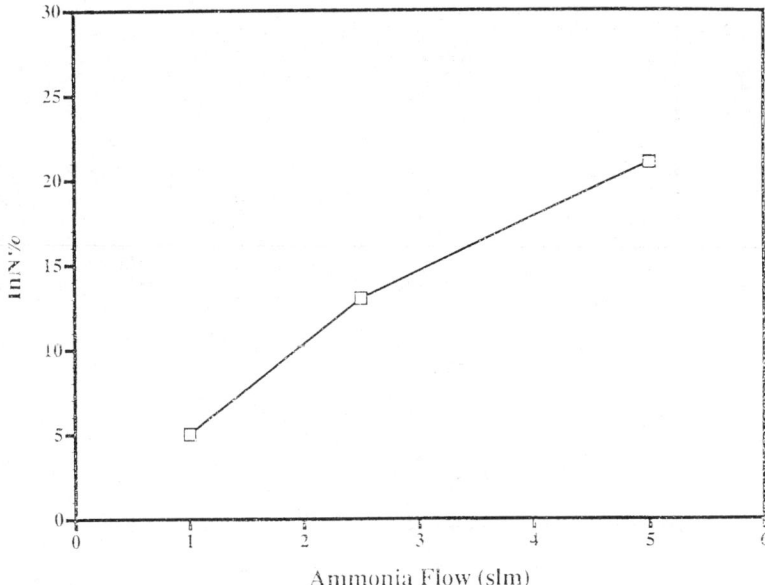

FIG. 20. InN% in InGaN as determined by XRD as a function of ammonia flow for high EDMIn flow.

hydrogen flowing at 2 sccm with OM and the 100 sccm with the ammonia gas stream, 17% InN in InGaN was observed. When the hydrogen flow was increased to 4 and 200 sccm, respectively, the indium incorporation dropped to 11%.

In an effort to determine the effect of the InN% due to the hydrogen generated during the decomposition of ammonia, a series of experiments were performed in which the overall ammonia flow was varied while keeping all other parameters constant. The results of that study are shown in Fig. 10 where the ammonia flow varied from 0.3 to 5 slm while using nitrogen as a make-up gas to keep the overall column V flow constant at 5 slm. The samples representing curve (a) in Fig. 10 were grown at 730°C with no hydrogen while those on curve (b) were grown at 780°C with 50 sccm hydrogen. Both were deposited by MOCVD with the same TMGa and EDMIn flows as those in Fig. 19. Curves (a) and (b) show little change in InN% for ammonia flows greater than 1 slm, indicating a sufficient supply of reactive nitrogen species at the growing surface with little effect from the hydrogen being generated by the decomposition of ammonia. This is an indication that the decomposition of ammonia into nitrogen and hydrogen at these temperatures is extremely low. If decomposition was occurring at a higher rate, the increased hydrogen being generated from ammonia would result in a decrease in InN% similar to that observed in Fig. 19. The data from Figs. 10 and 20 suggested an ammonia decomposition rate of less than 0.1% at growth temperatures in the 730–780°C range. Data shown in Fig. 10 are for low EDMIn flux, however, for higher values of EDMIn flux (270 sccm at $+10°C$) (Piner et al., to be published), the NH_3 flow seems to have more pronounced effect in the In incorporation as shown in Fig. 20.

1. THE PHASE SEPARATION ISSUE

The large difference in interatomic spacing between GaN and InN can give rise to a solid phase miscibility gap. Stringfellow and co-workers studied the temperature dependence of the binodal and spinodal lines in the $Ga_{1-x}In_xN$ system using a modified valence-force-field model where the lattice is allowed to relax beyond the first nearest neighbor. At a typical growth temperature of 800°C, the solubility of In in GaN was predicted to be less than 6% (I-hsiu and Stringfellow, 1996) as shown in Fig. 21.

Singh et al., (1997) reported further experimental evidence of the phase separation in MBE grown $In_xGa_{1-x}N$. For low values of x, XRD showed one single peak indicating a single phase on the macroscopic level. However, for values of x approaching 0.35, a weak broad peak at the diffraction angle corresponding to approximately pure InN was observed by XRD (Fig. 22).

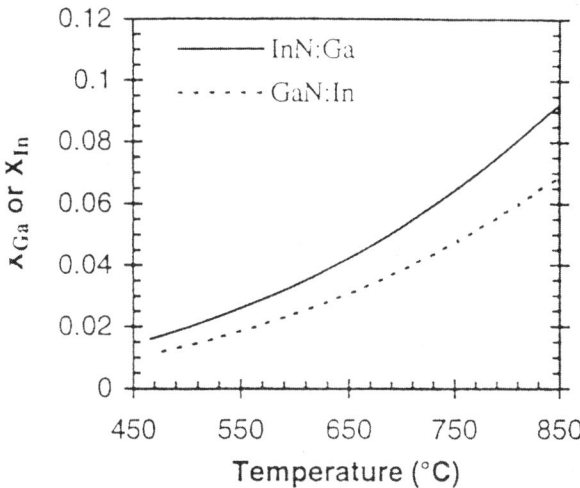

FIG. 21. Solubility of GaN in InN (InN:Ga) and InN in GaN (GaN:In) calculated using the valence-force-field model. (From I-hsiu and Stringfellow, 1996.)

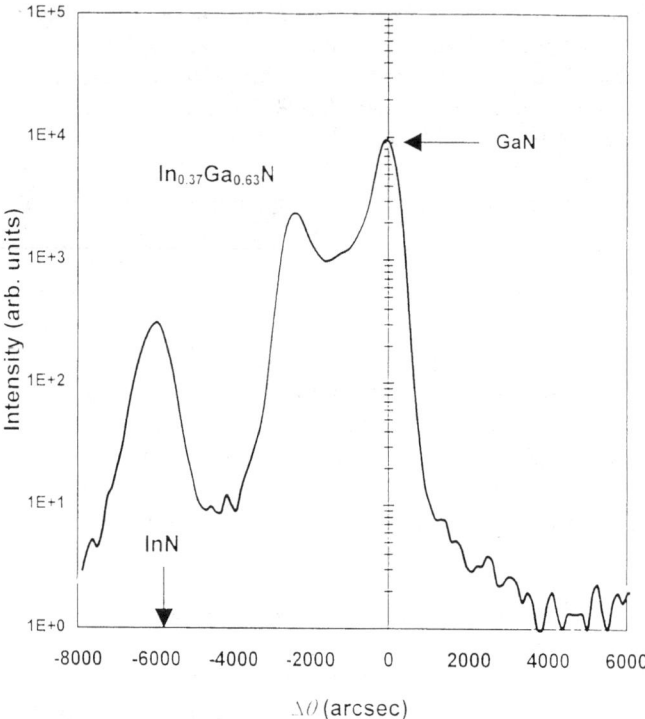

FIG. 22. XRD data for the $In_{0.37}Ga_{0.63}N$ film plotted in a logarithmic scale. This shows clear evidence of phase separated InN. (From Singh, et al., 1997.)

Further evidence of phase separation was also reported using optical absorption measurement where the optical bandgap was found to correspond to pure InN (Fig. 23). Similar results were also observed in MOCVD grown InGaN by Bedair et al. (unpublished). For compositions of approximately 40% InN, XRD showed only one single peak (Fig. 14). Efforts to increase the value of x beyond 40% were not successful under the growth conditions used. Reducing growth temperature to below 700°C resulted in extra peaks in the XRD caused by phase separation (Fig. 24). We have also found that PL emission close to that of InN was observed from these samples.

Previously mentioned data for MBE and MOCVD grown samples can be used to confirm the presence of phase separation in InGaN. The onset of phase separation for compositions higher than 35% for MBE and MOCVD samples, using XRD technique can be questionable. XRD is a macroscopic characterization approach, and separated phases have to constitute a sizable volume fraction in the grown films to be detected by this technique. Another technique such as selected area diffraction in the transmission electron microscope (TEM) can be more useful than XRD to determine the onset of phase separation. However, difficulties were encountered as the diffraction pattern changed as a function of beam exposure time. Coupling that with a high defect density makes (TEM) analysis of phase separation difficult, at least at this early stage of investigation (Singh et al., 1997).

Phase separation or composition modulation may also be present in InGaN as a part of quantum structures, such as AlGaN/InGaN and GaN/InGaN. For example, TEM and energy-dispersive X-ray microanalysis were used to identify regions with higher values of In composition in the InGaN films in blue laser structures (Narukawa *et al.*, 1997). Their claim is that emission is due to quantum dots with high In composition. It is not clear at this stage what effect strain induced in the InGaN QW has on the phase separation process. Added strain may result in stabilizing these alloys against phase separation (Singh *et al.*, 1997).

2. Doping of In-based Nitride Compounds

As-grown $In_xGa_{1-x}N$ films have high background carrier concentrations, n-type, that varies from 10^{20} cm^3 for InN to 10^{17} for $In_xGa_{1-x}N$ depending on the value of x. Doping these In n-type compounds to high levels does not seem to be a problem. Si doping using SiH_4 produces shallow levels and carrier concentrations in the $10^{19}/$cm^3 range (Nakamura, Mukai, and Senoh, 1993). It was also reported that Si doping improves the optical properties of InGaN and the band edge emission of Si-doped InGaN is approximately 36 times stronger than undoped films (Nakamura, Mukai, and Senoh, 1993). Cadmium doping was also reported by using TMCd as a means of introducing deep levels and to create a red shift in the emission of InGaN based devices (Nakamura, Iwasa, Nagahama, 1993). Deep levels, 0.5 eV from the band edge were reported. Zn doping was also reported for InGaN based double heterostructure to produce a red shift in LED structures (Nakamura, Mukai, and Senoh, 1994). Doping by ion implantation of InGaN using N and fluorine results in an increase in sheet resistance of these films (Zolper *et al.*, 1995). p-type doping using Mg in 10^{17} cm^3 range of $In_{0.09}Ga_{0.91}N$ was reported by Yamasaki *et al.* (1995). The activation energy of a Mg acceptor is estimated to be 204 meV.

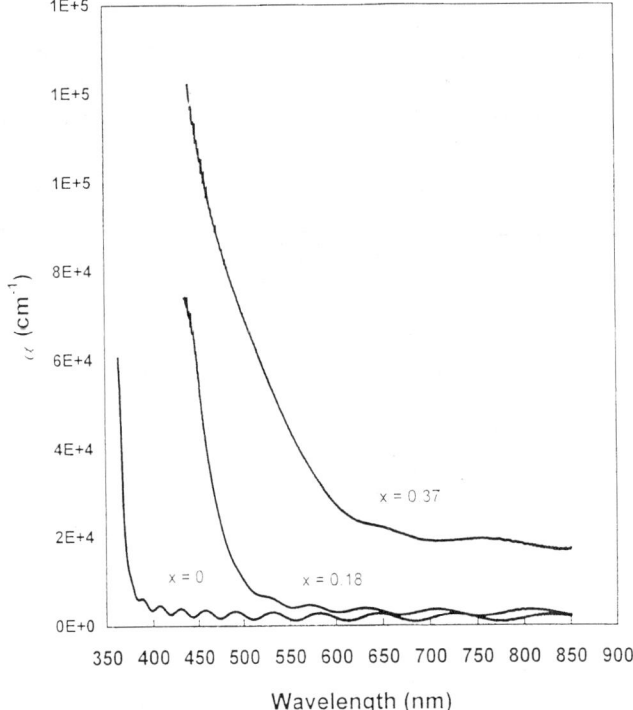

FIG. 23. Optical absorption versus wavelength for (a) GaN, (b) $In_{0.18}Ga_{0.82}N$, and (c) $In_{0.37}Ga_{0.63}N$ films. (From Singh et al., 1997.)

V. InGaN Based Heterostructures

The growth and characterization of strained superlattices and QW structures for high In-containing compounds is lacking their counter parts in the arsenide and phosphide compounds. The lack of a compatible growth temperature, phase separation and In segregation at the growing surface can be responsible for the paucity of information in the literature. Nakamura et al. (1993) reported growth and properties of $In_xGa_{1-x}N/In_yGa_{1-y}N$ superlattices with periods ranging from 200 to 600 Å. XRD of these superlattice structures (SLS) showed zero order and satellite peaks due to periodicity of the structure. Composition of the SLS layers were determined by growing thick InGaN films (600 Å) and characterizing them by XRD and PL to estimate the value of x (barrier) and y (well) as 22% and 6%, respectively (Nakamura et al., 1993). Satellite peaks were broad and weak probably due

6 INDIUM-BASED NITRIDE COMPOUNDS 157

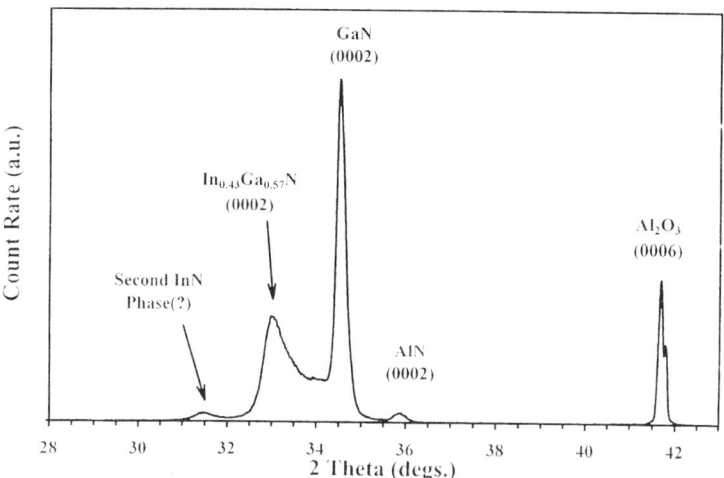

FIG. 24. XRD data of MOCVD InGaN at indicating the presence of separated phases. (From Bedair, unpublished.)

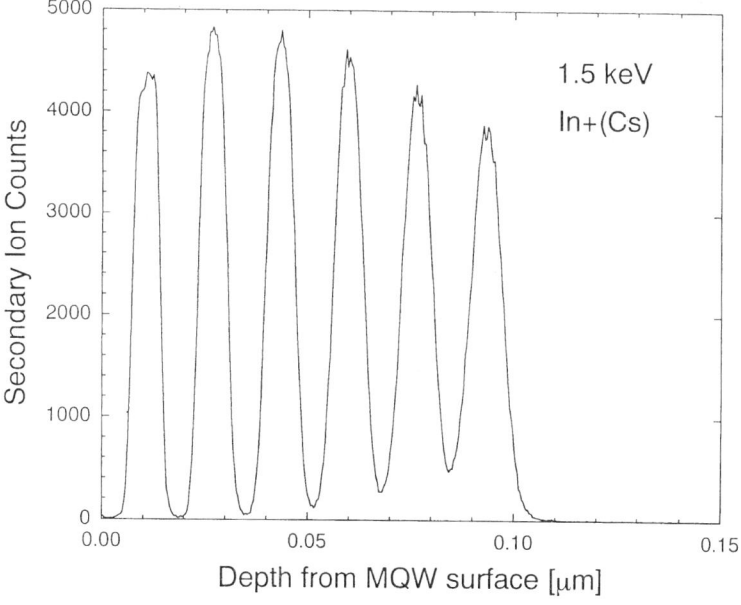

FIG. 25. In-depth profile in GaInN-GaN MQW by SIMS analysis showing the six-well 27.GaInN structure. (From Koike et al., 1996.)

to lack of sharp interfaces. It is not clear why $In_yGa_{1-y}N/In_xGa_{1-x}N$ SLSs were tried rather than GaN/InGaN which can be easily characterized. It is possible that for these high values of InN, poor interfaces were obtained and GaN may form an intermediate InGaN layer due to In segregation. The optical properties of these SLSs are excellent and a quantum size effect was observed that depends on the well width. For example, a shift from the band edge of the InGaN of 42 and 100 meV were observed for well widths of 100 and 30 Å, respectively. It should be mentioned that the growth of SLS is an excellent way to characterize the nitride compounds, determine layer thicknesses, and give indications about the quality of interfaces, rather than relying on electron or ion beam techniques.

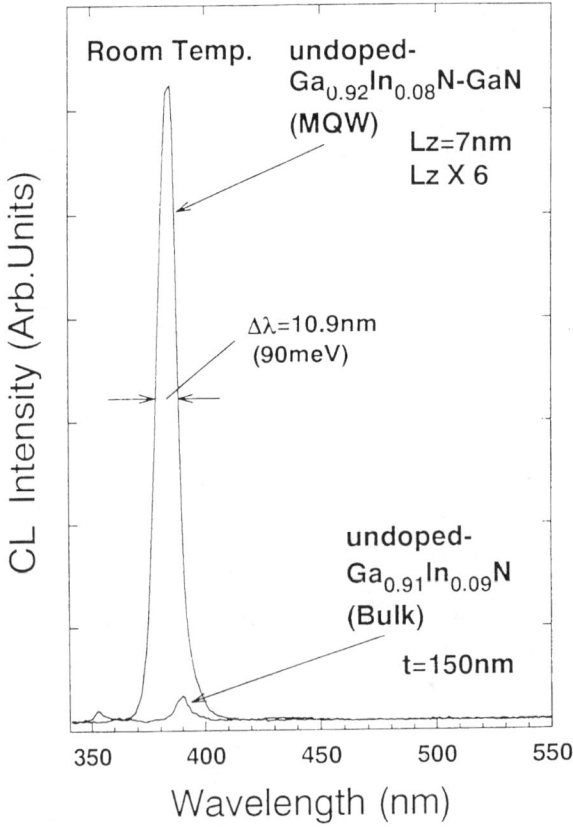

FIG. 26. Cathodoluminescence spectra of undoped $Ga_{0.92}In_{0.08}$-GaN MQW at room temperature with comparison of undoped $Ga_{0.91}In_{0.09}N$ bulk layer. (From Koike, 1996.)

GaN/InGaN heterostructures with low values of x have been reported by several groups as a part of single quantum wells (SQW) (Amano et al., 1994; Koike et al., 1996; Nakamura, Senoh, and Mukai, 1993; Nakamura et al., 1995), multiple quantum wells (MQW), and heterostructures for stimulated emission by optical pumping. For example, a GaN/$In_{0.08}Ga_{0.92}$N MQW was reported by Koike et al. (1996) with a well width of approximately 70 Å, and characterized by secondary ion mass spectroscopy (SIMS) shown in Fig. 25. They also reported that the MQW enhances the cathodoluminescence intensity by two orders of magnitude compared to bulk, $In_{0.08}Ga_{0.92}$N bulk films (Fig. 26). Keller et al. (1996) also reported the growth of GaN/$In_{0.16}Ga_{0.84}$N SQW structures with a well width ranging from 17 Å to 50 Å. Emission wavelength of 409 nm measured for 50 and 30 Å thick wells is the same as for InGaN bulk layers grown under the same conditions. Peak wavelength shifted slightly for a QW thickness below 30 Å (Fig. 27). Similar results were also reported by Scholz et al. (1997).

In our laboratory a variety of double heterostructures with $In_xGa_{1-x}N$ as the active layer have been grown with AlGaN cladding layers having ~10% AlN (Bedair et al., 1997; Roberts et al., 1996; Joshkin et al., to be published). InGaN active layers were typically grown on graded InGaN prelayers, which seem to improve the optical characteristics of the structure. The InGaN prelayer is grown in the temperature range 780–810°C, and the

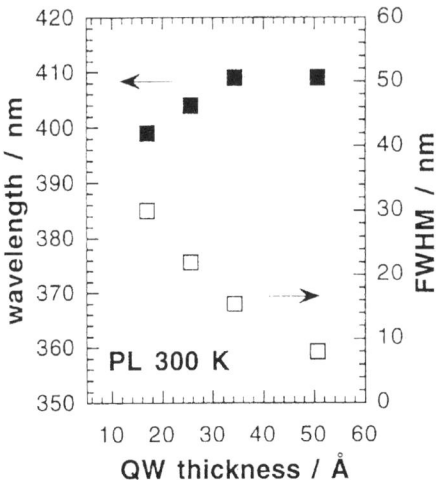

FIG. 27. 300 K PL emission wavelength (■) and the FWHM (□) of the SQWs of various thickness. (From Keller et al., 1996.)

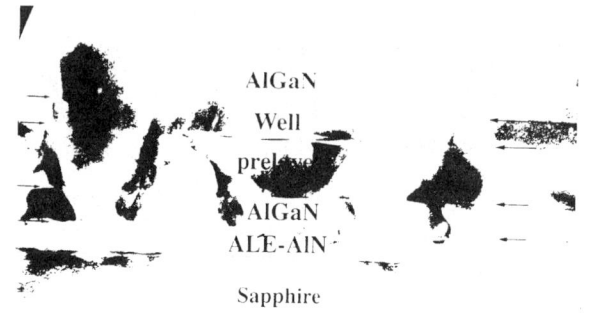

Fig. 28. Cross-sectional TEM of AlGaN/InGaN double heterostructure for (a) low InN% in the well and (b) high InN% in the well. (From Bedair et al., 1997.)

active layer grown in the temperature range 750–780°C. We found that for $In_xGa_{1-x}N$ double heterostructures (DH) with low values of x and emission wavelength less than 410 nm, graded layers are not critical and the InGaN active layer can be directly grown on GaN or AlGaN. Cross section Transmission Electron Microscopy (XTEM) of an AlGaN/InGaN/AlGaN double heterostructure is shown in Fig. 28(a) and the corresponding PL showed peak emission at approximately 400 nm. For green emission structures (~500 nm) interfaces are poor and strain contrast is apparent in Fig. 28(b). To achieve this longer wavelength emission (i.e., 400–560 nm), several growth parameters need to be carefully considered. The growth

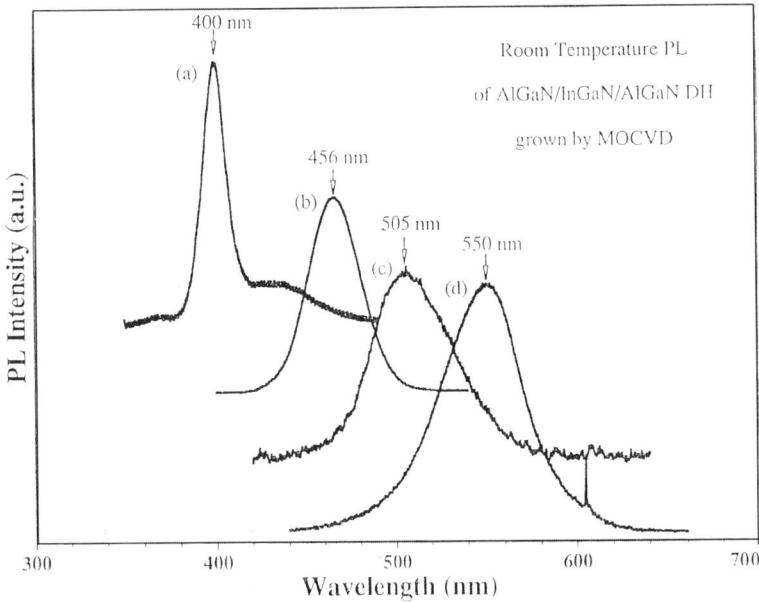

FIG. 29. Room temperature PL spectra of AlGaN/InGaN double heterostructures with emission from UV to the green and yellow part of the spectrum. (From Bedair et al., 1997; Robert et al., 1996.)

temperature of the active InGaN layer plays a critical role. For example, the emission peak shifts from 400 to 435 nm when the growth temperature is reduced from 780 to 750°C, while all other growth parameters were fixed. Also, increasing the absolute value of EDMIn results in a red shift in the emission wavelength. Other groups have also reported the growth and characterization of structures with emission in the green and yellow regions (Amano et al., 1994; Nakamura, Senoh, and Mukai, 1993; Nakamura et al., 1995). Growth and fabrication of LED structures emitting in green and yellow regions were reported by Nakamura et al., in 1995. In our laboratory, growth and optical properties of these structures emitting in the wavelength 400–560 nm were achieved (Bedair et al., 1997; Roberts et al., 1996). The room temperature PL spectra for some of these heterostructures (Roberts et al., 1996) is shown in Fig. 29. Data displayed in Fig. 29, curve (a) corresponds to a structure having a thicker active layer while the active layers of structures corresponding to curves (b) through (d) are each successively thinner (Bedair et al., 1997; Roberts et al., 1996). While accurate thickness measurements for these samples are not currently available, TEM data from

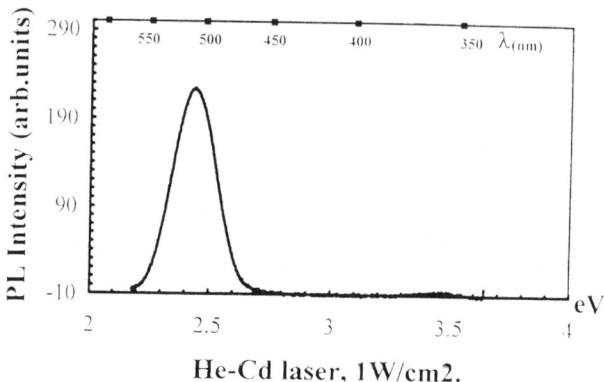

He-Cd laser, 1W/cm2.

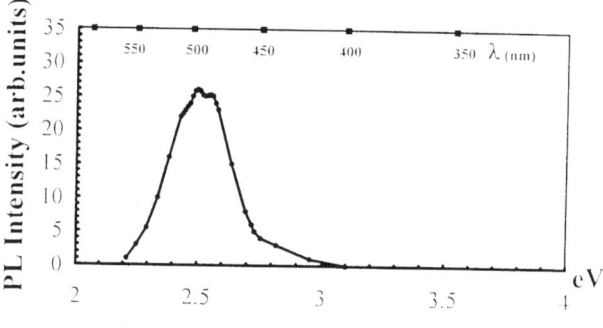

Pulsed Ti-supphire laser, 10 MW/cm2.

FIG. 30. Room temperature of AlGaN/InGaN with green emission at (a) low excitation power density in the order of watts/cm² and (b) high excitation power density in the order of molwt/cm². (From Bedair et al., 1997; Joshkin et al., to be published.)

other similar samples coupled with appropriate scaling of the active layer growth times suggest that thickness of the InGaN well is less than 100 Å. FWHMs of the spectra are broad and increase with emission wavelength. The question whether these emissions shown in Fig. 29, or reported by Nakamura et al. (1995) are band–band or impurities and defect related, is not clear. We are currently investigating the dependence of PL peak emission on temperature and intensity of the excitation sources (Joshkin et al., to be published) and initial results are shown in Figs. 30 and 31.

The AlGaN/InGaN double heterostructures shown in Section V suffer from the high lattice mismatch between the AlGaN and InGaN. For example, the mismatch can be close to 6% for the structure with peak emission at 550 nm, assuming it is band edge emission, that will require

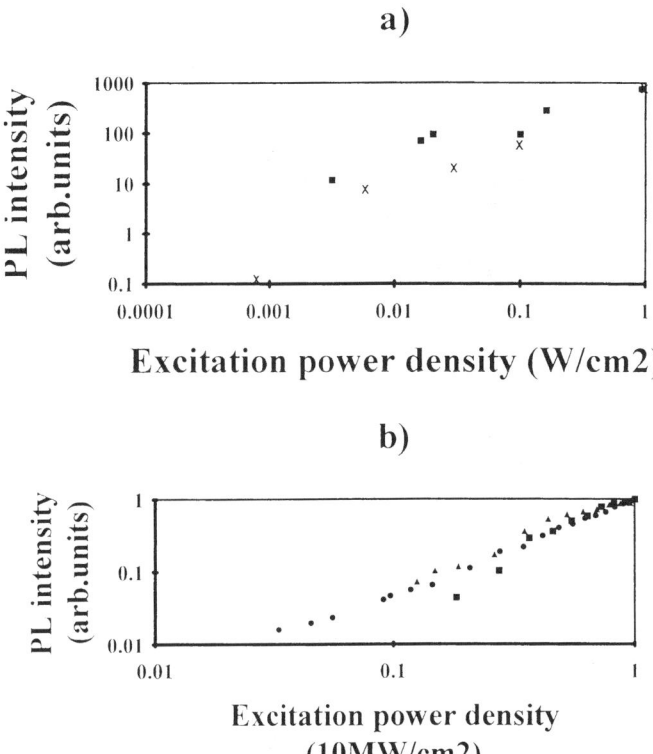

FIG. 31. Relation between room temperature PL intensity and excitation power density for green emitting double heterostructures. (a) for low laser power density and (b) for high laser power density. (From Bedair et al., 1997; Joshkin et al., to be published.)

InGaN films with thickness less than 20 Å. It should be possible to obtain a lattice matched heterostructure if an $Al_y In_x Ga_{1-x-y} N$ layer can be grown with the correct molar compositions x and y to obtain a lattice match to the InGaN active layer (Fig. 17). As a first attempt to address this problem, we deposited an $Al_y In_x Ga_{1-x-y} N$ layer at 780°C on AlGaN and then proceeded to grow the active InGaN layer, at 780°C which in turn was capped by the upper AlInGaN cladding layer also grown at 780°C. The active layer was grown by simply turning off the TMA flow and increasing the EdMIn flow while keeping all other growth conditions constant. The PL spectra for this structure is shown in Fig. 32, which exhibits sharp band edge emission. The quality of this quarternary/ternary should be the subject of further investigations.

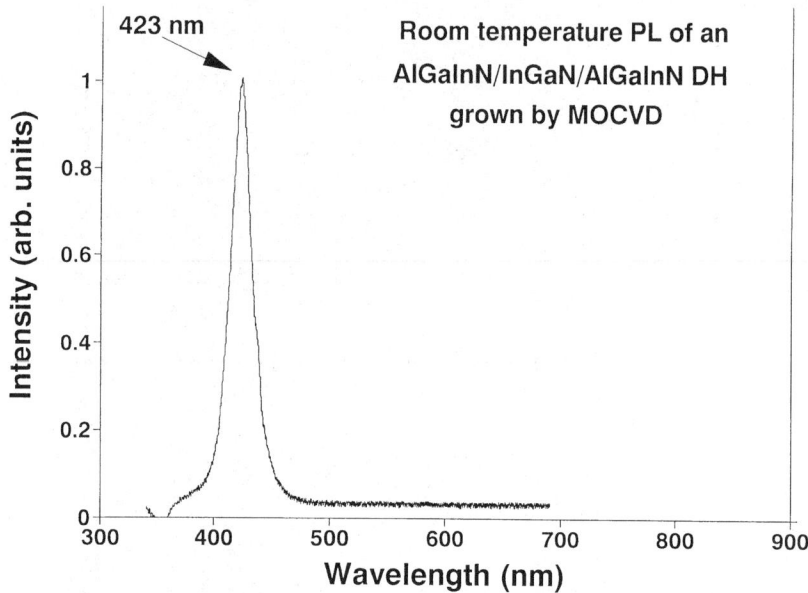

FIG. 32. Room temperature PL of AlInGaN/InGaN/AlInGaN double heterostructure. (From Bedair et al., 1997.)

VI. Conclusions

In-based nitride compounds are lagging behind their Ga and Al counterparts. This may be due to the weak In–N bond and the demand for a fairly high nitrogen over-pressure. Indium segregation at the growing surfaces and interfaces also represents another difficulty. More work is needed to adjust growth parameters to achieve device quality films and double heterostructures.

Acknowledgments

This work was supported by the office of the Naval Research Contract No. N00014-92-J-1477 and the Army Office research (ARPA) Contract No. DAAH04-96-1-0173. The author would like to thank Dr. N. El-Masry, J. Roberts, G. McIntosh, and E. Piner for their valuable contributions.

References

Abernathy, C. R., MacKenzie, J. D., Bharatan, S. R., Jones, K. S., and Pearton, S. J. (1995). *Appl. Phys. Lett.* **66**, 1632.
Amano, H., Tanaka, T., Kunii, Y., and Akasaki, I. (1994). *Appl. Phys. Lett.* **64**, 1377.
Arthur, J. (1968). *J. Appl. Phys.* **39**, 4032.
Bedair, S. M., and co-workers (1997). To be published.
Bedair, S. M., and co-workers Unpublished work.
Bedair, S. M., McIntosh, F. G., Roberts, J. C., Piner, E. L., Boutros, K. S., and El-Masry, N. A. (1997). *J. Cryst. Growth* **178**, 32.
Boutros, K. S. (1995). Ph.D. thesis, North Carolina State University.
Boutros, K. S., Robert, J. C., McIntosh, F. G., Piner, E. L., El-Masry, N. A., and Bedair, S. M. (1995). *MRS Proc.* **395**, 273.
Boutros, K. S., McIntosh, F. G., Roberts, J. C., Bedair, S. M., Piner, E. L., and El-Masry, N. A. (1995). *Appl. Phys. Lett.* **67**, 1856.
Bryden, W., Ecelberger, S., and Kistenmacher, T. (1994). *Appl. Phys. Lett.* **64**, 2864.
Doverspike, K., Rowland, L., Gaskill, D. K., and Freitas, J. A. (1995). *J. Elec. Mat.* **24**, 269.
Gaskill, D. K., Bottka, N., and Lin, M. C. (1986). *Appl. Phys. Lett.* **48**, 1449.
Gerard, J.-M. (1992). *Appl. Phys. Lett.* **61**, 2096.
Gong, J. R., Nakamura, S., Bedair, S. M., and El-Masry, N. A. (1992). *J. Elec. Mat.* **21**, 965.
Guo, Q., Ogawa, H., and Yoshida, A. (1995). *J. Cryst. Growth* **146**, 462.
I-hsiu, and Stringfellow, G. B. (1996). *Appl. Phys. Lett.* **69**, 2701.
Joshkin, V., Robert, J. C., McIntosh, F. G., Piner, E. L., Kolbas, R., Shmagin, I., Krishnankatty, S., Bedair, S. M., Wang, L., and Lin, S. To be published. MRS Spring Meeting, 1997.
Karam, N., Parados, T., Rowland, W., Schetzina, J., El-Masry, N. A. , and Bedair, S. M. (1995). *Appl. Phys. Lett.* **67**, 94.
Keller, S., Keller, B. P., Kapolnek, D., Abare, A. C., Masiri, H., Colden, L. A. Mishra, U. K., and DenBaars, S. P. (1996). *Appl. Phys. Lett.* **68**, 3147.
Koike, M., Yamasaki, S., Nagai, S., Koide, N., Asami, S., Amano, H., and Akasaki, I. (1996). *Appl. Phys. Lett.* **68**, 1403.
Kubota, K., Kobayashi, Y., and Fujimoto, K. (1990). *J. Appl. Phys.* **66**, 2984.
Massies, T., Turco, F., SaRates, A., and Contour, T. P. (1987). *J. Cryst. Growth* **80**, 307.
Matsuoka, T., Tanaka, H., Sasaki, T., and Katsui, A. (1989). International Symposium on GaAs and Related Compounds, *Inst. Phys. Conf.* **106**, p. 141.
Matsuoka, T., Yoshimoto, N., Sasaki, T., and Katsui, A. (1992). *J. Elec. Mat.* **21**, 157.
Matthews, L. W., and Blakeslee, A. E. (1974). *J. Cryst. Growth* **27**, 118.
McIntosh, F. G., Piner, E. L., Boutros, K. S., Roberts, J. C., He, Y., Moussa, M., El-Masry, N. A., and Bedair, S. M. (1995). *MRS Proc.* **395**, 219.
McIntosh, F. G., Boutros, K. S., Roberts, J. C., Bedair, S. M., Piner, E. L., and El-Masry, N. A. (1996). *Appl. Phys. Lett.* **68**, 40.
McIntosh, F. G., Piner, E. L., Roberts, J. C., Behbehani, M. K., Aumer, M. E., El-Masry, N. A., and Bedair, S. M. To be published.
Moison, T. M., Guille, C., Houzy, F., Barthe, F., and Van Rompay, M. (1989). *Phys. Rev.* **40**, 6149.
Morishita, Y., Nomura, Y., Goto, S., and Katayama, Y. (1995). *Appl. Phys. Lett.* **67**, 2500.
Moustakas, T. D. (1996). Third Nitride Workshop, St. Louis, MO.
Singh, R., Doppalapudi, D., Moustakas, T. D. and Romano, L.T., (1997). *Appl. Phys. Lett.* **70**, 1089.

Nagle, T., Landesman, L. P., Larive, M., Mottet, C., and Bois, P. (1993). *J. Crys. Growth* **127**, 550.
Nakamura, S., and Mukai, T. (1992). *Jpn. J. Appl. Phys.* **31**, L1457.
Nakamura, S., Iwasa, N., and Nagahama, S. (1993). *Jpn. J. Appl. Phys.* **32**, L338.
Nakamura, S., Mukai, T., and Senoh, M. (1993). *Jpn. J. Appl. Phys.* **32**, L16.
Nakamura, S., Mukai, T., Senoh, M., and Nagahama, S. (1993). *J. Appl. Phys.* **74**, 3911.
Nakamura, S., Senoh, M., and Mukai, T. (1993). *Appl. Phys. Lett.* **62**, 2390.
Nakamura, S. (1994). *J. Microelectron* **25**, 651.
Nakamura, S., Mukai, T., and Senoh, M. (1994). *Appl. Phys. Lett.* **64**, 1687.
Nakamura, S., Senoh, M., Iwasa, N., and Nagahana, S. (1995). *Jpn. J. Appl. Phys.* **34**, L797.
Narukawa, Y., Kawakami, Y., Fujita, S., and Nakamura, S. (1997). *Appl. Phys. Lett.* **70**, 981.
Nata Rajan, B. R., Etoukhy, A. H., Green, J. E., and Barr, T. L. (1980). *Thin Solid Films* **69**, 201.
Oberman, D., Lee, H., Gotz, W., Soloman, G., and Harris, J. (1994). *EMC Session Q-10*, Boulder, CO.
Parker, E. H. (ed.) (1985). *The Physics and Technology of Molecular Beam Epitaxy*, Plenum, New York.
Piner, E. L., Behbehani, M. K., El-Masry, N. A., McIntosh, F. G., Roberts, J. C., Boutros, K. S., and Bedair, S. M. (1997). *Appl. Phys. Lett.* **70**, 461.
Piner, E. L., McIntosh, F. G., Roberts, J. C., Aumer, M. E., Joshkin, V., Bedair, S. M., and El-Masry, N. A. (1996). *MRS Internet J. Nitride Semicond. Res.* **1**, 43.
Piner, E. L., and co-workers. To be submitted.
Ren, F., Abernathy, C. R., Chu, S. N. G., Lothian, J. R., and Pearton, S. T. (1995). **66**, 1503.
Roberts, J. C., McIntosh, F. G., Boutros, K. S., Bedair, S. M., Moussa, M., Piner, E. L., He, Y., and El-Masry, N. A. (1995). *MRS Proc.* **395**, 273.
Roberts, J. C., McIntosh, F. G., Aumer, M., Joshkin, V., El-Masry, N. A., and Bedair, S. M. (1996). *MRS Proc.* **449**, 1161.
Seki, H., and Koukitu, A. (1989). *J. Cryst. Growth* **98**, 118.
Scholz, F., Harle, V., Steuber, F., Bolay, H., Dormen, A., Kaufmann, B., Syganow, V., and Hangleiter, A. (1997). *J. Cryst. Growth* **170**, 321.
Shimizu, M., Hiramatsu, K., and Sawaki, N. (1994). *J. Cryst. Growth* **145**, 209.
Starosta, K. (1981). *Phys. Status Solidi* (a) **68**, K55.
Suntola, T. (1989). *Acta Polytechnica Scand.* **64**, 242.
Tansley, T. L., and Foley, C. P. (1986). *J. Appl. Phys.* **59**, 3241.
Tichler, M. A., and Bedair, S. M. (1986). *Appl. Phys. Lett.* **49**, 1199.
Trainer, J. W., and Rose, K. (1974). *J. Electron Mat.* **3**, 831.
Williams, C. K., Glisson, T. H., Hauser, J. R., and Littlejohn, M. A. (1978). *J. Electron Mat.* **7**, 639.
Yamasaki, S., Asami, S., Shibata, N., Koike, M., Manabe, K., Tanaka, T., Amano, H., and Akasaki, I. (1995). *Appl. Phys. Lett.* **66**, 1112.
Zolper, J. C., Pearton, S. J., Abernathy, C. R., and Vartuli, C. B. (1995). *Appl. Phys. Lett.* **66**, 3042.

CHAPTER 7

Crystal Structure of Group III Nitrides

A. Trampert, O. Brandt, and K. H. Ploog

PAUL-DRUDE-INSTITUT FÜR FESTKÖRPERELEKTRONIK
HAUSVOGTEIPLATZ 5-7, 10117 BERLIN, GERMANY

I. INTRODUCTION 167
II. CRYSTAL STRUCTURES 168
   1. *Polarity of the Structures* 172
III. LATTICE CONSTANTS AND MECHANICAL AND THERMAL PROPERTIES 172
   1. *Aluminum Nitride* 173
   2. *Gallium Nitride* 174
   3. *Indium Nitride* 175
   4. *Nitrides with Zinc-blende Structure* 176
IV. PHASE STABILITY, PHASE TRANSITIONS, AND POLYTYPISM 178
V. REAL STRUCTURES AND IMAGING 179
   1. *Defects in Nitrides* 181
VI. EPITAXIAL GROWTH 184
   1. *Application to Nitride Epitaxy* 185
   2. *Outlook* 190
   References 190

I. Introduction

The elements B, Al, Ga, and In are forming compounds with N having the composition $A^{III}N$. The chemical bond of these compounds is predominantly covalent, that is, the constituents develop four tetrahedral bonds for each atom. Because of the large the differences in electronegativity of the two constituents, there is a significant ionic contribution to the bond which determines the stability of the respective structural phase. The group III nitrides AlN, GaN, and InN can crystallize in the following three crystal structures: (1) wurtzite, (2) zinc-blende, and (3) rock-salt. At ambient conditions, the thermodynamically stable phase is the wurtzite structure. A phase transition to the rock-salt structure takes place at high pressure. In contrast, the zinc-blende structure is metastable and may be stabilized only by heteroepitaxial growth on substrates reflecting topological compatibility.

Boron nitride (BN), however, has properties unlike all other group III nitrides. In fact, crystal structures and related physical properties (such as the indirect bandgap) are analogous to modifications of carbon. Thus, BN exists in tetrahedral like and graphite like modifications. The former includes the stable zinc-blende and metastable wurtzite crystal structures, whereas graphite like modifications include a hexagonal and a less common rhombohedral lattice type, respectively. No high pressure phase transition to the rock-salt structure has been observed. Hexagonal BN has outstanding mechanical properties, but is less interesting for electronic devices. Since there are already excellent review articles about BN (Edgar, 1992) and also about boron compounds in general (Meller, 1988), we will restrict ourselves to a review of the three other group III nitrides, that is, AlN, GaN, and InN. In Sections I and II we describe their crystal structures and their mechanical and thermal properties in more detail.

II. Crystal Structures

The wurtzite structure has a hexagonal unit cell with two lattice parameters a and c in ratio $c/a = \sqrt{8/3} = 1.633$. As shown in Fig. 1, this structure is composed of two hexagonal close-packed (hcp) sublattices, which are shifted with respect to each other along the three-fold c axis by the amount of $u = 3/8$ in fractional coordinates. Both sublattices are occupied by one atomic species only, resulting in four atoms/unit cell.

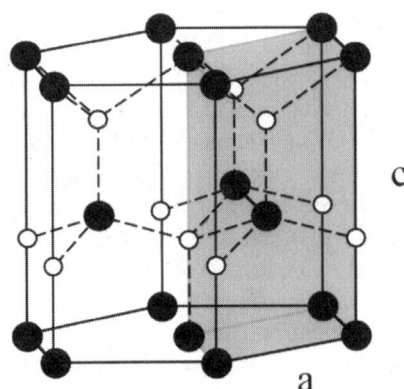

FIG. 1. The wurtzite crystal structure: atoms of one species are black, those of the other species are white, the unit cell is shaded.

Locally, every atom of one kind is surrounded by four atoms of the other kind which are arranged at the edges of a tetrahedron (coordination number 4). The symmetry of the wurtzite structure is given by space group $P6_3mc$, and the two inequivalent atom positions are $(\frac{1}{3}, \frac{2}{3}, 0)$ and $(\frac{1}{3}, \frac{2}{3}, u)$. For actual nitrides, axial ratios deviate considerably from the ideal value for hexagonal close packing of spheres, and hence, each lattice is distorted from the ideal geometry to some extent. As a result, there are two slightly different bond lengths defined by Yeh et al. (1992)

$$R(1) = u \cdot c \quad \text{and} \quad R(2) = a \cdot [1/3 + (1/2 - u)^2 \cdot (c/a)^2]^{1/2},$$

where u is called the dimensionless cell internal structure parameter. For an ideal wurtzite structure, $R(1)$ is equal to $R(2)$. The degree of deviation for the different nitrides is related to their phase stability in the sense that a critical deviation exists when the wurtzite structure becomes unstable (Lawaetz, 1972).

At high external pressures, a phase transformation of the wurtzite towards the rock-salt (NaCl) structure takes place which is predominantly found in ionic bonded crystals. In this structure, anions form a face-centered cubic (fcc) lattice where open sites in between are occupied by ions of opposite charge (Fig. 2). These cations in turn build an fcc lattice, which is displaced by half a unit cell dimension along the cubic axis. Thus, every cation is surrounded by six nearest neighbor anions and vice versa, which are forming a regular octahedron (coordination number 6). The space group symmetry of the NaCl type structure is $F m3m$.

The zinc-blende structure belongs to space group $F\bar{4}3m$ and consists of two the interpenetrating and identically oriented fcc lattices offset from one another by one-quarter of a body diagonal as shown in Fig. 3. There are four molecules/unit cell, the coordinates of one atom type being $(0, 0, 0)$,

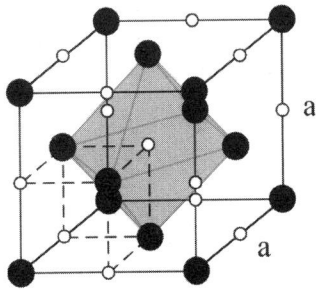

FIG. 2. The rock–salt structure: every ion is surrounded by a regular octahedron.

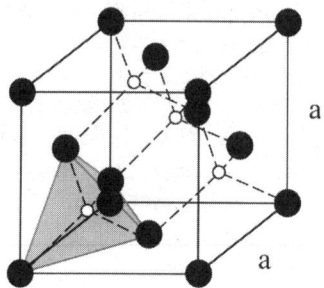

FIG. 3. The zinc-blende structure (a regular tetrahedron is inserted).

$(0, \frac{1}{2}, \frac{1}{2})$, $(\frac{1}{2}, 0, \frac{1}{2})$, $(\frac{1}{2}, \frac{1}{2}, 0)$, and those of the other type being $(\frac{1}{4}, \frac{1}{4}, \frac{1}{4})$, $(\frac{1}{4}, \frac{3}{4}, \frac{3}{4})$, $(\frac{3}{4}, \frac{1}{4}, \frac{3}{4})$, $(\frac{3}{4}, \frac{3}{4}, \frac{1}{4})$. As a result, each atom of one kind is surrounded by four atoms of the other kind forming a regular tetrahedron (coordination number 4). The overall equivalent bond length amounts to $R = (a/4)\sqrt{3}$.

Because of the tetrahedral coordination of both wurtzite and zinc-blende structures, the short range order of these structures is virtually identical. Neglecting slight deviations from ideal spherical closed packing in the wurtzite structure, each atom is tetrahedrally surrounded by four nearest neighbors with the same bond distance as in the zinc-blende modification. Also the twelve next-nearest neighbors are at the same distance in both structures. Significant differences are found only for the third next-nearest neighbors. This difference can simply be described by regarding the handedness of the fourth interatomic bond along the $\langle 111 \rangle$ closed packed chain corresponding to the third nearest neighbor relation in both structures (Fig. 4(a)). In the zinc-blende structure, triangularly arranged atoms in the closed-packed $\{111\}$ plane are in "staggered" configuration with the triangularly arranged unlike atoms in third nearest neighbor distance, that is, the latter are rotated around an angle of 60° with respect to the former (Fig. 4(b)). In contrast, triangularly arranged atoms in the $\{0001\}$[1] closed packed basal plane are in an "eclipsed" (i.e., identical) configuration in the wurtzite structure. Taking these different arrangements in closed-packed planes as the building principle for both crystal structures, one finally ends up with the following stacking sequences:

...AaBbAaBbAaBbAaBb...

---

[1] Note here the use of the four index notation $\{hkil\}$ which is sometimes omitted in literature $\{hk.l\}$ for simplicity because of $i = -(h + k)$.

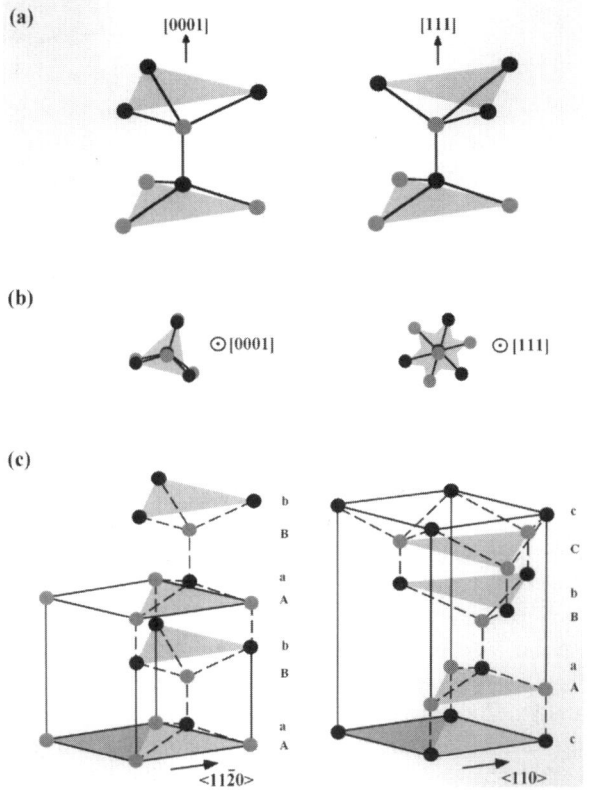

FIG. 4. (a) Handedness of the fourth interatomic bond along the closed-packed direction for wurtzite (*left*) and zinc-blende (*right*). (b) Scheme of the layer stackings and (c) the corresponding three-dimensional representations.

for the wurtzite, and

$$...AaBbCcAaBbCcAaBbCc...$$

for the zinc-blende type, which completely characterize both structure types as shown in Fig. 4. Small and large letters stand for the two different kinds of constituents. A fault in the stacking sequence transforms one structure into the other, as we will discuss later. Hence, the respective stacking fault energy corresponds to the free energy difference between the wurtzite and zinc-blende modifications. As a consequence, the density of such structural defects is expected to be high in nitrides exhibiting a low energy difference between these phases.

1. POLARITY OF THE STRUCTURES

None of the three structures described previously possess a center of inversion. This means that, when defining an atom position on a closed packed plane with coordinates (x, y, z), it is not invariant to the position $(-x, -y, -z)$ since inversion results in replacement of group III atoms by nitogen atoms and vice versa. As a result of the lack of inversion symmetry, the crystal exhibits crystallographic polarity: closed-packed {111} planes in zinc-blende and rock-salt structures and corresponding {0001} basal planes in the wurtzite structure differ from $\{\bar{1}\bar{1}\bar{1}\}$ and $\{000\bar{1}\}$ planes, respectively. In general, group III terminated planes are denoted as (111)A and group V terminated planes are denoted as (111)B. Beyond these primary polar planes many other secondary ones exist, which, however, have been historically ignored since they do not show up in macroscopic facetting of crystallites. Since the geometric structure factor is different in opposite directions along the polar axis, the A or B exposed faces of the crystal may be identified by anomalous scattering of X-rays (Warekois *et al.*, 1962). More recently, identification has been achieved also by transmission electron microscopy (TEM), employing convergent beam electron diffraction techniques (Taftø and Spence, 1982; Ponce *et al.*, 1996). A variety of properties of the material depends on its polarity, for example, growth, etching, defect generation and plasticity, and piezoelectricity. It is interesting to note here that wurtzite nitride layers epitaxially grown on hexagonal (0001) substrates are potentially piezoelectric, and become indeed piezoelectrically active if they are under residual (thermal) stress.

III. Lattice Constants and Mechanical and Thermal Properties

In general, single crystals are required for accurate measurements of lattice constants and associated mechanical and thermal properties, such as thermal expansion and thermal conductivity. However, growth of bulk crystals in GaN has proved to be exceedingly difficult, and virtually impossible in AlN and InN. Because of the consequential lack of single crystals, most of the data acquired to date stems from nitride layers grown heteroepitaxially. More often, nitride layers are grown in their wurtzite modification on hexagonal sapphire (α-Al$_2$O$_3$) and 6H-SiC substrates. Only a few studies have been reported on the metastable zinc-blende modification. In either case, density and type of structural defects (dislocations, planar faults) as well as point defects (native and extrinsic) are often unknown, although these defects may profoundly influence mechanical properties of

TABLE I

SOME PROPERTIES OF THE WURTZITE POLYTYPES. (Numbers are calculated by semiempirical (left) and first-principle (right) methods)

Parameter	AlN	GaN	InN
Bond energy (eV/bond)	2.88[a]/—	2.20[a]/—	1.93[a]/—
Bond length (Å)[a]	1.89[a]/1.89[d]	1.94[a]/1.94[d]	2.15[a]/2.14[d]
Bulk modulus (GPa)[b]	199[b]/205[d]	176[b]/202[d]	122[b]/139[d]
Melting point (K)[c]	3487[c]/—	2791[c]/—	2146[c]/—
Bandgap (eV)[d]	—/4.41[d]	—/2.04[d]	—/−0.04[d]

[a]Harrison, 1988.
[b]Azuhata, Sota, and Suzuki, 1996.
[c]van Vechten, 1973.
[d]Wright and Nelson, 1995.

the crystal (Lagerstedt and Monemar, 1979). Comparisons with structural properties predicted by first-principle calculations in fact reveal inconsistencies, which we will not discuss further (for an overview, see Kim, Lambrecht, and Segall, 1996; Wright and Nelson, 1994, 1995; Yeh et al., 1992). In this respect, it is instructive to examine the order within the homological chemical sequence AlN → GaN → InN as reflected by bond energies, bond lengths, elastic moduli, melting points, and bandgap energies. All these properties show the same trend within the homological sequence (Table I) which is explained by the characteristic bond nature (Harrison, 1988).

In the following sections, we review some of the mechanical and thermal properties of each of the wurtzite nitrides, and also give results for their zinc-blende modification, if available.

1. ALUMINUM NITRIDE

AlN is a ceramic and refractory material and has a combination of attractive physical properties, such as low thermal expansion, high thermal conductivity, high hardness, and high melting point (Table II). Furthermore, AlN attracts interest for its insulating properties (typical resistivities are around $10^{12}\,\Omega$ cm). However, AlN is a difficult material to grow for various reasons. The same properties which make AlN an interesting material also intricate growth, and the high reactivity of Al with O is yet another obstacle to overcome. The lattice constants mentioned here should be treated with care, as the appreciable amount of O present in today's AlN samples also affects the lattice constants (Slack, 1973). The most common

TABLE II

SOME PROPERTIES OF AlN

Wurtzite Polytype		
Bandgap energy (eV)	6.2[a]	
Lattice constants (nm)	0.311(2) $(a)^a$	0.498(2) $(c)^a$
Thermal expansion (K^{-1})	$4.2 \cdot 10^{-6}$ $(\alpha_{\|})^b$	$5.3 \cdot 10^{-6}$ $(\alpha_\perp)^b$
Thermal conductivity (W cm^{-1} K^{-1})	2.8[c]	
Melting point (K)	3487[d]	

[a]Yim et al., 1973.
[b]Yim and Paff, 1974.
[c]Slack et al., 1987.
[d]van Vechten, 1973.

values measured at room temperature by X-ray diffraction have been reported by Yim et al. (1973). Other reported values for the lattice parameters range from 0.3110–0.3113 nm for a and 0.4978–0.4982 nm for c (Schulz and Thiemann, 1977; Huseby, 1983), while the c/a ratio varies from 1.600 to 1.602 and deviates from the ideal value of 1.633. The temperature dependence of the lattice constants is characterized by the mean thermal expansion coefficient $\alpha = (1/\Delta T)(\Delta a/a)$ in a temperature interval ΔT, where Δa is calculated from the temperature dependence of lattice constant $a(T) = a_0(1 + A \cdot T + B \cdot T^2)$. Here, $a(T)$ and a_0 are the lattice constants at temperature T and 0°C and A and B are fitting parameters. The values in Table II are valid for the temperature range between 20 and 800°C (Yim and Paff, 1974). In comparison to sapphire ($\alpha_\| = 7.28 \cdot 10^{-6}$ K^{-1}, $\alpha_\perp = 8.11 \cdot 10^{-6}$ K^{-1}), both AlN and GaN (Table III) have much smaller thermal expansion coefficients which suggests that thermal stress resides in nitride layers grown on sapphire substrates.

All nitride materials are distinguished by their high thermal conductivity with AlN reaching the highest values of ~ 2.85 W cm^{-1} K^{-1} at room temperature (Slack et al., 1987). In comparison, Al$_2$O$_3$ and SiC have values of 0.25 W cm^{-1} K^{-1} and 4.9 W cm^{-1} K^{-1} respectively, at the same temperature.

2. GALLIUM NITRIDE

Wurtzite GaN with a direct bandgap of 3.43 eV (Maruska and Tietjen, 1969; Pankove, Berkeyheiser, and Maruska, 1970) is, in principle, ideally suited for ultraviolet and blue light-emitting diode and laser devices, and is

TABLE III

SOME PROPERTIES OF GaN

Wurtzite Polytype		
Bandgap energy (eV)	3.39[a]	
Lattice constants (nm)	0.318(8) (a)[b]	0.518(5) (c)[b]
Thermal expansion (K^{-1})	5.59·10^{-6} ($\alpha_{\parallel}$)[a]	3.17·10^{-6} ($\alpha_{\perp}$) (layer)[a]
	4.44·10^{-6} (α_a)[c]	4.03·10^{-6} (α_c) (bulk)[c]
Thermal conductivity (W cm^{-1} K^{-1})	1.3[d]	
Melting point (K)	2791[e]	

[a]Maruska and Tietjen, 1969.
[b]Detchprohm et al., 1992; Leszczynski et al., 1994.
[c]Mean linear coefficient calculated after: Leszczynski et al., 1994.
[d]Sichel and Pankove, 1977.
[e]van Vechten, 1973.

the best studied material among all group III nitrides. These devices, which have become commercially available recently, are prepared on sapphire substrates despite the large lattice and thermal mismatch. In fact, thermal stresses generated in the layers during heteroepitaxy may result in bending and even cracking the samples (Itoh et al., 1985; Hiramatsu, Detchprohm, and Akasaki, 1993). As a consequence, the lattice constants reported for such samples were found to exhibit significant scattering. Detchprohm et al. (1992) determined the intrinsic lattice constants of GaN in a strain-free state using layer thicknesses up to 1200 μm. The results were $a = 0.31892 \pm 0.00009$ nm and $c = 0.51850 \pm 0.00005$ nm with an axial ratio $c/a = 1.626$ very close to the value of ideal hcp lattice. Leszczynski et al. (1994) measured the lattice constants and the thermal expansion of bulk GaN crystals and compared them to those of heteroepitaxial layers. The values for the lattice parameter c are, within experimental error, identical for bulk and layer, and are also in agreement to values mentioned previously. However, lattice constant a of the layer is slightly smaller when compared with the bulk. This is discussed in terms of relaxation effects with regard to the sapphire substrate (Leszczynski et al., 1994). The expansion coefficients of bulk as well as layer material are listed in Table III.

3. INDIUM NITRIDE

InN has not received the same attention as GaN or AlN. Its bandgap of 1.89 eV (Zetterstrom, 1970) is in a spectral range where alternative material systems are available, also the mechanical and thermal properties of InN are

TABLE IV

SOME PROPERTIES OF InN

Wurtzite Polytype		
Bandgap energy (eV)	1.89[a]	
Lattice constants (nm)	0.3533...0.3548 (a)[b]	0.5963...0.5760 (c)[b]
Thermal expansion (K^{-1})	2.7...3.7·10^{-6} $(\alpha_{\|\|})$[c]	3.4...5.7·10^{-6} $(\alpha_{\perp})$[c]
Thermal conductivity (W cm^{-1} K^{-1})	0.8[d]	
Melting point (K)	2146[e]	

[a]Zetterstrom, 1970.
[b]See Tansley and Foley, 1986; Wakahara and Yoshida, 1989
[c]Madelung (ed.), 1982.
[d]Slack *et al.*, 1987.
[e]van Vechten, 1973.

not as outstanding as those of AlN. Preparation of InN is thus considered to be of scientific interest rather than technological importance. Note, however, that for obtaining blue light emission from nitride based structures, a significant amount (40–50%) must be alloyed into GaN. Pure InN has not yet been grown as single crystal by any growth technique, rather, fine grained polycrystalline layers are obtained. The dissociation temperature of InN is approximately 500°C (Trainor and Rose, 1974) and nonstoichiometry at temperatures required for crystallization severely affects the quality of these InN samples. No accurate measurements exist for either the lattice constants or the mechanical and thermal properties. Values given in Table IV serve as a rough guideline only.

4. NITRIDES WITH ZINC-BLENDE STRUCTURE

Zinc-blende III–V nitrides have been synthesized by heteroepitaxy on cubic substrates, such as Si, GaAs, MgO, and cubic 3C-SiC. All these substrates share the handicap of a very large lattice mismatch to the nitrides. Only a few groups have succeeded in growing zinc-blende nitrides and have reported measurements of fundamental structure properties of these cubic layers. Lattice constants reported so far are compiled in Table V. Measurements of other mechanical and thermal properties are still lacking.

Cubic AlN has been prepared by epitaxial growth on Si substrates using pulsed laser ablation (Lin *et al.*, 1995a) and by solid-phase epitaxy between Al and TiN (Petrov *et al.*, 1992), respectively. The lattice constant

TABLE V

PROPERTIES OF ZINC-BLENDE NITRIDES

Zinc-blende Polytype	AlN	GaN	InN
Bandgap energy (eV)	5.11^a	$3.2–3.3^b$	2.2^c
Lattice constant (nm)	$0.43(8)^d$	0.452^e	$0.49(8)^f$
Thermal expansion	—	—	—
Thermal conductivity	—	—	—

[a] Lamprecht and Segall, 1991.
[b] Lei et al., 1991.
[c] Jenkins, Hong, and Dow, 1987.
[d] Petrov et al., 1992.
[e] Strite et al., 1991.
[f] Strite et al., 1993.

is measured to be $a = 0.438$ nm for a thin film by applying electron diffraction.

Zinc-blende GaN has been synthesized by several groups (Mitzuta et al., 1986; Paisley et al., 1989; Lei et al., 1991, 1992; Strite et al., 1991; Powell et al., 1993; Kuwano et al., 1994; Lin et al., 1995b) on different substrates. An average lattice parameter of 0.452 nm fits the reported values within experimental errors. The structural quality of layers was found to be poor. A rough substrate/epilayer interface, high density of planar defects, and the tendency of phase transformation into the wurtzite structure are reported to be the basic problems. Also, the optical quality of zinc-blende GaN cannot compete with that of wurtzite GaN because only broad and weak emission is detected which is quenched already at low temperatures (Ramírez-Flores et al., 1994). Recently, however, progress has been made in understanding the nucleation stage of GaN on GaAs (Yang et al., 1996; Brandt et al., 1996) resulting in a much improved interface and a higher crystal perfection (Trampert et al., 1997). These results will be discussed in a later section of this chapter. The optical quality of these samples, which exhibit intense luminescence at room temperature and above, give rise to the hope that zinc-blende nitrides may eventually mature (Yang, Brandt, and Ploog, 1996; Brandt et al., 1997).

Strite et al. (1993) reported the existence of a zinc-blende InN polytype prepared by molecular beam epitaxy (MBE). The lattice constant $a = 0.498 \pm 0.001$ nm was measured by X-ray diffraction. TEM microstructural analysis revealed a high density of stacking faults from which wurtzite domains were nucleated. Abernathy and MacKenzie (1995) obtained InN layers containing only the cubic phase, but also with a high density of stacking faults and microtwins.

IV. Phase Stability, Phase Transitions, and Polytypism

Pressure induced transformation from wurtzite to rock-salt structures results in a transition from four-to six-fold coordination of atoms corresponding to a change in bond character from covalent to more ionic like. This transition indicates that nitrides with higher ionic bond contribution should transform more readily under pressure. The structural phase transition was experimentally observed at the following pressure values (Xia, Q., Xia, H., and Ruoff, 1993; Perlin et al., 1992; Ueno et al., 1994): 22.9 GPa, 52.2 GPa, and 12.1 GPa for AlN, GaN, and InN, respectively. At first sight, this result seems surprising because the trend of transition pressures does not follow the trend in Phillips ionicity (Phillips, 1973). The experimental trend is instead in agreement with the ionicity scale derived by first-principle calculations of electronegativity (Garcia and Cohen, 1993). However, the expected correlation between transition pressure and deviation from ideal axial ratio c/a was not observed. This ratio decreases with pressure in AlN and InN, deviating further from the ideal value up to a critical value at the phase transition, whereas it stays constant in GaN (Ueno et al., 1994).

Of more practical interest is the polytypism between wurtzite and zincblende structure as previously mentioned. First-principle calculations predict the ground state to be the wurtzite structure (Yeh et al., 1992) for all nitrides. The energy differences ΔE between both structure modifications follow the sequence

$$\Delta E\,(\text{GaN}) < \Delta E\,(\text{InN}) < \Delta E\,(\text{AlN})$$

suggesting that the problem of phase purity is most severe in GaN. Furthermore, these energy differences are correlated to the stacking fault energy in the wurtzite structure. Suzuki, Ichihara, and Takeuchi (1994) measured the stacking fault energy of several wurtzite type materials including nitrides by high-resolution TEM. The authors found an increase in the reduced stacking fault energy[2] with an increasing deviation Δ of c/a ratio from the ideal value of 1.633: $\Delta c/a\,(\text{GaN}) < \Delta c/a(\text{InN}) < \Delta c/a\,(\text{AlN})$. This is, in fact, on the same order as given previously for the energy differences ΔE.

---

[2]The *reduced* stacking fault energy is defined by the energy per atomic bond which is introduced to eliminate the influence of the lattice constant.

V. Real Structures and Imaging

Any real crystal contains structural imperfections, which are — according to their dimension — defined as point, line, interface, and volume defects. It is common to all defect types that they locally disturb the strict periodicity of the crystal lattice. It is, however, just this periodicity which underlies the physical properties of crystals, such as its bandstructure. In general, we thus may conclude that the physical properties of crystals will be modified through the presence of defects. The exact knowledge of their atomic structure and interaction with crystal properties is therefore of major importance. An experimental tool for investigating defects on an atomic scale is provided by high resolution TEM (HRTEM) with a point-resolution up to approximately 1 Å (Phillipp *et al.*, 1994). Using HRTEM, it becomes possible to image crystal lattice structure and local deviations from perfect periodicity in real space. At the same time, we may determine the precise location of defects within the crystal matrix. It is, however, important to note that the naive interpretation of an HRTEM image, which is based on the presumption that the contrast observed directly corresponds to the position of atoms in the crystal, may lead to erroneous conclusions. It is, in general, incorrect to explain this contrast as being due to projection of the observed structure.

The electron microscope in its high-resolution mode produces an interference pattern, which is the Fourier transformation (FT) of the amplitude distribution in the diffraction plane. The contrast, often observed as bright and dark dots, is related to the real three-dimensional structure of the sample. In a first order approximation, the image formation process can be described by the following flowchart:

$$\text{Sample} \xrightarrow{\text{FT}} \text{Fourier amplitude} \xrightarrow{\text{FT}} \text{image}$$

Fourier amplitudes in the back focal plane of the diffraction lens are given by FT of the transmission function which describes the interaction between electrons and sample. Only under special conditions ("weak-phase approximation", see Buseck, Cowley, and Eyring, 1992; Spence, 1989) is this process directly connected to the projected potential of the transmitted sample. During the back transformation of the Fourier amplitudes to the interference image, contrast transfer through the microscope must be considered by a convolution of amplitude distribution and contrast transfer function. Interpretation of HRTEM images thus requires the exact knowledge of electron-matter interaction as well as the contrast transfer function of the specific microscope used for imaging.

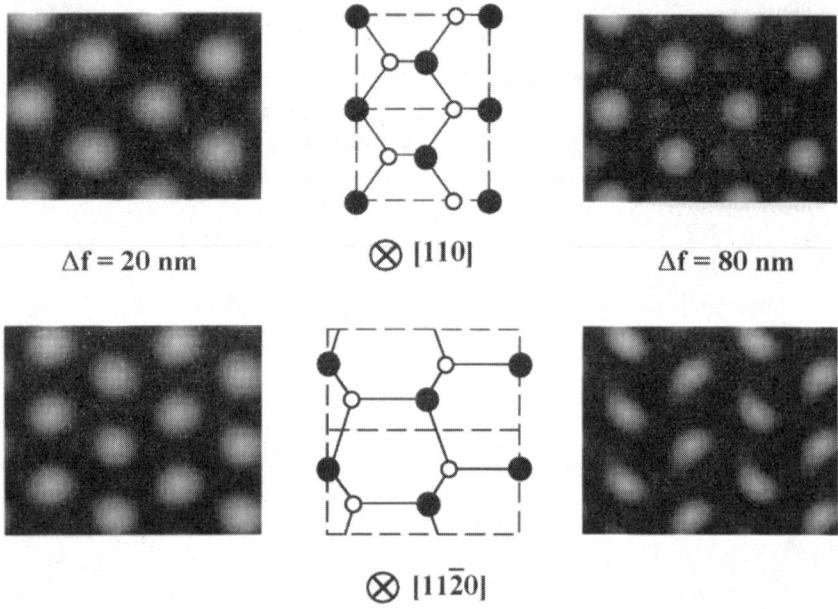

FIG. 5. Simulated HRTEM images of the wurtzite (*bottom*) and zinc-blende (*top*) GaN structure for two different defocus values (Δf = 20, and 80 nm). The underlying structural models are shown in the center.

These parameters in turn may serve as input parameters for a simulation of the image formation process. Strictly speaking, we may obtain a reliable determination of atomic positions of the investigated structure only by comparison between simulated and experimental images. In Fig. 5 we show simulations of HRTEM images of GaN in wurtzite and zinc-blende modification in comparable projection, that is, [110] for zinc-blende and [11$\bar{2}$0] for wurtzite (Stadelmann, 1987). For a certain defocus value of the microscope, simulated micrographs exhibit a contrast of similar symmetry and shape, and two different crystal symmetries are difficult to distinguish. Only when selecting another value for the defocus does the interference pattern become distinctly different. However, in neither of the micrographs does maxima of contrast (white spots) correspond to actual positions of atom rows, as is seen in comparison with structural models underlying these simulations.

1. DEFECTS IN NITRIDES

In spite of the enormous effort to improve crystal quality, homoepitaxial and heteroepitaxial nitride films still contain high densities of structural defects. Growth of bulk crystals is particularly difficult because of the high melting points of nitrides in connection with the extremely high nitrogen vapor pressure. Therefore, most III–V nitrides are prepared by epitaxial growth despite the absence of substrates with a low lattice mismatch to the nitride epilayer. A detailed discussion of the defects which appear during epitaxial growth is presented in a separate chapter of this book. Here, we will describe briefly the grown-in structural defects which originate from the bistability of the crystal structures.

The most prevalent structural defects observed in III–V nitride epilayers are threading dislocations in wurtzite materials, and stacking faults and micro-twins in zinc-blende material. The HRTEM micrograph in Fig. 6 is an example of a highly imperfect part of zinc-blende GaN grown on (001) GaAs which reveals planar defect types running along the ⟨111⟩ directions. Similar defect structures have been observed in GaN grown on MgO (Powell et al., 1993) and 3C-SiC (Paisley et al., 1989). These planar defects are best described in zinc-blende and wurtzite structures by regarding the sequence of the closed packed planes (Hull and Bacon, 1984). As was mentioned previously, the stacking sequence of a perfect material with

FIG. 6. HRTEM micrograph of zinc-blende GaN grown on (001) GaAs. Numerous planar defects which propagate along the {111} planes are apparent.

zinc-blende structure is

$$\ldots AaBbCcAaBbCcAa\ldots$$

A change of this sequence resulting from the removal or introduction of an extra layer will result in

$$\ldots AaBb\underline{CcAaCcAa}BbCcAa\ldots (intrinsic\ type)$$
$$\downarrow$$

$$\ldots AaBbCc\underline{BbAaBb}CcAa\ldots (extrinsic\ type)$$
$$\uparrow$$

which contains units (underlined) of an ideal wurtzite structure

$$\ldots AaBbAaBbAaBbAa\ldots.$$

The arrows indicate where an extra layer has been removed or inserted. It follows that stacking faults, which are introduced in a perfect zinc-blende structure, form an ideal wurtzite like lattice at the location of the planar defect. Thus, a certain equidistant sequence of stacking faults may transform the lattice structure from zinc-blende to wurtzite. This geometrical argument also applies to the reverse case of a wurtzite structure with stacking faults acting as zinc-blende units:

$$\ldots AaBb\underline{AaBbCc}BbCcBb\ldots (intrinsic\ type)$$

$$\ldots AaBb\underline{AaBbCcAa}CcAa\ldots (extrinsic\ type)$$

The intrinsic fault is formed by removal of a basal plane

$$(\ldots AaBbAaBb\uparrow BbAaBbAa\ldots)$$

resulting in an unstable situation of two similar atomic layers coming into contact with each other, followed by a slip of $1/3\langle 10\bar{1}0\rangle$ of the crystal above the fault to reduce fault energy (Hull and Bacon, 1984). The main contribution of the stacking fault energy arises from the second nearest neighbor sequence of double layers (Aa) which corresponds to the energy difference of both phases. However, in real crystals, the deviation from ideal geometry must be taken into consideration as discussed previously.

In lattice mismatched heterostructures, stacking faults can additionally play an important role for strain relief. These latter faults are generated by dissociation of the strain-relieving perfect dislocation into two partial

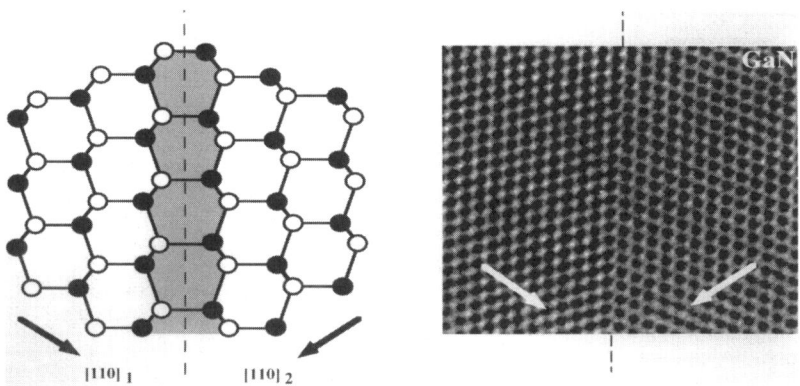

FIG. 7. A model of the rotational twin boundary (from Hold, 1964) and the corresponding HRTEM image of a GaN twin. The shaded area in the model represents a wurtzite like unit.

dislocations by way of the following reaction:

$$\mathbf{b} = a/2[110] \rightarrow a/6[112] + a/6[2\bar{1}1] = \mathbf{b}_{P1} + \mathbf{b}_{P2}$$

which is applied to a perfect 60° dislocation in a fcc-type lattice.[3] Because $\mathbf{b}_{P1}$ and $\mathbf{b}_{P2}$ (the so-called Shockley partial dislocations) are not translational vectors of the lattice, they produce a stacking fault in-between which is geometrically identical to the intrinsic type mentioned previously. One partial dislocation is placed at the interface forming a misfit dislocation and the second partial dislocation is terminating within the epilayer. In general, every stacking fault ending in the material is bound by partial dislocations.

The second type of planar defects (twin boundary) can also be regarded as a two phase region. In the $\langle 111 \rangle$ rotational twin crystal, the atomic arrangement on the $\{111\}$ plane is rotated 180° around the $\langle 111 \rangle$ axis at the twin boundary. Rotation of separate domains of 250° 32' relative to each other about the $\langle 110 \rangle$ tilt axis is an alternative method of generating such a twin. Note that a monolayer of the wurtzite lattice is generated at the twin boundary (Fig. 7).

In addition, the defect types introduced in nitride epilayers by strain relaxation are of a different nature than for the two polytypes. In wurtzite layers, a high density of dislocations is found and can be separated into two parts, (1) a threading arm running through the layer and (2) a misfit arm lying in the interface to relieve the misfit strain. In zinc-blende layers, a high

---

[3]Frank's rule shows that splitting reactions are energetically favorable, if $\mathbf{b}^2 > \mathbf{b}_{P_2}^2 + \mathbf{b}_{P_2}^2$ (Hull and Bacon, 1984).

density of planar defects are found instead and their generation will be discussed in the following section.

VI. Epitaxial Growth

Epitaxy-induced growth of thin films is, in general, based on specific interface structures defined by a minimum of interfacial energy. These interfaces correspond to preferred orientation relationships between the crystal lattices of the deposit (a_{epi}) and substrate (a_{sub}) which are usually characterized by a low mismatch f, defined by:

$$f = \frac{a_{sub} - a_{epi}}{a_{sub}}$$

with regard to the substrate reference lattice. Continuum elasticity theory has been used to predict that for small misfits epitaxy can occur by formation of layers which are initially pseudomorphic (van der Merwe and Ball, 1975; Matthews, 1975). As the epilayer thickness increases, its elastic strain increases until it is sufficient to activate misfit dislocation sources, after which plastic relaxation occurs. This critical thickness at which misfit dislocations are generated varies inversely with f. The interface is then characterized by extended regions of excellent lattice fit, separated by regions of poor fit located at misfit dislocations. The epitaxial growth of (In, Ga)As on GaAs with low In content may serve as an example for this low mismatch range. For large misfit systems, the assumptions of elastic theory are no longer valid and are expected to result in a breakdown of epitaxial growth (Royer, 1928). Indeed, for systems with a mismatch higher than about 15%, polycrystalline growth is frequently observed. Exceptions for this rule are, however, also found, particularly for purely ionic and metallic deposits.

Many of these exceptions can be explained by a coincidence site lattice (CSL). True coincidence between the epilayer lattice a_{epi} and substrate lattice a_{sub} would occur when $a_{epi}/a_{sub} = m/n$, where m and n are integers. If $m = n \pm 1$, there is one extra lattice plane (i.e., a simple edge dislocation) in each unit cell of the coincidence lattice. In general, however, the epitaxial system is not expected to be at true coincidence, and the coincidence lattice mismatch δ expresses this deviation from true coincidence as

$$\delta = \frac{ma_{sub} - na_{epi}}{ma_{sub}}.$$

This deviation introduces elastic strain at the interface in addition to strain accommodated by misfit dislocations. Therefore, the energy of heteroboundaries is expected to be small if and only if δ does not deviate substantially from true coincidence. In the next section we want to demonstrate the application of the CSL model to the epitaxy of GaN grown on different substrate types.

1. APPLICATION TO NITRIDE EPITAXY

Wurtzite GaN with the highest structural quality is grown on (0001) oriented hexagonal sapphire (α-Al_2O_3) substrates using a very thin AlN buffer layer (Yoshida, Misawa, and Gonda, 1983; Amano et al., 1986) providing nucleation centers which retain the crystallographic information of the substrate and transmit it to the growing GaN layer. Without the buffer layer, hexagonal pyramids form which can be several microns in diameter. Thus, the AlN buffer layer may promote wetting of the sapphire substrate surface and slightly reduces the mismatch between both lattices. The lattice mismatch between AlN and sapphire lattices is given by $f = (a_{AlN} - a_{Al_2O_3})/a_{Al_2O_3} = 0.345$. The actual epitaxial relationship between AlN and sapphire however, is characterized by:

$$(0001)AlN \parallel (0001)Al_2O_3 \quad \text{with} \quad [2\bar{1}\bar{1}0] \parallel [1\bar{1}00]Al_2O_3$$

as shown in Fig. 8. The 30° rotation of both lattices against each other leads to a significant reduction of lattice mismatch to 13.29%, a value which still is not expected to allow epitaxial growth. However, the epitaxial relationship depicted above leads simultaneously to a coincidence site lattice where 8 Al-Al interatomic distances in AlN fit to 9 Al-Al interatomic distances in sapphire (Sun et al., 1994) with a CSL mismatch of only $\delta = 0.70\%$. This decrease in strain is primarily responsible for epitaxial growth of AlN on sapphire. Experimental confirmation of this CSL-assisted epitaxy is found by the regular distance of misfit dislocations of 2.0 nm (Ponce et al., 1994) which is very close to the expected 8/9 lattice plane relation.[4] In addition to this high number of misfit dislocations at the AlN/sapphire interface, there are approximately 10^{10} cm^{-2} threading dislocations in the active nitride layer. In spite of this high defect density, the optical quality of these heterostructures is surprisingly good (Lester et al., 1995). The apparent inertness of defects with regard to the electronic and optical properties of nitrides is not yet understood and is under discussion.

[4]The ideal distance of misfit dislocations is equal to $8 \cdot b = 2.07$ nm with the Burgers vector $\mathbf{b} = 1/3\langle 11\bar{2}0 \rangle$.

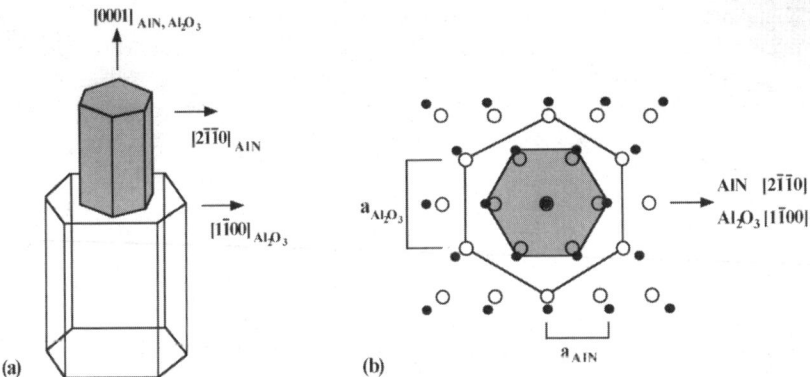

FIG. 8. (a) Shows a schematic illustration of the epitaxial relationship between AlN and Al_2O_3 (0001) representing the 30° rotation of the AlN basal plane and (b) shows the top view.

Epitaxial growth of the metastable zinc-blende structures has proved to be even more difficult than that of its wurtzite counterpart. The results reported so far indicates excessive interface roughness and huge defect densities within the layer which additionally contains wurtzite domains. Recently, GaN with relatively high structural perfection has been grown on (001) GaAs by plasma-assisted MBE (Trampert et al., 1997). This improvement has been made possible by an accurate control of the nucleation stage. The epitaxial quality has been found to be most affected by the crystallinity of the first few atomic layers, and the structural characteristics of the interface.

The cross-sectional HRTEM image of a sample with 0.5 monolayer (ml) of GaN deposited (Fig. 9a)) proves that nucleation occurs by formation of nanoscale three-dimensional islands. All of these islands exhibit a well defined epitaxial relationship. Even the smallest island observed, which has a volume of less than 10 nm^3, is relaxed by misfit dislocations. Specifically, the island shown in Fig. 9(a) contains an edge-type dislocation in the center and a second edge-type dislocation which is about to form at the edge of the island. The atomic structure close to the interface within the GaN island is highly distorted because the extension of lattice plane bending near the dislocation core is of the same order of magnitude as the radius of the observed island. Additionally, elastic relaxation processes at the island surface may influence the lattice image of the interface. Figure 9(b) shows the cross-sectional HRTEM image of a sample with 5 ml of GaN deposited. Strikingly, this sample exhibits a connected film like morphology with a thickness comparable to the height of the initial nuclei (Fig. 9(a)). The initial

7 Crystal Structure of Group III Nitrides

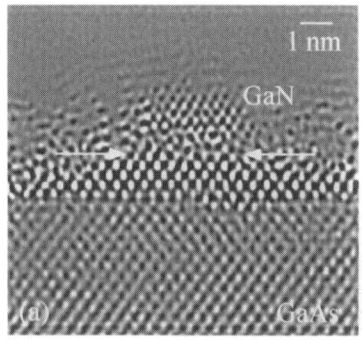 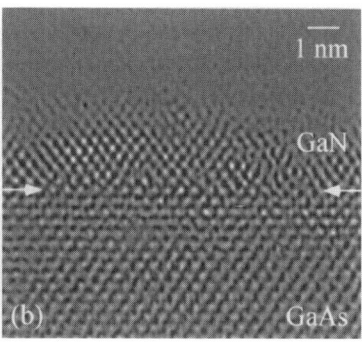

FIG. 9. The nucleation stage of zinc-blende GaN grown on (001) GaAs as observed by cross-sectional HRTEM: GaN after deposition of nominally 0.5 ml (a) and 5 ml (b). The interfaces are marked by arrows (the images are Fourier filtered).

nuclei thus seem to grow almost exclusively laterally, with little growth at this stage along the [001] direction. Moreover, misfit dislocation cores are detected with a distance of 5 {111} GaN lattice planes. The density of these edge-type dislocations is just sufficient to relieve the entire misfit strain of 20%.

These results directly visualize that pure edge-type misfit dislocations are instantaneously formed during the initial nucleation of islands by way of the incorporation of extra {110} lattice planes into the edge of the growing island without any climb or glide mechanism. However, it is clear that the epitaxial cubic interface has to be energetically more favorable than an interface involving the thermodynamically stable hexagonal lattice. Being thus confident that strain energy represents the most important part of total interfacial energy (the chemical portion is assumed to be comparable between cubic and hexagonal phase), these results can be explained by a CSL occurring between cubic GaN and GaAs. For the heterosystem investigated, the CSL mismatch is calculated to be $\delta = -0.0002 \pm 0.0020$ taking $m/n = 4/5$ and the most accurate values for lattice constants available at growth temperature, namely, $a_{GaN} = 0.455 \pm 0.01$ nm and $a_{GaAs} = 0.568886$ nm. As a result, this system is close to true coincidence ("magic mismatch") and an array of edge dislocations with a period of 5 GaN lattice planes will indeed account for the entire misfit. The occurrence of a "magic mismatch" between cubic GaN and GaAs provides an explanation of the phenomenon of epitaxial growth of a metastable phase for a strain at which epitaxy of covalently bonded materials is usually no longer achieved.

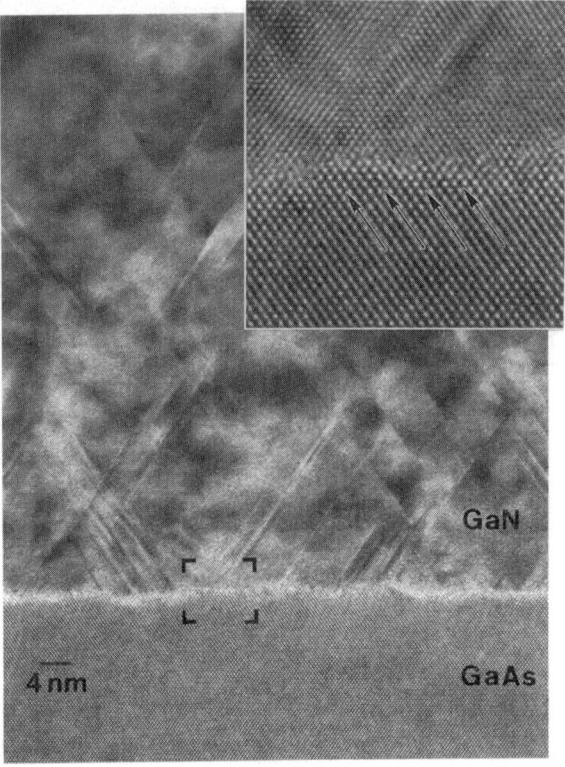

Fig. 10. A cross-sectional HRTEM image of the GaN/GaAs (001) heterostructure taken along a ⟨110⟩ direction. An atomically flat part of the interface containing regular arranged misfit dislocations is magnified in the inset (arrows indicate misfit dislocations).

The subsequent growth and the evolution of the defect microstructure detected in the TEM micrographs (Fig. 10) are determined by the initial nucleation of the islands. No planar defects are observed in the isolated nuclei (Fig. 9a)), but stacking faults and micro-twins are generated during the coalescence stage of the nuclei as observed in HRTEM images of the 5 ml sample. In fact, since the spacing of the individual nuclei will not necessarily be in phase with respect to their dislocation array the periodicity of the dislocation array will, in general, be broken upon their coalescence. These locations are centers of high local strain (a 4:3 ratio between GaN and GaAs lattice planes, for example, corresponds to a residual misfit δ of -6.7%). We thus believe that these local strain concentrations are responsible for the generation of secondary defects in the layers, namely, stacking faults and micro-twins, which are able to fit the coincidence lattice in the region of coalesced islands. The interface imaged in Fig. 10 is abrupt on an atomic scale with an overall vertical roughness of only a few monolayers. Careful inspection of the interface structure (see inset in Fig. 10) reveals that the boundary contains atomically straight parts. Within these regions no planar defects originate from the interface, and, simultaneously, regularly arranged pure edge-type dislocations appear at the interface. As expected, the cores of these dislocations are observed at every fifth GaN {111} lattice plane from the nucleation state described previously (Fig. 9(b)).

Surprisingly, the optical quality of cubic GaN films grown under optimized nucleation conditions does not appear to be affected by the presence of defects, as intense photoluminescence (PL) is detected at room temperature and above (Brandt *et al.*, 1997). Figure 11 shows PL spectra at three different temperatures taken from a 1 μm thick GaN layer. The spectra are dominated by near bandgap emission which exhibits a low intensity tail to lower energies modulated by Fabry–Perot interferences. This emission line is determined to be of exitonic origin up to room temperature (Klann *et al.*, 1995; Menniger *et al.*, 1996). The existence of optical interference indicates the macroscopic homogeneity of the layer and its transparency below the bandgap. Absolute PL efficiency of this layer is quite comparable to that of considerable thicker GaN layers on (0001) sapphire substrates measured side-by-side.

In conclusion, the defects observed in such samples which exhibit a high PL intensity seem to be electrically inactive. It is noteworthy that we do not observe 60° dislocations in these samples. It can be speculated that the defects observed, namely, pure edge-type dislocation at the GaN/GaAs

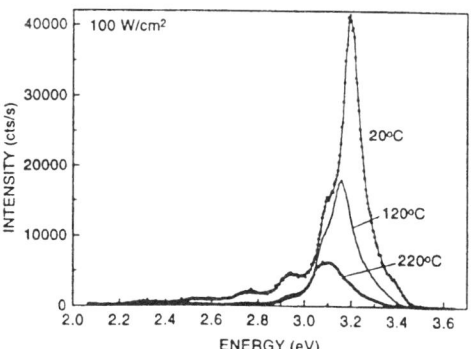

FIG. 11. Photoluminescence spectra at three different temperatures taken from a 1 μm thick cubic GaN layer on GaAs (001). (From Brandt *et al.*, 1997.)

interface and partial dislocations associated with stacking faults, reconstruct at their cores and are then electrically neutral (Petroff, Logan, and Savage, 1980). This speculation also applies to hexagonal GaN despite its different structure and might be a result of the fundamental properties of GaN such as its high ionicity.

2. Outlook

In order to overcome possible negative effects originating from lattice mismatch during heteroepitaxy, substrates are required to have smaller mismatch, therefore suggesting that hexagonal 6H-SiC substrates are more suitable than sapphire. The similarity of the lattices of GaN or AlN, respectively, and SiC, combined with a closer lattice mismatch and a similar thermal expansion parallel to the growing basal plane is expected to give rise to improved structural and optoelectronic properties. However, in spite of the very small lattice mismatch of $f = (a_{AlN} - a_{SiC})/a_{SiC} = 0.010$, the density of structural defects detectable in such nitride layers is on the same order of magnitude as in comparable films grown on sapphire. The dominant defects observed are threading dislocations and planar domain boundaries running perpendicular to the interface (Chien et al., 1996; Davis, Tanaka, and Kern, 1996; Ponce et al., 1995; Smith et al., 1995). Their origins are still under investigation, but defects in the SiC substrate surface and surface steps are held responsible for their occurrence.

Acknowledgments

The authors would like to thank the Bundesministerium für Bildung und Forschung (BMBF) of the Federal Republic of Germany for financial support. Finally, we are indebted to H. Yang for crystal growth and invaluable discussions.

References

Abernathy, C. R., and MacKenzie, J. D. (1995). *Appl. Phys. Lett.* **66**, 1632.
Amano, H., Sawaki, N., Akasaki, I., and Toyoda, Y. (1986). *Appl. Phys. Lett.* **48**, 353.
Azuhata, T., Sota, T., and Suzuki, K. (1996). *J. Phys.: Condens. Mat.* **8**, 3111.
Brandt, O., Yang, H., Jenichen, B., Suzuki, Y., Däweritz, L., and Ploog, K. H. (1996). *Phys. Rev. B* **52**, R2253.
Brandt, O., Yang, H., Müllhäuser, J., Trampert, A., and Ploog, K. H. (1997). *Mat. Sci. Eng.* **343**, 215.

Buseck, P., Cowley, J., and Eyring, L. (1992). *High-Resolution Transmission Electon Microscopy and Associated Techniques*, Oxford University, New York.
Chien, F. R., Ning, X. J., Pirouz, P., Bremser, M. D., and Davis, R. F. (1996). *Appl. Phys. Lett.* **68**, 2678.
Davis, R. F., Tanaka, S., and Kern, R. S. (1996). *J. Cryst. Growth* **163**, 93.
Detchprohm, T., Hiramatsu, K., Itoh, and Akasaki, I. (1992). *Jpn. J. Appl. Phys.* **31**, L1454.
Edgar, J. A. (1992). *J. Mat. Res.* **7**, 235.
Gracia, A., and Cohen, M. (1993). *Phys. Rev. B* **47**, 4221.
Harrison, W. A. (1988). *Electronic Structure and the Properties of Solids*, Dover, New York.
Hiramatsu, K., Detchprohm, T., and Akasaki, I. (1993). *Jpn. J. Appl. Phys.* **32**, 1528.
Hold, D. B. (1964). *J. Phys. Chem. Solids* **25**, 1385.
Hull, D., and Bacon, D. J. (1984). *Introduction to Dislocations*, Pergamon, Oxford.
Huseby, I. (1983). *J. Am. Ceram. Soc.* **66**, 217.
Itoh, N., Rhee, J. C., Kawabata, T., and Koike, S. (1985). *J. Appl. Phys.* **58**, 1828.
Jenkins, D. W., Hong, R.-D., and Dow, J. D. (1987). *Superlat. Microstruct.* **3**, 365.
Kim, K., Lambrecht, W. R. L., and Segall, B. (1996). *Phys. Rev. B* **53**, 16310.
Klann, R., Yang, H., Grahn, H. T., Ploog, K. H., Trampert, A. (1995). *Phys. Rev. B* **52**, R11615.
Kuwano, N., Nagatomo, Y., Kobayashi, K., Oki, K., Miyoshi, S., Yaguchi, H., Onabe, K., and Shiraki, Y. (1994). *Jpn. J. Appl. Phys.* **33**, 18.
Lagerstedt, O., and Monemar, B. (1979). *Phys. Rev. B* **19**, 3064.
Lambrecht, W. R. L., and Segall, B. (1991). *Phys. Rev. B* **43**, 7070.
Lawaetz, P. (1972). *Phys. Rev. B* **5**, 4039.
Lei, T., Fanciulli, M., Molnar, R., Moustakas, T. D., Graham, R., and Scanlon, J. (1991). *Appl. Phys. Lett.* **59**, 944.
Lei, T., Moustakas, T. D., Graham, R. J., He, Y., and Berkowitz, S. J. (1992). *J. Appl. Phys.* **71**, 4933.
Lester, S., Ponce, F., Craford, M., and Steigerwald, D. (1995). *Appl. Phys. Lett.* **66**, 1249.
Leszczynski, M., Suski, T., Teisseyre, H., Perlin, P., Grzegory, Jr., I., Jun, J., Porowski, S., and Moustakas, T. (1994). *J. Appl. Phys.* **76**, 4909.
Lin, W.-T., Meng, L.-C., Chen, G.-J., and Liu, H.-S. (1995a). *Appl. Phys. Lett.* **66**, 2066.
Lin, X. W., Behar, M., Maltez, R., Swider, W., Lilental-Weber, Z., and Washburn, J. (1995b). *Appl. Phys. Lett.* **67**, 2699.
Madelung, O. (ed.) (1982). Landolt-Börnstein, Bd 17a, *Halbleiter*, Springer Verlag, Berlin.
Maruska, H., and Tietjen, J. (1969). *Appl. Phys. Lett.* **15**, 327.
Matthews, J. W. (1975). "Coherent Interfaces and Misfit Dislocations" In *Epitaxial Growth*, Part B (ed. J. W. Matthews), Academic New York, p. 529.
Meller, A. (1988). *Gmelin Handbuch der Anorganischen Chemie, Boron Compounds*, 3rd Suppl., Vol. 3, Springer Verlag, Berlin, p. 1–91.
Menniger, J., Jahn, U., Brandt, O., Yang, H., and Ploog, K. H. (1996). *Phys. Rev. B* **53**, 1881.
Mizuta, M., Fujieda, S., Matsumoto, Y., and Kawamura, T. (1986). *Jpn. J. Appl. Phys.* **25**, L945.
Paisley, M., Sitar, Z., Posthill, J., and Davis, R. (1989). *J. Vac. Sci. Technol. A* **7**, 701.
Pankove, J. I., Berkeyheiser, J. E., and Maruska, H. P. (1970). *Appl. Phys. Lett.* **22**, 303.
Perlin, P., Jauberthie-Carillon, C., Itie, J. P., San Miguel, A., Grzegory, I., and Polian, A. (1992). *Phys. Rev. B* **45**, 83.
Petroff, P. M., Logan, R. A., and Savage, A. (1980). *Phys. Rev. Lett.* **44**, 287.
Petrov, I., Mojab, E., Powell, R., Greene, J., Hultman, L., and Sundgren, J. (1992). *Appl. Phys. Lett.* **60**, 2491.
Phillipp, F., Höschen, R., Osaki, M., Möbus, G., and Rühle, M. (1994). Ultramicroscopy **56**, 1.
Phillips, J. (1973). *Bonds and Bands in Semiconductors*, Academic, New York.

Polian, A., Grimsditch, M., and Grzegory, I. (1996). *J. Appl. Phys.* **79**, 3343.
Ponce, F. A., Major, Jr., J. S., Plano, W., and Welch, D. (1994). *Appl. Phys. Lett.* **65**, 2302.
Ponce, F. A., Krusor, B. S., Major, Jr., J. S., Plano, W. E., and Welch, D. F. (1995). *Appl. Phys. Lett.* **67**, 410.
Ponce, F. A., Bour, D. P., Young, W. T., Saunders, M., and Steeds, J. W. (1996). *Appl. Phys. Lett.* **69**, 337.
Powell, R., Lee, N., Kim, Y., and Greene, J. (1993). *J. Appl. Phys.* **73**, 189.
Ramırez-Flores, G., Navarro-Contreras, H., Lastras-Martínez, A., Powell, R. C., and Greene, J. E. (1994). *Phys. Rev. B* **50**, 8433.
Royer, L. (1928). *Bull. Soc. Fr. Mineral. Crist.* **51**, 7.
Schulz, H., and Thiemann, K. (1977). *J. Am. Ceram. Soc.* **23**, 815.
Sichel, E., and Pankove, J. (1977). *J. Phys. Chem. Solids* **38**, 330.
Slack, G. (1973). *J. Phys. Chem. Solids* **34**, 321.
Slack, G., Tanzilli, R., Pohl, R., and Vandersande, J. (1987). *J. Phys. Chem. Solids* **48**, 641.
Smith, D. J., Chandrasekhar, D., Sverdlov, B., Botchkarev, A., Salvador, A., and Morkoç, H. (1995). *Appl. Phys. Lett.* **67**, 1830.
Spence, J. C. H. (1989). *Experimental High-Resolution Transmission Electron Microscopy*, Clarendon Press, Oxford.
Stadelmann, P. (1987). Ultramicroscopy **21**, 131.
Strite, S., Ruan, J., Li, Z., Salvador, A., Chen, H., Smith, D., Choyke, W., and Morkoç, H. (1991). *J. Vac. Sci. Technol. B* **9**, 1924.
Strite, S., Chandrasekhar, D., Smith, D. J., Sariel, J., Chen, H., Teraguchi, N., Moroç, H. (1993). *J. Cryst. Growth* **127**, 294.
Sun, C., Kung, P., Saxler, A., Ohsato, H., Haritos, K., and Razeghi, M. (1994). *J. Appl. Phys.* **75**, 3964.
Suzuki, K., Ichihara, M., and Takeuchi, S. (1994). *Jpn. J. Appl. Phys.* **33**, 1114.
Taftø, J., and Spence, J. (1982). *Acta Cryst.* **15**, 60.
Tansley, T. L., and Foley, C. P. (1986). *J. Appl. Phys.* **59**, 3241.
Trainor, J., and Rose, K. (1974). *J. Elec. Mat.* **3**, 821.
Trampert, A., Brandt, O., Yang, H., and Ploog, K. H. (1996). *Appl. Phys. Lett.* **70**, 583.
Ueno, M., Yoshida, M., Onodera, A., Shimommura, O., and Takemura, K. (1994). *P4hys. Rev. B* **49**, 14.
van der Merwe, J. H., and Ball, C. A. B. (1975). "Energy of Interfaces between Crystals" In *Epitaxial Growth*, Part B (ed. J. W. Matthews), Academic, New York, p. 493.
van Vechten, J. (1973). *Phys. Rev. B* **7**, 1479.
Wakahara, A., and Yoshida, A. (1989). *Appl. Phys. Lett.* **54**, 2984.
Warekois, E., Lavine, M., Mariano, A., and Gatos, H. (1962). *J. Appl. Phys.* **33**, 690.
Wright, A., and Nelson, J. (1994). *Phys. Rev. B* **50**, 2159.
Wright, A., and Nelson, J. (1995). *Phys. Rev. B* **51**, 7866.
Xia, Q., Xia, H., Ruoff, A. L. (1993). *J. Appl. Phys.* **73**, 8198.
Yang, H., Brandt, O., and Ploog, K. H. (1996). *Phys. Stat. Sol.* (b) **194**, 109.
Yang, H., Brandt, O., Trampert, A., and Ploog, K. H. (1996). *Appl. Surf. Sci.* **104/105**, 461.
Yeh, C., Lu, Z., Froyen, S., and Zunger, A. (1992). *Phys. Rev. B* **46**, 10086.
Yim, W., and Paff, R. (1974). *J. Appl. Phys.* **45**, 1456.
Yim, W., Stofko, E., Zanzucchi, P., Pankove, J., Ettenberg, M., and Gilbert, S. (1973). *J. Appl. Phys.* **44**, 292.
Yohshida, S., Misawa, S., and Gonda, S. (1983). *Appl. Phys. Lett.* **42**, 427.
Zetterstrom, R. B. (1970). *J. Mat. Sci.* **5**, 1102.

CHAPTER 8

Electronic and Optical Properties of III–V Nitride based Quantum Wells and Superlattices

H. Morkoç, F. Hamdani, and A. Salvador

UNIVERSITY OF ILLINOIS AT URBANA-CHAMPAIGN
MATERIALS RESEARCH LABORATORY AND COORDINATED SCIENCES LABORATORY
URBANA, IL

I. INTRODUCTION	193
II. OPTICAL TRANSITIONS IN BULK GaN	195
III. CALCULATION OF CONFINED STATES	199
IV. EXPERIMENTAL RESULTS	214
1. Band Discontinuity Determination	215
2. Optical Properties of Quantum Well Structures	221
V. AlGaN/GaN QUASI TRIANGULAR QUANTUM WELLS	230
1. Theoretical Method	231
2. 2DEG Concentration as a Function of ΔE_{Fi} and Calculation Δd	234
3. Band Diagrams for Normally on and Quasi Normally Off Modulation Doped Field Effect Transistors	237
4. Modulation Doped Field Effect Transistors Utilizing Quasi Triangular Wells	239
VI. InGaN/GaN QUANTUM WELLS	241
1. Light Emitting Diodes	246
VII. InGaN/InGaN QUANTUM WELLS	248
1. Use of $In_xGa_{1-x}N/In_yGa_{1-y}N$ Quantum Wells in Laser Diodes	250
VIII. CONCLUSIONS	253
References	254

I. Introduction

Wide bandgap III–V nitrides such as GaN, AlN, InN and their alloys exhibit considerable hardness, high thermal conductivity, large bandgap energies, and both conduction and valence band offsets for carrier confinement. Consequently, III–V nitrides are attractive for potential optoelectronic devices applications in the blue-green or near ultraviolet spectrum, and for high power and high temperature microelectronic devices. Recent advances in the growth technology of these semiconductors by metalorganic chemical vapor deposition (MOCVD) method and modified molecular beam epitaxy (MBE), such as reactive ammonia MBE, electron cyclotron resonance (ECR), and radio frequency (RF) assisted MBE methods, have made possible the demonstration of violet and ultraviolet single quantum well (SQW) based light emitting diodes (LEDs), multiple quantum wells (MQWs) based laser diodes (LDs), and modulation doped field effect transistors (MODFET). For a review of semiconductor properties and advances made on the materials as well as device front in this class of materials, see Mohammad, Salvador, and Morkoç (1995); Mohammad and Morkoç (in press); Morkoç and Mohammad (1995); Edgar (1994); Akasaki and Amano (1994); Strite, Lin, and Morkoç (1993); Strite and Morkoç (1992); Morkoç (in press); Morkoç et al. (1994); Davis (1991).

Since the pioneering work of Esaki and Tsu (1970) considerable effort has been directed toward the growth of semiconductors heterostructures such as quantum wells (QWs) and superlattices (SLs). Early demonstration of the feasibility of such systems has been focused on the lattice matched materials such as GaAs/GaAlAs, until Osbourn (1985) pointed out that lattice mismatched materials such as CdTe/ZnTe, Si/Ge can be grown in heterostructures systems pseudomorphically by means of elastic strain accommodation which prevents formation of dislocations in the layers grown up to a certain critical thickness (t_c). It has been demonstrated that t_c is inversely proportional to lattice mismatch between the two compounds forming the heterostructure. Quantum confinement occurs when the Debye length in the semiconductor become comparable to the size of the well layer. Large electron effective masses in GaN and related materials require the well thickness in a few tens of angstrom. Although few of the heterostructures reported so far really fit this criterion, the term QW will be interpreted liberally here and heterostructures approaching this criterion will be treated. We must also mention that quality of the QW like structures grown so far does not sufficiently resolve the confined states well. Additional complications occur because of the lack of native substrates and small critical thickness in the nitride system. Nevertheless, great strides have been made in a considerably short period of time, compelling us to begin investigation of such structures.

As mentioned previously, the mismatch between GaN and the commonly used substrate Al_2O_3, is 14%. The critical thickness (Matthews and Blakeslee, 1976; Dodson and Tsao, 1987a,b, 1988; People and Bean, 1985) is nearly zero which indicates that dislocations are introduced during the first stages of growth. Transmission electron microscopy (TEM) investigation on different GaN grown samples indicates that the density of dislocations generated in layers is in the range 10^9 to 10^{10} cm^{-2}. It is worth noting that

III–V nitride compounds exhibit some features which are: (i) the binary compound can be achieved in both hexagonal and cubic phase depending on the substrate used and growth conditions, (ii) the band discontinuities are much larger than those exhibited by any other III–V or II–VI wide bandgap materials (this leads to higher carrier localization), and (iii) the small spin–orbit interaction leads to mixing of valence bands and a high effective mass for the holes, which is of benefit in the sense that matrix elements for radiative recombination are large leading to high modal gain and eventually power, and not so beneficial in the sense that the transparency current is increased.

In this chapter we present an overview of the work devoted to III–V wide bandgap energy nitrides based pseudo QWs. Section II presents various theoretical approaches for calculating band structure modification in QWs with or without the strain effect. In Section III we discuss experimental results available in the literature concerning GaN/GaAlN, GaN/InGaN, InGaN/InGaN, and AlN/GaN heterostructures. Section IV presents a resume of the performance of low dimensional based devices such as MODFET, LDs, and LEDs, and in Section V we discuss the prospects and directions of future development in that field.

II. Optical Transitions in Bulk GaN

Before lunging into the confinement issues, we shall briefly discuss the unique nature of the band structure of wurtzitic GaN at the Γ point. The bottom of the conduction band of GaN is formed predominantly from the s levels of Ga, and upper valence band states are formed from the p levels of N (Dingle *et al.*, 1971; Reynolds *et al.*, 1996). Upper valence band states are constructed out of appropriate linear combinations of products of p_x, p_y, p_z like orbitals with spin functions. In the absence of both spin–orbit and crystal field effects, these states are degenerate. The crystal field of the wurtzite lattice partially removes the degeneracy, separating the p_z orbital from the p_x and p_y orbitals. The p_x and p_y bands are further split by spin–orbit coupling. If one assumes an approximately spherical potential in the neighborhood of the N atoms, the higher energy of two spin states is the one where electron spin and orbital angular momentum are parallel. This result is anticipated also from the atomic spin–orbit splitting, where the $P_{3/2}$ state is known to have a greater energy than the $P_{1/2}$ state.

The contributions of spin–orbit interaction and crystal field perturbation to experimentally observed splittings, $E_{1,2}$ and $E_{2,3}$, have been calculated in various linear combinations of atomic orbitals (LCAO) approximations by several investigators (Kawabe *et al.*, 1967; Balkanski and de Cloizeaux, 1960). In treatments where wurtzite energy levels are treated as a perturbation of those in the zinc-blende structure, Hopfield and Thomas (1963) have derived the relations:

$$E_1 = 0 \tag{1a}$$

$$E_2 = \frac{\delta + \Delta}{2} + \sqrt{\left[\left\{\frac{\delta + \Delta}{2}\right\}^2 - \frac{2}{3}\delta\Delta\right]} \tag{1b}$$

$$E_3 = \frac{\delta + \Delta}{2} - \sqrt{\left[\left\{\frac{\delta + \Delta}{2}\right\}^2 - \frac{2}{3}\delta\Delta\right]} \tag{1c}$$

where Δ and δ represent the contributions of uniaxial field and spin–orbit interaction, respectively, to the splittings $E_{1,2}$ and $E_{2,3}$. The optical transition $\Gamma_7 \to \Gamma_9$ is allowed for $E \perp c$, where E is the electric field vector and c is the c axis of the crystal. Optical transitions $\Gamma_7 \to \Gamma_7$ are allowed for both $E \perp c$ and $E \parallel c$. The transitions mentioned previously have been observed in high quality GaN samples as shown in Fig. 1.

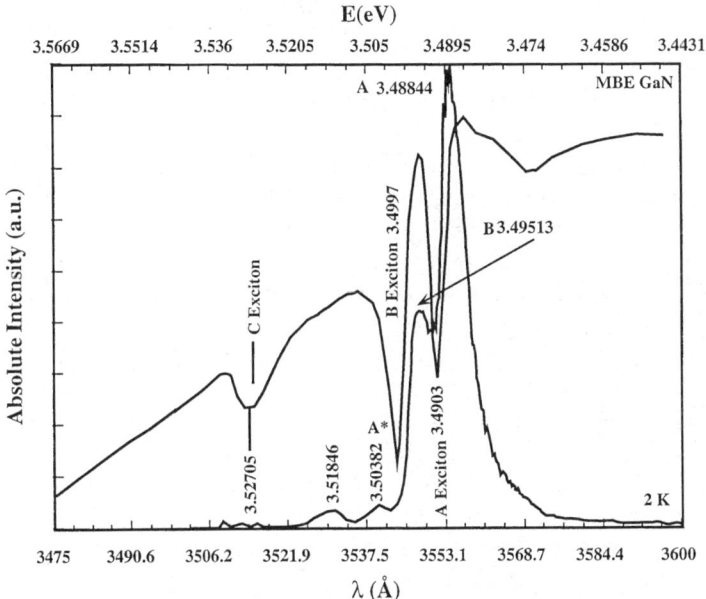

FIG. 1. Photoluminescence and reflectance spectra of a high quality GaN film prepared by MBE. (The scan was obtained by D. C. Reynolds.)

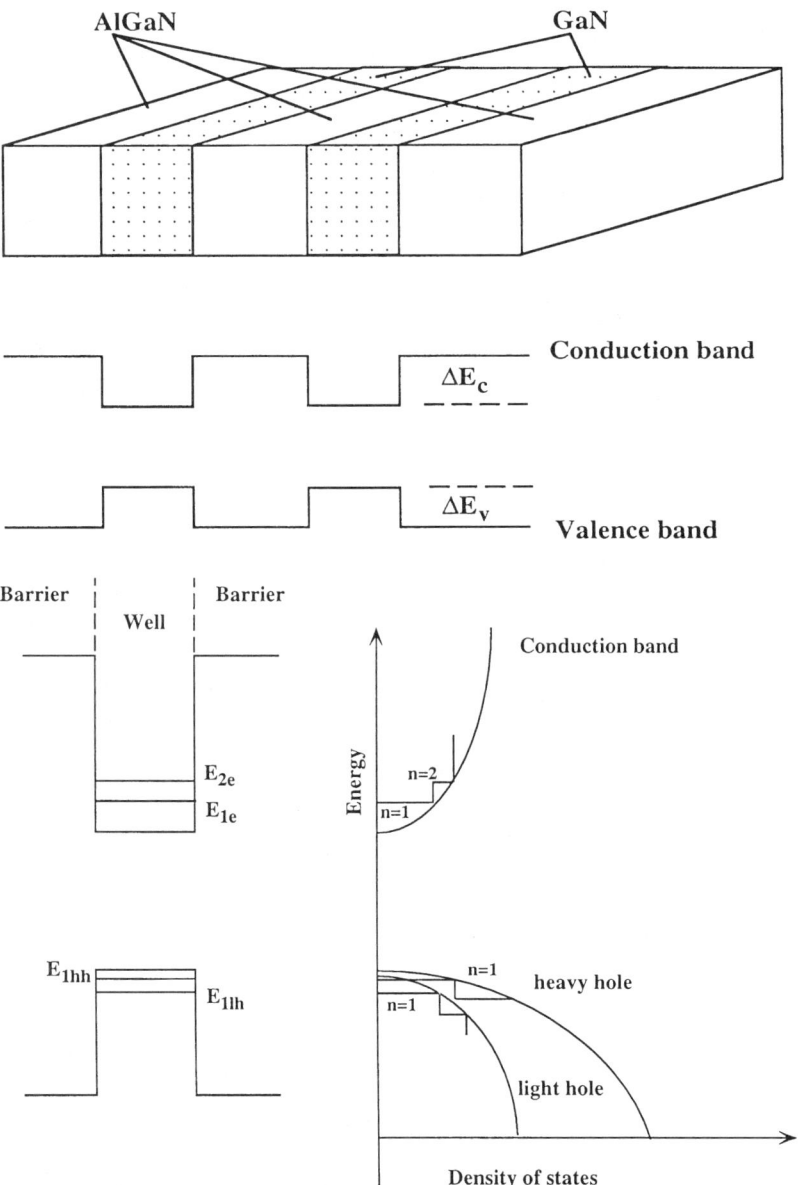

FIG. 2. Schematic representation of (a) a GaN/AlGaN MQW structure and the conduction and valence band edges at the Γ point, (b) valence and conduction band diagrams of the same with confined states indicated.

When the well size is comparable to the Debye length in the semiconductor forming the well, conduction and valence bands are modified so that the density states in both conduction and valence bands are discretized. This is depicted schematically in Fig. 2(a) and (b) for a wurtzitic semiconductor QW. Moreover, equally important is the modification to these energies in the semiconductor caused by strain effects. In short, eigenstates of strained layer SLs require consideration of both strain and quantum size effects. Eigenstates applicable to early versions of compound semiconductor based QWs without strain were modeled by envelope function formalism of Bastard (1982). If this method is to be used, first we have to calculate the strain dependent bandgap. Bastard's formalism can then be used to determine the transition energies in QWs which must be added to the strain component as shown by Marzin (1982) who developed a method for calculating the bandgap of a strained cubic semiconductor that when used

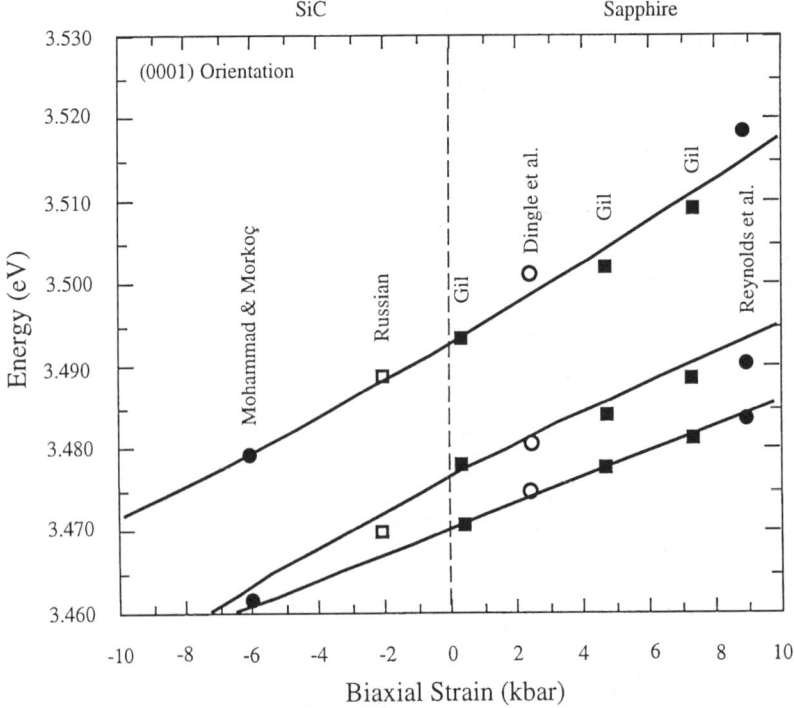

FIG. 3. Position of A, B, and C excitons in bulk GaN with respect to strain with experimental results corresponding to differing degrees of strain indicated. (From Gil, Briot, and Aulombard, 1996.)

in conjunction with envelope function approximation leads to eingenstates in quantized structures in strained systems.

In Marzin's revision of Bastard's method, an 8 × 8 Kane hamiltonian matrix (Pollak and Cardona, 1968) is used to give an accurate description of a strained quantum structure. The size of the matrix is justified because in addition to the conduction band there are three valence bands with spin up and spin down for each band. Similar to cubic semiconductors, wurtzitic phase also has one conduction band, three valence bands, heavy and light hole states, and a spin–orbit band, each one having spin up and spin down, leading to the necessity of using a 8 × 8 hamiltonian matrix. The strain effects in the GaN system has been treated by Gil, Briot and Aulombard (1995); Gil, Hamdani and Morkoç (1996), the results of which are shown in Fig. 3.

III. Calculation of Confined States

For simplicity, we now discuss briefly how to go about calculating the confined states and give examples of confined states in GaN/AlGaN, InGaN/GaN, and InGaN/InGaN QW of popular compositions and dimensions.

The ability to grow alternating thin layers of slightly dissimilar semiconductors, with each individual layer maintaining it's crystallinity, results in a modulated conduction band profile along the growth direction. This, in turn, will lead to confinement of the lowest level conduction electron (holes) within the well region. Depending on how the conduction bands (and valence bands) between low and high bandgap material align, well regions for the conduction electron and hole can either occur in the same layer (type I) or in alternating layers (type II) (Fig. 4). Band alignment is critical in determining whether the particular QW systems can be efficiently used for optical emitters. In general, intersubband optical transition is proportional to the overlap between the electron and hole wavefunction. Thus, stronger photoluminescence (PL) emission intensity is expected from structures yielding type I QWs, because of a larger overlap between the electron and hole wavefunction. A wide variety of experiments have been performed to accurately determine the band offset between GaN and AlN as well as GaN and InN. Recent X-ray photoelectron spectroscopy (XPS) measurements indicate that the conduction band to valence band offsets between GaN and AlN is 67:33 and that GaN/Al(Ga)N QWs are type I (Martin et al., 1994).

The conduction band minimum for GaN and AlN is at the zone center and two fold degenerate, and the confinement energies for GaN/AlGaN

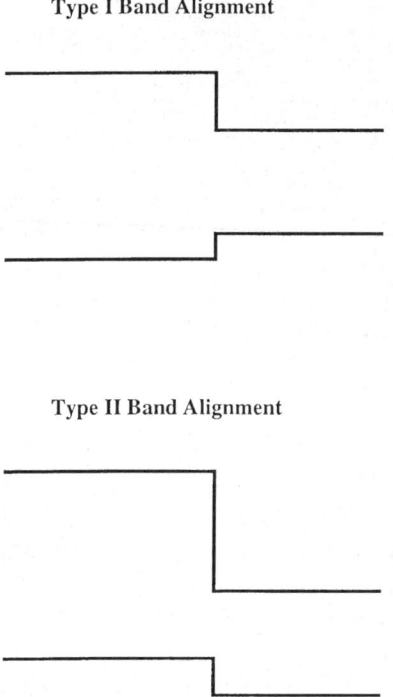

FIG. 4. Schematic description of type I and type II band alignments.

QWs can be reasonably estimated using envelope function approximation in the same manner as that extensively used for the GaAs/AlGaAs material system. Following Weisbuch's notation, the low lying conduction electron state can be represented by (Weisbuch and Vinter, 1991; Bastard and Brum, 1986)

$$\Psi(r) = \sum_{j=A,B} e^{ik_\perp \cdot r} u_c^j(r) f_n(z) \qquad (2)$$

where $u_c^j(r)$ is the conduction band zone center Bloch wavefunction of GaN(AlGaN) and $f_n(z)$ is a slowly varying envelope function, $k_\perp$ is the in-plane wave vector with the growth direction along the z axis. Since $u_c^j(r)$ is the same for GaN and Al(Ga)N, the Schrödinger's equation reduces to

$$\left[\left(\frac{-\hbar^2}{2m^*(z)} \cdot \frac{d^2}{d^2 z} + V(z)\right)\right] f_n(z) = E_n f_n(z) \qquad (3)$$

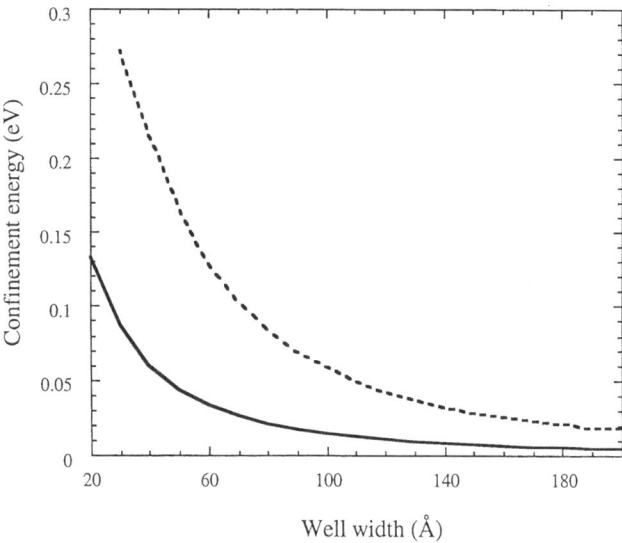

FIG. 5. Plot of the conduction band groundstate (solid line) and excited state (dashed line) confinement energies for a $GaN/Al_{0.15}Ga_{0.85}N$ QW as a function of well width.

In the above equation, $m^*(z)$ is the corresponding effective mass of the conduction electron, $V(z)$ represents the profile of the minimum of the conduction band along growth direction, and $E_n(z)$ are the so-called confinement energies. Assuming no doping in either regions, $V(z)$ has a rectangular well like profile and solutions for confinement energies is similar to a particle in a box problem. The boundary conditions are that $f_n(z)$ and $(1/(m^*(z))(df_n(z)/dz)$ are continuous at the interface. The latter is necessary to ensure conservation of particle current. In Fig. 5, plots of the conduction band groundstate (solid line) and excited state (dashed line) confinement energies for a $GaN/Al_{0.15}Ga_{0.85}N$ QW as a function of well width are displayed. GaN/AlGaN QWs will figure prominently in its use as the active layer in GaN based injection mode lasers. In generating the plot, a linear interpolation was used, with the bandgap energies of AlN and GaN being 6.2 eV and 3.42 eV, respectively, to calculate the AlGaN bandgap. The conduction electron effective mass, $0.19m_e$, was assumed to be the same in GaN and AlGaN regions because of the uncertainty in the effective mass for AlN. In Fig. 6, plots of the conduction band groundstate (solid line) and excited state (dashed line) confinement energies for a $GaN/Al_{0.15}Ga_{0.85}N$ QW as a function of well width are displayed.

By using the reported bandgap versus composition in InGaN (see Section VI), we have also calculated the confined states in $In_{0.20}Ga_{0.80}N/$

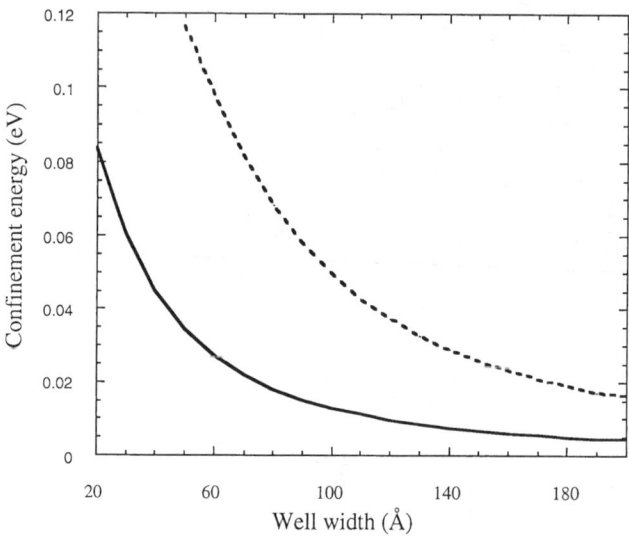

FIG. 6. Plot of the conduction band groundstate (solid line) and excited state (dashed line) confinement energies for a GaN/Al$_{0.07}$Ga$_{0.93}$N QW as a function of well width.

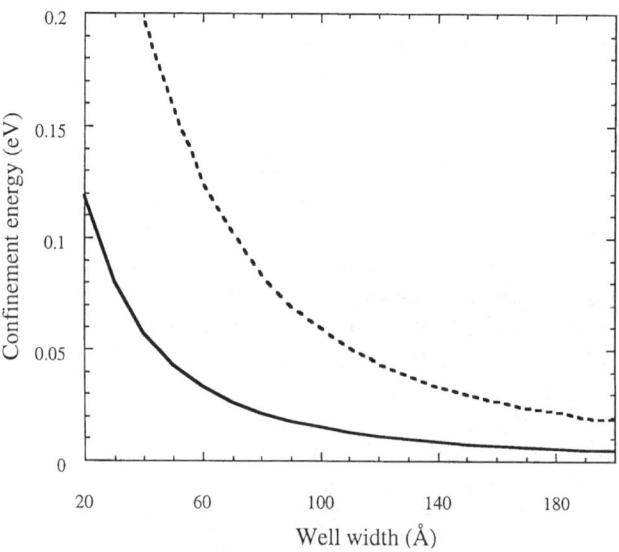

FIG. 7. Plot of the conduction band groundstate (solid line) and excited state (dashed line) confinement energies for a In$_{0.20}$Ga$_{0.80}$N/In$_{0.05}$Ga$_{0.95}$N QW as a function of well width.

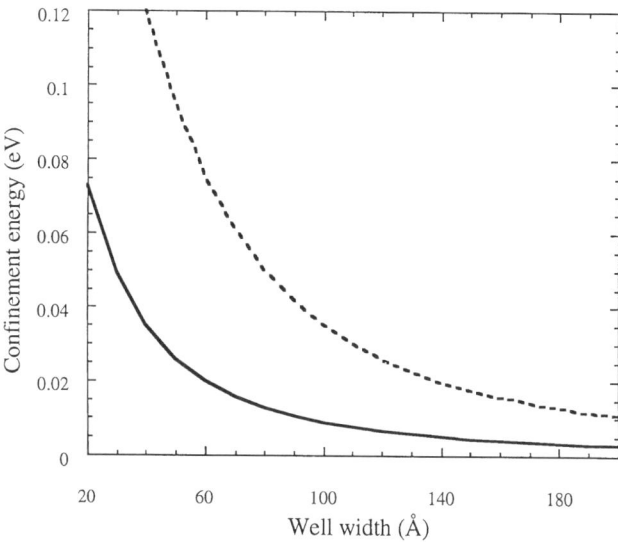

FIG. 8. Plot of the valence band heavy hole groundstate (solid line) and excited state (dashed line) confinement energies for a GaN/Al$_{0.15}$Ga$_{0.85}$N QW as a function of well width assuming a hole relative effective mass of 0.3.

In$_{0.05}$Ga$_{0.95}$N QWs. Figure 7 shows the plots of conduction band groundstate (solid sline) and excited state (dashed line) confinement energies for a In$_{0.20}$Ga$_{0.80}$N/In$_{0.05}$Ga$_{0.95}$N QW as a function of well width.

The calculation of the confined states for the holes in GaN/AlGaN systems is more complex due to the proximity of the bands near the zone center. For wurtzitic GaN, the energy separation between the so-called A, B, and C valence bands at the zone center is only 8.5 and 25 meV, respectively (Mohammad, Salvador, and Morkoç, 1995). In general, for the case of $k_\perp \neq 0$, case band mixing between these three bands are likely to occur and has to be included in the calculation (Sanders and Chang, 1985; Masselink, Chang, and Morkoç, 1983). However, for $k_\perp = 0$ and the lowest confined states, energy level can be calculated by using equations similar to that described for the conduction electron case. It is also worth noting that there is still controversy regarding the hole effective mass for GaN, with reports of values ranging from 1.0–0.4m_e (Khan et al., 1990; Orton, 1995). With improved GaN film quality, it is expected that a more accurate determination of hole effective mass can be made. Recently, low temperature PL spectroscopy of high quality undoped GaN film, wherein all the observed emission peaks were due solely to groundstate and excited

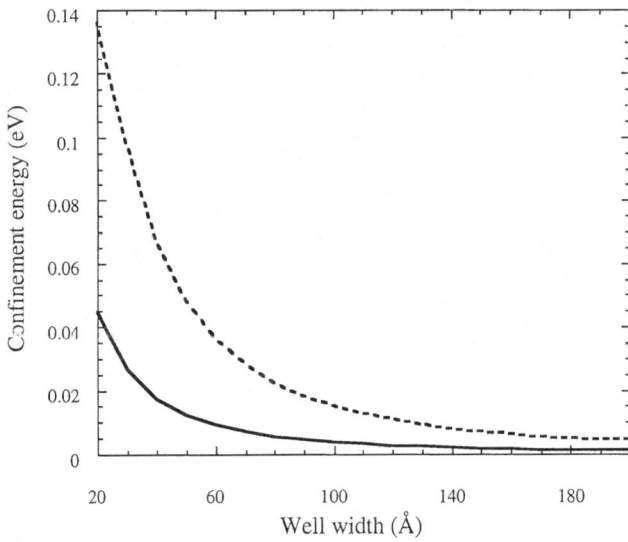

FIG. 9. Plot of the valence band heavy hole groundstate (solid line) and excited state (dashed line) confinement energies for a GaN/Al$_{0.15}$Ga$_{0.85}$N QW as a function of well width assuming a hole relative effective mass of 0.8.

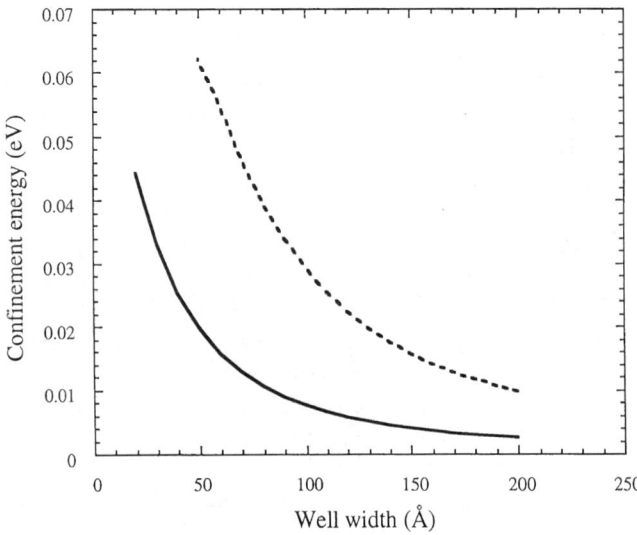

FIG. 10. Plot of the valence band heavy hole groundstate (solid line) and excited state (dashed line) confinement energies for a GaN/Al$_{0.07}$Ga$_{0.93}$N QW as a function of well width assuming a hole relative effective mass of 0.3.

excitonic states, reveal that the A exciton binding energy to be 20 meV (Mohammad, Salvador, and Morkoç, 1995). This implies a reduced mass of $0.129m_e$ for the A exciton and a hole effective mass of $0.4m_e$. In Fig. 8 a plot of the hole confined states, from the A valence band, GaN/AlGaN QW is shown as a function of well width. Figure 8 also shows the plots of valence band heavy hole groundstate (solid line) and excited state (dashed line) confinement energies for a GaN/$Al_{0.15}Ga_{0.75}$N QW as a function of well width assuming a hole relative effective mass of 0.3. Figure 9 shows the same as Fig. 8 but with a heavy hole relative effective mass of 0.8.

Figure 10 shows the plots of valence band heavy hole groundstate (solid line) and excited state (dashed line) confinement energies for a GaN/$Al_{0.07}Ga_{0.93}$N QW as a function of well width assuming a hole relative effective mass of 0.3. Figure 11 shows the same as Fig. 10 but with a heavy hole relative effective mass of 0.8. Using the composition dependent bandgap of InGaN we generated the confined states in that material. Figure 12 shows the plots of valence band heavy hole groundstate (solid line) and excited state (dashed line) confinement energies for a $In_{0.20}Ga_{0.80}$N/$In_{0.05}Ga_{0.95}$N QW as a function of well width.

An indirect method of confirming hole effective mass is in comparing the sum of calculated confinement energies for both electron and hole, with the

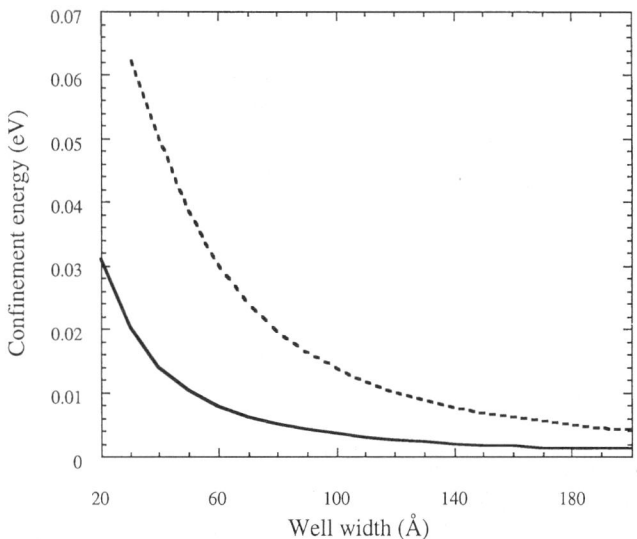

FIG. 11. Plot of the valence band heavy hole groundstate (solid line) and excited state (dashed line) confinement energies for a GaN/$Al_{0.07}Ga_{0.93}$N QW as a function of well width assuming a hole relative effective mass of 0.8.

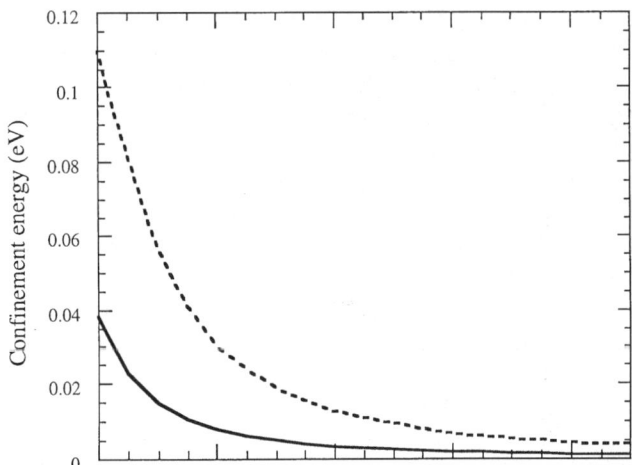

FIG. 12. Plot of the valence band heavy hole groundstate (solid line) and excited state (dashed line) confinement energies for a $In_{0.20}Ga_{0.80}N/In_{0.05}Ga_{0.95}N$ QW as a function of well width.

energy position of PL spectra of GaN/AlGaN QW. Strictly speaking, residual strain effects as well as the contribution of exciton binding energy has to be considered and calculated band edge transition is given by

$$E = E_g + E_{1c} + E_{1h} - E_{xb} + E_s \qquad (4)$$

where E_g is the bandgap of bulk GaN. E_{1c}, E_{1h} are groundstate confinement energies of the conduction electron and hole, respectively, E_{xb} is the exciton binding energy while E_s is the contribution of the strain. Strain effects due to lattice mismatch between the AlGaN barrier and GaN well region lead to an increase in bandgap energy of the well material provided that presence of a large concentration of defects does not prevent formation of a coherent strain. Exciton binding energy, on the other hand is dependent on the well width. However, it is expected that this variation becomes important only in thin wells since the calculated exciton Bohr radius in bulk GaN is 36 Å. Initial attempts on the optical properties of GaN/AlGaN QWs indicated that effective mass of the hole is still in the 1 to $0.8m_e$ range. These studies were hampered by uncertainty in band offsets as well as lack of high quality samples. A recent study on the PL spectrum of a silicon (Si) doped GaN/AlGaN QW embedded in a SCH laser structure show that a good fit to the PL spectra can be obtained by assuming 0.19 and $0.3m_e$ effective masses for the conduction electron and hole, respectively (Salvador et al.,

1995). We should point out that this is an indirect means of determining effective masses and relies on accurate knowledge of many other parameters.

As was mentioned previously, wide energy bandgap III–V nitrides can exist in different crystal polytypes depending on growth conditions and substrates used. Nevertheless, calculations specific to wurtzitic GaN and related structures have recently begun to emerge. It has recently been demonstrated theoretically and experimentally that the wurtzite structure represent the most stable phases of binary nitride compounds GaN, AlN, and InN. A number of approaches, such as envelope function formalism for valence bands in wurtzitic QWs (Sirenko et al., 1996), k.p. model for calculating the valence band structure for wurtzitic GaN including strain (Chuang and Chang, 1996a,b), first principle calculation of effective mass parameters using full potential linearized augmented plane wave method (Suzuki and Uenoyhama, 1996a,b). In short, methods such as *ab initio*, tight-binding, muffin-in, linear combination of atomic orbitals (LCAO), and linearized augmented plane wave methods, have been used to calculate energy bands for both wurtzite and zinc-blende GaN, InN, and AlN bulk materials. It should be noted that all these binary materials including alloy compounds obtained by combination of these binaries are wide direct bandgap semiconductors in both crystal phases, except zinc-blende AlN, which is expected to have an indirect bandgap with conduction band minimum being at X valley. Due to a lack of available experimental data, many details of these studies must be improved in order to provide an accurate band description. Recently, many theoretical investigations have been devoted to the study of band structure in GaN/GaAlN QWs with impact on the performance of optoelectronic and electronic devices based on these QWs.

We again emphasize that there exist many discrepancies concerning the determination of some important parameters for the relevant bulk material such as hole masses, bandgap energy bowing parameters, shear and deformation potentials, and band offsets, which are necessary for determining and understanding laser performance such as gain and transparency, as well as the performance of electronic devices. Calculations, using *ab initio* methods, have been used to estimate unknown but necessary parameters for band structure calculations. Sirenko et al. (1996) have performed envelope function calculations of the valence band in wurtzite QWs following the forming of Rashda-Sheka-Pi kus (RSP), developed early in bulk wurtzite semiconductors. The band structure calculation in bulk materials is presented in Chapter 11. We discuss in this section only the calculation results obtained for the nitride based heterostructures.

Due to the lack of accuracy in important physical parameters, such as effective masses, Luttinger parameters, spin–orbit, and crystal field parameters, several authors have calculated the band structure modifications of GaN heterostructures using variants of *ab initio* calculations, in order to estimate the gain and transparency of GaN based lasers. Since the devices are covered in Chapters 9 and 10, we shall present in this section only modifications to band structures and their influence on electronic transitions. Particular attention is paid to the influence of strain induced shift and modification of the effective masses and the symmetries of ground electronic levels.

Employing a 6 × 6 Luttinger-Kohn model, Ahn (1994) studied the effects of a very strong spin–orbit split-off band coupling on the valence band structure of GaN based materials. Considering that spin–orbit band is extremely important for GaN because of its very narrow spin–orbit splitting (10 meV), the spin–orbit split-off coupling was taken into consideration for the calculations. Also, it was assumed that electrons in QWs are confined by the conduction band offset ΔE_c and the holes by the valence band offset ΔE_v.

Using a k.p. method where the parameters are derived by first principle calculations, Uenoyama and Suzuki (1995) studied the valence subband structures of wurtzite GaN/AlGaN QWs. Since wurtzite GaN and AlN have

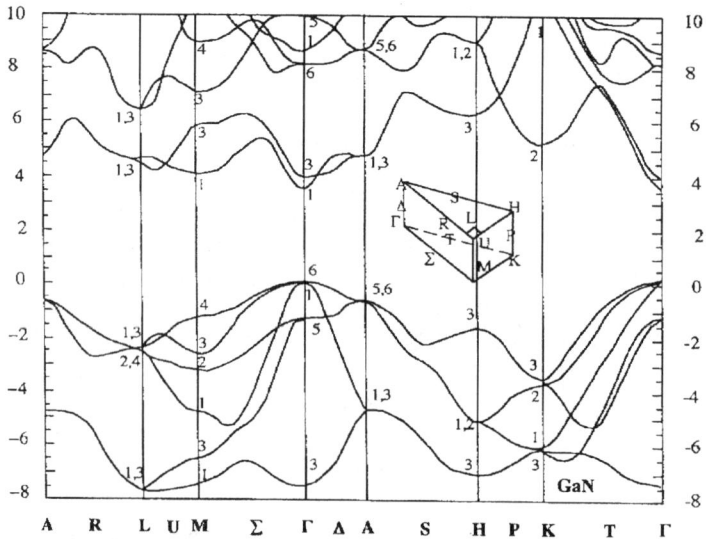

FIG. 13. Calculated band structure of GaN. Note that the GaN LVB has considerable structure as a result of the Ga 3d levels hybridizing with the other valence band orbitals. (From Fiorentini, Methfessel, and Scheffler, 1993.)

small spin–orbit splitting energies (<20 meV), the mixing among the six bands including the spin–orbit split band was considered to obtain the subband states. When the heterojunction is perpendicular to the c axis, the subband mixing is strong, particularly between HH and LH subbands with the same subband indices. Consequently, energy dispersion has a weak nonparabolicity. Furthermore, hole masses in two dimensional systems are not reduced from bulk masses because of a small spin–orbit splitting energy. Even in the case where herterojunction is parallel to c axis, average hole masses in the k space are as heavy as those in the bulk. It was noted that the hole carrier confinement does not lead to a significant reduction of hole masses, and so the threshold current density of wurtzite GaN/AlGaN QWs is not reduced.

Figure 13 shows the calculated band structure of wurtzitic GaN (Fiorentini, Methfessel, and Scheffler, 1993). On the other hand, Fig. 14 shows the band structure around the Γ point of bulk wurtzitic GaN without strain, whereas Fig. 15 shows the band structure around the Γ point for cubic GaN

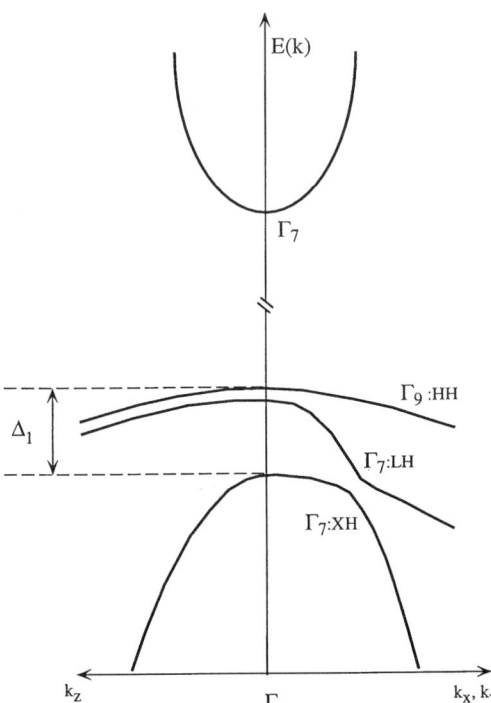

FIG. 14. Schematic band structure near the Γ point for wurtzitic GaN without strain. (From Suzuki, and Uenoyama, 1996a.)

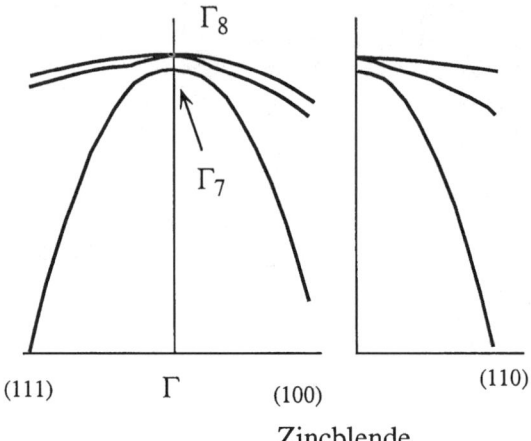

FIG. 15. Schematic band structure of cubic GaN near the Γ without strain. (From Uenoyama, and Suzuki, 1996.)

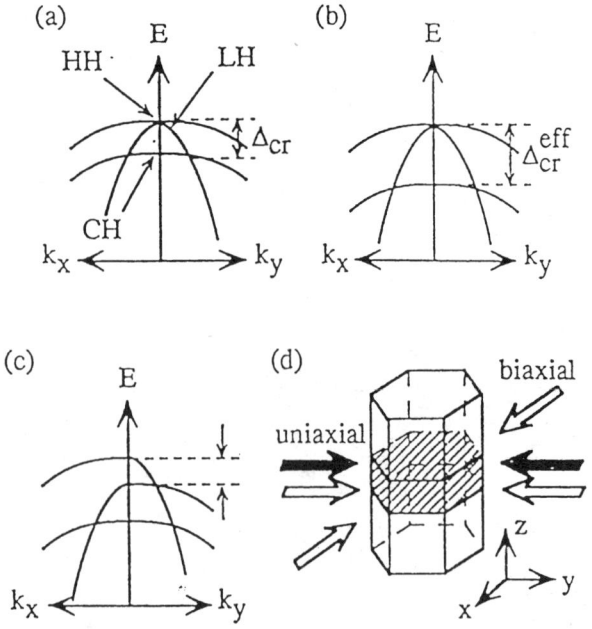

FIG. 16. Schematic valence band structure for wurtzitic GaN, (a) without any strain, (b) with biaxial strain, and (c) with uniaxial strain. Schematic on the right shows the particulars of the strain applied. (From Uenoyama, and Suzuki, 1996.)

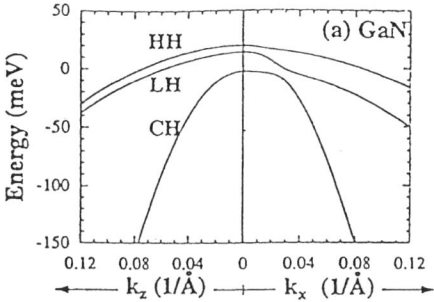

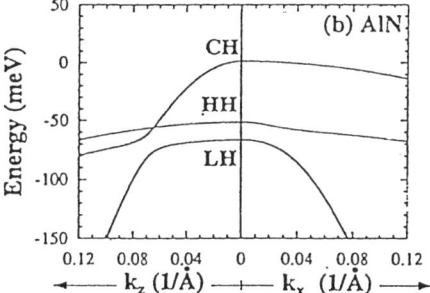

FIG. 17. Valence band dispersion around the Γ point for GaN and AlN calculated. (From Chuang, and Chang, 1996b.)

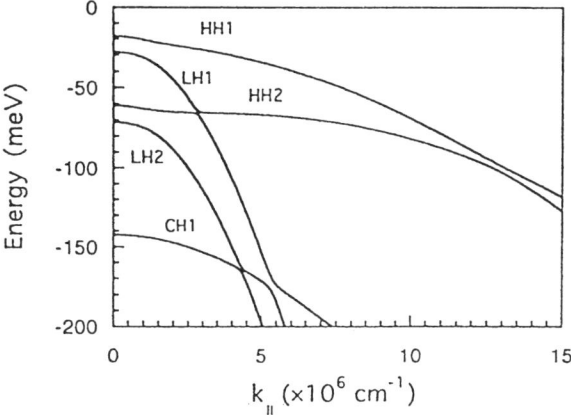

FIG. 18. Valence band structure of a 50 Å thick GaN QW surrounded by two $Al_{0.2}Ga_{0.8}N$ layers. The conduction and valence band discontinuities used in calculations were 0.425 and 0.108 eV, respectively. (From Kamiyama et al., 1995.)

TABLE I

CALCULATED EFFECTIVE MASS PARAMETERS, AND SPIN ORBIT AND CRYSTAL FIELD SPLITTING PARAMETERS

	Δ_{cr}	Δ_{so}		m_e	m_{hh}	m_{lh}	$m_{ch(sh)}$
WZ	72	16	k_x	0.18	1.61	0.14	1.04
			k_y	0.18	1.44	0.15	1.03
			k_z	0.20	1.76	1.76	0.16
ZB	—	20	[100]	0.17	0.86	0.86	0.17
			[110]	0.17	>2	0.84	0.15
			[111]	0.18	1.74	1.74	0.15

The terms m_e, m_{hh}, m_{lh}, $m_{ch}(m_{sh})$ denote the electron, heavy hole, light hole, crystal field (spin–orbit) split off hole masses, respectively, for the wurtzitic and cubic phases. (From Suzuki, and Uenoyama, 1996.) We should mention that all the recent experimental investigation indicate the conduction band mass to be 0.23 in wurtzitic GaN.

TABLE II

THE VALUE OF THE PHYSICAL PARAMETERS SUCH AS THE EFFECTIVE MASSES OBTAINED FROM THE *ab-initio* CALCULATIONS. (From Kamiyama *et al.*, 1995.)

Effective mass parameter	A_1	−6.56
	A_2	−0.91
	A_3	5.65
	A_4	−2.83
	A_5	−3.13
	A_6	−4.86
	A_7	0
	m_c	0.18
Deformation potential (eV)	D_1	−13.87
	D_2	−10.95
	D_3	2.92
	D_4	−5.84
Splitting energy (meV)	Δ_1	72.9
	Δ_2	5.17
Elastic stiffness constant (10^{11} dyn/cm^2)	C_{13}	12.0
	C_{33}	39.5

without strain. Figure 16 attempts to show that shown in Fig. 14 with biaxial and uniaxial strain ignoring the spin–orbit splitting. Figure 17 shows the band structure again near the Γ point in wurtzitic GaN and AlN. Uenoyama and Suzuki (1996) calculated effective mass parameters as well as spin–orbit and crystal field splitting in GaN are shown in Table I.

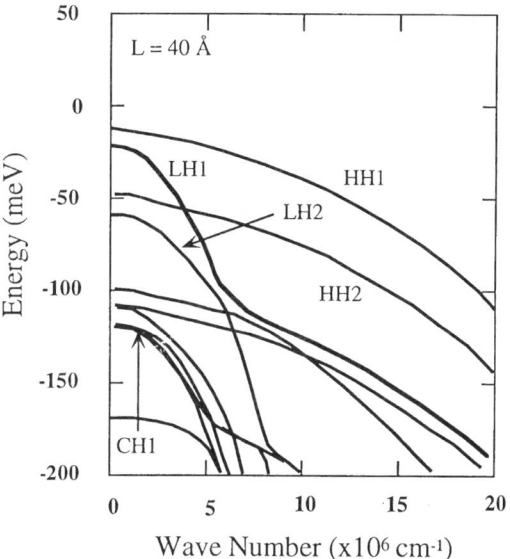

FIG. 19. Valence band structure of a 40 Å QW with barrier layers being $Al_{0.2}Ga_{0.8}N$ in the wurtzitic form. (From Uenoyama, and Suzuki, 1996.)

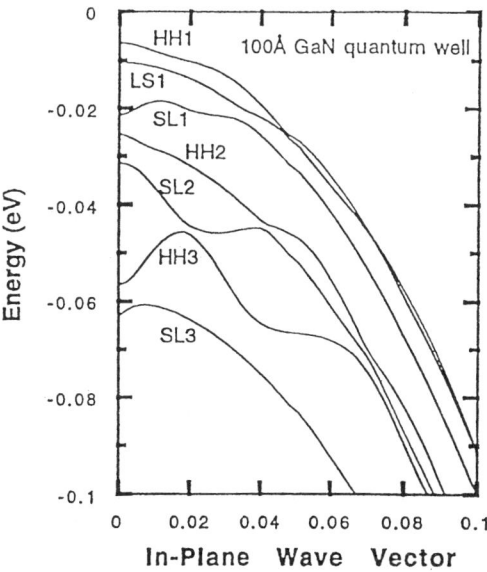

FIG. 20. Valence band structure of a 100 Å QW with barrier layers being $Al_{0.2}Ga_{0.8}N$ in the cubic form. (From Ahn, 1994.)

Figure 18 shows valence subband structure of 50 Å thick wurtzite GaN QW calculated by Kamiyama *et al.* (1995), using a k.p. perturbation theory and value of physical parameters such as effective masses obtained from *ab initio* calculations. These values are reported in Table II. Valence band structure of a 40 Å QW with barrier layers being $Al_{0.2}Ga_{0.8}N$ in the wurtzitic form is shown in Fig. 19. The results for valence subband calculations by Ahn (1994) for a 100 Å cubic GaN QW surrounded by assumed lattice matched AlGaN barriers, an assumption for computational convenience versus in-plane wavevectors are shown in Fig. 20.

IV. Experimental Results

GaAlN/GaN QWs were grown on sapphire substrate for the first time by Krishnakutty *et al.* (1992a,b) using low pressure metalorganic vapor deposition (MOVD). Itoh *et al.* (1991) have reported high quality GaAlN/GaN SLs grown by metalorganic vapor phase epitaxy (MOVPE) on sapphire substrate, using appropriate buffer layers. Sitar *et al.* (1991) were able to grow good quality AlN/GaN SLs on both 6H-SiC and sapphire substrate using ECR source MBE. The band engineering of these heterostructures and the quantum confinement effect caused by reduced dimensionality have been clearly demonstrated and the improvement needed for the devices performance has been carried out. Good quality QWs and SLs based on InGaN/InGaN grown by MOVPE demonstrated the feasibility of devices such as LED and lasers based on nitride compounds.

Nevertheless, many issues still need to be addressed in order to optimize device performance such as those for III–V (GaAs,InP...) and II–VI (ZnSe, CdTe...) materials. For example, excitonic binding enhancement with decreased dimensionality has not yet been demonstrated in III–V nitride heterostructures. It is worth noting that the binding energy of exciton in bulk GaN is approximately 20 meV, which is some four times higher than that in GaAs. One can expect that piezoelectic effect relative to the wurtzitic QW could be similar to (111) cubic QW. This effect is enhanced in strained heterostructures such as InGaN/AlGaN. The piezoelectric field in these heterostructures has been considered to explain the forward–backward asymmetry of the band discontinuities with respect to growth direction in InN/AlN and InN/GaN heterojunctions. Presence of piezoelectric effect in QW structures should influence the variation of exciton binding energy and the potential of these structures in optoelectronic devices.

1. BAND DISCONTINUITY DETERMINATION

Among the most important physical parameters for any heterojunction system are conduction and valence band discontinuities at the interface between the adjacent two semiconductors (also called band offsets); indeed, quality and even feasibility of heterojunction device concepts often depend crucially on values of these band discontinuities. An exhaustive review of band offsets in semiconductor heterojunctions is given by Yu *et al.* (1992). Epitaxial crystal growth techniques such as MBE and MOCVD of II–VI and III–V heterostructures have proved capable of producing abrupt band edges/discontinuity. Moreover, theoretical calculations indicate that the electronic structure in each layer of a heterojunction becomes very nearly bulk like even a single atomic layer away from the interface, lending credence to the idealized notion of an abrupt band edge discontinuity. Various types of band alignments can arise in semiconductor interfaces depending on the relative adjustment of energy bands with respect to each other. Figure 4 shows two types of possible alignments which occur most commonly in semiconductor heterojunctions. It is worth noting that device concepts which can be implemented successfully in a given heterojuncton system will depend very strongly on the type of band alignment of that heterojunction, and heterojunction device performance will often depend critically on exact values of band discontinuities. Type I alignment, in which the bandgap of one semiconductor lies completely within the bandgap of the other, is the most useful one for optoelectronic devices. A type II alignment occurs when the bandgaps of two materials overlap but not completely covering the other, as shown in Fig. 4. Type II ZnSe/ZnTe heterojunctions have been used to overcome crucial problems relating to difficulties in performing p-ohmic contact for blue ZnSe based lasers.

First we note that both theoretical and experimental works devoted to determination of band discontinuities in semiconductor heterojunction present large discrepancies between measured and calculated values. The origin of controversy between different values of band discontinuity for a given heterojunction can be related to different factors, among them are:

(i) technical difficulty and often indirect nature of measurements,
(ii) possible dependence of band discontinuity on detailed conditions of interface preparation,
(iii) strain dependence of band discontinuities.

The theoretical treatment of semiconductor heterojunction discontinuities can be divided into three categories: (1) empirical rules such as electron affinity rule and common anion rules which give an indication of band alignment type and band discontinuity values, (2) pseudopotential and LCAO theories are utilized to extract band discontinuities from electronic properties of bulk semiconductors, and (3) self-consistent calculations for specific interfaces using a super cell geometry such as linear muffin-tin orbital (LMTO). It is worth noting that capabilities of current theoretical treatments for band discontinuities are such that consistently reliable predictions of band discontinuity values in novel semiconductor heterojunctions cannot yet be made; band offsets must, therefore, be determined experimentally for each new material system of interest. XPS and ultraviolet photoelectron spectroscopy (UPS) are primarily used to determine band discontinuities by means of electron core level energies. The energies of core levels are obtained after an extensive band structure modelization in order to obtain the valence band edge. Optical techniques, such as excitation PL and reflectivity, present a more accurate tool to determine band discontinuity values due to the higher experimental resolution compared to previous X-ray techniques. Measurement of electrical characteristics such as capacitance-voltage and current-voltage has also been used with considerable success to determine GaAs/GaAlAs and HgTe/CdTe band discontinuities, but requires an accurate knowledge of charge density and its distribution throughout the region of the structure that is sampled.

Early attempts at determining band discontinuities by cathodoluminescence measurements for a GaN/AlN SL (Sitar *et al.*, 1991) and by PL measurements for GaN/Al$_{0.14}$Ga$_{0.86}$N QWs (Khan *et al.*, 1990) led to conduction- to valence-band discontinuity ratios of 50:50 and 60:40, respectively.

The first detailed investigation of band discontinuity in GaN/AlN heterojunction was reported by Martin *et al.* (1994) by using *in situ* XPS method. Further work by the same authors has been extended to other binary nitride heterojunctions such as InN/GaN, AlN/InN. All experimental and theoretical estimates of band discontinuities indicate the occurrence of type I alignment between nitride materials InN, GaN, and AlN. We compare the results obtained by Martin *et al.* (1994) with other works available and the estimated values obtained using different theoretical models.

a. *GaN/AlN Heterostructures*

In Table III we present all band discontinuity values obtained by various experimental methods and the relevant experimental details. Martin and co-workers made systematic *in situ* XPS studies of both GaN/AlN and

TABLE III

VALENCE BAND DISCONTINUITY VALUES GIVEN IN eV

	GaN/AlN	AlN/GaN	Substrate
Martin et al. (1994)	0.8 ± 0.3	0.8 ± 0.3	SiC, Al$_2$O$_3$
Baur et al. (1994)	0.5 ± 0.5		(polycrystal)
Waldrop and Grant (1996)	1.36 ± 0.07		SiC
Martin et al. (1996)	0.60 ± 0.24	0.57 ± 0.22	SiC, Al$_2$O$_3$

AlN/GaN heterostructure. Values obtained using Ga$_{2d}$ and Al$_{2p}$ core levels for both heterostructure is merely the same within the experimental errors ($\Delta E_v = 0.67 \pm 0.02$ eV). Waldrop and Grant (1996) reported a higher value of the valence band discontinuity in GaN/AlN grown on SiC ($\Delta E_v = 1.36 + 0.07$ eV) from Ga$_{3d}$ and Al$_{2p}$ core levels measured by XPS in both bulk and heterostructure. The large discrepancy between values obtained from these two studies is related to the discrepancy in determining core level binding energies which depend crucially on how the valence band edge is fitted using a density of state calculations.

In a parallel investigation, Baur et al. (1994) estimated the valence band for GaN/AlN by applying the concept of Langer and Heinrich (1985) who postulated that transition metal impurity levels act as a common reference level to predict the band alignment in semiconductor heterojunctions. In the nitride system, the fact that a characteristic infrared luminescence spectrum for both polycrystalline AlN and GaN is dominated by a zero phonon line at 1.3 eV of the iron level, was utilized to extract the band discontinuities. It was also observed that by employing PL excitation spectroscopy the (–/0) acceptor level of iron is located at $E_v + 3.0$ eV for AlN, and at $E_v + 2.5$ eV for GaN. The observed zero phonon line was assigned to the spin-forbidden internal 3d-3d transition $^4T_1(G) \rightarrow {}^6A_1(S)$ of Fe$^{3+}_{Al}$(3d^5). Based on this observation the valence band offset was determined to be $\Delta E_v = 0.5$ eV, and the conduction band offset to be $\Delta E_c = 2.3$ eV for GaN/AlN heterostructure (Baur et al., 1994).

Albanesi et al. (1994) have performed self-consistent calculation by means of linear muffin-tin orbital method to estimate the valence band discontinuity at the zinc-blende AlN/GaN interface, in the [001] direction. The value obtained ($\Delta E_v = 0.85$ eV) is close to the value obtained by Martin and co-workers using XPS, while the experimental value of ΔE_v obtained by Waldrop and Grant is in accordance with the affinity model prediction.

b. *GaN/InN and AlN/InN*

In situ XPS study by Martin *et al.* (1996) revealed a large asymmetry of band discontinuity values depending on the order in which binary layers are grown, that is, GaN on InN versus InN on GaN. The experimental values are reported in Table IV. It is interesting to note that the mean values obtained for both heterostructures obey the transitivity law when compared to the experimental value obtained for GaN/AlN. Piezoelectric effect, as described later, has been evoked to explain the asymmetry of band discontinuities. Dangling bonds unique to one polarity of interface may also induce such an asymmetry (Martin *et al.*, 1996). Though this concept accounts for major features of the observed data, deterministic investigation must be undertaken before a more definitive statement can be made.

Note that InN/GaN–GaN/InN and InN/AlN–AlN/InN heterojunctions show a significant forward–backward asymmetry, while AlN/GaN–GaN/AlN heterojunctions give almost identical values. All heterojunctions were in standard type I heterojunction alignment. Insight into the asymmetrical nature may be provided by strain induced piezoelectric fields (Smith, 1986; Kuech *et al.*, 1990). The lattice constant of GaN is larger than that of AlN by approximately 2.5%, while the lattice constant of InN is larger than that of GaN by approximately 11% (Strite and Morkoç, 1992). Heterojunction underlayers were thick enough to be relaxed to their natural lattice constants at growth temperature with residual strain remaining due to the thermal mismatch between substrate and layers, so the heterojunction thin overlayers used were at least partially strained. The nitrides are piezoelectric materials, so the strain induce static electric fields via the piezoelectric effect. Figure 21 shows how a strain induced piezoelectric field changes the apparent valence band discontinuity. For a wurtzite material grown in the (0001) direction, the strain induced electric field is entirely longitudinal. MBE produces N terminated layers, so in the language of Smith (1986) each grown layer ends with a B (anion N) face and begins with an A (cation In/Ga/Al) face. The nitrides have negative piezoelectric coefficients just like III–V zinc-blende materials, and furthermore have decreasing bandgaps

TABLE IV

VALENCE BAND DISCONTINUITY VALUES

InN/GaN = 0.93 ± 0.25 eV
GaN/InN = 0.59 ± 0.24 eV
InN/AlN = 1.71 ± 0.20 eV
AlN/InN = 1.32 ± 0.14 eV

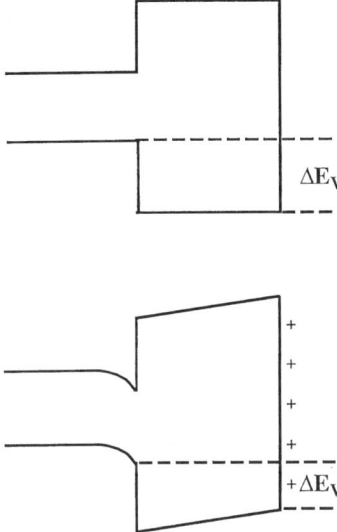

FIG. 21. Schematic representation of how piezoelectric effects alter the measured band discontinuities.

with increasing lattice constants, so that the field directions of Smith (1986) lead to the result that strain induced piezoelectric fields always tend to decrease apparent valence band discontinuities for nitride materials. The field magnitude is given by (Bykhovski, Gelmont, and Shur, 1993)

$$E = \frac{2d_{31}}{\varepsilon}\left(c_{11} + c_{12} - \frac{2c_{13}^2}{c_{33}}\right)u_{xx} \qquad (5)$$

where d_{31} is the relevant piezoelectric constant (of five for wurtzite materials), ε is the static dielectric constant, c_{ij} are elastic constants, and u_{xx} is the fractional strain from lattice mismatch given by $-1 + a_1/a_2$. Several of these constants are not well known for InN, GaN, and AlN; the values and sources used in this work are given in Table V. The calculated electric fields in the strained regions are:

$$\text{InN (on GaN)} = 5.5 \times 10^8 \text{ V/m}$$

$$\text{GaN (on InN)} = 2.1 \times 10^9 \text{ V/m}$$

$$\text{GaN (on AlN)} = 4.7 \times 10^8 \text{ V/m}$$

TABLE V

VALUES USED FOR CALCULATING STRAIN-INDUCED PIEZOELECTRIC EFFECTS

	InN	GaN	AlN
d_{31} (cm/V)	$-1.1 \times 10^{-10 a}$	$-1.7 \times 10^{-10 a}$	$-2.0 \times 10^{-10 b}$
ε (static)	15.3^c	10.0^c	8.5^c
c_{11} (GPa)	271^d	396^d	398^d
c_{12} (GPa)	124^d	144^d	140^d
c_{13} (GPa)	94^d	100^d	127^d
c_{33} (GPa)	200^d	392^d	382^d
a (Å)	3.548^d	3.189^d	3.112^d

[a]Estimated from the AlN value in the ratio $1/\varepsilon$.
[b]Landolt-Börnstein, 1971.
[c]V. W. L. Chin et al., 1994.
[d]Kwiseon et al., 1995.

$$\text{AlN (on GaN)} = 6.0 \times 10^8 \text{ V/m}$$

$$\text{InN (on AlN)} = 5.5 \times 10^8 \text{ V/m}$$

$$\text{AlN (on InN)} = 2.7 \times 10^9 \text{ V/m}$$

The important factor to note is the strong asymmetry for heterojunctions containing InN, where InN on the bottom leads to much bigger piezoelectric fields than does InN on the top. Assuming typical heterojunction overlayer thickness of 10 Å, these fields lead to voltage changes of 0.47–2.7 V, values much larger than the observed discrepancies. The critical thickness for GaN/AlN pseudomorphic growth is approximately 30 Å (Sitar et al., 1990), but critical thickness for GaN/InN and AlN/InN are estimated at 6 Å, even less than the heterojunction overlayer thickness. Even if lattice matched substrates were available and employed, epitaxial layers containing large fractions of InN would practically contain defects unless the critical thickness is not challenged. In the current approaches, because the lattice mismatch substrates dense networks of threading defects extending from substrates to surfaces are present they provide strain relief mechanisms. Therefore, the simple model for pseudomorphic strained layers is not quantitatively applicable even for thin heterojunction overlayers. We now proceed with the supposition that heterojunction overlayers are partially strained to an unknown degree. The presence of the strain induced piezoelectric effect accounts for the observed asymmetries and leads to preference for measurements with the smallest piezoelectric effects. After rejecting the cases with smaller valence band discontinuities (based on

TABLE VI

VALENCE BAND DISCONTINUITY VALUES AFTER PIEZOELECTRIC CORRECTION

InN|GaN: $\Delta Ev = 1.05 \pm 0.25$ eV
GaN|AlN: $\Delta Ev = 0.70 \pm 0.24$ eV
InN|AlN: $\Delta Ev = 1.81 \pm 0.20$ eV

---

having larger piezoelectric fields) and applying rough corrections for small strain induced piezoelectric fields to cases with larger valence band discontinuities, the results shown in Table VI are obtained:

$$\text{InN} \mid \text{GaN}: \Delta E_v = 1.05 \pm 0.25 \text{ eV}$$
$$\text{GaN} \mid \text{AlN}: \Delta E_v = 0.70 \pm 0.24 \text{ eV}$$
$$\text{InN} \mid \text{AlN}: \Delta E_v = 1.81 \pm 0.20 \text{ eV}$$

These results obey transitivity to well within the experimental accuracy. Ratios of conduction band discontinuities to valence band discontinuities are roughly 30:70, 75:25, and 60:40, respectively. We briefly note that strain induced piezoelectric fields lead to a small broadening of photoemission signals, since piezoelectric fields create spatial variations in the potential in near surface regions where photoelectrons originate. This effect was observed throughout our nitride work. We also note that Fermi levels for GaN and InN were observed just below the conduction band edges; both materials typically have large n-type background doping when grown by MBE. The Fermi level for AlN was observed at roughly 4 eV above the valence band edge, although considerable variation was seen, likely caused by charging. Having gone through this treatment, we should also point out that asymmetric interfacial defect can also contribute to the observed order dependence which was briefly considered by Martin *et al.* (1996).

2. OPTICAL PROPERTIES OF QUANTUM WELL STRUCTURES

Optical probing of QWs and SLs have been decidedly successful in compound semiconductor samples since their inception. As is the case in almost all semiconductor research, the better the quality, the better the spectroscopic clarity. Early GaAs based QW exhibited only one blue shifted transition which was automatically attributed to free excitons. As the quality improved, it was discovered that transitions observed in bulk GaAs, such as free and bound excitons are also observed in GaAs QWs with the added feature that the valence band degeneracy is removed, therefore light hole related transitions also occur (Reynolds et al., 1984).

The nitride system is no exception, and already efforts, though in their embryonic stage, are underway to probe nitride semiconductor heterostructures by means of optical probes which are also nondestructive. Later we will review the work that has been performed on the topic, although quality of structures prepared so far is not sufficiently adequate. Characteristically, we will subdivide this section by the type of heterostructures utilized so far and discuss the important results obtained. We should mention that all heterostructures measured are accompanied by some degree of strain caused by thermal mismatch with the substrates and variation of lattice parameters with Al and In content. This causes transition energies to be dependent on the substrate on which the structure is grown. Even on sapphire, residual strain is not always the same, thus the spread in the data concerning energy transitions is present, further complicating the determination of confinement effect. Additional uncertainties are brought to bear because of the dearth of accurate knowledge of pertinent parameters (Martin et al., 1994), nevertheless, a set of representative data points on sapphire, SiC, and ZnO are shown in Fig. 3.

a. GaN/AlGaN Heterostructures

Krishnakutty et al. (1992a) carried out PL analysis of GaN/AlGaN QWs at 77 K. They noted that emission energies from the well change with well width and aluminum composition of barriers in a manner characteristic of type I heterojunctions. The PL spectra of AlGaN/GaN QWs are shown in Fig. 22. An inspection of these two figures indicates that peak emission energy from GaN QW is shifted to a higher energy. A finite square well model was used to calculate the confined particle transitions in QWs. The observed energy shift between peak emission energies of bulk GaN and GaN/AlGaN QW was attributed to different physical phenomena: Quantum size effects and the strain induced bandgap shifts in the AlGaN/GaN QWs. With new data gathered since then, it is clear that epitaxial layer quality must be sufficiently good for the intrinsic transitions to be delineated. In addition, the strain effect must be accounted for, which is not simple as each sample appears to be under varying degrees of strain even when grown on the same kind of substrate. In short, the initial investigation of optical properties and confinement energies of GaN/AlGaN QWs were severely hampered by lack of high quality samples, uncertainty in band offsets, and a wide range of reported values for hole and conduction electron

Optical Characterization of AlGaN-GaN-AlGaN Quantum Wells

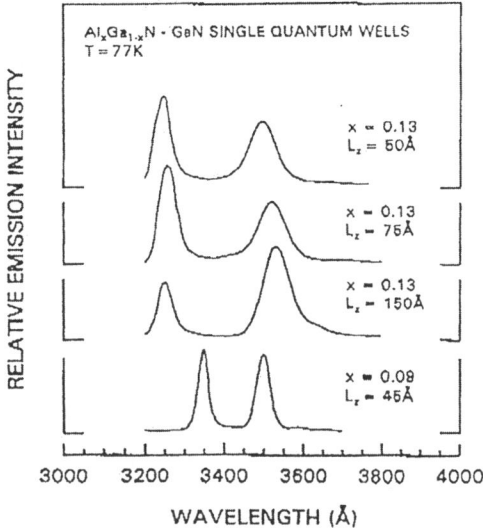

FIG. 22. Emission spectra of $Al_xGa_{1-x}N/GaN/Al_xGa_{1-x}N$ SQWs with well thicknesses of 45, 50, 75, and 150 Å and with 0.13 and 0.09 AlN mole fractions in the barrier layers. (From Krishnakutty et al., 1992a.)

effective masses, not to mention varying strain effects (Gil, Briot, and Aulombard, 1995; Gil, Hamdani, and Morkoç, 1996).

Recent and improved samples have paved the way more accurately for the determination of parameters mentioned previously, but variation of strain from sample to sample still presents a problem (Reynolds et al., 1996). Coupled with improved GaN based heterostructure layers, GaN/AlGaN QWs are now understood better. Salvador et al. (1995) employed PL emanating from an AlGaN/GaN/AlGaN single well heterostructure. A good fit to the observed room temperature band edge luminescence was obtained using a 67:33 conduction to valence band offset, and $0.19m_e$ and $0.3m_h$ as the conduction electron and hole effective masses. The GaN/AlGaN QW layer used was grown on the c plane of sapphire by reactive MBE. The well width was 60 Å and only the well region was Si doped ($5 \times 10^{17} cm^{-3}$). Photoluminescence measurements were performed using the 325 nm line of a 27 mW HeCd laser as the excitation source and the emitted PL signal was collected from the epilayer front surface and was directed to a Spex monochromator. The signal was detected using a GaAs photomultiplier tube. For temperature dependent PL measurements the sample was mounted on the cold finger of a Janis cryostat.

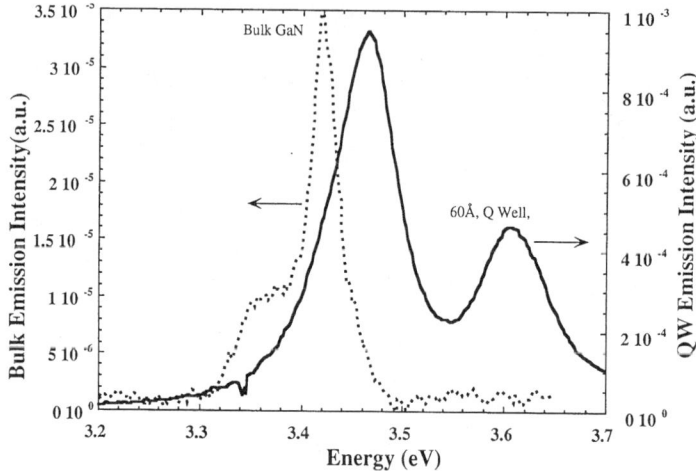

FIG. 23. Room temperature PL spectrum of a 60 Å GaN/AlGaN SQW. The peak at 3.6 eV is due to emission in the $Al_{0.07}Ga_{0.93}N$ barrier. Superimposed for reference is the PL spectrum for a bulk GaN layer. (From Salvador et al., 1995.)

The room temperature PL spectrum of a 60 Å GaN/AlGaN SQW is shown in Fig. 23. The peak at 3.6 eV is due to emission in the $Al_{0.07}Ga_{0.93}N$ barrier. Also superimposed for reference is the PL spectrum from undoped bulk GaN. Of the two salient features seen in PL spectra of single GaN QW, the peak at 3.465 eV is assigned to the band edge transition in QW, while that at 3.6 eV is assigned to emissions in the AlGaN barrier. From this it was inferred that the AlN composition in the barrier was 0.07 which is consistent with that determined by X-ray in a thick AlGaN of the same composition as that forming the barriers. In contrast, the band edge signal for the bulk GaN sample is seen at 3.42 eV.

To account for the blue shift in PL spectra of the GaN QW, confinement energies for the groundstate heavy hole and conduction electron must be determined. Reported X-ray photoemission spectroscopy results indicate that ratio of conduction band to valence band offset between AlN and GaN is 67:33 (Martin et al., 1994). Extrapolating this band alignment to the case of the AlGaN/GaN with the composition mentioned, and using $0.19m_e$ for the conduction electron effective mass, the calculated groundstate conduction band confinement energy is 27 meV. The heavy hole effective mass was chosen to be $0.3m_e$, consistent with recent reports that determine the heavy hole effective mass to be between $0.3-0.4m_h$, resulting in a confinement energy of 16 meV (Orton, 1995). The sum of calculated confinement energies thus indicate that band edge transition for 60 Å GaN/Al$_{0.07}$Ga$_{0.93}$N QW would blue shift by 43 meV compared to bulk GaN and this matches well with the observed 45 meV shift in our PL spectrum.

In principle, excitonic effects and residual strain also must be considered to properly account for the band edge transition. Excitonic effects at room temperature can be neglected in our study, since the GaN well is doped intentionally with Si to a level of 1×10^{18} cm^{-3} and screening effects would most likely minimize the excitonic binding energy. Thus, transitions at room temperature will be interband in character. As will be shown later, the room temperature PL peak assigned to the GaN QW follows a temperature dependence typical of band-to-band transition.

Strain effects caused by lattice mismatch between GaN well material and AlGaN barrier lead to an increase in bandgap energy of well material, providing that a presence of large concentration of defects in the material does not prevent formation of coherent strain. This shift in energy can be estimated using the relation $\Delta E_g = b(\Delta V/V)$, where b is the deformation potential and V the volume of material. Recent work relating the shift in bandgap energy to hydrostatic pressure in GaN indicates a value of -9.2 eV for the deformation potential (Shan *et al.*, 1995). The in-plane lattice mismatch ($\Delta a/a$) between Al$_{0.07}$Ga$_{0.93}$N, and GaN is 0.169%. Assuming that the GaN well is strained, a blue shift of $\Delta E_g = 25$ meV is obtained. This implies that the expected band edge emission for the GaN QW should be at 3.495 eV which is much higher than the observed value. Inspection of the film under optical microscopes, however, reveal cracks on the film. The density of cracks did not change after the sample was cooled to 4 K. Moreover, GaN films grown on sapphire characteristically show extended defect densities of greater than 10^8 cm^{-2}. We therefore conclude that the film is relaxed and that strain can be neglected.

Temperature dependence PL spectra of the sample was also investigated. Figure 24 displays the PL spectra at temperatures of 4, 35, and 68 K. There are two PL peaks of interest, one centered at 3.495 eV and the wide PL signal in the region of 3.4–3.47 eV. The origin of the latter is attributed to transitions from Si (intentional) donor level to GaN valence band. Features seen in the broad energy region are caused by interference patterns resulting from multiple reflections between the AlGaN/air interface and the epitaxial layer substrate interface. This is supported by the agreement of the observed spacing of peaks with that calculated from measured film thickness. As the temperature is increased from T = 4 K to 120 K, this PL signal blue shifts and is quenched while another PL feature develops on its high energy side. This temperature behavior is typical of the onset of conduction band to valence band transition, from donor to valence band transition, and is brought about by thermal ionization of donors. As shown in Fig. 25, PL

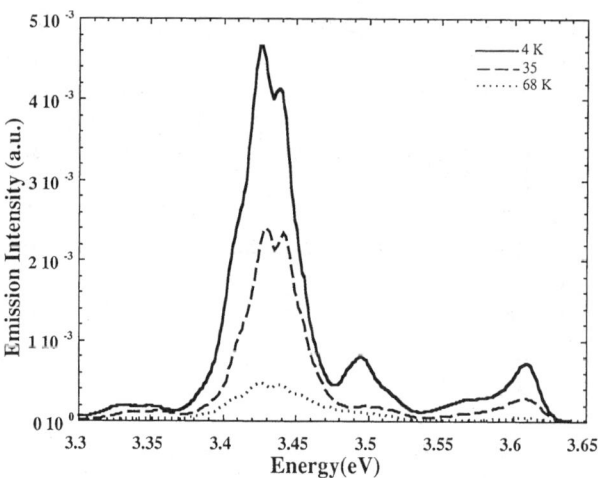

FIG. 24. Photoluminescence spectra of the 60 Å GaN/AlGaN SQW at $T = 4$ K, 35, and 68 K. (From Salvador et al., 1995.)

spectra of the GaN/AlGaN QW at $T = 68$ K, 120 K, and 170 K. The PL signal attributed to Si donor level to valence band transitions are quenched as the temperature is raised while band-to-band transition is enhanced, the PL feature on the high energy side develops into PL peak seen at room temperature.

As mentioned earlier, the PL peak centered at 3.495 eV is accompanied with a shoulder like feature at 3.51 eV. As the temperature is increased to 35 K, this shoulder like feature becomes distinct and the PL peak observed at 3.495 eV is quenched. From the temperature dependence of these two features, peak at 3.495 eV is assigned to transitions involving donor bound excitons while the feature at 3.51 eV is assigned to interband transitions. It is likely that at $T = 4$ K, excitonic effects may not be neglected, since not all Si donors are ionized, leading to reduced screening effects. If we take into consideration exciton binding energy of 20 meV for GaN (Reynolds et al., 1996) it indicates that the GaN QW band edge is at 3.53 eV.

The expected location of the interband transition at $T = 5$ K can be determined using the Varshni empirical relation $E_g(T) = E_0(0) - \alpha T^2/(\beta + T)$ (Shan et al., 1995; Varshni, 1967). We recently have investigated PL dependence on temperature of high quality bulk GaN, wherein at $T = 5$ K both A (Γ_9^v-Γ_7^c) and B (Γ_7^v (upper band)-Γ_7^c) groundstate excitons as well as their excited states are seen. A best fit to the plot of energy location of

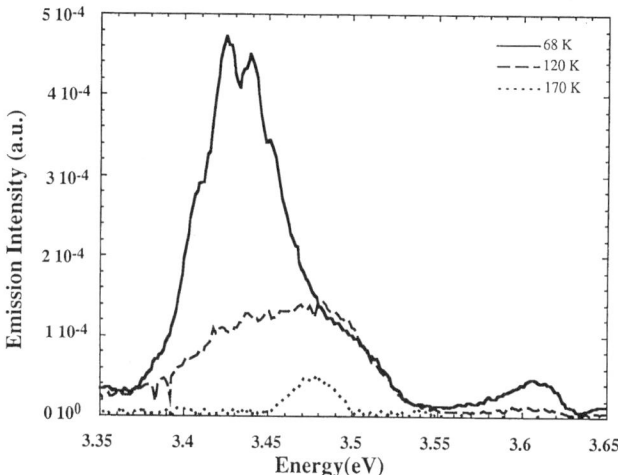

FIG. 25. Photoluminescence spectra of the GaN/AlGaN QW at $T = 68$ K, 120 K, and 170 K. The PL signal attributed to Si donor level to valence band transitions are quenched as the temperature is raised while band-to-band transition is enhanced (From Salvador et al., 1995.)

interband Γ_9^v-Γ_7^c transition as a function of temperature in this high quality layer yields $\alpha = 7.32 \times 10^{-4}$ eV/K and $\beta = 700$ K. Using these values and the observed PL peak location of QW at room temperature, the expected location of QW band-to-band transition at $T = 4$ K is calculated to be at 3.53 eV which compares favorably with observed PL spectrum.

Following the work of Salvador et al. (1995), Smith et al. (1996) carried out an experimental investigation of optical transitions in GaN/AlGaN MQWs employing time resolved PL measurements. For the study, wurtzite GaN epitaxial layers of approximately 4 µm thick and GaN/AlGaN MQWs were grown by the reactive MBE method on sapphire substrates with 50 nm AlN buffer layer. MQWs consisted of 10 periods of alternating 25 Å GaN wells and 50 Å $Al_{0.07}Ga_{0.93}N$ barriers. GaN layers were nominally undoped and insulating. Low temperature time resolved PL spectra were measured by using a picosecond laser spectroscopy system with an average output power of 20 mW, a tunable photon energy up to 45 eV, and a spectral resolution of about 0.2 meV. A microchannel plate photomultiplier tube, together with a single photon system were used to collect time resolved PL data, and the overall time resolution of the detection system was about 20 picoseconds.

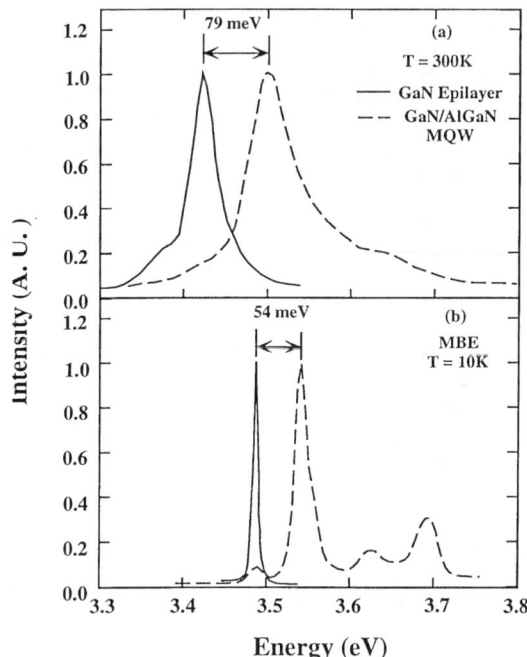

FIG. 26. Continuous wave PL spectra of a GaN epilayer (solid dots) and GaN/Al$_x$Ga$_{1-x}$N MQW measured at (a) $T = 300$ K and (b) $T = 10$ K. Note that the blue shift of the excitonic transition in the MQW at $T = 10$ K (54 meV) is approximately 25 meV less than that at $T = 300$ K (79 meV). (From Smith et al., 1996).

Figure 26 shows the plot of the continuous wave (CW) PL spectra of the GaN/Al$_x$Ga$_{1-x}$N MQW sample obtained at (a) $T = 300$ K and (b) $T = 10$ K. For comparison, PL spectra of a GaN epilayer are also shown. In the GaN epilayer, the dominating transition line at $T = 10$ K is caused by recombination of the groundstate of A exciton. In the MQW, excitonic transition peak position is blue shifted because of the well known effect of quantum confinement of electrons and holes. Blue shift at room temperature (79 meV) is what we expected for our MQW structure with a 67% (33%) conduction (valence) band offset. One of the interesting features shown in Fig. 26 is that the blue shift observed at 10 K is 54 meV, approximately 25 meV less than the shift of 79 meV seen at 300 K. We attribute this difference to the fact that PL emission in the MQW measured at low and room temperatures resulted from the recombination of localized excitons and free excitons, respectively. The exciton localization at low temperatures is caused by the interface roughness of the MQW. As the temperature increases, localization energy is no longer efficient to localize excitons. Thus, the difference in blue shift seen at 10 and 300 K of 25 meV measures the exciton localization energy, which gives a well thickness fluctuation of about ±4 Å. This interpretation is further supported by the time resolved PL data to be discussed later. On the other hand, one could also argue that low and room temperature dominating transition peaks in MQW correspond to free exciton and band-to-band transitions, respectively. In such a context, the 25 meV difference would then correspond to exciton binding energy in MQW. However, the band-to-band transition was not observed in the GaN epilayer under the same experimental condition. Moreover, the enhancement of the exciton binding energy in MQW also makes the band-to-band recombination less likely.

In addition to the main exciton emission band resulting from GaN wells, Fig. 26(b) also shows that there are more features in the higher emission energy side. For a clearer illustration, we have replotted the 10 K CW emission spectrum of the MQW sample in Fig. 27. Four transition peaks at 3.692, 3.625, 3.558, and 3.489 eV are clearly resolved. We assign these to 0–3

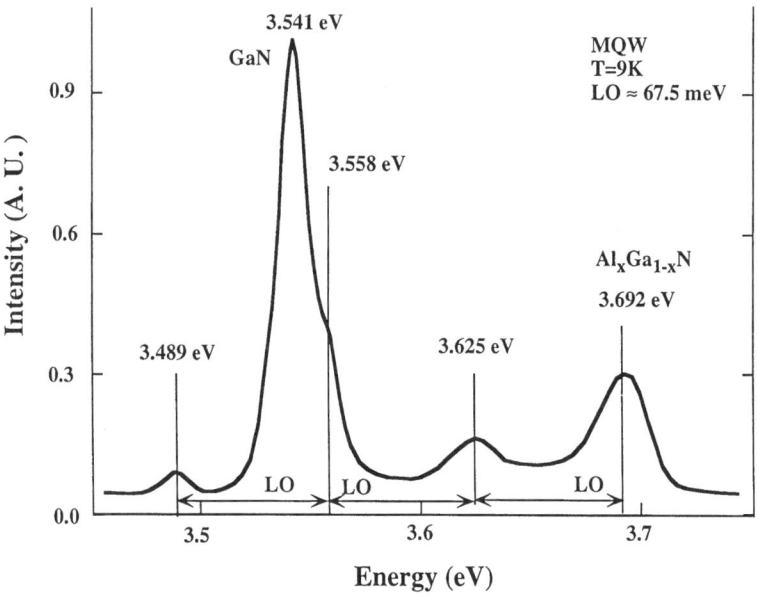

FIG. 27. Continuous wave PL spectra of GaN/Al$_x$Ga$_{1-x}$N MQW measured at $T = 10$ K, showing several phonon replicas of the excitonic transition at 3.692 eV resulting from the Al$_x$Ga$_{1-x}$N barriers with a modified LO phonon energy of 67.5 meV. The excitonic transition resulting from the GaN wells at 3.541 eV is also shown. (From Smith et al., 1996).

phonon replicas of the excitonic transition resulting from $Al_xGa_{1-x}N$ barriers. This assignment is based on the fact that these transition lines are separated by an equal energy space (67.5 meV). Moreover, relative emission intensities of these transitions also support our assignment. Excitonic transitions from $Al_xGa_{1-x}N$ barriers are easily seen here because the AlN mole fraction in the barrier material is relatively low ($x \sim 0.07$), which makes the energy difference between excitons in wells and barriers relatively small. The emission intensities of the excitonic transition at 3.692 eV and its phonon replicas decrease with an increase of temperature, which is due to an increased rate of exciton transfer from $Al_xGa_{1-x}N$ barriers to GaN wells. Our results shown in Fig. 27 seem to suggest that exciton–phonon interactions are quite strong in $Al_xGa_{1-x}N$ barrier regions. Another surprising feature is that energy separation between these transition lines is 67.5 meV. In GaN epilayers, phonon replica emissions are usually associated with longitudinal optical (LO) phonons, which have an energy of 91 meV Landolt-Börnstein, 1971). The measured phonon energy in the MQW of 67.5 meV corresponds to the transverse optical (TO) phonon energy in GaN epilayers.[57] Similar phonon replica emissions with TO phonon frequency in $GaAs/Al_xGa_{1-x}As$ MQW and with LO phonon frequency in GaAs epilayers have been observed previously (Klein, 1990). Such a behavior has been theoretically understood. In MQW, the LO phonons involved in optical transitions vibrate at the bulk TO phonon frequency due to the symmetry properties of MQW. Thus, phonon replica emissions observed in $GaN/Al_xGa_{1-x}N$ MQW shown in Fig. 27 are also associated with the LO phonons, except that they vibrate at GaN bulk TO frequency. It is well known that the ratio of exciton binding energy to energy of optical phonons involved in optical transitions is one of the crucial factors that determine gain mechanisms for lasing. The modified optical phonon frequency seen in MQW may have an important consequence in the design of laser structures.

V. AlGaN/GaN Quasi Triangular Quantum Wells

Quantum well structures are not limited to double heterostructures but can also appear in structures involving single heterojunction. The most common example are MODFETs. In this structure the higher bandgap material (AlGaN) is doped n-type. The carriers fall into the lower bandgap (GaN) material until an equilibrium is reached where the Fermi level is the same on both sides of the interface. The resulting band bending creates a triangular well like potential at the GaN/AlGaN interface. In this well the electrons accumulate forming a two dimensional electron gas similar to that obtained in a GaN/AlGaN double heterojunction QW. The confinement energy levels and the amount of band bending in the conduction band are determined in a self-consistent manner (Stengel, Mohammad, and Morkoç, (1996).

1. THEORETICAL METHOD

The structure considered for the present study is a wurtzite $Al_xGa_{1-x}N$ layer structure grown on wurtzite GaN. A metal contact is deposited on $Al_xGa_{1-x}N$ for source and drain contacts. Because of conduction band discontinuity, a triangular QW is created at the $Al_xGa_{1-x}N$/GaN interface. In this well, electrons accumulate, forming a 2DEG. In what follows, we present the treatment of Stengel et al. (Stengel, Mohammad, and Morkoç, 1996).

a. Two-dimensional Electron Gas

In order to obtain the 2DEG concentration at the $Al_xGa_{1-x}N$/GaN interface Poisson's Eq. (1) and Schrödinger's Eq. (2) are solved self-consistently. In reduced units (energy units being Rydberg and length unit being Bohr radius) these equations are given by

$$\frac{d^2 E_c}{dz^2} = -8\pi n(z) \tag{6}$$

$$-\frac{\hbar^2}{2m^*}\frac{d^2\Psi_i}{dz^2} + E_c(z)\Psi_i = E_i\Psi_i \tag{7}$$

where z is the coordinate in a direction perpendicular to the channel, $E_c(z)$ is the conduction band position, Ψ_i is the wavefunction of electrons, m^* is the effective electron mass in GaN, $n(z)$ is the electron concentration at the point z, h is the Planck's constant, and $\hbar = h/2\pi$. When unintentionally doped, GaN shows a donor implanted behavior. However, these donors do not affect the 2DEG concentration, because most of them are not ionized at the interface, owing to the Fermi level lying above the conduction band. If, for the sake of convenience, the origin of the energy is chosen at the edge of the conduction band level of GaN at the $Al_xGa_{1-x}N/GaN$ interface, and if the unintentional donor concentration in the bulk of GaN is accounted for, then the first boundary condition may be written as

$$E_c(z \to +\infty) = E_F + \frac{1}{2}E_g \tag{8a}$$

or

$$E_c(z \to +\infty) = E_F + k_B T \ln\left(\frac{N_c}{N_d}\right), \tag{8b}$$

where N_c is the effective electron density of states, N_d is the donor impurity density in the bulk GaN, E_F is the Fermi level, E_g is the energy bandgap, all for GaN, k_B is the Boltzmann constant and T is the absolute temperature. Note that, in practical situations, this boundary condition merely affects the 2DEG concentration. The two other boundary conditions are

$$\left(\frac{dE_c}{dz}\right)_{z=0} = 8\pi n_{2D} \tag{9a}$$

$$\Psi_i(z \to \pm\infty) = 0 \tag{9b}$$

where n_{2D} is the density of two-dimensional electron gas. The value of the electron concentration at the point z, $n(z)$, is given by

$$n(z) = \sum_{i=1}^{\infty} N_i |\Psi_i(z)|^2 \tag{10}$$

where

$$N_i = \frac{k_B T}{2\pi} \ln\left[1 + \exp\left(\frac{E_F - E_i}{k_B T}\right)\right] \tag{11}$$

Equations (10) and (11) are solved using trial wavefunctions of the form:

$$\phi_n(z) = (z + z_0)e^{-b_n z} \quad \text{for } z \geq 0 \tag{12a}$$

$$\phi_n(z) = z_0 e^{(1/z_0 - b_n)z} \quad \text{for } z < 0 \tag{12b}$$

The wavefunction is obtained by Eq. (8) with C_n^i as the variational parameters.

$$\Psi_i = \sum_{n=1}^{\infty} C_n^i \phi_n \tag{13}$$

Although theoretically the summation series of Eqs. (10) and (13) can extend up to ∞, a value of $n = 10$ is sufficient for self-consistency. Correlation and exchange terms are also considered in this self-consistent calculation (Masselink, (1986). This calculation, for a given 2DEG concentration, yields the conduction band, Fermi level (which is also the quasi-Fermi level of the electrons), energy levels of the electrons in the triangular well, and electron concentration as a function of z. Another important parameter extracted from these calculation, is $\Delta E_{Fi} = E_F - E_c$ at the $Al_xGa_{1-x}N/GaN$ interface.

b. $Al_xGa_{1-x}N$ LAYER

The Poisson's equation and the current continuity equation for the $Al_xGa_{1-x}N$ region may be given by

$$\frac{d^2 E_c}{dz^2} = 8\pi[N_D^+(z) - n(z)] \tag{14}$$

$$\frac{dE_F}{dz} = \frac{J}{e\mu n(z)} \tag{15}$$

where J is the constant current density, μ the constant mobility of electrons, $N_D^+(z)$ the ionized donor concentration, and $n(z)$ the electron concentration, all in $Al_xGa_{1-x}N$. For the solution of the Poisson's Eq. (14) and the current continuity Eq. (15), which involve the quasi Fermi level of the electrons (Ponce, Masselink and Morkoç, 1985), the origin of energy is taken at the Fermi level at the $Al_xGa_{1-x}N/GaN$ interface.

The boundary conditions necessary for the solutions of Eqs. (14) and (15) are

$$E_c(0) = \Delta E_c - \Delta E_{Fi} \tag{16a}$$

$$\left(\frac{dE_c}{dz}\right)_{z=0} = 8\pi n_{2D} \tag{16b}$$

$$E_F(0) = 0 \tag{16c}$$

The current density is obtained in a self-consistent calculation involving the value of the Fermi level E_{Fm} at the semiconductor metal interface (E_{cg} is the conduction band at $Al_xGa_{1-x}N$ metal contact, and ϕ_b is the Schottky barrier height of the $Al_xGa_{1-x}N$ metal contact).

$$E_{Fm} = E_{cg} - q\phi_b \tag{17}$$

As a result of this calculation, the gate bias needed to achieve the originally given 2DEG concentration is calculated as:

$$-qV_G = E_{Fm} \tag{18}$$

By performing this numerical calculation for different values of the 2DEG concentration, a curve of this concentration as a function of the gate bias is obtained.

a. Default Parameters

For our calculations, the default parameters used for both GaN and $Al_xGa_{1-x}N$ (for values of x lower than 0.4) are listed in Table VII. Donor level in GaN depends on the impurity atom (Strite and Morkoç, 1992) for the present calculations it is assumed to be 45 meV. The energy bandgap of $Al_xGa_{1-x}N$ is modeled as

$$E_g(Al_xGa_{1-x}N) = xE_g(AlN) + (1-x)E_g(GaN). \tag{19}$$

2. 2DEG CONCENTRATION AS A FUNCTION OF ΔE_{Fi} AND CALCULATION Δd

Calculated results for ΔE_{Fi} as a function of n_{2D} are shown in Fig. 28. These are compared with a similar curve for GaAs and with a curve obtained by data fitting. The approximation used for the curve for GaAs is

$$n_{2D} = \left(\frac{\Delta E_{Fi}}{18.38}\right)^{1.5} \tag{20}$$

for which the length units are Bohr radii, and the energy unit is in Rydberg (5.31 meV in GaAs). As a result, the density of 2DEG, n_{2D}, is in $10^{12}\,cm^{-2}$ (as Bohr radius is approximately 100 Å in GaAs). Considering ΔE_{Fi} to be in meV, and using the value of the Bohr radius (27.51 Å) and of the energy unit

TABLE VII
Values of Various Default Parameters used for the Present Calculations

Parameter	Symbol for Parameter	Value
Temperature	T	300 K
Boltzmann constant	k_B	1.38066×10^{-23} J/K
Effec. electron mass in GaN	m^*	0.2
Dielectric constant (GaN)	ε_r	10.4
Effec. density of states (Elec)	N_c (GaN)	2.25×10^{18} cm^{-3}
Energy bandgap (GaN)	E_g	3.43 eV
Energy bandgap (AlN)	E_g	6.0 eV
E_g(AlGaN)-E_g(GaN)	ΔE_g	—
Conduction band discontinuity	ΔE_c	$0.82 \Delta E_g$
Metal/AlGaN barrier height	ϕ_b	1.10 eV
Electron mobility in GaN	μ_0	500 cm^2/Vs
Saturation velocity in GaN	v_{sat}	2.0×10^7 cm/s
AlGaN layer thickness	d	200 Å
Spacer layer thickness	W_{sp}	20 Å
Al mole fraction in AlGaN	x	0.25
Donor conc. in AlGaN	N_d (AlGaN)	5.0×10^{18} cm^{-3}
Channel width	W_D	40 μm
Impurity level in GaN	E_d	45 meV

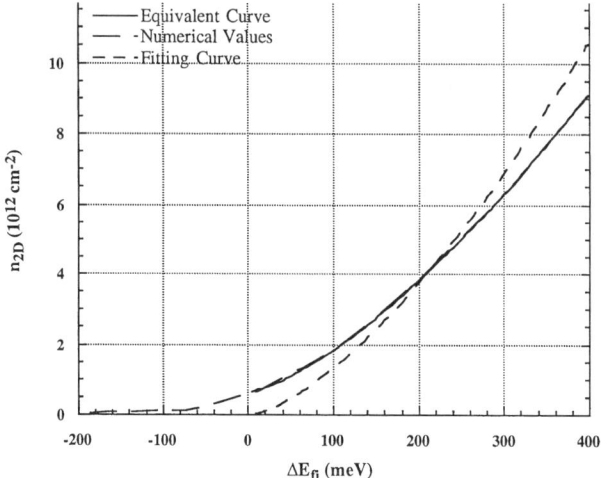

FIG. 28. Position of the Fermi level with respect to the conduction band edge, ΔE_{Fi}, in the GaN side of the heterointerface as a function of the 2DEG concentration. Curves 1, 2, and 3 correspond respectively to the equivalence with GaAs (equivalent curve), numerical simulation, and simple analytical fitting of this calculation. (From Stengel, Mohammad, and Morkoç, 1996.)

of Rydberg (25.15 meV) for GaN, the formula for n_{2D} can be simplified as

$$n_{2D} = 13.21 \times 10^{12} \left(\frac{\Delta E_n}{462}\right)^{1.5}, \text{cm}^{-2} \qquad (21)$$

For the sake of convenience and for analytical modeling, the values of n_{2D} for the range of 0.65×10^{12} cm^{-2} to 9.0×10^{12} cm^{-2} may be fitted to

$$n_{2D} = [0.65 + 1.72 \times 10^{-3} \Delta E_{Fi}^{1.42}] \times 10^{12}, \text{cm}^{-2} \qquad (22)$$

From Eq. (22) it is evident that, in order to obtain a significantly large 2DEG concentration, conduction band discontinuity must also be quite large. For example, this conduction band discontinuity should be larger than 400 meV for n_{2D} and approximately 9×10^{12} cm^{-2}.

The value of Δd, which is the average distance of 2DEG from the heterointerface, can also be obtained from numerical simulations, the result of which is presented in Fig. 29. A simple analytical formula for Δd obtained as a function of n_{2D} ranging between 0.2×10^{12} cm^{-2} and 10^{13} cm^{-2} may

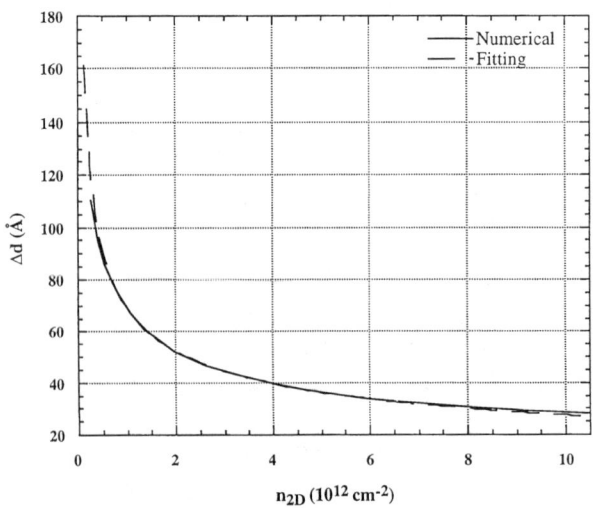

FIG. 29. Average distance (Δd) of electrons of the 2DEG from the heterointerface as a function of the 2DEG concentration. The solid curve corresponds to the numerical simulation, and the dashed curve to analytical fitting. (From Stengel, Mohammad, and Morkoç, 1996.)

be given by

$$\Delta d = \frac{69}{(n_{2D})^{0.4}}, \text{Å} \tag{23}$$

For the sake of comparison, the results from Eq. (22) are also presented in Fig. 2. From this figure it may be noted that the fitted results compare very well with the numerical ones. This suggests that, for simple analytical models, a value of Δd ranging between 30 and 60 Å would be quite reasonable, which would correspond to n_{2D} varying from 10 to 1 times 10^{12} cm^{-2}.

3. Band Diagrams for Normally on and Quasi Normally Off Modulation Doped Field Effect Transistors

Energy band diagrams for a quasi normally off (QN-OFF) and a normally on (N-ON) MODFET are shown in Fig. 30 for $x = 0.25$, $N_d = 10^{19}$ cm^{-3}, and $W_{sp} = 20$ Å. For the calculations for QN-OFF MODFET, $V_G = 0.03$ V, and $d = 130$ Å, and for calculations for N-ON MODFET, $V_G = 0.04$ V, and $d = 200$ Å. Calculations indicate that for both MODFETs, the 2DEG does not extend to the AlGaN region due to a high $Al_xGa_{1-x}N/GaN$ conduction band discontinuity (more than 500 meV as compared to 142 meV for $Al_xGa_{1-x}As/GaAs$ at $x = 0.3$). Because of this, and the fact that probability of electron wavefunctions extending to the $Al_xGa_{1-x}N$ is very low, a thinner spacer would be needed to achieve the optimal mobility in 2DEG due to a lower alloy scattering (Weisbuch and Vinter). Despite this, the effect of Coulombic scattering could presumably be opposite, especially because of the lower dielectric constant of GaN and AlGaN which leads to a higher scattering potential. Very precise calculations or experiments would be needed to resolve this matter. As is apparent from Fig. 30(a), for QN-OFF MODFET, the calculated 2DEG concentration is 2×10^{12} cm^{-2}, and maximum electron concentration in GaN is approximately 4×10^{18} cm^{-3}. The energy band diagram shows that quasi Fermi level (origin of energy axis) in GaN is very close to the first energy level. Donor atoms in $Al_xGa_{1-x}N$ are all ionized, and yet, because the conduction band is far above the Fermi level, there is no electrons present in the $Al_xGa_{1-x}N$ region. The gate bias can thus be raised to increase 2DEG concentration, and be lowered to decrease it.

The situation is, however, different in the case of the N-ON MODFET, for which the density of 2DEG is 3.8×10^{12} cm^{-2} (see Fig. 30(b)). For this

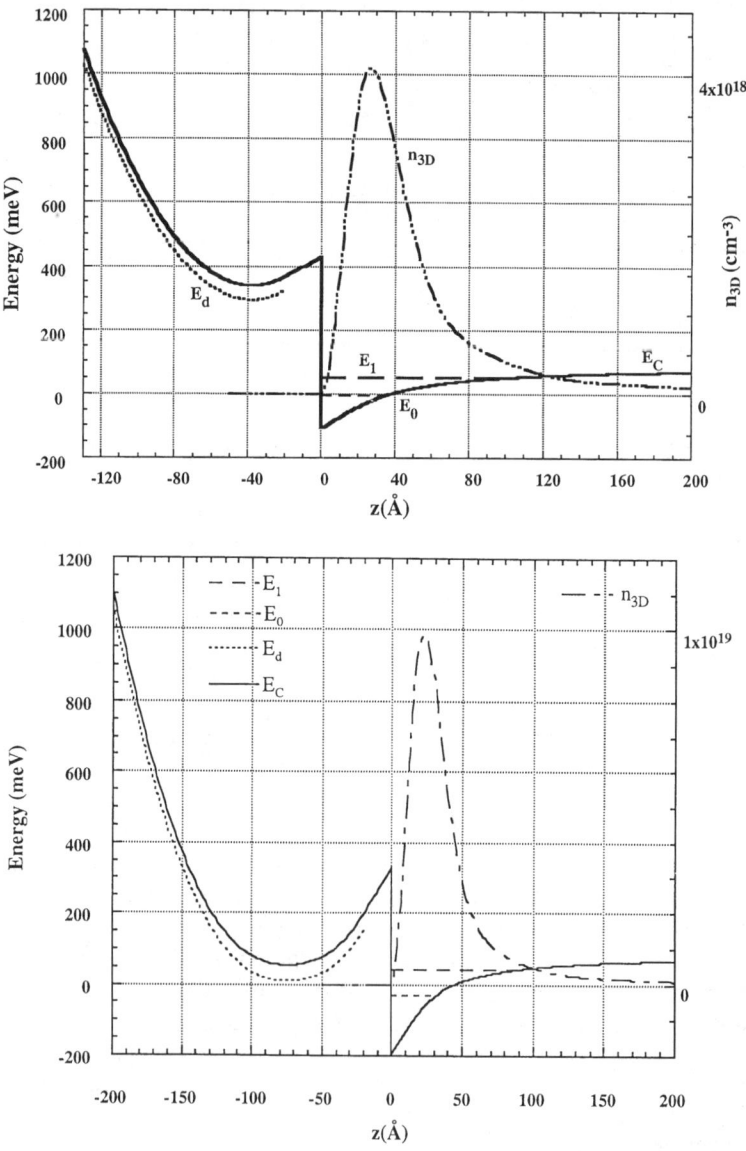

FIG. 30. Band diagram for (a) a quasi normally off and (b) a normally on MODFET. The origin of energy for these band diagrams is the Fermi level. The left side of $z = 0$ line corresponds to the AlGaN region, and the right side to the GaN. The conduction band is represented on both side by curve 1, donor level E_d in AlGaN by curve 2, the energy level E_0 in GaN by curve 3, and energy level E_1 in GaN by the curve 4. (From Stengel, Mohammad, and Morkoç, 1996.)

MODFET, the quasi Fermi level in GaN is far above the lowest energy level, and peak concentration of electron in 2DEG is 10^{19} cm^{-3}. Also, some of the donor atoms (for z between -100 Å and -50 Å) are now neutralized, and some electrons start to appear in the $Al_xGa_{1-x}N$ region. Because of these, a further rise of gate bias causes not only an increase in donor neutralization, but also an increase in electron concentration in $Al_xGa_{1-x}N$. However, the 2DEG concentration remains unaltered.

4. Modulation Doped Field Effect Transistors Utilizing Quasi Triangular Wells

Modulation doped structure, by virtue of selective doping form a quasi triangular potential well deserving a short discussion here. Following a succinct mathematical description of the problem, we will discuss the performance of devices exploiting this structure.

GaN based field-effect transistors (FETs) are projected to be highly useful for amplification and switching in high temperature and/or high power environment. The research activity, however, is still in its embryonic state. Notwithstanding, a few reports have been put forward describing fabrication and electrical characterization of some GaN FETs during the past several years, the activity is on a rapid rise. As in the case of conventional III–V area, the MODFETs based on this new heterostructure system are expected to exhibit the best performance. Consequently, only MODFETs will be treated here.

MODFET utilize a two dimensional carrier gas confined at an interface between two layers with an interfacial energy barrier such as AlGaAs/GaAs (Drummond, Masselink and Morkoç, 1986) and AlGaAs/InGaAs (Morkoç, Sverdlov and Gao, 1993). A GaN MODFET taking advantage of the background donors in the AlGaN layer is not a controllable or desirable situation and has been reported by Khan et al. (1994). Congruent with early stages of development and defect laden nature of early GaN and AlGaN layers, MODFETs exhibited a low resistance and a high resistance state before and after application of a high drain voltage (20 V). As in the case of GaAs/AlGaAs MODFETS (Fischer et al. 1984), hot electron trapping in gate insulator layer at the drain side of the gate is primarily responsible for the current collapse. Negative electron charge accumulated from this trapping causes a significant depletion of the channel layer, more probably a pinch-off, leading to a drastic reduction of channel conductance and decrease of drain current. This continues to be effective until the drain source bias is substantially increased leading to a space charge injection and giving rise to an increased drain source current.

Very recently some high performance n-channel normally off and normally on GaN/AlGaN MODFETs utilizing a doped AlGaN donor layer have been successfully fabricated in the author's laboratory (Aktas et al. 1995; Mohammad et al. 1996; Fan et al. 1996). For fabrication of these MODFETs, the GaN based heterostructure was grown epitaxially on basal plane sapphire substrate by MBE. The Hall measurements of the GaN layer yielded the electron mobility to be over $1000 \text{ cm}^2/\text{Vs}$ at 40 K, although it was found to decrease first slowly and later rapidly with temperature reaching approximately $490 \text{ cm}^2/\text{Vs}$ at 310 K. The sheet carrier concentration in the layer was 10^{13} cm^{-2}, which was effectively independent of temperature over the temperature range of 40 and 200 K. It rose to approximately $1.2 \times 10^{13} \text{ cm}^{-2}$ at 300 K. These sheet carrier concentrations are at least one order of magnitude higher than those measured for GaAs/AlGaAs and InGaAs/InAlAs systems.

The dc drain characteristics of our MODFETs (Fan et al. 1996) with a gate length of $2\,\mu\text{m}$, gate width of $40\,\mu\text{m}$, and source drain separation of $4\,\mu\text{m}$, indicate a maximum transconductance of approximately 220 mS/mm which is comparable to that obtained in GaAs. As expected the extrinsic transconductance of the MODFETs increases with decreasing gate length. Binari et al. (1994) reported GaN MODFETs with $L = 0.7\,\mu\text{m}$ and $W = 150\,\mu\text{m}$ with a room temperature extrinsic transconductance $g_m = 45 \text{ mS/mm}$ at $V_{GS} = -2\text{ V}$ and $V_{DS} = 10\text{ V}$. Similar MODFETS (Khan et al. 1995) with $L = 0.23\,\mu\text{m}$ and $W = 100\,\mu\text{m}$ exhibited, however, $g_m = 23 \text{ mS/mm}$ at $V_{GS} = -0.6\text{ V}$ and $V_{DS} = 10\text{ V}$. The transconductance performance of both of these MODFETs appears to be inferior to the ones reported by Aktas et al. (1995); Mohammad et al. (1996); Fan et al. (1996).

Continued effort in the author's laboratory led to much enhanced performance in terms of transconductance and breakdown voltages, the hallmark of what GaN has to offer. MODFETs with gate lengths of $2\,\mu\text{m}$ exhibited transconductances of approximately 180 mS/mm and drain breakdown voltages of approximately 100 V for a gate to drain spacing of $1\,\mu\text{m}$ (Fan et al., 1996). The fact that similar results, in terms of the breakdown voltage (Wu et al., 1996a,b) are being obtained at other laboratories as well is indicative of the potential of this material system for high power applications.

Using MODFET structures similar to those employed in the author's laboratory, Khan et al. (1996a,b) achieved sheet carrier concentrations of 10^{13} cm^{-2} which led to MODFETs with $1\,\mu\text{m}$ gate lengths exhibiting transconductances between 80 and 120 mS/mm. Normally on devices with $1\,\mu\text{m}$ gate lengths, exhibited current gain cut off frequencies of approximately 20 GHz which is larger than that expected from GaAs devices. Moreover, devices with quarter micron gate lengths exhibit current gain cut off and power gain cut off frequencies of approximately 40 and 90 GHz, respectively (Eastman, 1996). Preliminary simulations conducted by the author indicate that GaN should outperform GaAs for short gate length devices (Stengel, Mohammad, and Morkoç, 1996).

These preliminary devices fabricated in the authors' laboratory were tested for microwave power at Wright Laboratory (Aktas *et al.*). A normalized power level of 1.5 W/mm was obtained at 4 GHz for a device with approximately 1.5–2.0 gate length, having a gate width of 78 μm and gate to drain distance of 1 μm. This power level compares to the already exciting performance of 1 W/mm obtained at 1 GHz (Wu, *et al.*, 1996a,b). The current power level is very encouraging and compare favorably in other materials with 0.25 μm gate lengths, and much better thermal conductivity than sapphire used for the GaN/AlGaN MODFET. The small signal S parameter measurements indicated a current gain cut off frequency, meaning that frequency at which the short circuited current gain is one or zero dB is approximately 6 GHz and the maximum frequency of oscillation is approximately 11 GHz.

VI. InGaN/GaN Quantum Wells

InGaN represents an important corner piece in the triad of nitride based semiconductor heterostructures. With its smaller bandgap compared to GaN and robustness against defect propagation, it is firmly established as QW or active region of many devices based on this important class of semiconductor. At present however, layers grown with any technique are characterized with high background doping concentration and poorer overall structural quality as compared to GaN and to some extent AlGaN. Despite the above, great strides have been made on the path to GaN and InGaN films that have led to demonstration of p-n junction GaN LEDs ((Amano, Hiramatsu, and Akasaki, 1988) and followed by their commercialization (Nakamura, 1993) and demonstration of pulsed operation of injection lasers at room temperature (Nakamura, 1996).

Binary compound InN, which represents the binary end point of InGaN alloy has not received the experiment attention given to GaN and AlN, probably because of difficulties associated with growth of high quality crystalline InN samples, and existence of alternative well characterized semiconductors, such as InGaAlP and AlGaAs, which have energy bandgaps close to that (1.89 eV) for InN. The energy bandgap of InN corresponds to a portion of the electromagnetic spectrum in which alternative and well developed semiconductor technology is available. Consequently, practical applications of InN are restricted to its alloys with GaN and AlN. Growth of high quality InN and the enumeration of its fundamental physical properties remains for the present a purely scientific enterprise. In contrast though and as indicated above, InGaN is already an integral part of important device designs. $In_xGa_{1-x}N$ is not less important than $Al_xGa_{1-x}N$ for the fabrication of electrical and optical devices, such as LEDs and lasers, which can emit in violet or blue wavelengths. It can be a promising strained QW material for these devices. Based on a significant progress in growth and characterization of this material, particularly the solid solution, during the last several years, bandgap dependence of $In_xGa_{1-x}N$ on In mole fraction has been studied by a number of researchers (Osamura, Nakajima, and Murakami, 1972; Osamura, Naka, and Murakami, 1975; Nagatomo et al., 1989; Yoshimoto, Matsuoka, and Katsui, 1991). Among them, Osamura, Naka, and Murakami (1975), suggested that

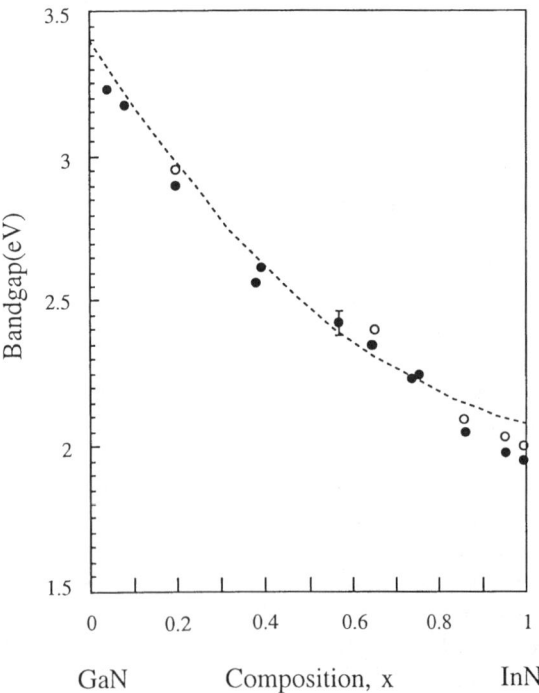

FIG. 31. Variation of the bandgap energy of $In_xGa_{1-x}N$ with InN mole fraction. The dashed line indicates the best fit to the various data points collected by Osamura, Nakajima, and Murakami (1972) and Dr. S. Nakamura.

the energy bandgap of $In_xGa_{1-x}N$ across $0 \leqslant x \leqslant 1$ would be represented by

$$E_g(x) = (1 - x)E_g(\text{InN}) + xE_g(\text{GaN}) - bx(1 - x) \qquad (24)$$

where $E_g(\text{GaN}) = 3.40 \text{ eV}$, $E_g(\text{InN}) = 2.07 \text{ eV}$, and $b = 1.0 \text{ eV}$. The observed and calculated dependencies of InGaN energy bandgap with InN mole fraction are displayed in Fig. 31. This figure demonstrates that experimental dependence of energy bandgap on In mole fraction is rather insensitive to the choice of growth chambers, and that it tends to follow a regular trend despite variations, for example, in growth conditions. The calculated results from Eq. (24) agree well with experiments as indicated in Fig. 31.

Nagatomo et al. (1989) have observed that the InN lattice constant is 11% larger than GaN. Such a large difference in lattice constants places strict limits on the In content in $In_xGa_{1-x}N$, particularly on thickness of

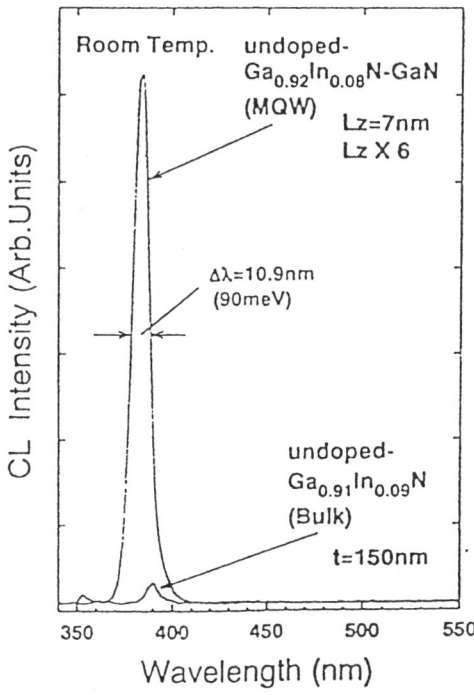

FIG. 32. Cathodoluminescence spectra of bulk $In_{0.09}Ga_{0.91}N$ and $In_{0.08}Ga_{0.92}N$/GaN MQWs indicating the superior properties of the QW structure. (From Koike et al., 1996.)

InGaN layers on GaN. Yoshimoto, Matsuoka, and Katsui (1991) studied the effect of growth conditions on carrier concentration and transport properties of $In_xGa_{1-x}N$. They observed that if deposition temperature is increased from 500 to 900°C, $In_xGa_{1-x}N$ with ($x \approx 0.2$) and grown on sapphire and later on ZnO (Matsuoka *et al.*, 1992), will suffer from a reduction in carrier concentration from 10^{20} to 10^{18} cm^{-3}, but will gain from an increase in carrier mobility from less than 10 cm^2/Vs to 100 cm^2/Vs. The work on ZnO led to good InGaN with In mole fractions as large as 23%. Nakamura and Mukai (1992) discovered that film quality of $In_xGa_{1-x}N$ can be significantly improved if these films are grown on high quality GaN films. Thus, from reports cited above it may be concluded that the major challenge of obtaining high mobility InGaN is to find a compromise in growth temperature, since InN is unstable at typical GaN deposition temperatures. This growth temperature would undoubtedly be a function of

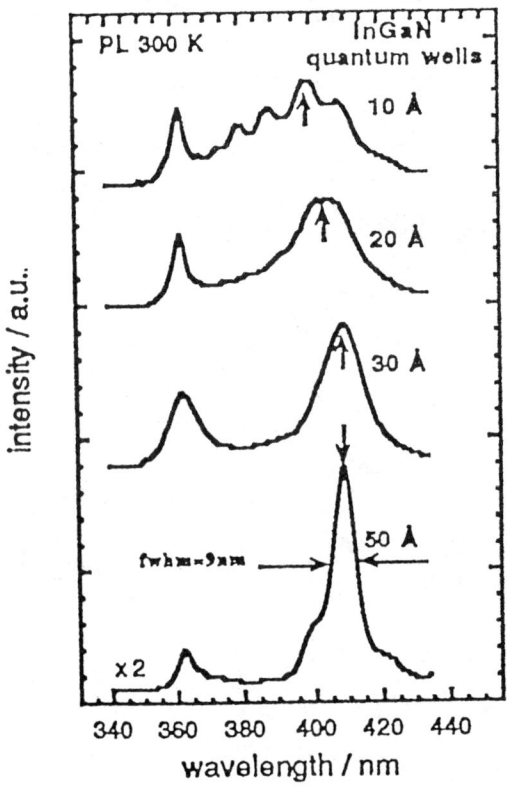

FIG. 33. The emission spectra for QWs of 10, 20, 30, and 50 Å. (From Keller *et al.*, 1996.)

dopant atoms, as well as the method (MBE, MOCVD, etc.) used for growth. This is evident from a study by Nakamura and co-workers who have since expanded the study of InGaN employing Si (Nakamura, Mukai, and Seno, 1993) and Cd (Nakamura, Iwasa, and Nagahama, 1993) dopant atoms. A review of various transport properties of GaInN and AlInN is given by Bryden and Kistenmacher (1994), and growth and mobility of p-GaInN is presented by Yamasaki et al. (1995).

The high equilibrium pressure of InGaN and its tendency to form balls following the initial quasi two dimensional growth does not lend itself to layers with reasonable quality. However, much of these restrictions are bypassed when thin layers of InGaN are sandwiched between GaN layers. This is evident in the work of Koike et al. (1996) who reported much improved cathodoluminescence in $In_{0.08}Ga_{0.92}N/GaN$ MQWs as shown in Fig. 32, compared to bulk InGaN. $GaN/In_xGa_{1-x}N/GaN$ SQWs, SQWs have also been investigated by Keller et al. (1996). Using PL as the probe for gauging expected red shift, the author noted that as the QW thickness is reduced, the luminescence intensity decreases and is accompanied by a line broadening. Shown in Fig. 33 are the emission spectra for QWs of 10,

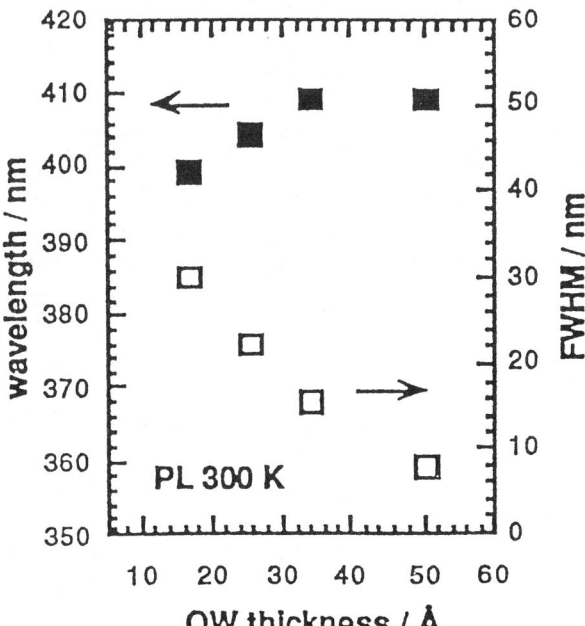

FIG. 34. The peak wavelength of emission and the line width as a function of well thickness. (From Keller et al., 1996.)

20, 30, and 50 Å. Figure 34 shows the peak wavelength of emission and the line width as a function of well thickness. Moreover, the authors indicated evidence for vertical gradient in the solid solution within the QW.

1. LIGHT EMITTING DIODES

Although the details of LEDs are discussed in the device related sections of this book, it is beneficial to treat LEDs performance briefly as high performance LEDs utilize InGaN/GaN QW structures. Perhaps the most anticipated application that has been exploited already, is InGaN QW like structures and their use in violet, blue, and green LEDs. In recent years, LEDs have undergone a tremendous advancement in performance, and are now used in just every aspect of life. The future of many technologies including printing, communications, displays, and sensors, depend profoundly on the development of compact, reliable and inexpensive light sources. Currently, conversion efficiencies of commercial LEDs emitting in the red (650–660 nm) stand at 16%, and lasers at 75% (Crawford, 1992). A primary goal of GaN research is to efficiently harness its direct energy bandgap for optical emission. Though the band edge emission in GaN occurs at approximately 362 nm, which is ultraviolet, by appropriately alloying GaN with its cousins AlN and InN the energy bandgap of the resulting Al(In)GaN can be altered for emission in the range of ultraviolet to yellow or may even be red. The first GaN LEDs were reported some twenty-five years ago by Pankove, Miller, and Berkeyheiser (1977). Due to difficulties in doping GaN p-type at the time, these LEDs were MIS LEDs, rather than p-n junction LEDs. The electroluminescence (EL) of these LEDs could be varied from blue to yellow, depending on doping of the insulator layer. Measured efficiencies of these preliminary MIS LEDs were not unfortunately sufficient for competition in the commercially available LEDs of that time.

The first GaN p-n junction LED was demonstrated by Amano *et al.* (1989). The fabricated device consisted of a Mg doped GaN layer grown on top of an undoped *n*-type ($n = 2 \times 10^{17}$ cm^{-3}) GaN film with the chemical Mg concentration estimated to be 2×10^{20} cm^{-3}. The EL of the devices was dominated by near band edge emission at 375 nm, which was attributed to transitions involving injected electrons and Mg associated centers in the p-GaN region. Additionally, a small shoulder extending to 420 nm, due to defect levels, was also observed.

One of the timely advancements in nitride effort has been exploitation of double heterostructures (DH) for light emission devices (Nakamura *et al.*, 1991; Amano *et al.*, 1989; Akasaki and Amano, 1994). The advantage of DH

LEDs over homojunction LEDs is that the entire structure outside of the active region where light is generated appears transparent. This reduces the internal absorption losses. Furthermore, this cladding region serves as an interface for scattering of light, thus minimizing the probability of total internal reflections within the device. These two factors together enhance the probability of escape of the light out of the device.

In order to achieve other desired colors, InGaN alloys for emission media are required. While increased InN mole fraction in GaN red shifts the spectrum, it would be at the expense of working against the introduction of structural defects unless InGaN is made sufficiently thin so that it is not lattice matched to GaN. Lattice mismatched films can be grown to a certain thickness called "the critical thickness" for a given composition. The larger the composition, the smaller the critical thickness. In view of this, there should be a substantial effort devoted to optimization. In this vein, band edge emissions were also obtained for LEDs employing Si doped InGaN QW as the active region in a GaN/InGaN DH LED (Amano, Hiramatsu and Akasaki, 1988). The In mole fraction content of the active layer was varied, resulting in shift of peak wavelengths of the device's EL spectra from 411 to 420 nm. Impressively, researchers at Nichia Chemical (Nakamura *et al.*, 1995) were recently successful in reducing the thickness of InGaN emission layers to approximately 30 Å. With this achievement, InGaN QWs with InN mole fractions up to 70% have been obtained, and LEDs with commercial capabilities are now possible in blue, green, and yellow.

The blue and blue-green LEDs developed by Nichia Chemical initially relied on the transitions to Zn centers in InGaN. This was necessitated by the need to extend the wavelength to the desired values and the amount of In that could be added was limited while maintaining good crystalline quality. These LEDs suffered from wide spectral widths and saturation in output power with injection current at the intended wavelength of operation. The large spectral width spoiled color saturation with the undesirable outcome that not all the colors could be obtained through color mixing. However, with the QW approach, In mole fraction can be extended to approximately 70%, paving the way for excellent ultraviolet, violet, blue, green, and yellow LEDs. These LEDs exhibit power levels of 5, 3, and 1 mW at 20 mA of injection current for the wavelengths of 450, 525, and 590 nm, respectively.

Very important is the fact that the full width at half maximum (FWHM) of the spectrum was 20, 30, and 80 nm, for blue, green, and yellow LEDs, respectively, owing to the fact that these new LEDs take advantage of band-to-band transitions. The In mole fractions used are 20, 43, and 60%, for the 450, 525, and 590 nm emission, respectively. At the wavelength corresponding to green, the 3 mW power level which corresponds to 12 cd in a 10 degree viewing cone, was obtained with an efficiency of 6.3%. Most recent results for GaN and ZnTeSe material systems are over 20 cd level. In short, at an injection current of 20 mA, blue, green, and yellow LEDs produce 5 mW, 2.5 cd, (efficiency = 9.1%), 3 mW (efficiency = 6.3%), and 1 mW (efficiency = 2.3%) at blue, green, and yellow wavelengths. The ultraviolet LEDs operating at 400 nm exhibit efficiencies of 10% at $I = 20$ mA. At the current injection level of 100 mA, 400, and 450 nm, LEDs exhibit power levels of 13 and 12 mW, respectively.

Continuing on their long term activity, Koike *et al.* (1995) at Toyota Gosei, using an asymmetric GaN/InGaN/AlGaN DH design, obtained blue InGaN/GaN/AlGaN LEDs with excellent performance. For example, with injection currents of 20 mA (efficiency = 3.9%) and 100 mA, the output power is 3 and 9 mW, respectively, for blue LEDs. Since LEDs rely on the Zn centers in InGaN, the spectral width is approximately 70 nm. Although not utilizing InGaN, for the sake of completeness, we mention that Cree Research (Edmond, 1995), also has marketed blue GaN based LEDs grown on SiC which take advantage of extrinsic transitions involving Mg induced defect levels in GaN. As of September 1995, the Cree devices exhibit 1.4 mW with an external efficiency of 2.4%. One purported advantage of the Cree device is that it is on conducting SiC substrate in a vertical geometry.

While blue and green InGaN based LEDs are finding increasing applications in displays and traffic lights, the attention is turning onto white light generation and illumination. One approach is to use the best red, green, and blue LED technologies together in a primary color mixing configuration. In this scheme the red will be either AlGaAs or InGaAlP based, while the green and blue will be provided by InGaN. Another approach for production of white light utilizes a bright violet LED which pumps a medium that has in it dyes corresponding to the three primary colors as was recently demonstrated by Nichia Chemical. The success of this approach will depend on the efficiency of conversion. At present, violet LEDs are about 10% efficient and further enhancements can be expected with the development of new substrates.

VII. InGaN/InGaN Quantum Wells

The ternary semiconductor compound, InGaN, has a direct bandgap tuning in the range from 1.95 ($x = 1$) to 3.4 eV ($x = 0$) at room temperature. Thin InGaN layers, in the thickness range which could be considered the QW region, straddled by GaN and/or AlGaN layers for LEDs. Moreover,

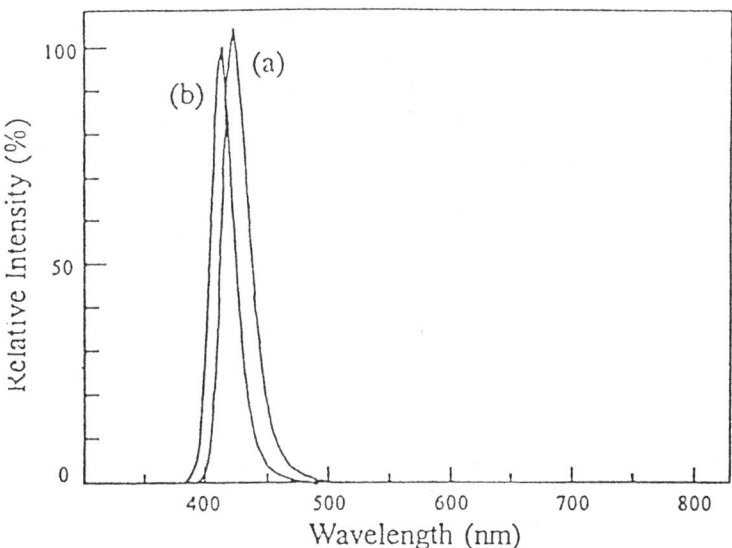

FIG. 35. The PL emission spectra for two $In_xGa_{1-x}N/In_yGa_{1-y}N$ periodic heterostructures samples having periodicity of 60 and 200 Å. (From Nakamura et al., 1993.)

InGaN/InGaN MQW layers where the mole fraction is varied from the wells to the barriers were used to produce injection lasers for short wavelength operation. Accurate spectroscopic work however, is lacking to extract confinement energies. Nevertheless, we will discuss available data on the topic followed by a succinct treatment of lasers which utilize InGaN/InGaN QWs.

Nakamura et al. (1993) has undertaken a study of $In_xGa_{1-x}N/In_yGa_{1-y}N$ heterostructures which, as indicated above, form active layers of lasers to follow. Structural as well as emission characteristics were analyzed by X-ray diffraction and PL. The former was used to determine the periodicity and the latter for determining the emission intensity and energy. The PL emission spectra for two such samples having periodicity of 60 and 200 Å, as determined by X-ray diffraction, also show the expected satellite peaks, (Fig. 35). Using a Kronig-Penny formulation with the applicable and available carrier effective masses of 0.2 and 0.8 for electron and hole effective masses, respectively, in GaN, and 0.11 and 1.6 for electron and hole relative effective masses, respectively, in InN (Strite and Morkoç, 1992) the well layer was determined to be $In_{0.22}Ga_{0.78}N$. We should note that since publication of these papers, preponderance of available data indicate the relative electron effective mass in GaN to be 0.23. What is technologically relevant here is that structures of this kind were used to optimize growth conditions which laid the ground work for high performance LEDs to be produced. It should be noted that as part of this particular study, the PL emission peak position in bulk InGaN was also investigated with varying layer thickness resulting in the observation that peak position is thickness independent. This would imply that any In surface segregation is minimized. What is striking is that the structures investigated and optimized are in many ways similar to the active layers used by Nakamura and his colleagues in their laser structures.

1. Use of $In_xGa_{1-x}N/In_yGa_{1-y}N$ Quantum Wells in Laser Diodes

As was the case of InGaN/GaN QWs, while the details of lasers are covered in the device sections of this book, we will, however, mention laser results here as the active laser used in those devices are of QW nature. Unlike display and lighting applications, digital information storage and reading require coherent light sources, namely, lasers. The output of these coherent light sources can be focused to a diffraction limited spot, permitting an optical system in which bits of information can be recorded and read with ease and uncommon accuracy (Asthana, 1994). As the wavelength of the light gets shorter, the focal diameter becomes smaller. The interim approach adopted by the industry relies on red lasers with which pit dimensions of approximately 0.4 μm can be read. Using a two layer scheme in what has been dubbed as the digital versatile disk (DVD), the density can be increased from today's 1 Gb to approximately 17 Gb for each compact disk (Bell, 1996). The cycle in the consumer electronics is rather short in that even if the red laser based DVDs are implemented, the violet or the blue laser can be implemented two years after the red lasers. For consumer applications, CW operation lifetimes on the order of 10,000 h at 60°C are required.

Significant gains in recording density can be achieved by using potential GaN based lasers that operate at shorter wavelength(s). The nitride based materials system, when adapted to semiconductor lasers in blue and ultraviolet wavelengths, offers increased data storage capacity, possibly in excess of 40 Gb for each compact disk. This prospect is luring numerous corporations across the globe to rush to build lasers for an affordable compact digital information storage system. Present approaches rely on wavelength doubling of high power red lasers, and are bulky and expensive. Consequently, compact coherent sources emitting at the wavelength of interest are preferable. Although II–VI compound semiconductors, such as zinc-selenide ($E_g = 2.67$ eV) and its pseudomorphic ternary and lattice matched quaternary alloys led to the demonstration of 490 nm diode lasers, they suffer from several weaknesses (Haase et al., 1991; Salovatke et al., 1993). For example, the ZnSe has low thermal conductivity, poor thermal stability, large ohmic contact resistance, and a low damage threshold. Though tremendous strides have been made at Sony Laboratories and elsewhere, operating lifetimes of Sony devices are apparently only on the order of 200 hours.

The injection level required for achieving the lasing condition or for rendering the active layer nonabsorbing at the wavelength of emission in GaAs is approximately 1.6×10^{18} cm^{-3}. At this injection level the separation of the quasi Fermi levels equals the bandgap of the active layer (Morkoç, Sverdlov, and Gao, 1993). The lower the injection level, the lower the injection current requirement for lasing condition or transparency. It then follows that in semiconductors with large density of states (joint density of states), the transparency condition is reached at higher injection levels. Using the hexagonal band parameters, injection level for transparency in GaN based lasers with GaN active layers is approximately 6×10^{18} cm^{-3}. Simply put, all things being equal, the transparency current for GaN lasers is approximately 3 to 4 times larger than in GaAs based lasers.

Uenoyama and Suzuki (1996) considered the gain in bulk GaN and 40 Å thick QW lasers utilizing the wurtzitic and zinc-blende phases. With the joint density of states, polarization effects/matrix elements, and the occupation probabilities taken into account, they calculated the gain versus normalized injection current. The unique valence band nature of wurtzitic GaN where the spin–orbit and crystal field splitting lead to three bands, results in extremely large polarization effects and large gain, which is much larger than that for the zinc-blende case. The zinc-blende structure, however, exhibits much enhanced performance in the QW case because of its enhanced matrix elements. Nevertheless, a laser based on the zinc-blende polytype of GaN would still be inferior to a wurtzitic laser. This exercise leads one to conclude that when, and if, quality of GaN based heterostructures is improved to the point where scattering losses and nonradiative recombination losses are lowered and smooth facets are formed, excellent lasers can be expected.

As anxiously anticipated, Nakamura et al. (1996) reported the observation of laser oscillations at room temperature in InGaN QWs utilizing GaN waveguides and AlGaN cladding layers. Although the initial laser reported utilized 26 periods of $In_{0.2}Ga_{0.8}N/In_{0.05}Ga_{0.95}N$ MQW structures consisting of 25 Å thick $In_{0.2}Ga_{0.8}N$ well layers and 50 Å thick $In_{0.05}Ga_{0.95}N$ barrier layers, recent structures incorporate MQWs with as little as 7 periods. A 200 Å thick p-type $Al_{0.2}Ga_{0.8}N$:Mg layer was employed to prevent disassociation of the InGaN layers during growth of subsequent

GaN and AlGaN layers which require much higher substrate temperatures. The 0.1 μm thick layer of n-type $In_{0.1}Ga_{0.9}N$ was imbedded in the buffer layer to prevent cracking. Since it is difficult to cleave the c face sapphire substrates, reactive ion etching (RIE) was employed to form the cavity facets. High reflectivity facet coatings (60–70%) were used to reduce the threshold current. A Ni/Au contact was evaporated onto the entire area of the p-type GaN layer, and a Ti/Al contact onto the n-type GaN layer. Akasaki et al. (1996) also reported lasing action in a single $In_{0.1}Ga_{0.9}N$ at 376 nm, (shortest of any injection semiconductor laser) in separate confinement structure (SCH) with a threshold current of $2.9 \, kA/cm^2$ at room temperature under pulsed operation, duty cycle of 1% and pulse width of 0.3 μs.

The lasers utilizing films grown on the c plane of sapphire had etched cavities as cleavage is not practical because sapphire does not cleave well. This is complicated further by GaN epilayers rotating with respect to the underlying sapphire making it impossible to align the cleavage plane, a plane, of GaN and sapphire. It is for this reason, Nakamura et al. (1996) explored lasers grown on the a plane of sapphire. Even though material quality does not compare to that on the c plane, improved cavity formation along the R plane ($1\bar{1}02$) outweighs its reduced materials quality. Moreover, lasr structures on (111) $MgAl_2O_4$ spinel substrates, led to wurtzitic GaN along the c plane being explored for optical pumping (Kuramata et al., 1995) and injection laser experiments (Nakamura et al., 1996). In this scheme, spinel cleaves along the (100) plane inclined to the surface with cleavage following the R plane of GaN to approximately where the epilayers are reached. Even though the facet quality in this scheme is the best among the aforementioned approaches, material quality degradation is too severe to push it ahead of other approaches. It is, however, very clear that a substrate with good cleavage characteristics and on which GaN can be grown without rotation is desperately needed. While SiC meets some of these criteria, poor surface quality of SiC as far as growth is concerned is making its implementation difficult although optically pump stimulated emission has been reported (Zubrilov et al., 1995; Song). Until such time as native substrates are available, ZnO meets the above criteria. Again, this approach too has been hampered by the scarcity of high quality substrates prepared by the sublimation technique.

Room temperature pulsed operation of injection lasers having an InGaN MQW laser with 7 periods have been reported by Nakamura (unpublished). The threshold current is slightly over 300 mA and voltage across the device at threshold is approximately 30 V. Needless to say, the voltage drop which is largely attributed to large contact resistance on p-type GaN must be reduced for reliable CW operation. Nakamura (unpublished) recently lowered the forward voltage drop to under 10 V with CW operation for 5 h at $-40°C$ and by approximately 1 h at room temperature. For injection levels above the threshold current, the device shows strong cavity modes. Far field pattern of the device for both in-plane and out-plane directions obtained with polarization sensitive power measurements indicates that TE mode dominates over TM mode above the threshold current. Although the sum of all these figures culminates in the conclusion that devices exhibit lasing oscillations there are still issues to be worked out about the spectral purity and lack of consistent agreement between the cavity length and mode spacing. The fact that GaN epitaxial layer is not commensurate with the c face of sapphire substrates and it is rotated with respect to the substrate to minimize strain, the weak cleavage plane of sapphire and cleavage plane of GaN do not align. To circumvent this problem, Nakamura et al. (1995) has grown laser structures on a plane of sapphire where the substrate and GaN cleavage planes align, leading to reduced threshold current densities.

VIII. Conclusions

Even though the GaN based structures in general and QWs in particular are characterized with many structural and electrical defects, optical and electrical properties of these structures as well as the devices based on them have been impressive and to some extent unpredictable. This is because a whole body of work indicating clearly that defects of this extent in other semiconductors proved to be insurmountable barriers. Obviously, nitrides differ from other semiconductors, noticeably in that they are very robust and additional defects are not easily created during device operation even when such device operation are, for example, injection lasers. All indicators point to a future where heterostructures constructed in GaN will push the edge of the envelope in device performance—optoelectronic emitters and detectors and high power electronic devices—with wide spread use.

ACKNOWLEDGMENTS

The work at the University of Illinois was funded by grants from ONR and AFOSR and monitored by Drs. G. L. Witt, Y. S. Park, C. E. C. Wood, and Mr. M. Yoder. The authors are indebted to Donna Guzy for editorial assistance as well as preparing some of the figures. One of us, H.M. would like to thank his colleagues at Wright Laboratory, namely, Drs. C. W. Litton, D. C. Reynolds, D. C. Look, P. Hemenger, W. Mitchel, and A. Garscadden for many useful discussions and support.

REFERENCES

Ahn, D. (1994). *J. Appl. Phys.* **76**, 8206 (1994).
Akasaki, I., and Amano, H. (1994). *J. Electrochem. Soc.* **141**, 2266–2271 (1994).
Akasaki, I., Sota, S., Sakai, H. Tanaka, T., and Amano, H. (1996). *Electron. Letts.* **32**, 1105.
Aktas, Ö., Fan, Z., Mohammad, S. N., Roth, M., Jenkins, T., Kehias, L., and Morkoç, H. *IEEE Electron Dev. Lett.* (submitted).
Aktas, Ö., Kim, W., Fan, Z., Botchkarev, A. E., Salvador, A., Mohammad, S. N., Sverdlov, B., and Morkoç, H. (1995). *Electron. Lett.* **31**, 1389.
Albanesi, E. A., Lambrecht, W. R. L., and Segall, B. E. (1994). *Proc. Mater. Res. Soc. Symp.*, Vol. 339.
Amano, H., Hiramatsu, K., and Akasaki, I. (1988). *Jpn. J. Appl. Phys.* **27**, L1384.
Amano, H., Hiramatsu, K., and Akasaki, I. (1988). *J. Appl. Phys.* **27**, L1384.
Amano, H., Kito, M., Hiramatsu, K., and Akasaki, I. (1989). *Jpn. J. Appl. Phys.* **28**, L2112.
Asthana, P. (1994). IEEE Spectrum, **31**, 60.
Bastard, G. (1982). *Phy. Rev. B.* Vol. 25, pp. 7584–7597.
Bastard, G., and Brum, J. A. (1986). *IEEE J. Quantum Elect.* QE-**22**, 1625.
Baur, J., Maier, K., Kunzer, M. Kaufmann, U., and Schneider, J. (1994). *Appl. Phys. Lett.* **65**, 2211.
Bell, A. (1996). Scientific American, pp. 42–46.
Binari, S. C., Rowland, L. B., Kelner, G. Kruppa, W., Dietrich, H. B., Doverspike, K., and Gaskill, D. K. (1994). *21st Int. Symp. Compound Semicond. Proc.*
Landolt-Börnstein (1971). Numerical Data and Functional Relationships in Science and Technology (Eckerlin, P. and Kandler, H., eds.), Springer-Verlag, Berlin.
Bryden, W. A., and Kistenmacher, T. J., (1994). In *Properties of Group III Nitrides*, (Edgar, J. H., ed.), (Published by INSPEC, Institution of Electrical Engineers, London, U.K.), pp. 117–124.
Bykhovski, A., Gelmont, B., and Shur, M. (1993). *J. Appl. Phys.* **74**, 6734.
Chin, V. W. L., Tansley, T. L., and Osotchan, T. (1994). *J. Appl. Phys.* **75**, 7365.
Chuang, S. L., and Chang, C. S. (1996a). *Phys. Rev B* **54**, 54.
Chuang, S. L., and Chang, C. S. (1996b). *Appl. Phys. Lett.* **68**, 1127.
Crawford, M. G. (1992). IEEE Circuits and Devices Mag. **8**, 24; Bhargava, R. M. (1992). Optoelectronics, Devices and Technol. **7**, 19–47.
Davis, R. F. (1991). *Proc. IEEE*, **79**, 702.
Dingle, R., Sell, D. D., Stokowski, S. E., and Ilegems, M. (1971). *Phys. Rev. B.* **4**, 1211.
Drummond, T. J., Masselink, W. T., and Morkoç, H. (1986). *Proc. of IEEE*, **74**, 773.
Eastman, L. F. private communication. Burm, A. U., Schaff, W. J. Eastman, L. F., Amano, H., and Akasaki, I. (1996). *Appl. Phys. Lett.*, **68**, 2849.
Edgar, J. H. (1994). "Properties of Group II Nitrides", Ed. London, U.K.: Inspec IEE.
Edmond, J. (1995). Topical Meeting on II-Nitrides, Nagoya, Japan.
Esaki, L., and Tsu, R. (1970). *IBM J. Res. Dev.* **14**, 61.

Fan, Z., Mohammad, S. N., Aktas, Ö., Botchkarev, A., Salvador, A., and Morkoç, H. (1996). *Appl. Phys. Letts.* **69**, 1229.
Fiorentini, Y., Methfessel, M., and Scheffler, M. (1993). *Phys. Rev. B.* **47**, 13353.
Fischer, R., Drummond, T. J., Klem, J., Kopp, W., Henderson, T., Perrachione, D., and Morkoç, H. (1984). *IEEE Trans. Electron. Dev.* ED-**31**, 1028.
Gil, N., Briot, O., and Aulombard, R. L. (1995). *Phys. Rev. B.* **52**, R17028.
Gil, B., Hamdani, F., and Morkoç, H. (1996). *Phys. Rev. B.* **54**, 7678.
Haase, M. A., Qui, J., Depuydt, J. M., and Cheng, H. (1991). *Appl. Phys. Lett.* **59**, 1272.
Hopfield, J. J., and Thomas, D. G. (1963). *Phys. Rev.* **132**, 563.
Itoh, K., Kawamoto, T., Amano, H., Hiramatsu, K., and Akasaki, I. (1991). *Jpn. J. Appl. Phys.* **30**, 1924.
Kamiyama, S., Ohnaka, K., Susuki, M., and Uenoyama, T. (1995). *Jpn. J. Appl. Phys.* **34**, L821.
Kawabe, K., Tredgold, R. H., and Inuishi, Y. (1967). *Elect. Eng. Jpn.* **87**, 62; Balkanski, M., and de Cloizeaux, J. (1960). *J. Phys. Radium* **21**, 825.
Keller, S., Keller, B. P., Masui, H., Kapolnek, D., Abare, A., Mishra, U., Coldren, L., and Denbaars, S. (1996). Proceeding of the Symposium on Blue Laser and Light Emitting Diodes, Chiba Univ. Japan., p. 54; *Appl. Phys. Letts.*, **68**, 3147.
Khan, M. A., Chen, Q., Shur, M. S., Dermott, B. T., Higgins, J. A., Burm, J. Schaff, W., Eastman, L. F. (1996a). *Elect. Lett.* **32**, 357; Khan, M. A., Chen, Q., Wang, J. W., Shur, M. S., Dermott, B. T., and Higgins, J. A. (1996b). *IEEE Elect. Dev. Lett.*, **17**, 325.
Khan, M. A., Shur, M. S., Chen, Q. C., and Kuznia, J. N. (1994). *Electron. Lett.* **30**, 2175.
Khan, M. A., Shur, M. S., Kuznia, J. N., Cheng, Q. C., Burm, J., and Schaff, W. (1995). *Appl. Phys. Lett.* **66**, 1083.
Khan, M. A., Skogman, R. A., van Hove, J. M., Krishnankutty, S., and Kolbas, R. M. (1990). *Appl. Phys. Letts.* **56**, 1257.
Klein, M. V. (1986). *IEEE, Quantum Electronics*, QE-**22**, 1760.
Koike, M., Yamasaki, S., Nagai, S., Koide, N., Asami, S., Amano, H., and Akasaki, I. (1995). Topical Meeting on III-Nitrides, Nagoya, Japan.
Koike, M., Yamasaki, S., Nagai, S., Koide, N., Asami, S., Amano, H., and Akasaki, I. (1996). *Appl. Phys. Lett.* **68**, 1403.
Krishnakutty, S., Kolbas, R. M., Khan, M. A., Kuznia, J. N., Van Hove, J. M., and Olson, D. T. (1992a). *J. Electron Mater.* **21**, 437.
Krishnakutty, S., Kolbas, R. M., Khan, M. A., Kuznia, J. N., Van Hove, J. M., and Olson, D. T. (1992b). *J. Electron Mater.* **21**, 609.
Kuech, T. F., Collins, R. T., Smith, D. L., and Mailhiot, C. (1990). *J. Appl. Phys.* **67**, 2650.
Kuramata, A., Horino, K., Domen, K., Shinohora, K. and Tanahashi, T. (1995). *Appl. Phys. Letts.* **67**, 2521.
Langer, J. M., and Heinrich, H. (1985). *Phys. Rev. Lett.* **55**, 1414.
Martin, G. A., Botchkarev, A., Agarwal, A., Rockett, A., Morkoç, H. (1996a). *Appl. Phys. Lett.* **68**, 2541; Martin, G. A. (1996). Ph.D. Thesis, Univ. of Illinois.
Marzin, J. Y. (1982). In *Heterojunction and Semiconductor Superlattices*, Springer Verlag, pp. 161–176.
Martin, G., Strite, S., Botchkarev, A., Agarwal, A., Rockett, A., Morkoç, H., Lambrecht, W. R. L., and Segall, B. (1994). *Appl. Phys. Lett.* **65**, 610.
Masselink, W. T. (1986). Ph.D. Thesis, University of Illinois at Urbana-Champaign.
Masselink, W. T., Chang, Y. C., and Morkoç, H. (1983). *Phys. Rev. B.* **28**, 7373.
Matsuoka, T., Yoshimoto, N., Sasaki, T., and Katsui, A. (1992). *J. Electron. Mater.* **21**, 157–163.
Matthews, J. W., and Blakeslee, A. E. (1976). *J. Crystal Growth*, Vol. 32, pp. 265–283; Dodson, B. W., and Tsao, J. (1987a) *Appl. Phys. Lett.*, Vol. 51, pp. 1325–1327; Dodson, B. W., and

Tsao, J. (1987b). *Phys. Lett.*, **51**, 1325; *Appl. Phys. Lett.*, Vol. 52, p. 852; People, R., and Bean, J. C. (1985). *Appl. Phys. Lett.*, Vol. 47, pp. 322–324.
Mohammad, S. N., Fan, Z., Aktas, Ö., Botchkarev, A., Salvador, A., and Morkoç, H. (1996). *Appl. Phys. Letts.* **69**, 1420.
Mohammad, S. N., and Morkoç, H. "Progress and Prospects of Group III-V Nitride Semiconductors", *Progress in Quantum Electronics*, in press.
Mohammad, S. N., Salvador, A., Morkoç, H. (1995). "Emerging GaN Based Devices", *Proc. IEEE* **83**, 1306.
Morkoç, H., and Mohammad, S. N. (1995). "High Luminosity Gallium Nitride Blue and Blue-Green Light Emitting Diodes", *Science Magazine*, Vol. 267, pp. 51–55.
Morkoç, H., Strite, S., Gao, G. B., Lin, M. E., Sverdlov, B., and Burns, M. (1994). *J. Appl. Phys. Reviews*, Vol. 76, No. 3. pp. 1363–1398.
Morkoç, H., Sverdlov, B., and Gao, G. B. (1993). *Proceeding of IEEE*, **81**, 492.
Morkoç, H. "III-V Nitrides and Silicon Carbide for Photonic Materials and Devices". In CRC Handbook of Photonics (Gupta, M. S. (ed.)), in press.
Nagatomo, T., Kuboyama, T., Minamino, H., and Omoto, O. (1989). *Jpn. J. Appl. Phys.* **28**, L1334–L336.
Nagatomo, T., Kuboyama, K., Minamino, H., and Omoto, O. (1989). *Jpn. J. Appl. Phys.* **28**, L1334–L1336.
Nakamura, S. unpublished.
Nakamura, S., Iwasa, N., and Nagahama, S. (1993). *Jpn. J. Appl. Phys.* **32**, L338–341.
Nakamura, S., and Mukai, T. (1992). *Jpn. J. Appl. Phys.* **31**, L1457–L1459.
Nakamura, S., Mukai, T., and Seno, M. (1993). *Jpn. J. Appl. Phys.* **31**, L16–L19.
Nakamura, S., Mukai, T., and Senoh, M. (1991). *Jpn. J. Appl. Phys.* **30**, L1998.
Nakamura, S., Mukai, T., Senoh, M., Nagahama, S. I., and Iwasa, N. (1993). *J. Appl. Phys.* **74**, 3911.
Nakamura, S., Senoh, M., Isawa, N., and Nagahama, S. (1995). *Jpn. J. Appl. Phys.* **34**, L797.
Nakamura, S., Senoh, M., Isawa, N., and Nagahama, S. (1995). *Jphn. J. Appl. Phys.* **34**, L797.
Nakamura, S., Senoh, M., Nagahama, S., Iwasa, N., Yamada, T., Matsushita, T., Kiyoku, H., and Sugimoto, Y. (1996). *Appl. Phys. Letts.* **68**, 2105.
Nakamura, S., Senoh, M., Nagahama, S., Iwasa, N., Yamada, T., Matsushita, T., Kiyoku, H., and Sugimoto, Y. (1996). *Jpn. J. Appl. Phys.* **35**, L74.
Nakamura, S., Senoh, M., Nagahama, S., Iwasa, N. Yamada, T., Matsushita, T., Kiyoku, H., and Sugimoto, Y. (1996). *Jpn. J. Appl. Phys.* **35**, L74.
Nakamura, S., Senoh, M., Nagahama, S., Iwasa, N., Yamada, T., Matsushita, T., Kiyoku, H., and Sugimoto, Y. (1996). *Jpn. J. Appl. Phys.* **35**, L217.
Orton, J. M. (1995). *Semicond. Sci. Technol.* **10**, 101.
Osamura, K., Naka, S., and Murakami, Y. (1975). *J. Appl. Phys.* **46**, 3432–3438.
Osamura, K., Nakajima, K., and Murakami, Y. (1972). *Solid State Commun.* **11**, 617–621.
Osburn, G. C. (1985). *J. Vac. Sci. & Technol.*, **A3**, 826; Morkoç, H., Sverdlov, B. and Gao, G. B. (1993). *Proceeding of IEEE*, **81**, 492.
Pankove, J. I., Miller, E. I., and Berkeyheiser, J. E. (1971). *RCA Rev.* **32**, 383.
Pollak, F. H., and Cardona, M. (1968). *Phys. Rev. B.* **172**, 816.
Ponce, F., Masselink, W. T., and Morkoç, H. (1985). *IEEE Trans. Electron Devices* ED-**32**, 1017.
Reynolds, D. C., Bajaj, K. K., Litton, C. W., Yu, P. W., Masselink, W. T., Fischer, R., and Morkoç, H. (1984). *Phys. Rev. B, Rapid Commun.*, Vol. 29, pp. 7038–7040.
Reynolds, D. C., Look, D. C., Kim, W., Aktas, O., Botchkarev, A. E., Salvador, A., Morkoç, H., and Talwar, D. N. (1996). *J. Appl. Phys.* **80**, 594.
Salovatke, A., Jeon, H., Hoviven, M., Kelkar, P., Nurmikko, A. V., Grillo, D. C., He, L., Han, J. Fan, Y., Ringle, M., and Gunshor, R. L. (1993). *Electron. Lett.* **29**, 2041.

Salvador, A., Liu, G., Kim, W., Aktas, Ö., Botchkareve, A., and Morkoç, H. (1995). *Appl. Phys. Lett.* **67**, 3322.
Sanders, G. D., and Chang, Y. C. (1985). *Phys. Rev. B.* **31**, 6892.
Shan, W., Schmidt, T. J., Haustein, R. J., Song, J. J., and Goldenberg, B. (1995). *Appl. Phys. Lett.*, **66**, 3492.
Sirenko, Yu. M., Jeon, J. B., Kim, K. W., Littlejohn, M. A., and Stroscio, M. A. (1996). *Phys. Rev. B.* **53**, 1997.
Sitar, Z., Paisley, M. J., Yan, B., Davis, R. F., Ruan, J., and Choyke, J. W. (1991). *Thin Solid Films* **200**, 311.
Sitar, Z., Paisley, M. J., Yan, B., Ruan, J., Choyke, W. J., and Davis, R. F. (1990). *J. Vac. Sci. Technol. B.* **8**, 316.
Smith, D. L. (1986). *Solid State Communications* **57**, 919.
Smith, M., Lin, J. Y., Jiang, H. X., Salvador, A. Botchkarev, A., and Morkoç, H. (1996). *Appl. Phys. Letts.* **69**, 2453.
Song, J.-J. private communication.
Stengel, F., Mohammad, S. N., and Morkoç, H. (1996). *J. Appl. Phys.* **80**, 3031.
Strite, S. T., Lin, M. E., and Morkoç, H. (1993). In *Thin Solid Films*, 231(1–2), pp. 197–210. Special Issue: "Properties and Preparation of Emerging Device Heterostructures" (Morkoç, H. (ed.)).
Strite, S. T., and Morkoç, H. (1992). "GaN, AlN, and InN: A Review", *Journal of Vacuum Science and Technology*, Vol. B10, pp. 1237–1266.
Suzuki, M., and Uenoyama, T. (1996a). *Jpn. J. Appl. Phys.* **35**, 1421.
Suzuki, M., and Uenoyama, T. (1996b). *Phys. Rev. B* **52**, 8132.
Uenoyama, T., and Suzuki, M. (1995). *Appl. Phys. Lett.* **67**, 2527.
Uenoyama, T., and Suzuki, M. (1996). *Proc. of the International Symposium on Blue Light Emitting Diodes and Lasers*, Chiba, Japan; *Inst. Phys. Conf. Ser. No.* 142, 915.
Uenoyama, T., and Suzuki, M. (1996). *Proceeding of the Symposium on Blue Laser and Light Emitting Diodes*, Chiba Univ. Japan. p. 271.
Varshni, Y. P. (1967). *Physica* **34**, 149.
Waldrop, J. P., and Grant, R. W. (1996). *Appl. Phys. Lett.* **68**, 2879.
Weisbuch, C., and Vinter, B. *Quantum Semiconductor Structures*, Academic Press, San Diego.
Weisbusch, C., and Vinter, B. *"Quantum Semiconductor Structures: Fundamentals and Applications"*, Academic Press, San Diego, CA 1991, p. 11.
Wu, Y. F., Keller, B. P., Keller, S., Kapolnek, D., Kozodoy, P., Denbaars, S. P., and Mishra, U. K. (1996a) *Appl. Phys. Lett.* **69**, No 10; Wu, Y. F., Keller, B. P., Keller, S., Kapolnek, D., Denbaars, S. P., and Mishra, U. K. (1996b). *IEEE Elect. Dev. Lett.* **17**, 455.
Yamasaki, S., Asami, S., Shibata, N., Koike, M., Manabe, K., Tanaka, T., Amano, H., and Akasaki, I. (1995). *Appl. Phys. Lett.* **66**, 1112–1114.
Yoshimoto, N., Matsuoka, T., and Katsui, A. (1991). *Appl. Phys. Lett* **59**, 2251–2253.
Yu, E. T., McCaldin, J. O., and McGill, T. C. (1992). *Solid State Physics*, Vol. 46, (Ehrenreich, H., and Turnbull, D. (eds.)), Academic Press.
Zubrilov, A. S., Nikolaev, V. I., Tsevetkov, D. V., Dmitirev, V. A., Irvine, K. G., Edmond, J. A., and Carter, C. H. (1995). *Appl. Phys. Letts.* **67**, 533.

CHAPTER 9

Doping in the III-Nitrides

K. Doverspike

CREE RESEARCH, INC.
2810 MERIDIAN PARKWAY
DURHAM, NC

J. I. Pankove

ASTRALUX, INC
BOULDER, CO
AND UNIVERSITY OF COLORADO AT BOULDER
BOULDER, CO

I.	INTRODUCTION	259
II.	UNDOPED GaN	260
III.	COMMON DOPANTS FOR III-NITRIDES	262
	1. Donors	262
	2. Acceptors	265
IV.	DOPING OF THE ALLOYS	269
	1. InGaN	269
	2. AlGaN	271
V.	DOPING TECHNIQUES	271
	1. Doping During Growth	271
	2. Post-Growth Doping (Ion Implantation)	272
VI.	FUTURE DIRECTIONS	274
	References	275

I. Introduction

The control of doping in semiconductors is extremely important for the fabrication of devices. Doping determines the position of the Fermi level in a semiconductor making the material n-type if the Fermi level is close to the conduction band, p-type if it is closer to the valence band. Doping controls the electron (n-type) or hole (p-type) concentrations. Doping also affects the mobility of carriers: high doping reduces the mobility by impurity scattering. Most devices need a p-n junction, because minority carrier injection from a p-n junction is more efficient than injection from a Scottky barrier. The breakdown voltage of a p-n junction depends on the concentration profile of the doping on either side of the p-n junction. For optoelectronic applications, such as light emitters, the impurities can be the luminescent centers. Thus a deep acceptor provides tight localization for the electron it captures. Spacial localization corresponds to a state that is spread out in momentum space. Hence, momentum conservation in radiative recombination is easily achieved even if the band structure is indirect (e.g., in GaP).

In general, wide bandgap semiconductors are difficult to dope both n- and p-type, possibly due to native defects. When the enthalpy for defect formation is lower than the bandgap energy, the probability of generating a defect increases with the bandgap because, in general, the energy released by donor-to-acceptor transitions increase with the energy gap. In the present chapter, doping of the Al-Ga-In-N material system will be discussed. However, before discussing doping issues in this material system, various characteristics of unintentionally doped GaN will be examined. This is important since the characteristics of doped material, in general, is strongly influenced by the characteristics of the host material. The bulk of the chapter will deal with both n- and p-type doping of GaN, while the current status of doping of the alloys InGaN and AlGaN will also be discussed. Doping techniques, including ion-implantation, will then be discussed.

II. Undoped GaN

Prior to the 1980s, reports of GaN growth by vapor phase epitaxy illustrated the necessity to grow very thick films ($>100\,\mu$m) in order to reduce the background electron concentration to $10^{17}\,\text{cm}^{-3}$ levels (Ilegems, and Montgomery, 1973). However, currently with the use of either GaN (Nakamura, 1991) or AlN (Akasaki *et al.*, 1989) buffer layers, most researchers in the GaN community have reported their background electron concentration in unintentionally doped GaN to be approximately mid $10^{16}\,\text{cm}^{-3}$ to low $10^{17}\,\text{cm}^{-3}$ as measured by Hall measurements. There has been much speculation as to what species is responsible for this relatively high background carrier concentration. Probably the oldest speculation is that the unknown donor is due to nitrogen vacancies, but other candidates include oxygen or silicon acting as shallow donors, anti-site defects, and gallium vacancies (Boguslawski *et al.*; Jenkins, Dow, and Tsai, 1992; Neugebauer, and Van de Walle, 1994). Whatever the source of this background carrier concentration, for many applications including light emitting diodes and lasers, this background donor may not be detrimental since both the n-type and p-type regions in these devices are fairly heavily doped (Si and Mg concentrations mid $10^{18}\,\text{cm}^{-3}$ or higher). However, for electronic device applications, such as various types of field effect transistors (FETs), one needs the ability to grow a high mobility channel layer either on an insulating film, or directly on an insulating substrate. If the underlying film or substrate is conducting, then it will be detrimental to the high frequency performance of the device.

Because of the large lattice mismatch between GaN and sapphire, it is difficult to grow high quality, high mobility, GaN directly on sapphire. Therefore, being able to grow high quality highly resistive GaN is of great importance for FET applications. The other substrate of choice is SiC, which can be grown either insulating or conducting. The lattice mismatch between GaN and SiC is approximately 3%, compared to roughly 15% between sapphire and GaN. However, even with a closer lattice match, this material system has historically still required the use of a buffer layer (using the Al-Ga-N system) onto which GaN is then grown (Weeks et al., 1995). Therefore, whether SiC or sapphire is used as the substrate, one still needs the ability to grow highly resistive GaN that is of high quality, so high mobilities upon doping the channel layers can be obtained. Although several groups have reported the growth of high quality, highly resistive undoped GaN (Chen et al., 1996; Doverspike et al., 1996; Shan et al., 1995), many of the specific growth conditions, and correlations illustrating control of the background doping level in GaN, have not yet been reported. However, optimizing the growth conditions, (e.g., buffer layer, V/III ratio, and NH_3/H_2) appear to be important variables.

Since it is difficult to make contacts to GaN when the carrier concentration is below 10^{16} cm^{-3}, Hall measurements cannot be used to characterize these highly resistive films, although resistivities of 10^{10} Ω cm were obtained by using large area In contacts. In addition, the highly resistive GaN layers could withstand > 1000 V without exhibiting breakdown using a two point probe (Doverspike et al., 1996). X-ray rocking curves are also not very useful in characterizing these highly resistive films since the full width at half maximum (FWHM) of the (0002) peak does not show a direct correlation with the electrical properties, and may not be able to distinguish between low mobility compensated films or high mobility films. The mobility of doped channel layers grown on highly resistive layers is a powerful characterization tool and gives some indication of the quality of, not only the doped GaN layer, but also the highly resistive underlying GaN film. Another characterization tool that can be used for undoped GaN films is photoluminescence (PL), since in high quality material one should observe free exitons in the PL spectra.

In summary, the characterization of undoped GaN is an important part of characterizing and understanding the nature of doped GaN. If the undoped GaN is of high quality, this should result in high mobilities >400 cm^2 V^{-1} s^{-1} when doped with Si donors. However, highly resistive undoped GaN could also be the result of highly compensated GaN, which would result in low mobilities upon doping.

III. Common Dopants for III-Nitrides

1. DONORS

a. Nitrogen Vacancies

It was long believed that N-vacancies were the dominant donor in GaN, because GaN that was not deliberately doped was always n-type, and usually degenerately so. The donor nature of the N-vacancy was construed as a missing N atom surrounded by four Ga atoms that provide three valence electrons to complete the bonding octet with the five missing electrons of nitrogen. Two of these three electrons would be donated to the conduction band. Many attempts have been made to avoid N-vacancy formation by growing GaN at high pressure and high temperature (Karpinski and Porowski, 1984; Madar et al., 1975). As discussed in Section II, several groups have reported the growth of highly resistive GaN with low background electron concentrations simply by optimizing the growth parameters. This may imply that the N-vacancy is not the dominant donor in GaN since the unintentional background level was decreased without the use of high N_2 or NH_3 pressures.

b. Oxygen

Oxygen was proposed as a potential donor, because if it substitutes for nitrogen it can donate an extra electron. Although this concept was supported by theoretical analysis of reaction kinetics (Seifert et al., 1983), for several years there were no deliberate attempts at introducing oxygen into the growth chambers. Of course, there had been accidental introduction of oxygen due to air leaks into the reactor, but these usually resulted in reddish GaN:O that did not luminesce. A secondary ion mass spectrometry (SIMS) analysis revealed the presence of 30 atomic percent oxygen while the X-ray diffraction was identical to that of wurtzite GaN, indicating the crystallinity of $GaN_{1-x}O_x$ appears to be similar to undoped GaN (Pankove, 1992).

The donor nature of oxygen was demonstrated in a classical paper by Chung and Gershenzon who doped GaN with oxygen by introducing water vapor in the reaction chamber. Their results are shown in Fig. 1, demonstrating that the electron concentration increases with increasing water partial pressure. In addition, note that the electron concentration always decreases with increasing temperature, thus suggesting that nitrogen vacancies may not be present as donors, since one would expect a higher number of nitrogen vacancies with increasing temperature.

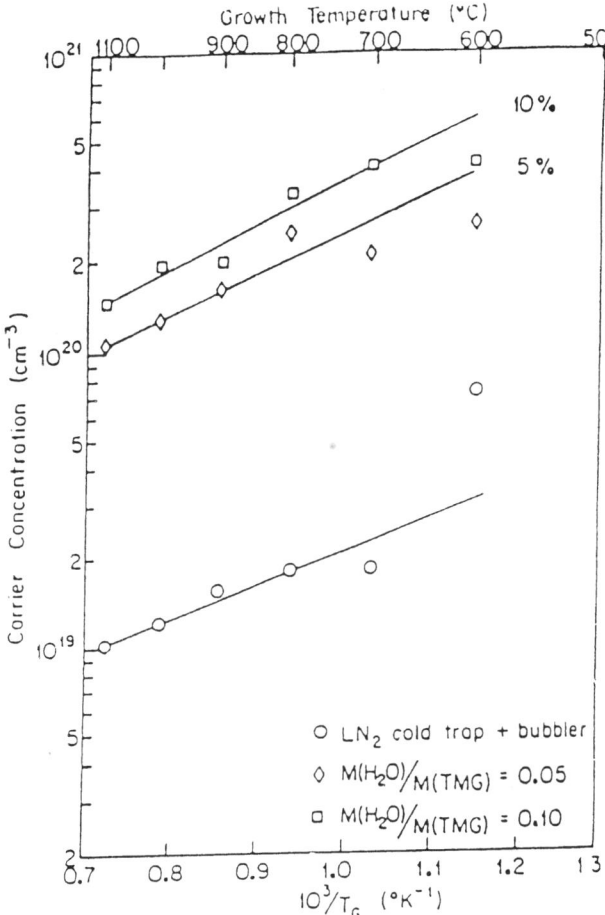

FIG. 1. Temperature dependence of carrier concentration for GaN layers prepared at different growth conditions. The data indicate the results of the layers grown with a gettering technique (circles) and with different amounts of intentional water injection during growth (diamonds and squares). The M(H$_2$O)/M(TMG) indicates the H$_2$O-to-TMG flux ratio entering the reactor. [After Chung and Gershenzon (1992)].

c. Silicon

The most common n-type dopant in GaN is silicon. Controllable silicon doping of GaN has been demonstrated over a wide range of concentrations (low 10^{17} cm^{-3} to mid 10^{19} cm^{-3}). Several groups have shown that the electron concentration measured at 300 K using van der Pauw Hall measurements increases linearly with the silicon/gallium ratio and is independent of the silane source (silane, disilane, or tetraethyl silane)

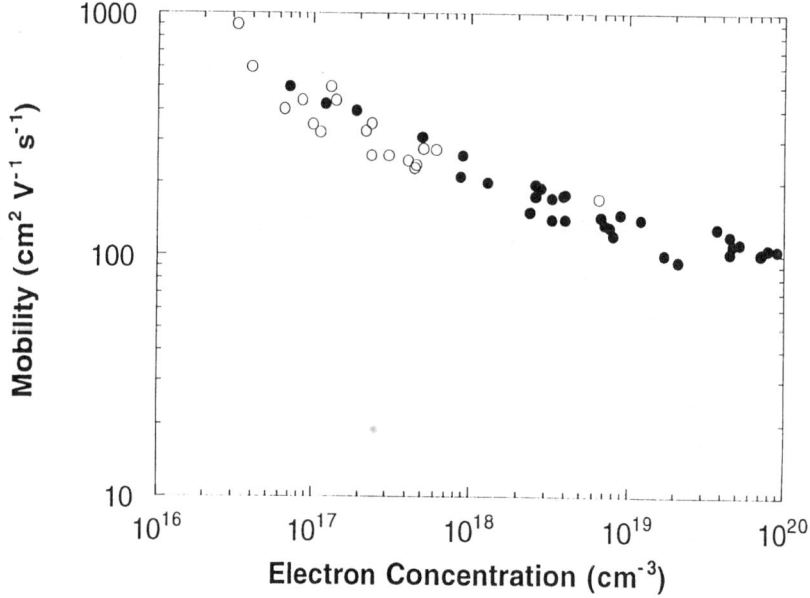

FIG. 2. Electron mobility vs. electron concentration in n-type GaN.

(Nakamura, Mukai, and Senoh, 1992; Rowland, Doverspike, and Gaskill, 1995; Kaneda et al., 1996). As the electron concentration is increased, there is also the expected decrease in the electron mobility as shown in Fig. 2 due to impurity scattering. The data for unintentionally doped samples (open symbols) are very similar to the data for intentionally doped samples (closed symbols) in the region where the electron concentrations overlap. This implies that the magnitude of compensation for doped and unintentionally doped samples is approximately the same (Gaskill et al., 1995).

Si-doped GaN is often characterized by variable temperature Hall measurements where donor activation energies of approximately 27 meV have been reported (Hacke et al., 1994; Wickenden et al., 1995). However, these values assume a temperature independent for the Arrhenius equation. Others (Gotz et al., 1996) report a range of 12–17 meV as the donor level from Hall effect analysis. Because the Si donor level is fairly shallow, nearly complete activation of Si in GaN has been observed by comparing Hall data to SIMS analysis (Wickenden et al., 1995).

Although electron concentrations up to mid 10^{19} cm^{-3} can be easily achieved with Si doping, at these concentrations, cracking of GaN films grown on sapphire has been observed (Murakami et al., 1991). There is also a GaN film thickness versus electron concentration relationship. The more lightly doped the GaN film is, the thicker it can be grown without observing cracking in the GaN film, but as the doping level is increased, the thickness that the GaN film can be grown without observing cracks is decreased. Although many researchers have observed this, a complete study showing the effects of specific growth parameters (e.g. growth temperature, cooling rate, etc.) on the cracking of GaN films at high electron concentrations has not yet been published. It is not known at this time if there is a similar trend in electron concentration and thickness of Si:GaN grown on SiC.

For light emitting diodes (LEDs) grown on sapphire, high Si doping levels are needed so a low resistance contact to the n-type GaN can be made. In addition, for LEDs grown on sapphire, the GaN:Si layer needs to be thick enough for low series resistance in the device. For LEDs grown on SiC, this is not as relevant, since a vertical device structure can be fabricated which takes advantage of the conducting nature of the SiC substrate.

More recently, it has been reported that Si doping of GaN to approximately 3×10^{18} cm^{-3} actually improves the layer quality by decreasing the threading dislocations from 5×10^9 cm^2 for undoped GaN to 7×10^8 cm^2 for GaN:Si (Ruvimov *et al.*, 1996). There was also a reduction in the X-ray diffraction FWHM of the (0002) peak from 8 arcmin to 4.2 arcmin for undoped and GaN:Si respectively. No cracking of the GaN films were observed at this electron concentration.

Recently Burm *et al.* have shown that a shallow Si implant at a dose of 1×10^{18} cm^{-2} to produce a doping density of 4×10^{20} followed by an 1150 C anneal for 30 sec results in very low contact resistance of 0.097 Ω mm and a specific contact resistance of 3.6×10^{-8} Ω cm^2 (Burm *et al.*, 1997).

d. Germanium

Although Si is predominantly used in the GaN community to control the electron concentrations, Ge can also be used. It has been shown that electron concentrations up to 1×10^{19} cm^{-3} have been obtained for Ge doping although the surface morphology began to degrade at these doping levels (Nakamura, Mukai, and Senoh, 1992). Ge doping can be easily controlled and is linear with the flow rate of GeH$_4$, however the doping efficiency is about one order of magnitude lower compared to Si doping.

2. ACCEPTORS

a. Divalent Atoms

The III-nitrides can be made p-type by inserting column II elements such as Zn, Cd, Mg, and Be substitutionally for Ga to form single acceptors. However all of these divalent elements form deep acceptors, the shallowest being Mg with an activation energy of 0.16 eV, which is still many kTs above the valence band edge of GaN.

Zinc. Although much work was done in the 1970's on GaN:Zn in the hope that it would lead to a shallow acceptor level and subsequently conducting p-type GaN, this was not to be the case. When GaN is Zn doped, the result is a highly resistive layer, due to the deep acceptor levels it creates (the shallowest level being ~ 0.21 eV above the valence band). Divalent atoms such as Zn can, not only substitute for Ga, but also substitute for nitrogen, in which case they form triple acceptors. Since divalent atoms may substitute for both Ga or N, one might expect four acceptor levels: one for the group II on Ga-sites, and three for the different charge states of a group II on N-sites. This property has been verified for the case of Zn doped GaN which gave four electroluminescence peaks (see Fig. 3) (Pankove, 1973). Four different Zn acceptor levels were also found at the following energies above the valence band by using low temperature PL, 0.34 eV, 0.65 eV, 1.02 eV, and 1.42 eV (Monemar, Lagerstedt, and Gislason, 1980). It has also been found experimentally that the nature of the Zn-related luminescence centers and their density are strongly affected by growth conditions (e.g., Zn concentrations and substrate temperature) (Amano et al., 1988). Therefore, these various Zn-related levels may be an example where the divalent Zn substitutes for Ga giving the shallowest level,

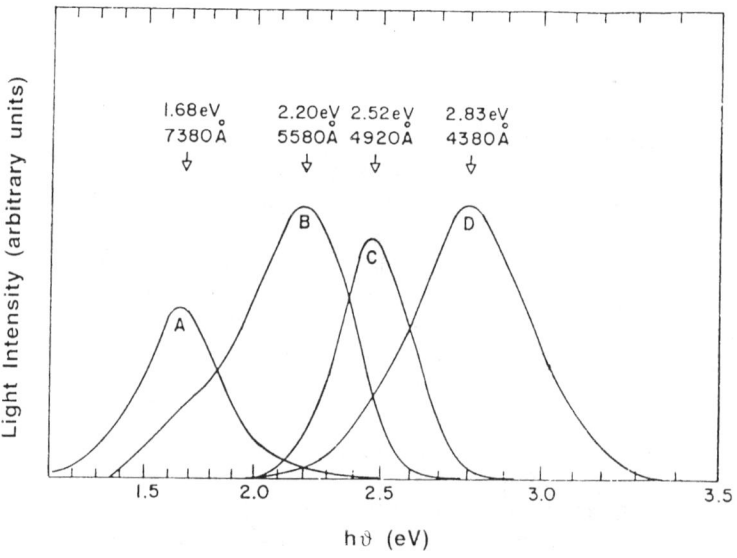

FIG. 3. Electroluminescence spectra from four M-i-n diodes where the i-layer is GaN:Zn.

and then at higher Zn concentrations and/or temperatures may substitute for N forming the deeper acceptor levels in which they are capable of binding up to three electrons giving the three different charge states. In general, the emission from GaN:Zn is very dependent on the growth parameters both when OMCVD or VPE was used as the growth technique (Amano et al., 1988; Jacob, Boulou, and Furtado, 1997; Pankove, 1992).

Magnesium. Although p-type GaN is now fairly easily achieved using Mg doping, this was one of the major breakthroughs in GaN research (Amano et al., 1989; Nakamura, Senoh, and Mukai, 1991). Although Mg is the shallowest acceptor that has been found, the acceptor level at approximately 0.16 eV (Akasaki, and Amano, 1994; Molnar, and Moustakas, 1993) above the valence band is still very deep. At this acceptor level, one should only expect approximately 1–5% of the Mg atoms to be ionized at room temperature, which means that the Mg concentration needs to be approximately two orders of magnitude larger than the desired hole concentration.

When OMCVD is used as the growth method, a post-growth treatment is necessary in order to convert the highly resistive as-grown GaN to lower resistance p-type GaN ($\sim 1\,\Omega\,cm$). A post growth treatment is necessary because it was found that hydrogen neutralizes the Mg acceptor. To activate this acceptor, H must be removed. This acceptor neutralization by H is similar to that observed earlier in silicon (Pankove, Zanzucchi, and Magee, 1985). The post-growth treatment can be either annealing in an N_2 atmosphere at temperatures typically between 700 and 900°C (Nakamura et al., 1992) or using a low energy electron beam treatment (LEEBI) technique (Amano et al., 1989). Earlier, some speculated that the success of the LEEBI treatment was simply due to local heating and, therefore, was similar in nature to a post-growth thermal anneal. However, recently a study was done in which the LEEBI treatment was done at low temperatures where local heating would not be a factor and p-type GaN was still achieved (Li and Coleman, 1996). Either post-growth treatment is believed to be a method of dissociating the H-Mg complexes that are formed during growth (Gotz et al., 1995). When molecular beam epitaxy (MBE) is used, as-grown GaN exhibits p-type behavior presumably due to the absence of H in the MBE growth environment.

Characterization tools used for p-type GaN include mobility and hole concentration, as determined by Hall measurements, Na-Nd as measured by the capacitance-voltage (C-V) technique (Huang et al., 1996), SIMS, and PL measurements. Since many device applications are dependent on the hole concentration and mobility, Hall measurements can be very useful. However, reliable Hall measurements assume the ability to make a contact to the p-type GaN. The highest reports of hole concentrations as measured by Hall measurements for OMCVD grown material are approximately

3×10^{18} cm^{-3} (Akasaki, and Amano, 1994; Nakamura, Senoh, and Mukai, 1991), while hole concentrations up to 1×10^{19} cm^{-3} have been reported recently for MBE grown material (Moustakas, 1993). It is still unclear as to whether these high hole concentrations can be achieved "routinely" in any laboratory. Typical mobilities in p-type GaN are very low, often 10 cm^2 V^{-1} s^{-1} or below for hole concentrations in the 10^{17} cm^{-3} range.

In the past few years, there has been much confusion over the interpretation of PL of GaN:Mg. In an early paper by Nakamura, a relationship between the resistivity of the GaN:Mg films and a 450 nm peak in the PL spectra was shown (Nakamura, Senoh, and Mukai, 1991). There were also reports in the literature (Amano et al., 1990) that showed a peak in the PL spectra around 390 nm when the Mg concentration was $<2 \times 10^{16}$ cm^{-3}, while for Mg concentrations of approximately 7×10^{19} cm^{-3} the PL peak was around 420–430 nm. Part of the confusion was the inability to explain why the conduction band to acceptor level transition would be at 420–430 nm or longer in the PL spectra if the Mg acceptor level is believed to be approximately 150 meV above the valence band (from variable temperature Hall measurements). A conduction band-acceptor level (150 meV above the valence band) transition would imply a peak approximately at 390 nm. It is possible that the PL peak around 420–430 nm is a conduction band-to-impurity related defect level transition. This implies the presence of a Mg related defect level approximately 500 meV above the valence band. GaN:Mg doped by ion beam implantation exhibited a cluster of levels 300–400 meV above the valence band according to PL data (Pankove and Hutchby, 1976). At this time, the nature of the Mg related defect level is still not known, but some of the possibilities include a defect composed of more than one Mg atom, a Mg substituting for N as opposed to substituting for Ga, or Mg sitting on an interstitial site. As mentioned in the section on GaN:Zn, deeper acceptor levels are also observed at higher Zn concentrations. Therefore, it is possible that both of these group II in GaN(Mg and Zn) may be examples of group II atoms substituting either for the Ga or the N atom. This also means that there may be no definitive relationship between the 420–430 nm peak in the PL spectra and the corresponding hole concentration, which has been experimentally confirmed by some laboratories. As more Mg is incorporated into the GaN film, the 420–430 nm peak may increase in intensity, but the hole concentration may not show the same increase. The growth conditions of Mg-doped GaN must be optimized so that the Mg that is being incorporated into the GaN film is substituting for the Ga atoms. Attempting to incorporate high Mg concentrations under non-optimized growth conditions, may lead to an increase in the presence of the Mg-related defect level.

Mg doped GaN films have also been studied using time-resolved PL (Smith et al., 1996) whose results yield an optical ionization energy of about

290 meV for the Mg acceptor level and approximately 550 meV for the Mg related defect complex. Because of the recombination lifetimes dependence on the excitation intensity and decay kinetics, the peak that is 290 meV below the bandgap energy is due to the conduction band-to-Mg acceptor transition, while the peak that is 550 meV below the bandgap energy is due to the transition conduction band to impurity related defect complex. The discrepancy between the thermal ionization energy (160 meV) and the optical ionization energy (290 meV) may be accounted for by the lattice relaxation associated with doping impurities in GaN.

In another recent study of luminescence from GaN:Mg Melton et al. (1977) found four PL peaks that are separated by 0.09 eV, the LO phonon energy of GaN. However the temperature dependence of the 3.19 eV peak is different from that of the other three peaks, indicating that the 3.19 eV peak is not a phonon replica of the zero-phonon line.

Berylium. Recently Prof. Ploog's group (Brandt et al., 1996) have found that Be forms a shallow band of acceptors in GaN. Furthermore they found that by compensating Be with O, a neutral dipole is formed that does not scatter the holes. Hence a record high hole mobility of 150 cm^2 V s is obtained. The transition to this metallic-like phase occurs at a Be concentration of 1.25×10^{20} cm^{-3}. This may be the ideal contact layer for a GaN-based injection laser. A similar low-scattering dipole pair is expected from Mg and O, though this has not been demonstrated yet.

b. Tetravalent Atoms

In GaAs, column IV elements can be amphoteric (i.e. can either be a donor or acceptor depending on the site they occupy). We have seen in Section IIIA that Si and Ge tend to be donors in GaN. Carbon is expected to be an acceptor if it substitutes for N in GaN. There has been one research group that has reported on carbon doping of GaN by MBE (Abernathy et al., 1995). P-type behavior was obtained by Hall measurements and the hole concentrations and mobility were 3×10^{17} cm^{-3} and 103 cm^2 V^{-1} s^{-1}, respectively. At this time, there has been no report of carbon doping leading to p-type behavior by using the MOCVD growth technique.

IV. Doping of the Alloys

1. InGaN

When InGaN layers are unintentionally doped, they show n-type conduction. Most researchers believe that this is due to a high number of N-vacancies. InGaN films can be easily doped with Si, and there are reports where researchers doped the InGaN layers in order to enhance the PL signal. It has been reported that the bandedge PL signal of InGaN:Si was 36 times stronger than that of undoped InGaN (Nakamura, Mukai, and Senoh, 1993).

When using InGaN layers as the active layer in light emitters, the InGaN can be simultaneously doped with Si and Zn in order to use the donor-acceptor transitions as the light emitting mechanism. Depending on the In concentration, LEDs made by co-doping with Si and Zn can emit anywhere from the purple, blue or green regions by controlling the In% and therefore the overall bandgap (Nakamura, 1995). The incorporation of Zn into InGaN layers is similar to Zn incorporation into GaN layers in that it acts as an acceptor level, but is too deep an acceptor to lead to p-type conduction. More recently, the LEDs being produced by Nichia are no longer based on Zn, Si-doped InGaN active layers, but rather quantum well InGaN layers (Nakamura et al., 1995).

It has also been reported that Cd can be used as a dopant in InGaN (Nakamura, Iwasa, and Nagahama, 1993). The blue emission from PL spectra had a peak wavelength at an energy level 0.5 eV lower than the bandgap energy of InGaN. Although Cd is a possible dopant for the active layer in InGaN LEDs, it is currently not being used in commercial LEDs.

One group has been successful in obtaining p-type $In_{0.90}Ga_{0.91}N$ (Yamasaki et al., 1995). This is of great interest to the GaN community, both for the active layer in light emitters, and also for use as a contact layer. If an InGaN active layer can be made p-type, this may contribute more to the improvement of the hole injection efficiency into the active layer

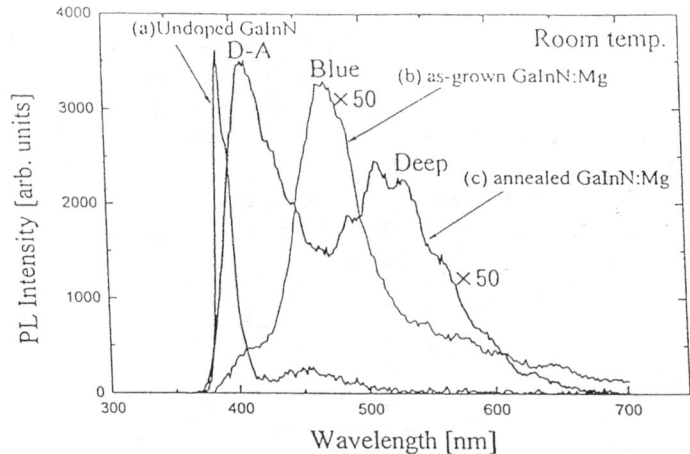

FIG. 4. PL spectra of (b) as-grown and (c) annealed $Ga_{0.91}In_{0.09}N$:Mg. Undoped $Ga_{0.91}In_{0.09}N$ (a) is also shown for comparison. [From Yamasaki et al. (1995)].

compared with n-type InGaN active layers, because the built-in potential for hole injection from the p-type AlGaN cladding layer into the InGaN active layer can be decreased (Yamasaki et al., 1995). Since InGaN has a smaller bandgap than either GaN or AlGaN, it is reasonable to expect that lower resistance contacts may also be made to Mg-InGaN as compared to Mg-GaN or Mg-AlGaN. As shown in Fig. 4, undoped InGaN shows a sharp emission in the PL spectra at 389 nm while the as-grown InGaN:Mg shows a broad peak centered around 470–490 nm (Yasasaki et al., 1995). After annealing, a peak believed to be conduction band to acceptor level transition is enhanced at approximately 405 nm. This would be similar to the peak observed at 390 nm in the PL spectra of GaN:Mg, but is shifted to longer wavelengths simply due to the smaller bandgap of the $In_{0.09}Ga_{0.91}N$ layer. In this paper, it was also reported that the hole concentration for InGaN:Mg as obtained from Hall measurements was approximately $7 \times 10^{17} cm^{-3}$.

2. AlGaN

The most common donor in AlGaN is Si, while the most commonly used acceptor is Mg, similar to GaN. The higher the Al% in AlGaN, the more difficult it is to controllably demonstrate p-type conduction, because in general the higher the bandgap of the semiconductor, the more difficult it is to dope both p- and n-type. The temperature dependence of the hole concentration and the Hall mobility of a $Al_{0.08}Ga_{0.92}N$ film indicates that the Mg acceptor is approximately 35 meV deeper than it is for GaN (Tanaka et al., 1994). Therefore, it is not surprising that the higher Al containing alloys (10–30%) are more difficult to make p-type. In one of the LED structures published by Nakamura, the p-type cladding layer was composed of $Al_{0.3}Ga_{0.7}N$ (Nakamura et al., 1995). This is probably the highest reported Al composition that has been made p-type, although the Howard University group has reported the ability to make AlN p-type by incorporating high concentrations of carbon into the film (Wongchotiqul et al., 1996).

V. Doping Techniques

1. Doping During Growth

The three main crystal growth methods are hydride vapor phase epitaxy (HVPE), molecular beam epitaxy (MBE) and metal organic vapor phase epitaxy (MOCVD). When using the HVPE technique, doping is achieved by

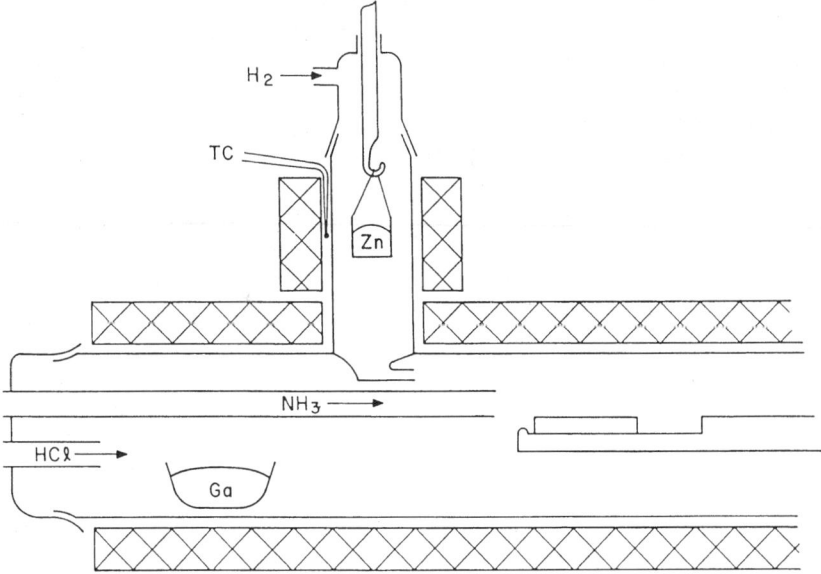

FIG. 5. Diagram of system for GaN growth by the chloride transport method showing a side arm for Zn-doping during growth.

a separately heated side arm containing a bucket of the desired impurity (e.g., Zn metal) (see Fig. 5) (Pankove, 1973). The bucket temperature is adjusted to obtain the desired vapor pressure, whereupon the dopant is flushed into the deposition chamber by a predetermined stream of H_2. For MBE growth, effusion cells containing the desired dopant are used. In gas-source MBE, the dopant is in the form of a volatile compound that decomposes at the heated substrate leaving behind the desired element. In MOCVD, the dopant is also a volatile compound that is dissociated at or near the surface of the growing crystal. See chapters on crystal growth at beginning of this volume.

2. Post-Growth Doping (Ion Implantation)

The ions to be implanted into the semiconductor films are derived either from elemental sources by evaporation, sputtering, or from compounds that are dissociated either thermally or by a plasma. The elements are ionized, accelerated, filtered in a magnetic field and implanted at a selected kinetic energy. The implantation energy determines the depth location of the maximum concentration of dopant below the surface of the crystal, and also the spread of the dopant distribution. To achieve a uniform distribution over a narrow range of depth (usually <1 mm) the implantation is done at several energies for different times.

Because of the high kinetic energy of the ions (up to 1 MeV) the crystal is severely damaged. Fortunately, the broken bonds can be reconstructed by a thermal anneal (950°C for GaN). To avoid the dissociation of GaN, it is prudent to perform the anneal in an ambient of NH_3 or atomic N. One advantage of doping by ion beam implantation is that several dopants can be implanted in different areas of the same crystal, then annealed simultaneously to assure that any variation in the subsequent analytical results is due to the implant rather than to the crystal or the annealing treatment. Much of the early information about the various acceptor levels in GaN was obtained by using ion implantation. A large number of candidate acceptors

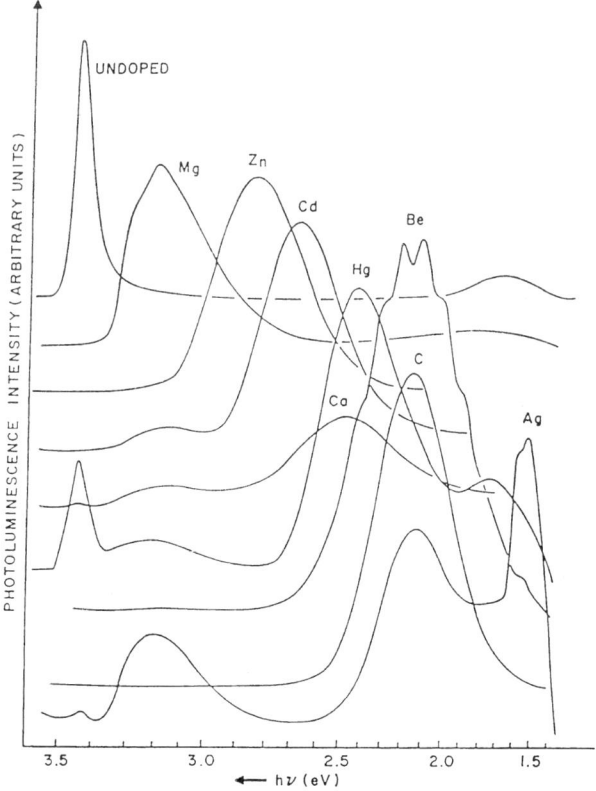

FIG. 6. Characteristic emission spectra at 78 K of an undoped GaN specimen and of several implanted dopants after identical anneal numerical. Pankove and Hutchby (1976).

was implanted in a few wafers of GaN deposited on sapphire that were identically annealed before the PL spectra were measured from each implanted area. There was also (on each wafer) a reference area that was not implanted (Pankove and Hutchby, 1976). The PL of the reference area was due to band to band recombination, while the peaks in the PL spectra of the implanted areas were due to the conduction band to acceptor transitions. By comparing these spectra one could evaluate the depth of the acceptor above the valence band edge (Fig. 6). Thus it appeared that all the candidate acceptor elements form deep acceptors, the shallowest being Mg. The peak in the PL spectra occurred about 0.3 eV above the valence band edge indicated by the reference area. Since this work was done before the awareness of the role of H, no effort was made to avoid H neutralization of the shallowest of the acceptor levels.

More recently, ion implantation has been used to produce both n- and p-type GaN. The p-type GaN was obtained by either implanting Ca alone or co-implanting Ca and P (Zolper et al., 1996). This results in an estimated ionization level of ∼170 meV. There is also evidence that the acceptors can be passivated with hydrogen forming Ca:H complexes similar to Mg:H complexes found in OMCVD grown GaN:Mg (Lee, Pearton, and Stall, 1996).

VI. Future Directions

Although impurities can be used as luminescence centers in GaN based devices, another luminescent center could be a rare earth (RE) atom. In RE atoms, the luminescent transition comes from the core electrons falling energetically from excited states to the ground state. To be luminescent, the RE must be in the 3+ state, which is usually achieved by co-doping with oxygen or fluorine. GaN:Er,O is of great interest for optical fiber communications because erbium's strongest emission is at 1.54 μm (0.8 eV) (Torvik et al., 1997). This 1.54 μm emission can be excited either optically by PL or by impact of hot electrons as in cathodoluminescence. The capture cross-section for hot electrons by Er is four to five orders of magnitude larger than the capture cross section for photons (Xu, Yu, and Zhong, 1986; Torvik et al., 1997). This observation is very attractive for an electrically pumped GaN:Er,O laser for optical fiber communications. Finally, the molecular doping a la Brandt et al. (Brandt et al., 1996) that led to a metallic-like degenerate acceptor band presents a great hope for more conducting p-type nitrides.

REFERENCES

Abernathy, C. R., MacKenzie, J. D., Pearton, S. J., and Hobson, W. S. (1995). CCl_4 doping of GaN grown by metalorganic molecular beam epitaxy. *Appl. Phys. Lett.* **66**(15), 1969–1971.

Akasaki, I., Amano, H., Koide, Y., Hiramatsu, K., and Sawaki, N. (1989). Effect of AlN buffer layer on crystallographic structure and on electrical and optical properties of GaN and $Ga_{1-x}Al_xN_{(0<x<0.4)}$ films grown on sapphire substrates by MOVPE. *J. Cryst. Growth.* **98**, 209—219.

Akasaki, I., and Amano, H. (1994). Widegap Column-III Nitride Semiconductors for UV/blue Light Emitting Devices. *J. Electrochem. Soc.* **141**(8), 2266–2271.

Amano, H., Hiramatsu, K., Kito, M., Sawaki, N., and Akasaki, I. (1988). Zn Related Electroluminescence Properties in MOVPE Grown GaN. *J. Cryst. Growth* **93**, 79–82.

Amano, H., Kito, M., Hiramatsu, K., and Akasaki, I. (1989). P-type Conduction in Mg-Doped GaN Treated with Low-Energy Electron Beam Irradiation. *Jpn. J. Appl. Phys.* **28**(12), L2112–L2114.

Amano, H., Kitoh, M., Hiramatsu, K., and Akasaki, A. (1990). Growth and Luminescence Properties of Mg-Doped GaN Prepared by MOVPE. *J. Electrochem. Soc.* **137**(5), 1639–1641.

Boguslawski, P., Briggs, E., White, T. A., Wensell, M. G., and Bernhol, J.

Brandt, O., Yang., H., Kostial, H., and Ploog, K. H. (1996). *Appl. Phys. Lett.* **69**, 2707–2709.

Burm, J., Chu, K., Davis, W. A., Schaff, W. J., Eastman, L. F., and Eustis, T. J. (1997). *Appl. Phys. Lett.* **70**, 464–466.

Chen, Q., Khan, M. A., Yang, J. W., Sun, C. J., Shur, M. S., and Park, H. (1996). High Transconductance Heterostructure Field-Effect Transistors Based on AlGaN/GaN. *Appl. Phys. Lett.* **69**(6), 794–796.

Chung, B. C., and Gershenzon, M. The Influence of Oxygen on the Electrical and Optical Properties of GaN Crystals Grown by Metalorganic Vapor Phase Epitaxy. *J. Appl. Phys.* **72**(2), 651–659.

Doverspike, K., Wickenden, A. E., Binari, S. C., Gaskill, D. K., and Freitas, J. A., Jr (1996). Growth of Silicon Doped and High Quality, Highly Resistive GaN for FET Applications (Ponce, F. A., Dupuis, R. D., Nakamura, S., and Edmond, J. A., eds.). *Mat. Res. Soc. Symp. Proc.* 395, 897–901.

Gaskill, D. K., Doverspike, K., Rowland, L. B., and Rode, D. L. (1995). Some Aspects of GaN Electron Transport Properties (Goronkin, H. ed.). *Int. Symp. on Cmpd. Semiconductors.* (IOP Publishing, Bristol), 425–431.

Gotz, W., Johnson, N. M., Walker, J., Bour, D. P., Amano, H., and Akasaki, I. (1995). Hydrogen Passivation of Mg Acceptors in GaN Grown by Metalorganic Chemical Vapor Deposition. *Appl. Phys. Lett.* **67**(18), 2666–2668.

Gotz, W., Johnson, N. M., Chen, C., Liu, H., Kuo, C., and Imler, W. (1996). Activation Energies of Si Donors in GaN. *Appl. Phys. Lett.* **68**(22), 3144–3146.

Hacke, P., Maekawa, A., Koide, N., Hiramatsu, K., and Sawaki, N. (1994). Characterization of the Shallow and Deep Levels in Si Doped GaN Grown by Metal-Organic Vapor Phase Epitaxy. *Jpn. J. Appl. Phys.* **33**, 6443–6447.

Huang, J. W., Kuech, T. F., Hongqiang, L. and Bhat, I. (1996). Electrical Characterization of Mg-Doped GaN Grown by Metalorganic Vapor Phase Epitaxy. *Appl. Phys. Lett.* **68**(17), 2392–2394.

Ilegems, M., and Montgomery, H. C. 1973). Electrical Properties of n-type Vapor-Grown Gallium Nitride. *J. Phys. Chem. Solids* **34**, 885–895.

Jacob, G., Boulou, M., and Furtado, M. (1977). Effect of Growth Parameters on the Properties of GaN:Zn Epilayers. *J. Cryst. Growth* **42**, 136–143.

Jenkins, D. W., Dow, J. D., and Tsai, M. (1992). N Vacancies in $Al_xGa_{1-x}N$. *J. Appl. Phys.* **72**(9), 4130–4133.

Kaneda, N., Detchprohm, T., Hiramatsu, K., and Sawaki, N. (1996). Si-Doping in GaN Grown by Metalorganic Vapor Phase Epitaxy Using Tetraethylsilane. *Jpn. J. Appl. Phys.* **35**, L468–L470.

Karpinski, J., and Porowski, S. (1984). High Pressure Thermodynamics of GaN. *J. Cryst. Growth* **66**, 11–20.

Lee, J. W., Pearton, S. J., and Stall, R. A. (1996). Hydrogen Passivation of Ca Acceptors in GaN. *Appl. Phys. Lett.* **68**(15), 2102–2104.

Li, X., and Coleman, J. J. (1996). Time-Dependent Study of Low Energy Electron Beam Irradiation of Mg-Doped GaN Grown by Metalorganic Chemical Vapor Deposition. *Appl. Phys. Lett.* **69**(11), 1605–1607.

Madar, R., Jacob, G., Hallais, J., and Fruchart, R. (1975). *J. Cryst. Growth* **31**, 197.

Pankove, J. I. (1973). Luminescence in GaN *J. Luminescence* **7**, 114–126.

Melton, W. A., Leksono, M. Qiu, C. H., and Pankove, J. I. (1977). Temperature Dependent Photoluminescence of Mg-Doped GaN. *Appl. Phys. Lett.* (to be submitted).

Molnar, R. J., and Moustakas, T. D. (1993). *Bull. Am. Phys. Soc.* **38**, 445.

Monemar, B., Lagerstedt, O., Gislason, H. P. (1980). Properties of Zn-Doped VPE-Grown GaN. Luminescence Data in Relation to Doping Conditions. *J. Appl. Phys.* **51**(1), 625–639.

Moustakas, T. D. (1993). 183rd Electrochemical Soc., 93-1 Meeting, Honolulu, 958.

Murakami, H., Asahi, T., Amano, H., Hiramatsu, K., Sawaki, N., and Akasaki, I. (1991). Growth of Si-Doped $Al_xGa_{1-x}N$ on (0001) Sapphire Substrate by Metalorganic Vapor Phase Epitaxy. *J. Cryst. Growth* **115**, 648–651.

Nakamura, S. (1991). GaN Growth Using GaN Buffer Layer. *Jpn. J. Appl. Phys.* **30**, L1705–1707.

Nakamura, S., Senoh, M., and Mukai, T. (1991). Highly p-Type Mg-Doped GaN Films Grown With GaN Buffer Layers. *Jpn. J. Appl. Phys.* **30**(10A), L1708–L1711.

Nakamura, S., Mukai, T., Senoh, M., and Iwasa, N. (1992a). Thermal Annealing Effect on p-Type Mg-Doped GaN Films. *Jpn. J. Appl. Phys.* **31**, L139–L142.

Nakamura, S., Mukai, T., and Senoh, M. (1992b). Si- and Ge-Doped GaN Films Grown with GaN Buffer Layers. *Jpn. J. Appl. Phys.* **31**, 2883–2888.

Nakamura, S., Mukai, T., and Senoh, M. (1993). Si-Doped InGaN Films Grown on GaN Films. *Jpn. J. Appl. Phys.* **31**, 2883–2888.

Nakamura, S., Iwasa, N., Nagahama, S. (1993). Cd-Doped InGaN Films Grown on GaN Films. *Jpn. J. Appl. Phys.* **32**, L338–L341.

Nakamura, S., Senoh, M., Iwasa, N., and Nagahama, S. (1995). High Power InGaN Single Quantum-Well-Structure Blue and Violet Light-Emitting Diodes. *Appl. Phys. Lett.* **67**(13), 1868–1870.

Nakamura, S., Senoh, M., Iwasa, N., Nagahama, S., Yamada, T., Mukai, T. (1995). Superbright Green InGaN Single-Quantum-Well-Structure Light-Emitting Diodes. *Jpn. J. Appl. Phys.* **34**, L1332–L1335.

Nakamura, S. (1995). InGaN/AlGaN Blue-Light-Emitting Diodes. *J. Vac. Sci. Technol. A* **13**(3), 705–710.

Neugebauer, J., and Van de Walle, C. G. (1994). Atomic Geometry and Electronic Structure of Native Defects in GaN. *Phys. Rev. B* **50**(11), 8067–8070.

Pankove, J. I., and Hutchby, J. A. (1976). Photoluminescence of Ion-Implanted GaN. *J. Appl. Phys.* **47**(12), 5387–5390.

Pankove, J. I., Zanzucchi, P. J., and Magee, C. W. (1985). Hydrogen Localization Near Boron in Silicon. *Appl. Phys. Lett.* **46**(4), 421–423.

Pankove, J. I. (1992). Stoichiometry Issues in Gallium Nitride and Other Wide Gap Semiconductors (Bachmann, K. J., Hwang, H-L., and Schwab, eds.). *Non-Stochiometry in Semiconductors* 143–153.

Rowland, L. B., Doverspike, K., and Gaskill, D. K. (1995). Silicon Doping of GaN Using Disilane. *Appl. Phys. Lett.* **66**(12), 1495–1497.

Ruvimov, S., Liliental-Weber, Z., Suski, T., Agger, J. W. III, Washburn, J., Krueger, J., Kisielowski, C., Weber, E. R., Amano, H., and Akasaki, I. (1996). Effect of Si Doping on the Dislocation Structure of GaN Grown on the A-Face of Sapphire. *Appl. Phys. Lett.* **69**(7), 990–992.

Seifert, W., Franzheld, R., Butter, E., Sobotta, H. and Riede, V. (1983). *Cryst. Res. Technol.* **18**, 383.

Shan, W., Schmidt, T. J., Yang, X. H., Hwang, S. J., Song, J. J., and Goldenberg, B. (1995). Temperature Dependence of Interband Transitions in GaN Grown by Metalorganic Chemical Vapor Deposition. *Appl. Phys. Lett.* **66**(8), 985–987.

Smith, M., Chen, G. D., Lin, J. Y., Jiang, H. X., Salvador, A. Sverdlov, B. N., Botchkarev, A., Morkoç, H., and Goldenberg, B. (1996). Mechanisms of Band-Edge Emission in Mg-Doped p-Type GaN. *Appl. Phys. Lett.* **68**(14), 1883–1885.

Tanaka, T., Watanabe, A., Amano, H., Kobayashi, Y., Akasaki, I., Yamazaki, S., and Koike, M. (1994). P-Type Conduction in Mg-doped GaN and $Al_{0.08}Ga_{0.92}N$ Grown by Metalorganic Vapor Phase Epitaxy. *Appl. Phys. Lett.* **65**(5), 593–594.

Torvik, J. T., Feurstein, R. J., Melton, W. A., and Pankove, J. I. (1997). Luminescence in Rare-Earth Doped Semiconductors. *Appl. Phys. Reviews* (submitted).

Torvik, J. T., Qiu, C.-H., Feuerstein, R. J., Pankove, J. I., and Namanar, F. (1997). Plate, Cathode and Electroluminescence in Erbium and Oxygen Co-Implanted GaN. *J. Appl. Phys.* **81**(a) 6343–6350.

Weeks, T. W., Jr., Bremser, M. D., Ailey, K. S., Carlson, E., Perry, W. G., and Davis, R. F. (1995). GaN Thin Films Deposited Via Organometallic Vapor Phase Epitaxy on α(6H)–SiC(0001) Using High-Temperature Monocrystalline AlN Buffer Layers. *Appl. Phys. Lett.* **67**(3), 401–403.

Wickenden, A. E., Rowland, L. B., Doverspike, K., Gaskill, D. K., Freitas, J. A., Jr., Simons, D. S., and Chi, P. H. (1995). Doping of Gallium Nitride Using Disilane. *J. Elect. Mat.* **24**(11), 1547–1550.

Wongchotiqul, K., Chen, N., Zhang, D. P., Tang, X., and Spencer, M. G. (1996). Low Resistivity Aluminum Nitride: Carbon (AlN:C) Films Grown by Metal Organic Chemical Vapor Deposition (Ponce, F. A., Dupuis, R. D., Nakamura, S., and Edmond, J. A. eds.). *Mat. Res. Soc. Symp. Proc.* **395**, 279–282.

Xu, X., Yu, J., and Zhong, G. (1986). The Impact Cross Section of Fr in ZnS. *J. Luminescence* **36**, 101.

Yamasaki, S., Asami, S., Shibata, N., Koike, M., Manabe, K., Tanaka, T., Amano, H., and Akasaki, I. (1995). P-Type Conduction in Mg-Doped $Ga_{0.91}In_{0.09}N$ Grown by Metalorganic Vapor Phase Epitaxy. *Appl. Phys. Lett.* **66**(9), 1112–1113.

Zolper, J. C., Wilson, R. G., Pearton, S. J., and Stall, R. A. (1996). Ca and O Ion Implantation Doping of GaN. *Appl. Phys. Lett.* **68**(14), 1945–1947.

CHAPTER 10

High Pressure Studies of Defects and Impurities in Gallium Nitride

T. Suski

UNIPRESS HIGH PRESSURE RESEARCH CENTER
POLISH ACADEMY OF SCIENCES
SOKOLOWSKA, WARSAW, POLAND

P. Perlin

CENTER FOR HIGH TECHNOLOGY MATERIALS
UNIVERSITY OF NEW MEXICO
ALBUQUERQUE, NM

I. INTRODUCTION	279
II. PRESSURE DEPENDENCE OF THE ELECTRONIC STATES OF DEFECTS	281
III. NATIVE VERSUS IMPURITY RELATED DONOR IN UNDOPED n-GaN	283
IV. OXYGEN AND SILICON IMPURITIES IN GaN	289
V. YELLOW LUMINESCENCE	291
VI. MECHANISM OF THE LUMINESCENCE IN GaN/InGaN/AlGaN QUANTUM WELLS	295
VII. SUMMARY	299
References	300

I. Introduction

In this review we demonstrate the usefulness of high pressure techniques in experimental studies of defects in wide bandgap semiconductor GaN. The discussed problems consist of the dominant donor in undoped GaN (Perlin et al., 1995; Wetzel et al., 1996), properties of oxygen (O) and silicon (Si) donors (Wetzel et al., 1997; Perlin et al., 1997a) and the mechanism of the parasitic effect of yellow luminescence observed often in GaN and related nitrides (Suski et al., 1995). Finally, we will present the high pressure test of the radiative recombination mechanism responsible for the luminescence in light emitting diodes (LEDs) based on quantum wells of InGaN (Perlin et al., 1997c).

Concentration of free electrons in undoped GaN achieves values 10^{19}–10^{20} cm^{-3} for bulk single crystals and 10^{16}–10^{17} cm^{-3} in epitaxial layers. The donor responsible for this situation has been commonly associated with the nitrogen vacancy, V_N. Recently, however, a number of theoretical predictions suggest that O or Si (Seifert et al., 1983; Koide et al., 1991; Neugebauer and Van de Walle, 1995a) can play a role of this unintentional donor. Questions concerning the presence of donors formed by V_N and possible coexistence of more than one donor in unintentionally doped GaN material (Perlin et al., 1997a) have not found a final answer.

Si and O are the natural candidates for donor impurities in group III nitrides. For this purpose Si substitutes cations (Ga, Al, In) and O replaces nitrogen. Si is commonly used as a donor impurity in device-oriented applications of nitrides giving electron concentrations up to 10^{20} cm^{-3} (Neugebauer and Van de Walle, 1995a). The natural question concerning the nature of states formed by Si and O donors in GaN appears when one recalls situations in other III–V semiconductors. In GaAs, for example, the donors (from group IV as well as from group VI) form resonant electronic states with different amounts of localization (Mooney, 1990). There are states with hydrogenic character as well as very localized states. They can coexist in the crystal. The most famous donor states, produced, for example, by Si, are known as metastable DX centers (Chadi and Chang, 1988). Oxygen in GaAs forms a metastable donor too (Skowronski, 1992a). Therefore, the question about the character of Si and O donors in GaN represent very interesting issue.

The problem of the so-called yellow luminescence in GaN has been discussed for many years (e.g., Suski et al., 1995; and references therein). A deep mid-gap localized state is involved in the mechanism of the parasitic luminescence which introduces an efficient radiative recombination channel which can compete with the band-edge luminescence. Controversial ideas have been proposed regarding not only the microscopic origin of this midgap defect state but also to mechanism of the radiative recombination leading to the yellow luminescence.

In spite of the successful commercialization of very efficient blue and green LEDs (Nakamura et al., 1995a), and the demonstration of the laser diode (Nakamura et al., 1995) based on the single or multiple quantum wells of the InGaN alloy, the mechanism of the optical emissions in these devices is not clear. There are several observations which suggest strongly that the radiative recombination does not have a standard near band edge (excitonic or band-to-band) character.

In the presented paper we will discuss ideas of the high pressure experiments which have been performed with the purpose of supplying information about the microscopic nature and/or models of the above mentioned phenomena.

II. Pressure Dependence of the Electronic States of Defects

Bond lengths and related lattice constants represent the most basic parameters characterizing a crystalline solid. The simplest method to change distances between atoms and thus, properties of the material, consists in applying hydrostatic pressure. In particular, the crystal structure and the energy spectrum of electrons are modified. Hydrostatic pressure as a symmetry preserving perturbation, has an important advantage with respect to electric or magnetic fields which lower crystal symmetry. Moreover, hydrostatic pressure has often been used as a superior method of lattice constant(s) variation in comparison with alloying which even in the simplest virtual-crystal approximation, introduces a disorder potential.

In the presented discussion, we concentrate on semiconductors. Defects and their particular representatives — impurities, control most of the semiconductor crystal properties. In the one-electron approximation, the defect/impurity states are obtained as the eigenvectors of the Hamiltonian $H = p^2/2m_0 + V_{cr} + V_{def}$ where p is the electron momentum and m_0 its mass, V_{cr} is the periodic crystal potential and V_{def} is the total potential introduced by defect/impurity. In general, V_{def} contains the long-range Coulomb potential (due to the effective charge on the defect), the short range potential (due to the difference in the host and defect/impurity cores) and the potential produced by the lattice deformation brought about by the defect. All these terms are affected by hydrostatic pressure. In particular, the modifications of V_{cr} through the decrease of the lattice constant, lead to the pressure dependence of the band structure of the perfect crystal. Changes of bandgap with pressure (dE_g/dp) result from a relative shift of the conduction band minimum (CBM) with respect to the valence band maximum (VBM). Magnitude of dE_g/dp differs significantly for different materials. For narrow gap compound InSb, dE_g/dp reaches 150 meV/GPa, it achieves 120 meV/GPa for medium gap GaAs and 40 meV/GPa for wide bandgap semiconductor–GaN. All the above compounds belong to direct bandgap semiconductors with the gap located at the Γ point of the Brillouin Zone. In general, variation of the individual bands and their extrema with pressure are described by values of the appropriate deformation potentials. They can have positive and negative values which results in "closing" or "opening" the bandgaps at different points of the Brillouin Zone. Concerning GaN crystal, the calculated values of dE_{CBM}/dp and dE_{VBM}/dp are 88 and 49 meV/GPa, respectively (Christensen and Gorczyca, 1994).

Electronic defect states in semiconductors and their pressure variation can be classified as belonging to one of the following classes:

(1) *Effective mass-like state (hydrogenic state)*. Coulombic, slowly-varying

potential characterizes a perturbation introduced by a charge defect center to a semiconductor matrix. Corresponding defect levels appear in the entire Brillouin Zone. However, their wave functions consist of the contributions from the close vicinity of the associated band structure extremum. The most important effective mass donors and acceptors introduce electronic levels close to the CBM and VBM, respectively. For the center with the donor character, the ground state of a bound electron appears at energy: $E_d \propto m^* \varepsilon^{-2}$, where m^* and ε are the electron effective mass and dielectric constant, and the CBM gives the level with zero of kinetic energy.

After applying hydrostatic pressure, the distance between the defect level and the parent extremum of the band structure changes weakly. In wide gap semiconductor GaN, changes of m^* and ε for pressures up to 10 GPa are very small. Therefore, hydrogenic state should practically follow the pressure induced shift of the parent extremum (subband).

(2) *Localized state (with small lattice relaxation).* The perturbation introduced by the defect corresponds to highly-localized potential, for example, to one elementary cell. Then, taking into account one band only, we may characterize the defect center by a single matrix element V_0 (between the Wannier functions centered at the impurity site (Callaway, 1974). Localized impurity levels E_{imp} are given by the condition

$$P \int \frac{N_0(E') \, dE'}{E_{imp} - E'} = \frac{1}{V_0}$$

where $N_0(E')$ denotes the density of states in the considered band and P stands for the principal value. The localized level can be located either in the forbidden gap or can represent the resonant state (with the conduction or valence band). The pressure dependence of E_{imp} follows from the pressure dependence of $N_0(E')$. Practically, the pressure coefficient of the impurity level is a weighted average of different subbands (e.g., Γ, L, and X). This coefficient depends on the position of the level with respect to the subbands (extrema) of the principal band. Usually, for the localized donors its magnitude differs significantly from the pressure coefficient of the CBM. More moderate difference is observed for relative shift of VBM and the localized acceptors.

(3) *Localized state (with large lattice relaxation).* Localized potential of the defects are often accompanied by large distortion of the surrounding lattice. The electron-lattice interaction becomes a relevant contribution in determination properties of the defect state. Experimental evidences of this interaction consist in thermal broadening of the optical spectrum, difference between optical and thermal ionization energies, thermally activated pro-

cesses of carrier capture and emission onto or from the defect center. Thermodynamic barriers responsible for these processes can be modified by applying a hydrostatic pressure. Moreover, value of thermal or optical ionization energies vary with the applied pressure too. Calculations of the above changes appear to be very complicated. In addition to the factors described in (2), one should include a contribution corresponding to pressure-induced changes of the electron–lattice interaction.

To complete the general description of the properties of electronic states produced by various defects/impurities one should mention a situation when different states of the same defect can coexist in the semiconductor. For example, for Ge-donor in GaAs one can observe coexistence of two localized states (one with the small, another with large lattice relaxation) (Suski, 1994). Depending on the applied pressure value and due to the fact that the pressure coefficients of these states are different, it becomes possible to modify their relative occupancy. At low pressure values ($>1\,\text{GPa}$), both states are resonant with the conduction band. For larger pressures they emerge to the bandgap of GaAs trapping ultimately all carriers. The latter situation causes the metal-insulator transition. Moreover, in GaAs the donor state with large lattice relaxation DX center, localizes two electrons (negative U center, where U is the Hubbard correlation energy). Whereas, the state with small lattice relaxation traps one electron only (positive U) (Baj, Dmowski, and Slupinski, 1993).

There are also situations when a hydrogenic state and a localized state of the same impurity can coexist (e.g., Suski, 1994).

III. Native Versus Impurity Related Donor in Undoped n-GaN

Despite impressive technological achievements of the last few years such as production of blue/green diodes and blue lasers (Nakamura et al., 1995; Nakamura et al., 1996a), there are still substantial gaps in our knowledge of the basic physical properties of GaN semiconductor. One of them has concerned till very recently, the origin of n-type conductivity in undoped GaN crystals. Experimentally, room temperature concentrations of conduction electrons in the GaN range from $10^{17}\,\text{cm}^{-3}$ in molecular beam epitaxy (MBE)- or metal-organic chemical vapor deposition (MOCVD)-grown epitaxial films to $10^{20}\,\text{cm}^{-3}$ in bulk crystals grown by high-pressure, high-temperature method (Porowski, Grzegory, and Jun, 1989). Maruska and Tietjen (1969), Ilegems and Montgomery (1973), and Perlin et al. (1995) suggested that the autodoping originates from native defects. Due to the lack of precise data on secondary ion mass spectroscopy (SIMS), it was a common belief that concentrations of contaminants should be lower by a few orders of magnitude than the highest concentrations of conduction electrons. The residual donor was tentatively identified with the nitrogen vacancy, V_N (Maruska and Tietjen, 1969; Ilegems and Montgomery, 1973; Monemar and Lagerstedt, 1979; Perlin et al., 1995). It has been argued that technologically, it was very likely that V_N appears during the growth because of the very high nitrogen equilibrium pressure at the growth temperatures; this implies that the growth occurs under Ga-rich conditions. So far, however, the dominant donor was not positively identified with the nitrogen vacancy.

Ab initio calculations of electronic structure of native defects in GaN were performed by Boguslawski, Briggs, and Bernholc (1995) and Neugebauer and Van de Walle (1994). They have shown that V_N introduces a resonant level inside the conduction band (cb) at about 0.8 eV above the cb minimum. In the neutral charge state, the one electron that should occupy this resonance autoionizes to bottom of the conduction bands and becomes bound by the Coulomb tail of the vacancy potential, forming a shallow level. Thus, at atmospheric pressure, the vacancy behaves like a shallow effective mass donor. When the concentration of vacancies exceeds the critical Mott value, which is about 10^{18} cm^{-3} for GaN, the sample should become metallic with free electrons in the conduction bands.

Recently, an alternative explanation for the dominant donor in undoped GaN material has been proposed. Namely, residual oxygen or silicon impurities (Neugebauer and Van de Walle, 1995a; Chung and Gershenzon, 1992). Both donors introduced intentionally form substitutional impurities leading to highly n-type GaN. O-donor substitutes N and Si replaces Ga ions. Matilla and Nieminen (1996) have performed calculations which strongly support this suggestion. Moreover, they showed that O introduces to GaN a localized donor state resonant with the conduction band. With increasing of the bandgap (by means of alloying with AlN) this state emerges in the bandgap, similar to the V_N, and exhibits properties similar to the donors forming metastable DX centers in GaAs.

In order to confirm the presence of a resonant level predicted by both theories, experiments under high hydrostatic pressures have been proposed. The corresponding method is based on the fact that the pressure coefficient of the bottom of the conduction band is usually higher than that of the resonance (see Section II). Consequently, at sufficiently high pressures, a crossover should occur between the two levels, and the resonance should become a genuine gap state that may trap electrons. After the crossover, free electrons would disappear from the conduction band and would occupy the deep defect-induced state.

In the case of V_N, to induce the crossover, the conduction band minimum should rise by at least 0.8 eV, assuming that the resonance does not move. Since the pressure coefficient of the gap is about 40 meV/GPa (Perlin *et al.*, 1992a), one has to apply a pressure of about 20 GPa. Such high pressures necessitate the use of diamond anvils, which practically excludes transport measurements. In this paragraph, we discuss the results of three optical experiments under pressure, which both probe for the existence of the resonant, deep state transition.

The first experiment measures the absorption due to free electrons in the infrared (IR) region of the spectrum (0.9–1.6 μm). In the second complementary experiment the far-infrared (2–20 μm) transmission is examined. The third one investigates the variation, with pressure, of the Raman spectrum of highly n-type GaN material.

In the performed experiments bulk GaN crystals grown at the nitrogen pressure of about 1.5 GPa and temperature of 1500°C (Porowski, Grzegory, and Jun, 1989) were used. These samples are characterized by free electron concentration of about 5×10^{19} cm^{-3}. The experiments were performed in diamond anvil cell. Details of the high pressure systems can be found in Perlin *et al.*, (1995). To measure the absorption, the light beam from a 150 W halogen lamp was focused on a 100 μm pinhole, whose image was projected on the sample. The outcoming light was dispersed by a Spex 500M spectrometer and detected by a Peltier-cooled PbS detector. Figure 1 shows the absorption spectra at various pressures. For the pressures lower than 15 GPa there is no substantial change in the shape of the absorption.

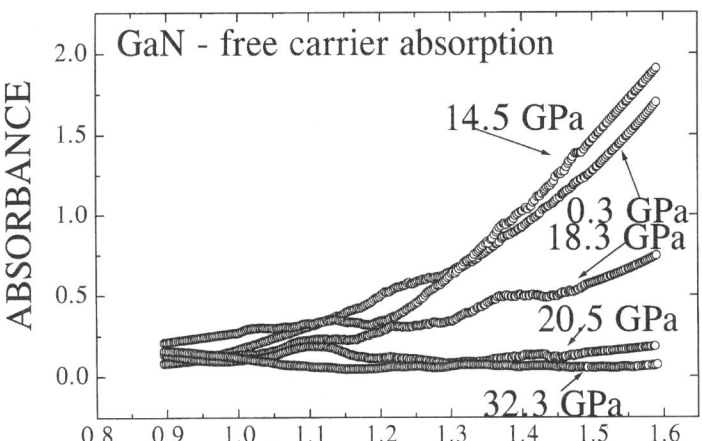

FIG. 1. Free carrier absorption spectra of a bulk GaN crystal for various pressures (at $T = 300$ K). (From Perlin *et al.*, 1995.)

However, between 15 GPa and 20 GPa the intensity of absorption decreases and practically disappears above 20 GPa.

The used experimental setup limited the available spectral range $\lambda \leqslant 1.6\,\mu$m. Since this represents only a very small part of the free carrier absorption, the examined spectral range was expanded by measuring the transmission. In this experiment a Perkin-Elmer 1600 Series Fourier spectrometer equipped with an IR microscope was used. The microscoping system allowed to determine very accurately, the ratio of the light intensity transmitted through the sample, to that passing through the pressure-transmitting medium close to the sample. The measurements consisted of the total IR transmission of the sample, for example, the ratio of intensities of the "white" light in the spectral range limited by the sensitivity of the employed IR detector (2–20 μm). Figure 2 shows the pressure dependence of the total transmission of a thin (15 μm) bulk single crystal. For pressures lower than about 20 GPa, the transmission is practically constant and close to 15%. The transmission increases rapidly starting at 20 GPa. At the highest pressures the transmission exceeds 100%, which simply means that the absorption of pressure-transmitting medium is higher than that of the sample.

The Raman scattering measurements were performed with the use of a triple Dilor-XY spectrometer. Details of this experiment are given elsewhere (Perlin *et al.*, 1992b). Figure 3 compares the Raman spectra of bulk GaN crystal at low (2 GPa) and very high (32.2 GPa) pressures. It shows that new

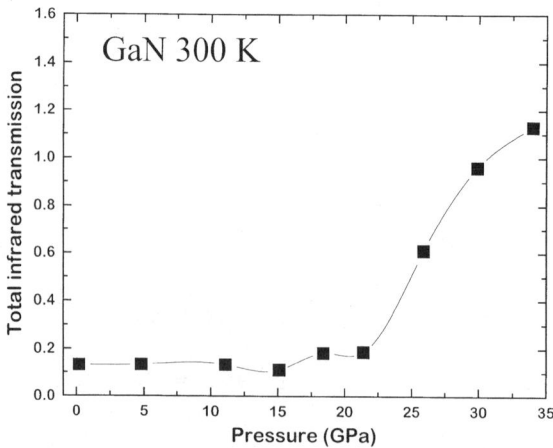

FIG. 2. Pressure dependence of the total infrared transmission of a bulk GaN crystal at $T = 300$ K. After Ref. 1.

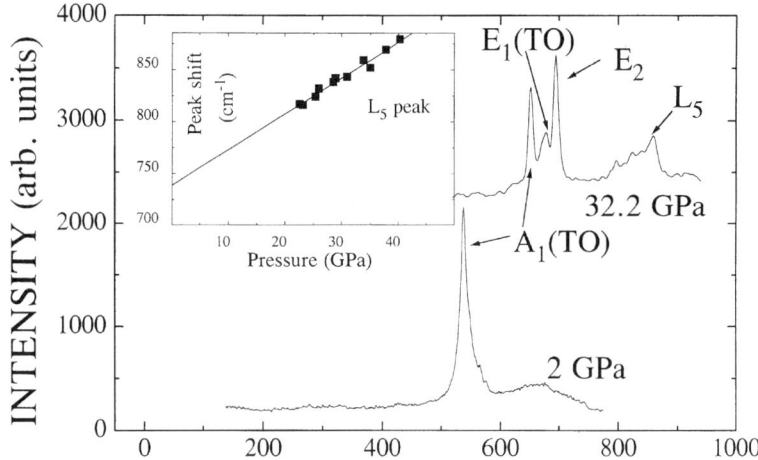

FIG. 3. Raman spectra of a bulk GaN crystal at low (2 GPa) and high (32.2 GPa) pressures. The peaks labeled A_1(TO), E_1(TO), and E_2 reflect the phonon modes of GaN. The peak L_5 appears above about 20 GPa. Insert: pressure dependence of the L_5 peak energy. The solid line represents a linear extrapolation of the peak energy to ambient pressure. (From Perlin et al., 1995.)

phonon peaks appear in the high pressure spectrum. Here we are interested in the peak labeled L_5 with wavenumber of about 850 cm^{-1} at 32.2 GPa. By extrapolating its position to the ambient pressure one obtains an energy of 740 cm^{-1} (see inset). Since this energy is very close to the energy 738–741 cm^{-1} of the LO phonon in GaN recently measured in epitaxial samples (Sobotta et al., 1992b), it is natural to assign the L_5 peak to the LO phonon. The presence of this peak was detected only for pressures higher than ≈ 22 GPa, as shown in the inset of Fig. 3.

We will now discuss each experiment in more detail. Both absorption and transmission measurements show that at pressures higher than about 20 GPa the IR absorption in the 0.9–1.5 μm range abruptly decreases. There are two main mechanisms of absorption in the IR region, phonon absorption and free carrier absorption. The former mechanism is important only in the far IR region, for example, for $\lambda \geqslant 0.18$ μm, and should not change substantially with pressure. In contrast, free carrier absorption can change considerably when the concentration of free electrons changes. Therefore, the quenching of the IR absorption is interpreted as due to a freeze-out of electrons from the conduction band. More precisely, one can estimate from the obtained data that the electron concentration decreases by a factor of about 15. This picture explains also the Raman experiment. The proposed interpretation is based on the work of Mooradian and Wright (1967), who observed and discussed the plasmon–phonon coupling in GaAs. They showed that the plasmon––phonon coupling shifts the LO mode to higher energies and reduces its oscillator strength. For GaN, this occurs at low pressures, when concentration of free carriers in epitaxial films is higher than $10^{18}\,\text{cm}^{-3}$ (Kozawa et al., 1994). In bulk GaN crystals with electron concentrations above $5 \cdot 10^{19}\,\text{cm}^{-3}$ the coupled plasmon LO-phonon has been clearly identified at energies in the range of $3000\,\text{cm}^{-1}$. In the latter case the pressure-induced freezout of electrons leads to a reappearance of the LO mode (see Fig. 3). More detailed description of the pressure variation of the phonon modes and the coupled LO phonon-plasmon modes in bulk GaN material can be found in Perlin et al. (1996).

The described experimental results can be consistently explained by the presence of a resonant state in the conduction band. Hydrostatic pressure induces a crossover of the resonance with the bottom of the conduction band at about 20 GPa. After the crossover, conduction electrons become trapped in the resonance-derived gap level. Consequently, since practically all electrons disappear from the conduction band, the concentration of defects giving rise to the resonance must be equal or higher than the concentration of dominant donors.

One can think of another explanation of the presented data. There is a possibility of a pressure-induced crossover of the conduction band minimum at Γ point with one of the secondary minima of the cb at the edge of the Brillouin zone. According to results presented in (Christensen and Gorczya, 1994), this crossover is expected to occur at about 100 GPa, which is too high.

At this point, it is interesting to recall the electrical properties of $Al_xGa_{1-x}N$. It turns out, that for $x \geqslant 20\text{--}30\%$, this alloy becomes an insulator (Yoshida, Misawa, and Gonda, 1982). A natural way of explaining this situation is to anticipate that for higher amount of AlN the energy gap of the alloy opens sufficiently to induce an emergence of the localized state of the donor in the band gap, and thus, to cause a metal-insulator transition. The situation is very similar to that occurring in GaAs and $Al_xGa_{1-x}As$, where for $x > 20\%$ the localized state related to DX center is formed in the bandgap of AlGaAs.

In summary, the high pressure optical experiments show that a freeze-out of electrons from the conduction band occurs at about 20 GPa. It has been proposed that this effect is due to an emergence of a conduction band resonance into the forbidden gap. The resonance-derived deep state captures free electrons, leading to the freeze-out of free carriers. At the present moment (in spite of the coincidence in the pressure value at which the electron freeze-out was observed with the estimated "critical" pressure for the emergency of V_N related levels to the GaN bandgap) both, the nitrogen vacancy as well as oxygen (or silicon), can be considered as probable origin of the localized, resonant donor state, studied in the described high pressure experiments.

IV. Oxygen and Silicon Impurities in GaN

One of the possible approaches to answer the question whether native defects or impurities play a decisive role in the formation of the localized, resonant donor state in GaN (and genuine localized gap state in $Al_xGa_{1-x}N$ with $x > 20$–30%) consists in performing the similar high pressure tests as those described in Section III, on samples which show both features, a high electron concentration, n_e, and (1) concentration of impurities clearly below n_e or (2) number of O or Si ions comparable or higher than n_e. Observation of the electron freeze-out for samples of type (1) would supply arguments against the involvement of impurities in the considered effect. The case related to (2), would support the hypothesis based on the impurity involvement.

It has become possible to prepare a set of highly conducting n-type GaN samples (Wetzel et al., 1977). It consists of two highly O doped films O1 and O2 grown by the hydride vapor phase epitaxy (HVPE) and a highly Si doped film grown by metalorganic vapor phase epitaxy (MOVPE). All the samples were deposited on (0001) sapphire substrates and they were characterized by room temperature Hall effect. The doping level was obtained from callibrated SIMS depth profiles using a Cs ion source. O1, O2 and GaN:Si samples exhibit concentrations of free carriers, n_e, of 3.5, 1, and $1 \cdot 10^{19} \, cm^{-3}$, respectively. Oxygen concentration of 2 and $0.8 \cdot 10^{19} \, cm^{-3}$ was found for O1 and O2 samples. In GaN:Si the Si concentration equal to n_e, for example, $1 \cdot 10^{19} \, cm^{-3}$ was detected. In all three samples, the described effect of the plasmon-LO phonon coupling led to the absence of LO phonon mode in the Raman spectrum measured at ambient pressure.

The idea of the high pressure experiment performed on these highly doped GaN epitaxial films consisted in the verification of the appearance of the LO phonon mode at high pressure (particularly in the vicinity of $P \sim 20 \, GPa$).

High pressure studies of GaN:Si sample led to the observation that, up to 25 GPa, no evidence of the LO mode could be observed. Thus, the Si donor is very unlikely to be the origin of the localized, resonant donor level responsible for the drop of n_e at $P \sim 20 \, GPa$. Hypothesis that Si impurity forms a donor state of the effective mass character only seems to be very probable (Boguslawski, Briggs, and Bernholc, 1995).

The consistently different situation has been found in GaN:O samples (Wetzel et al., 1997). The behavior of O1 and O2 samples resembles the results obtained for bulk single crystals of GaN. At pressures exceeding 20 GPa the band related to LO phonon mode appeared in the measured Raman spectra. Due to the similarities in application of hydrostatic pressure to GaN and alloying with AlN, the obtained result can be transferred to the $Al_xGa_{1-x}N$ alloy. Assuming a linear variation of the band gap with x, one can find a correspondence between application of 1 GPa pressure and an increase of AlN content by about 1.6%. It is very likely that O induces a strongly localized gap state at x higher than 0.2–0.3 (Wetzel et al., 1977; Matilla and Nieminen 1996). Moreover, AlN is always a good insulator, and due to strong affinity of Al and O an unintentional incorporation of O in AlN often occurs.

In this context, the question about the oxygen content in bulk GaN samples has been posed. Very recent SIMS measurements showed that the concentration of O was in the range of $1 \cdot 10^{20} cm^{-3}$, for example, very similar to the n_e in these samples (Perlin et al., 1997a). Thus, the idea that the oxygen impurity induces the donor state involved in the metal-insulator transition at about 20 GPa got strong supporting argument. As was mentioned in Perlin et al. (1997a), n_e drops by about one order of magnitude after removing the layer of 10–20 μm from a surface of bulk GaN crystals. The specific conditions of the high-pressure, high temperature synthesis of GaN single crystals lead likely to the formation of the oxygen-rich layer on the crystal surface. Since the "probing depth" for the Raman scattering experiment is comparable with the thickness of O-rich surface layer, the performed studies would favor regions with higher electron concentration, for example, very likely with high oxygen concentration.

It is necessary to remember, however, that the presence of impurity ions in the material, measured by, for example, SIMS method, do not prove unambiguously that they entirely become electrically active dopants. In particular, they can occupy improper position(s) in the matrix of parent crystal, possess an amphoteric character, or be passivated. Results of oxygen implantation to GaN crystal showed that only a few percent of the introduced O ions became electrically active (Zolper et al., 1996a).

As has been mentioned at the beginning of this section, the alternative way to examine a controversy concerning a role of native defects versus impurities (in problems related to the origin of n-type conductivity of undoped GaN) consists in performing high pressure studies of the samples characterized by significantly lower concentration of impurities with respect to n_e. Unfortunately, in spite of the intensive search, it was not possible to find such samples.

V. Yellow Luminescence

An important parasitic effect which strongly limits possible applications of GaN is the so-called yellow luminescence. Many of GaN samples exhibit the related broad luminescence band centered around 2.2–2.3 eV. This emission is a competitive radiative recombination channel with respect to the near band edge emission. Native defects and residual impurities represent natural candidates for the origin of the electronic level(s) participating in the considered luminescence.

A basic question which should be answered in studies of the yellow luminescence is a mechanism of the related radiative recombination. Two qualitatively different approaches have been proposed. According to the model discussed by Ogino and Aoki (1980) (model (a), Fig. 4), the initial state of the considered transition is a shallow donor (with the level depth of about 25 meV) and the final state is a deep acceptor (about 0.86 above the valence band maximum). A competitive model (b) has been introduced by Glaser *et al.*, (1995). These authors performed the photoluminescence and

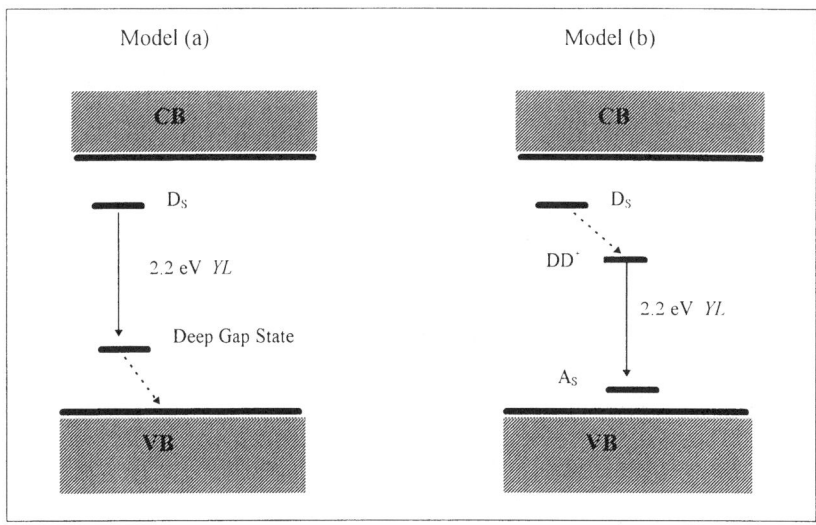

FIG. 4. Proposed models of the radiative recombination mechanisms which cause the yellow luminescence in GaN. Model (a) proposed in Ref. 35 and model (b) suggested in Ref. 36. D_s, DD^+, and A_s correspond to shallow donor, double (deep) donor, and shallow acceptor, respectively. YL stands for yellow luminescence and dotted arrows illustrate the nonradiative recombination processes.

optically detected magnetic resonance studies. They proposed a scheme of the radiative transition of electrons between a deep double donor of A_1 character (with a depth of about 0.8 eV below the bottom of the conduction band) and an acceptor state of the effective mass character. The transition is preceded by a nonradiative transfer of the electron between a shallow donor and the A_1 state (Fig. 4). It is worthwhile to mention, that both discussed approaches agree on the strong electron-lattice coupling interaction associated with the yellow-emission band.

The aim of the performed high pressure experiment (Suski et al., 1995) was to elucidate a mechanism of the radiative recombination leading to the yellow-luminescence band in GaN crystal, for example, to verify which model (a) or (b) is correct. The crucial idea of the used experiment consisted of a comparison of the pressure evolution of the yellow-band luminescence with the pressure variation of GaN energy gap, dE_g/dp.

Concerning model (a), application of pressure would induce a blue shift of the yellow band resembling dE_g/dp. According to Section II, the initial state of the radiative recombination, which is a shallow donor, follows the upward shift of the CBM at the Γ point of the Brillouin Zone. The final state, which is represented by the deep acceptor state is expected to move weakly with pressure. This pressure shift may be similar to the shift of the valence band maximum, VBM (at the Γ point of the Brillouin Zone) or to a weak negative shift of the acceptor level with respect to the VBM (Chadi, 1995).

Within the model (b), the pressure dependence of the yellow-band energy should be much smaller than dE_g/dp. The deep donor (A_1 state of double donor character) is formed by electronic states originating from the entire Brillouin Zone. Rough calculations of the pressure coefficient of this localized state led to the value of about 10 meV/GPa (Wetzel et al., 1996) (with respect to the VBM). Moreover, for the shallow acceptor state which is proposed as a final state of this recombination process, increasing pressure rises the binding energy by means of a decrease of the dielectric constant. It represents an additional factor lowering the pressure shift of the considered transition. Summarizing the above qualitative analysis, model (a) predicts a pressure coefficient of the yellow luminescence similar or even higher than the pressure dependence of the energy gap. On the contrary, much smaller modifications of the yellow-band energy with pressure can be expected within model (b).

Two types of samples were used in the performed experiment, bulk-single crystals ($n_e \sim 5 \cdot 10^{19}$ cm^{-3}) and epitaxial layers grown by molecular beam epitaxy on a sapphire substrate ($n_e \sim 4 \cdot 10^{17}$ cm^{-3}). The inset of Fig. 5 shows the yellow-luminescence spectra measured at different pressures for bulk crystal of GaN (at the highest pressures luminescence becomes blue-

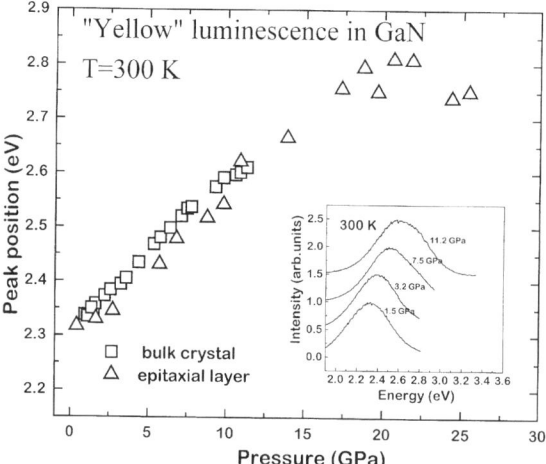

FIG. 5. Pressure dependence of the energy of the "yellow" luminescence (YL) for a bulk sample (squares) and an MBE epitaxial film (triangles) of GaN (at $T = 300$ K). The broken line is drawn to guide the eye. The insert shows the pressure evolution of the spectra of YL. (From Suski et al., (1995a).)

like). A Gaussian shape of the broad band related to this emission suggests an involvement of a strong electron-lattice coupling. The blue shift of the studied band is clearly seen. The pressure variation of the energy associated with the maximum of the examined luminescence is given in Fig. 5 (squares). It leads to the pressure coefficient of 30 ± 2 meV/GPa, which is similar to the following: (1) pressure dependence of the bandgap of GaN crystal (~ 40 meV/GPa) and (2) pressure coefficient of the yellow luminescence measured by other authors (Shan et al., 1995b). According to our earlier reasoning, this result strongly supports the model of the recombination consisting of the electron transition between a conduction band/shallow donor state and a deep-gap state.

The second experiment consists of measurements of pressure evolution of the yellow luminescence band in the MBE grown sample. The pressure dependence of this luminescence up to about 25 GPa is shown on Fig. 5 (triangles). The following important conclusion can be drawn. Pressure coefficients of the yellow luminescence are practically the same for both kinds of the used GaN samples. For MBE film, at $P \approx 20$ GPa, the pressure coefficient changes drastically. It starts to saturate or even to become negative. This behavior is associated with the effect of a resonant donor level entering the bandgap of GaN (Section III). A high degree of the localization of this state corresponds to its "deep" character and to its weak shift with

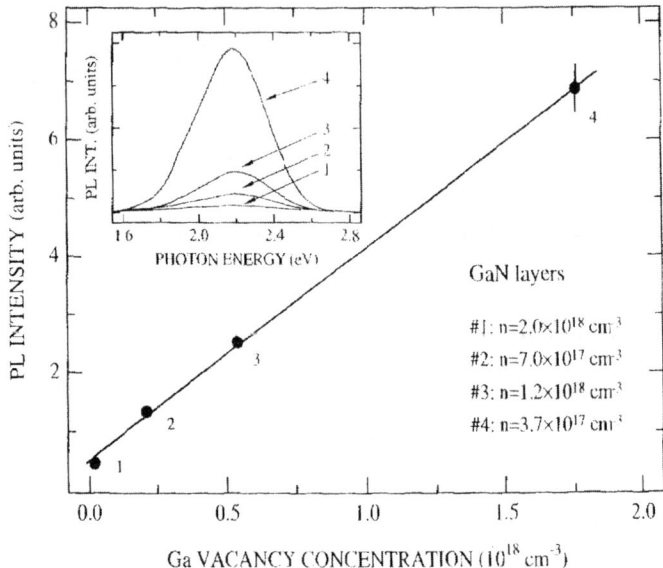

FIG. 6. The intensity of the yellow luminescence versus the Ga vacancy concentration (determined by the positron anihilation studies) in GaN MOCVD epitaxial layers. The insert shows the luminescence spectrum in the four studies layers, indexed according to the increasing Ga vacancy concentration. (From Saarinen et al., to be published.)

pressure. At pressures higher than about 20 GPa the radiative recombination can very likely occur between this localized donor and deep gap states. This result again supports the model (a), according to which the initial state of the analyzed luminescence is associated (at ambient pressure) with CBM/shallow donor state.

What is the chemical origin of the deep gap state? Ogino and Aoki (1980) suggested that a carbon atom is crucial for this luminescence. On the contrary, Molnar and Moustakas (1994) argued against the impurities as an origin of this effect. Moreover, Zhang et al. (1996b) used their experiment with varying the MOCVD growth conditions to anticipate a contribution of Ga vacancy, V_{Ga}, to the formation of the related localized gap state. This possibility was predicted also by the first principal theoretical calculations (Neugebauer and Van de Walle, 1996b). On the contrary, a suggestion concerning the nitrogen antisite involvement was proposed by Suski et al. (1995a). Results of the work by Sarrinen et al. (1996b) represent a very convincing proof that the origin of yellow luminescence consists in the Ga vacancy or complex defect involving the V_{Ga}. Accordingly, Fig. 6 shows the linear dependence between an intensity of the yellow luminescence and the concentration of V_{Ga} in different MOCVD grown GaN samples. Concentration of V_{Ga} was determined by means of the positron anihilation.

VI. Mechanism of the Luminescence in GaN/InGaN/AlGaN Quantum Wells

$In_xGa_{1-x}N$ alloys with x up to 0.45 are considered as the best materials for optoelectronic applications. Having bandgaps that are composition tunable in a large range of energies (1.9–3.4 eV) this material is well suited to form quantum wells in many optoelectronic devices based on group III nitrides. Also, compared to GaN, InGaN is commonly thought to provide better radiative recombination efficiency and to be free from parasitic yellow luminescence. Hence, it is not surprising that this material has been successfully employed in the single quantum well (SQW) LEDs manufactured by Nichia Chemical Industries (Nakamura et al., 1995b). Two types of SQW diodes have so far been fabricated: blue light emitting with the active layer built from $In_{0.15}Ga_{0.85}N$, and green light emitting in which the active layer consists of $In_{0.45}Ga_{0.55}N$. Also the first reported blue laser diode was based on a very similar structure (Nakamura et al., 1996a). In spite of these technological advances, the optical emission mechanism in InGaN QW diodes is not yet clear. Firstly, it was observed that the energy of emission in InGaN usually occurs at energies below the bandgap (Nakamura et al., 1995c). Secondly, it was shown that the linewidth of emission is characterized by a large nonthermal broadening (Nakamura et al., 1995b, 1995c). Finally, it has been found recently that the temperature shift of photo and electroluminescence in green LEDs (manufactured by Nichia Chemical Industries) is opposite to the shift of the bandgap (Perlin et al., 1997b). All these observations suggest strongly that the recombination in this material does not have a standard near-band-edge character (excitonic or band-to-band character). To account for this, several different recombination mechanisms have been put forward. Chichibu et al. (1997) suggested that the luminescence involves recombination of excitons localized on the potential fluctuations caused by variations in the indium content. Perlin et al. (1997b) proposed that the radiative transitions take place between uncorrelated electrons and holes in the band tails of this alloy material. From the measurements of the lifetime of the photoluminescence (PL) decay in InGaN alloys, it has also concluded that impurity states and alloy potential fluctuations are most likely responsible for the alloy PL signal (Shan et al., 1996). Lastly, it was asserted (only for quantum well systems) that the recombination could be related to quantum dots spontaneously formed in the InGaN confining layer due to indium content fluctuations (Chichibu et al., 1997; Shan et al., 1996).

In order to shed more light on the nature of the radiative recombination in InGaN QW's, hydrostatic pressure measurements of the electroluminescence and photoluminescence in the commercial Nichia LEDs were performed (Perlin et al., 1997c). The active layer of these devices consisted of a 3 nm thick $In_xGa_{1-x}N$ layer ($x = 0.45$ for the green diode, and $x = 0.15$ for the blue diode) sandwiched between n-GaN and p-$Al_{0.2}Ga_{0.8}N$ layers. The pressure coefficient of the emission peak can reveal whether the recombination has excitonic/band-to-band character (pressure coefficient should be close to that of the bandgap) or whether the recombination consists of different mechanisms, as for example, occurs with the contribution of the localized states (pressure coefficients much smaller than the bandgap are then expected — see Section II).

The presented studies (Perlin et al., 1997c) were performed at 300 K using a Unipress piston-cylinder cell equipped with sapphire window and electrical connections (electroluminescence (EL) studies), and diamond anvil cell (photoluminescence (PL) studies). Prior to the experiments, the diodes were de-encapsulated. For PL measurements, a part of the sapphire substrate of the diode structures was mechanically removed, and the samples were cleaved and placed in the diamond anvil cell. High pressure investigation of EL consisted in examination of the luminescence from the green diode ($In_{0.45}Ga_{0.55}N$ quantum well). Figure 7 shows the obtained spectra for few values of the applied pressure. Very weak changes of the EL band can be seen only. The pressure coefficient of the peak position was determined as equal to about 16 meV/GPa.

In order to prove that the mechanism of EL and PL is the same, the PL studies of pressure behavior of the green and blue ($In_{0.15}Ga_{0.85}N$ quantum well) diodes were performed. The luminescence was excited with an argon laser; the blue 458 nm line was used in the case of the green LED, and the UV 363 nm line was used for the blue LED. The obtained peak positions as a function of pressure are shown in Fig. 8. The linear pressure coefficients obtained for these PL peaks are again unusually low, about 16 meV/GPa and 12 meV/GPa for the blue and green diodes, respectively. These results are in agreement with the PL measurements.

One should compare these results with the value for the pressure coefficients of the bandgaps of GaN (40 meV/GPa from experiment (Perlin et al., 1992a)) and InN (33 meV/GPa from theory (Christensen and Gorczyca, 1994)). In order to ascertain why the pressure coefficients for the QW diodes are so low, one should estimate the pressure-induced change in the confinement energy. Application of pressure increases the bandgap and thus

10 DEFECTS AND IMPURITIES IN GALLIUM NITRIDE 297

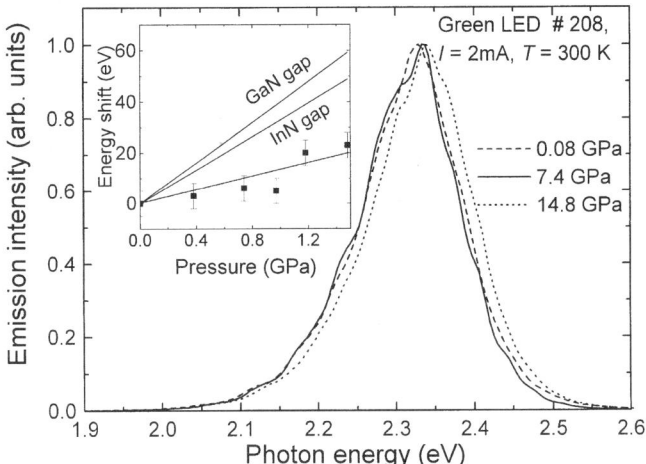

FIG. 7. Electroluminescence spectra of the green light emitting diode based on single quantum well of GaN/InGaN/AlGaN. The insert shows the energy value corresponding to the luminescence maximum as a function of pressure. (From Perlin et al., 1997c.)

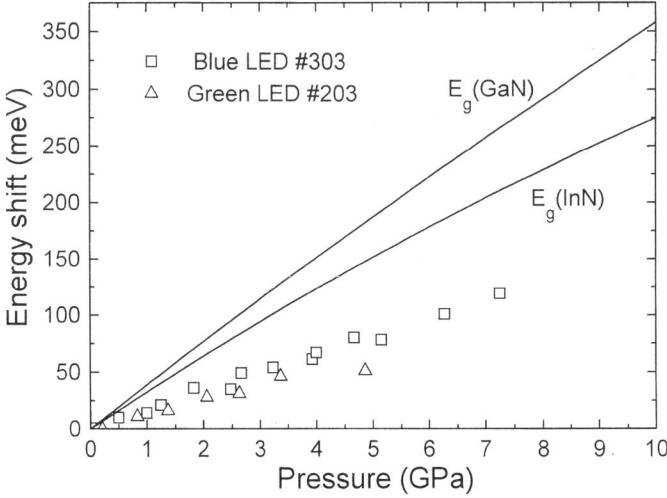

FIG. 8. Energy of the photoluminescence as a function of pressure for both the green (triangles) and the blue (squares) diodes with different InGaN single quantum wells. Solid lines show the pressure induced shift of the bandgaps in GaN and InN. (From Perlin et al., 1997c.)

tends to increase the effective mass, which in turn leads to a decrease of the pressure coefficient in the QW emission relative to that of the corresponding bulk alloy. These kinds of effects have been observed previously in GaAs QW's (Perlin *et al.*, 1994). Their magnitude can be estimated from first order k · p theory, in which the relative change in the effective mass is proportional to the relative change in the bandgap. Accordingly the pressure coefficient of the confinement energy should be $\sim 2\,\text{meV/GPa}$. Clearly, this does not make a very important contribution to the overall pressure coefficient of the luminescence peak. Another possible origin of the observed low value of the pressure coefficient is the influence of the substrate on the pressure behavior of the heterostructure. However, experimental evidence shows that bulk GaN crystals and epitaxial layers of GaN on sapphire both behave identically under applied hydrostatic pressure (see the case of the "yellow luminescence" described in Section V). Thus, since the observed pressure coefficients of the QW emission peaks are 2–3 times smaller than those of the bandgaps of the corresponding bulk alloys, it leads to the conclusion that the QW emission has not a standard near-band-edge character. The obtained pressure coefficients suggest the involvement of some kind of strongly localized states.

Let us assess this result in terms of the existing models of the luminescence. Considering the model based on highly localized excitons, the most familiar example is the nitrogen related luminescence in GaAs. In this material, nitrogen at an As site forms an isoelectronic center which can bind electrons via a short range potential. High pressure experiments (Leroux, Neu, and Verie, 1986) showed that the pressure coefficient of this emission can be much lower than that of the bulk GaAs bandgap. Although strongly localized excitons of this type usually produce sharp emission lines, the broadening seen in the present samples might arise from the effects of alloy fluctuations on the exciton localization in the InGaN QW's.

Small pressure coefficients are also expected if the recombination takes place between uncorrelated electrons and holes trapped in band-tail states (again arising from indium content fluctuations). Band-tail states become increasingly deep as they extend into the forbidden gap, and deep electronic states are known to have pressure coefficient determined by an average across the whole Brillouin zone. This average coefficient will be much lower than that of the Γ-point direct bandgap, in accord with the present results for the QW emission. This model can also account for the anomalous temperature shift of the QW peaks observed in these devices (Perlin, Osinski, and Eliseev, 1997b). The model of tail-related transitions has been successfully used to explain the light emission properties of amorphous silicon, including the effects of applied pressure (Perlin *et al.*, 1997c; and references therein).

Considering the third model based on recombination within quantum dots, it was demonstrated by Li *et al.* (1995) that the pressure coefficient of the luminescence from InAs quantum dots embedded in GaAs is not substantially different from the pressure coefficient of the bulk InAs bandgap. Comparing InAs and InGaN, one should remember that exciton radius becomes approximately five times smaller for InGaN in comparison with InAs, ~ 3 nm versus ~ 15 nm. The difference in the excitons radii may be responsible for the different pressure behavior. However, if we assume that the present recombination occurs inside quantum dots whose electronic levels are sufficiently unmixed by disorder that they behave like normal band edge states, we cannot account for the small measured value of the pressure coefficient for any possible local indium composition.

In conclusion, the pressure dependence of the electroluminescence and photoluminescence from two different GaN/InGaN/AlGaN LEDs was measured. Unusually low pressure coefficients were found indicating the involvement of highly localized states in the process of radiative recombination.

VII. Summary

The presented paper is a brief review of very recent results concerning defect and impurity centers in GaN. The discussed experiments were performed by means of various optical techniques under hydrostatic pressure. In the introduction, a general discussion of the pressure variation of different electronic states was given. Hydrostatic pressure offers a unique tool to examine the behavior of electronic levels associated with the localized states of defects. Contrary to the hydrogenic states, which under pressure follow the changes of their parent band extrema, a strong shift of the localized levels (with respect to the conduction band minimum) is usually observed. This tendency was used in the presented experiments to determine a nature of the electronic states of the studied defects or to verify models proposed for their description. In particular, the presented high pressure experiments showed that in undoped n-GaN bulk single crystals as well as in epitaxial layers highly doped with oxygen a freeze-out of electrons from the conduction band occurs at about 20 GPa. It has been proposed that this effect is due to an emergence of a conduction band resonance into the forbidden gap. The resonance-derived deep state captures free electrons, leading to the freezeout of the electrons. It is likely, that the oxygen forms the examined localized state in unintentionally doped n-GaN. On the contrary, Si donor leads preferentially to the effective mass-like state in GaN:Si.

In the case of parasitic yellow luminescence studies in GaN, hydrostatic pressure was used with the purpose of testing two models of the related radiative recombination. The obtained results support the model according to which the initial state of the analyzed luminescence is associated (at pressures lower than about 20 GPa) with conduction band minima or the shallow donor state. The model anticipating the involvement of a deep-duble donor and shallow acceptor level turned out to be much less probable.

Finally, the hydrostatic pressure was used to verify models of the luminescence in the commercial blue and green light emitting diodes consisting of GaN/InGaN/AlGaN single quantum well. Unusually low pressure coefficients were found, indicating the involvement of highly localized states in the process of radiative recombination (e.g., excitons localized by the alloy potential fluctuations in InGaN active layers).

ACKNOWLEDGMENTS

The authors acknowledge valuable discussions with Drs. M. Leszczynski, C. Wetzel, J. W. Ager III, S. Porowski, I. Grzegory, I. Gorczyca, P. Boguslawski, B. A. Weinstein, and J. Baranowski.

REFERENCES

Baj, M., Dmowski, L. H., and Slupinski, T. (1993). "Direct Proof of Two-Electron Occupation of Ge-DX Centers in GaAs Codoped with Ge and Te," *Phys. Rev. Lett.* **71**, 3592.

Boguslawski, P., Briggs, E., and Bernholc, J. (1995). "Native Defects in Gallium Nitride," *Phys. Rev. B* **51**, 17255; Boguslawski, P. and Bernholc, J. (1996). "Doping Properties of Amphoteric C, Si, and Ge Impurities in GaN and AlN," *Acta. Phys. Polon.* **A90**, 735.

Callaway, J. (1974). "Quantum Theory of the Solid State," part B, Academic Press, New York.

Chadi, D. J., and Chang, K. J. (1988). "Theory of the Atomic and Electronic Structure of DX Centers in GaAs and AlGaAs Alloys," *Phys. Rev. Lett.* **61**, 873.

Chadi, J. (1995). "Bond Instability and Impurity Compensation in II–VI Semiconductors: Theory," *Proceedings of 22nd Int. Conf. on the Physics of Semiconductors, Vancouver, 1994* (D. J. Lockwood, ed.), World Scientific Publishing Co., Singapore, p. 2311.

Chichibu, S., Azuhata, T., Sota, T., Nakamura, S. (1997). "Recombination of Localized Excitons in InGaN Single- and Multi-Quantum Well Structures," *Proceedings of MRS Fall Meeting* (T. Moustakas and I. Akasaki, eds.) Boston, December 1996.

Christensen, N. E., and Gorczyca, I. (1994). "Optical and Structural Properties of III–V Nitrides Under Pressure," *Phys. Rev. B* **50**, 4397.

Chung, B. C., and Gershenzon, M. (1992). "The Influence of Oxygen on the Electrical and Optical Properties of GaN Crystals Grown by Metalorganic Vapor Phase Epitaxy," *J. Appl. Phys.* **72**, 651.

Glaser, E. R., Kennedy, T. A., Doverspike, K., Rowland, L. B., Gaskill, D. K., Freitas, Jr. J. A.,

Asif Khan, M., Olson, D. T., Kuznia, J. N., and Wickenden D. K. (1995). "Optically Detected Magnetic Resonance of GaN Films Grown by Orhanometallic Chemical-Vapor Deposition," *Phys. Rev. B* **51**, 13326.

Ilegems, M., and Montgomery, M. C. (1973). "Electrical Properties of n-type Vapor-Grown Gallium Nitride," *J. Phys. Chem. Solids* **34**, 885.

Koide, N., Kato, H., Sassa, M., Yamasaki, S., Manabe, K., Hashimoto, M., Amano, H., Hiramatsu, K., and Akasaki, I. (1991). "Doping of GaN with Si and Properties of Blue GaN LED with Si-Doped Layer by MOVPE," *J. Cryst. Growth* **115**, 639.

Kozawa, T., Kachi, T., Kano, H., Taga, Y., Hashimoto, M., Koide, N., and Manabe, K. J. (1994). "Raman Scattering from LO Phonon-Plasmon Coupled Modes in Gallium Nitride," *J. Appl. Phys.* **75**, 1098.

Leroux, M., Neu, G., and Verie, C. (1986). "High Pressure Dependence of the Electronic Properties of Bound States in n-type GaAs," *Solid State Commun.* **58**, 289.

Li, G. H., Goñi, A. R., Syassen, K., Brandt, O., and Ploog, K. (1995). "High Pressure Study of Gamma-X Mixing in InAs/GaAs quantum dots," *J. Phys. Chem. Solids* **56**, 185.

Mooradian, A. A., and Wright, G. B. (1967). "Observation of the Interaction of Plasmons with Longitudinal Optical Phonons in GaAs," *Phys. Rev. Lett.* **16**, 999.

Maruska, H. P., and Teitjen, J. J. (1969). "The Preparation and Properties of Vapour-Deposited Single-Crystalline GaN," *Appl. Phys. Lett.* **15**, 327.

Monemar, B., and Lagerstedt, O. (1979). "Properties of VPE-Grown GaN Doped with Al and Some Iron-Group Metals," *J. Appl. Phys.* **50**, 6480; Tansley, T. L., and Egan, R. J. (1992). "Point-Defect Energies in the Nitrides of Aluminium, Gallium and Indium," *Phys. Rev. B* **45**, 10942.

Mooney, P. M. (1990). "Deep Donor Levels (DX Centers) in III–V Semiconductors," *J. Appl. Phys.* **67**, R1.

Molnar, R. J., and Moustakas, T. D. (1994). "Growth of Gallium Nitride by Electron-Cyclotron Resonance Plasma-Assisted Molecular-Beam Epitaxy: The Role of Charged Species," *J. Appl. Phys.* **76**, 4587.

Matilla, T., and Nieminen, R. M. (1996). "*Ab Initio* Study of Oxygen Point Defects in GaAs, GaN, and AlN," *Phys. Rev. B* **54**, 16676.

Neugebauer, J., and Van de Walle, C. G. (1994). "Atomic Geometry and Electronic Structure of Native Defects in GaN," *Phys. Rev. B* **50**, 8067.

Neugebauer, J., and Van de Walle, C. G. (1995a). "Defects and Doping in GaN," *Proceedings of 22nd Int. Conf. on the Physics of Semiconductors, Vancouver, 1994* (D. J. Lockwood, ed.) World Scientific Publishing Co., Singapore, p. 2327.

Nakamura, S., Senoh, M., Iwasa, N., and Nagahama, S. (1995b). "High Brightness InGaN Blue, Green, and Yellow Light-Emitting Diodes with Quantum Well Structures," *Jpn. J. Appl. Phys.* **34**, L797.

Nakamura, S., Senoh, M., Iwasa, N., Nagahama, S., Yamada, T., Mukai, T. (1995c). "Superbright Green InGaN Single-Quantum, Well-Structure Light-Emitting Diodes," *Jpn. J. Appl. Phys.* **34**, L1332.

Nakamura, S., Senoh, M., Nagahama, S., Iwasa, N., Yamada, T., Matsushita, T., Kiyoku, H., and Sugimoto, Y. (1996a). "InGaN-Based Multi-Quantum Well-Structure Laser Diode," *Jpn. J. Appl. Phys.* **35**, L74.

Neugebauer, J., and Van de Walle, C. G. (1996b). "Gallium Vacancies and the Yellow Luminescence in GaN," *Appl. Phys. Lett.* **69**, 503.

Ogino, T., and Aoki, M. (1980). "Mechanism of Yellow Luminescence in GaN," *Jpn. J. Appl. Phys.* **19**, 2395.

Porowski, S., Grzegory, I., and Jun, J. (1989). "Synthesis of Metal Nitrides Under High Nitrogen Pressure," In *High Pressure Chemical Synthesis* (J. Jurczak and B. Baranowski, eds.) Elsevier, Amsterdam, p. 21.

Perlin, P., Gorczyca, I., Christensen, N. E., Grzegory, I., Teisseyre, H., and Suski, T. (1992a). "Pressure Studies of Gallium Nitride: Crystal Growth and Fundamental Electronic Properties," *Phys. Rev. B* **45**, 13307; Teisseyre, H., Perlin, P., Suski, T., Grzegory, I., Porowski, S., Jun, J., Pietraszko, A., and Moustakas, T. D. (1994). "Temperature Dependence of the Energy Gap in GaN Bulk Single Crystals and Epitaxial Layer. *J. Appl. Phys.* **76**, 2429.

Perlin, P., Jauberthie-Carillon, C., Itie, J. P., San Miguel, A., Grzegory, I., and Polian, A. (1992b). "Raman Scattering and X-ray Absorption Spectroscopy in Gallium Nitride Under High Pressure," *Phys. Rev. B* **45**, 83.

Perlin, P., Trzeciakowski, W., Litwin-Staszewska, E., Muszalski, J., and Mikovic, M. (1994). "The Effect of Pressure on the Luminescence from GaAs/AlGaAs Quantum Wells. *Semicond. Sci. Technol.* **9**, 2239.

Perlin, P., Suski, T., Teisseyre, H., Leszczynski, M., Grzegory, I., Jun, J., Porowski, S., Boguslawski, P., Bernholc, J., Chervin, J. C., Polian, A., and Moustakas, T. D. (1995). "Towards the Identification of the Dominant Donor in GaN," *Phys. Rev. Lett.* **75**, 296.

Perlin, P., Knap, W., Camassel, J., Polian, A., Chervin, J. C., Suski, T., Grzegory, I., and Porowski, S. (1996). "Metal-Insulator Transition in GaN Crystals," *Phys. Stat. Sol. B* **198**, 223.

Perlin, P., Suski, T., Polian, A., Chervin, J. C., Knap, W., Camassel, J., Grzegory, I., Porowski, S., and Erickson, J. W. (1997a). "Coexistence of Shallow and Localized Donor Centers in Bulk GaN Crystals Studied by High Pressure Raman Spectroscopy," *Proceedings of the Fall MRS Meeting* (T. D. Moustakas and I. Abasaki, eds.) Boston 1996.

Perlin, P., Osinski, M., and Eliseev, P. G. (1997b). "Optical and Electrical Characteristics of Single-Quantum-Well InGaN Light Emitting Diodes," *Proceedings of MRS Fall Meeting* (T. Moustakas and I. Akasaki, eds.) Boston, December 1996.

Perlin, P., Iota, W., Weinstein, B. A., Wisniewski, P., Suski, T., Osinski, M., and Eliseev, P. G. (1997c). "Influence of Pressure on the Photoluminescence and Electroluminescence of GaN/InGaN/AlGaN Quantum Wells," *Appl. Phys. Lett.* **70**, 2993.

Seifert, W., Franzheld, R., Butter, E., Sobotta, H., and Riede, V. (1983). "On the Origin of Free Carriers in High-Conducting n-GaN," *Cryst. Res. Technol.* **18**, 383.

Skowronski, M. (1992a). "Complexes of Oxygen and Native Defects in GaAs," *Phys. Rev. B* **46**, 9476.

Sobotta, H., Neumann, H., Franzheld, R., and Seifert, W. (1992b). "Infrared Lattice Vibrations of GaN," *Phys. Stat. Solidi B* **174**, K57.

Suski T. (1994). "Hydrostatic Pressure Investigations of Metastable Defect States," *Mat. Sci. Forum* **143–147**, 975.

Suski, T., Perlin, P., Teisseyre, H., Leszczynski, M., Grzegory, I., Jun, J., Bockowski, M., Porowski, S., and Moustakas, T. D. (1995). "Mechanism of Yellow Luminescence in GaN," *Appl. Phys. Lett.* **67**, 2188.

Shan, W., Schmidt, T. J., Hauenstein, R. J., Song, J. J., and Goldenberg, B. (1995b). "Pressure-Dependent Photoluminescence Study of Wurtzite GaN," *Appl. Phys. Lett.* **66**, 3492: Kim, S., Herman, I. P., Tuchman, J. A., Doverspike, K., Rowland, L. B., and Gaskill, D. K. (1995). "Photoluminescence from Wurzite GaN under Hydrostatic Pressure," *Appl. Phys. Lett.* **67**, 380.

Shan, W., Little, B. D., Song, J. J., Feng, Z. C., Schurman, M., and Stall, R. A. (1996). "Optical Transitions in $In_xGa_{1-x}N$ Alloys Grown by Metalorganic Chemical Vapor Deposition," *Appl. Phys. Lett.* **69**, 3315.

Saarinen, K., Laine, T., Kuisma, S., Nassila, J., Hautojarvi, P., Dobrzynski, L., Baranowski, J., Pakula, K., Stepniewski, R., Wojdak, M., Wysmolek, A., Suski, T., Leszczynski, M.,

Grzegory, I., and Porowski, S. "Observation of Native Ga Vacancies in GaN by Positron Annihilation," to be published.

Wetzel, C., Walukiewicz, W., Haller, E. E., Ager III, J. W., Grzegory, I., Porowski, S., and Suski, T. (1996). "Carrier Localization of As-Grown n-type Gallium Nitride Under Large Hydrostatic Pressure," *Phys. Rev. B* **53**, 1322.

Wetzel, C., Suski, T., Ager III, J. W., Weber, E. R., Haller, E. E., Fischer, S., Meyer, B. K., Molnar, R. J., and Perlin, P. (1997). "Pressure Induced Deep Gap State of Oxygen in GaN," *Phys. Rev. Lett.* **78**, 3923.

Yoshida, S., Misawa, S., and Gonda, S. (1982). "Properties of $Al_xGa_{1-x}N$ Films Prepared by Reactive Molecular Beam Epitaxy," *J. Appl. Phys.* **53**, 6844.

Zolper, J. C., Wilson, R. G., Pearton, S. J., and Stall, R. A. (1996a). "Ca and O Ion Implantation Doping of GaN," *Appl. Phys. Lett.* **68**, 1945.

Zhang, X., Kung, P., Walker, D., Saxler, A., and Razeghi, M. (1996b). "Growth of GaN Without Yellow Luminescence," *Mat. Res. Soc. Symp. Proc.* **395**, 625.

CHAPTER 11

Optical Properties of GaN

B. Monemar

DEPARTMENT OF PHYSICS AND MEASUREMENT TECHNOLOGY
MATERIALS SCIENCE DIVISION, LINKÖPING UNIVERSITY
LINKÖPING, SWEDEN

I. INTRODUCTION 305
II. FUNDAMENTAL OPTICAL PROPERTIES 306
   1. Optical Properties Above the Bandgap Energy 306
   2. The Near Bandgap Region, Exciton Effects 311
   3. Exciton Recombination Dynamics 334
   4. Near Bandgap Optical Properties at High Carrier (Exciton) Densities . . . 339
III. BELOW BANDGAP OPTICAL PROPERTIES, REFRACTIVE INDEX 342
IV. OPTICAL PROPERTIES OF CUBIC GaN 345
V. DEFECT-RELATED OPTICAL PROPERTIES 348
   1. Bound Excitons in GaN 349
   2. Other Donor- or Acceptor-Related Optical Spectra 355
   3. Optical Spectra Related to Transition Metal Centers in GaN 361
   References 363

I. Introduction

The optical properties of a semiconductor are connected with both intrinsic and extrinsic effects. Intrinsic optical transitions are the band-to-band transitions, including exciton effects, but also phonon- or carrier-related absorption processes in the *ir* part of the spectrum. Extrinsic properties are related to dopants or defects, which usually create discrete electronic states in the bandgap, and therefore influence both optical absorption and emission processes, as well as many other material parameters. In this chapter, we shall first discuss the intrinsic optical properties of GaN, starting with the above bandgap energy region. Later, impurity and defect related effects will be treated. Similar properties in AlN, InN, and their alloys with GaN will be discussed only where relevant data exist.

II. Fundamental Optical Properties

1. Optical Properties above the Bandgap Energy

GaN (as well as AlN and InN) has a direct lowest bandgap, and therefore a sharp interband-band absorption edge with a large value of the absorption coefficient α at the edge, ($\sim 10^5$ cm^{-1}, see below). The absorption is even stronger above the edge and, therefore, direct transmission measurements of interband absorption are not practical, since sample thicknesses of the order 0.1 μm would be required. A common method to determine the optical properties in this high photon energy range is to measure reflectance over a large spectral range, which can be done quite accurately. With this information, it is possible to calculate the spectral dependence of other related optical quantities, such as the real and imaginary parts of the dielectric function, or the real and imaginary part of the refractive index via the so-called Kramers–Kronig relations (Klingshirn, 1995), here written for the relation between the real and imaginary part of the frequency dependent dielectric function ε:

$$\varepsilon_1(\omega) = 1 + 2/\pi P \int \omega' \varepsilon_2(\omega')\, d\omega'/(\omega^2 - \omega'^2) \tag{1}$$

where $h\nu = \hbar\omega$ is the photon energy, and P stands for the principal value of the integral. The integral here runs from 0 to infinity, but in practice, the integrand can be approximated with good accuracy, with a simple extrapolation outside the regions experimentally measured.

The refractive index **n** is related to the dielectric function $\varepsilon = \varepsilon_1 + i\varepsilon_2$ simply as

$$\mathbf{n} = n + ik = (\varepsilon_1 + i\varepsilon_2)^{1/2} \tag{2}$$

The quantity ε_2 is directly related to the absorption coefficient α via $\alpha = 2k\omega/c$, where $\omega = 2\pi c/\lambda$, (λ is the photon wavelength). The reflectance (normal to the surface) in turn is related to the refractive index via

$$R = \frac{(n-1)^2 + k^2}{(n+1)^2 + k^2} \tag{3}$$

Optical transitions discussed here are direct transitions, for example, they are vertical in k-space due to conservation of energy and momentum in the transition (Fig. 1(a)). The interband optical transition probability is related

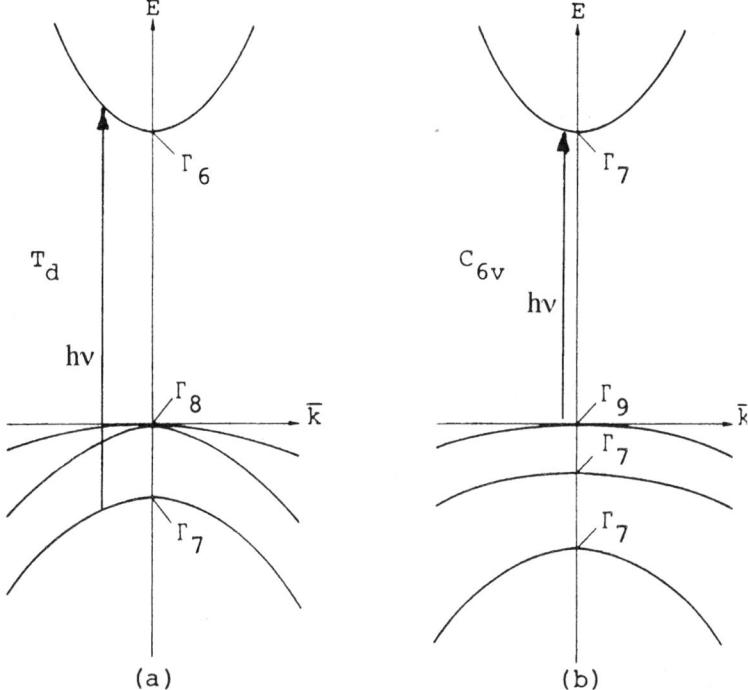

FIG. 1. Schematic picture of the near bandgap electronic structure of GaN: (a) cubic GaN (zincblende structure) and (b) wurtzite GaN. Examples of vertical optical interband transitions are also given.

to the interband optical matrix element M_{cv} (usually evaluated in the electric dipole approximation) and in addition to the density of initial states in the valence band and final states in the conduction band (Klingshirn, 1995). This can schematically be written as

$$\varepsilon_2 \sim \alpha \sim k \sim [M_{cv}]^2 \int_{BZ} 1/\nabla E_{cv} \quad (4)$$

where the integral should cover the entire Brillouin zone (BZ) (Klingshirn, 1995). It is therefore the joint density of states for the energy gradient ∇E_{cv} between valence band and conduction band states at each energy that is important for the absorption strength. Whenever extrema in this quantity occur, there is a singularity expected in the joint density of states (the

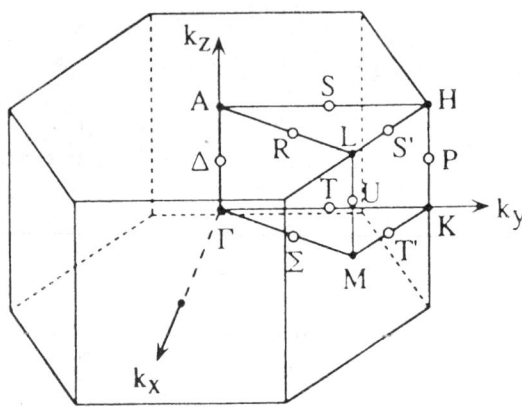

FIG. 2. The Brillouin zone of the hexagonal wurtzite structure. The standard notation for the high symmetry points and lines are also given.

different so-called Van Hove singularities (Yu and Cardona, 1996), and a corresponding feature will show up in optical spectra. It should be pointed out that the occurrence of such extrema is not necessarily connected with high symmetry points in the BZ, often the strongest peaks in interband spectra are related to other regions of the BZ, where the joint density of states is high. The BZ of the wurtzite structure is shown in Fig. 2, for easy reference.

An example of reflectance data of wurtzite GaN is shown in Fig. 3, which was obtained in the configuration $\mathbf{E} \perp \mathbf{c}$ between 3.5 eV and 30 eV, with a synchrotron radiation source. From the experimental data for R, the other quantities ε_1, ε_2, n and k are derived, via Kramers–Kronig analysis (Klingshirn, 1995; Yu and Cardona, 1996), as shown in the lower panels of Fig. 3.

In order to assign the observed peaks in the reflectance spectrum to features in the band structure, a theoretical band structure calculation has to be performed. Figure 4 shows an example of such a band structure from a recent calculation (Lambrecht et al., 1995) using the so-called density functional theory (DFT) in the local density approximation (LDA). Minor effects like exciton binding and spin-orbit splittings are ignored, since they are not important on the energy scale of these data. This technique underestimates the bandgap, and therefore, the conduction band states in Fig. 4 should be raised in energy by about 1 eV to get the correct bandgap. Assuming the same correction for all conduction band states, a theoretical reflectance spectrum can be calculated from the band structure, by first calculating ε_2 (Eq. (4)), further ε_1 via Kramers–Kronig analysis, and finally

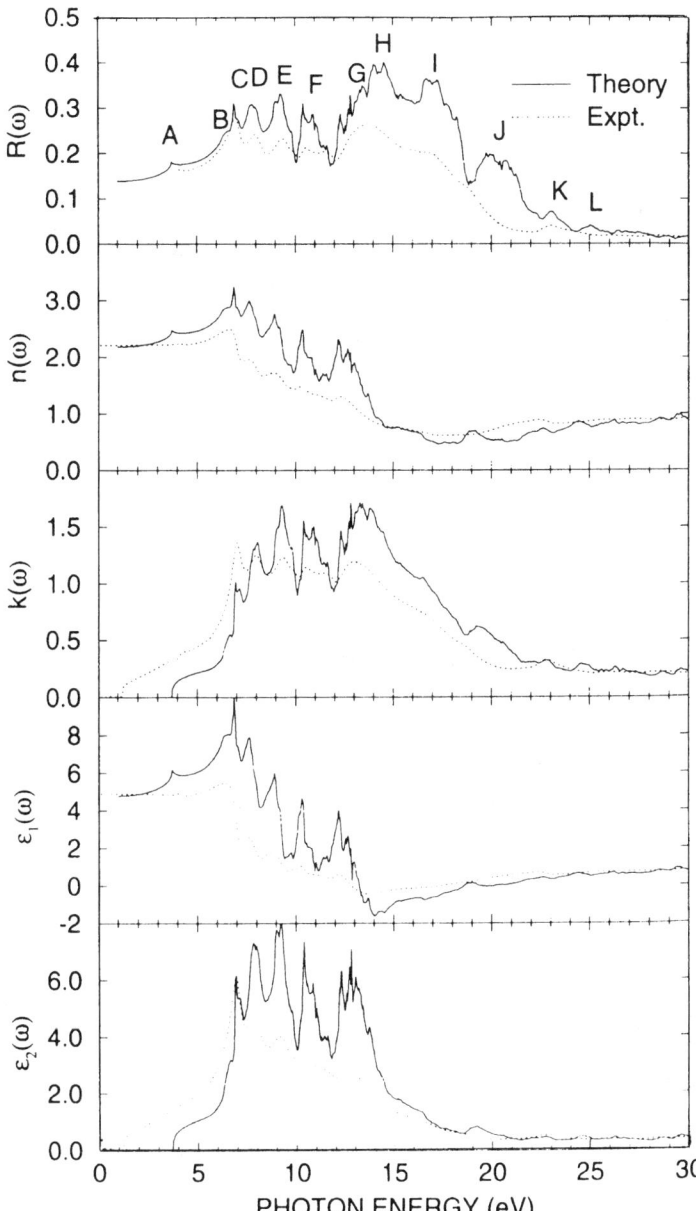

FIG. 3. The various optical response functions of GaN for polarization $\mathbf{E} \perp \mathbf{c}$, obtained from reflectance data using synchrotron radiation (.....), compared with theoretical calculations of the same spectra (—). (Reprinted from Lambrecht et al. (1995), with permission from The American Physical Society.)

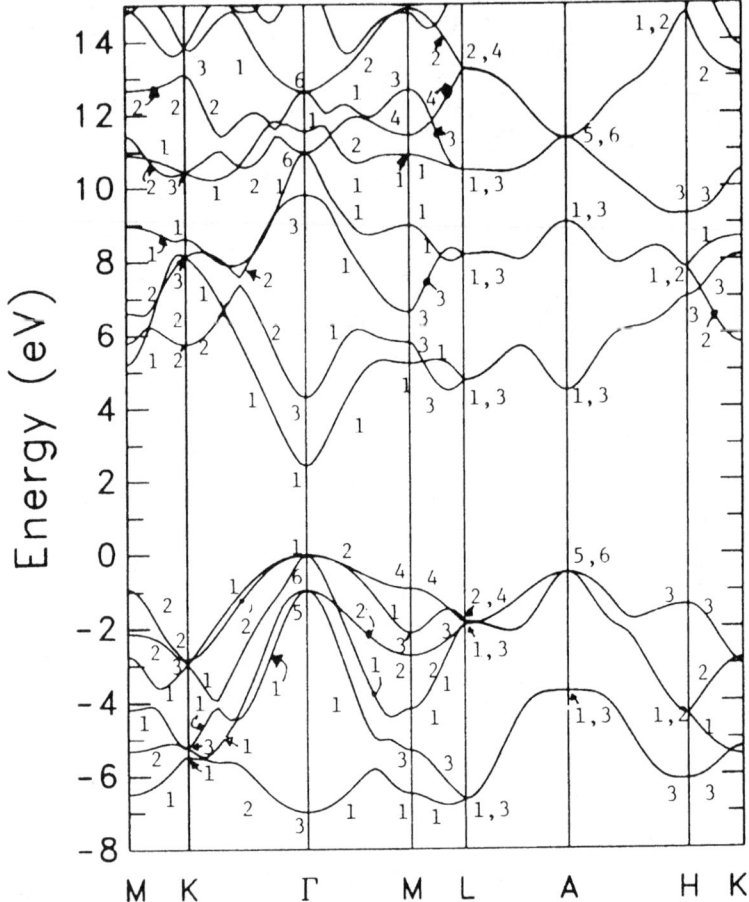

FIG. 4. Calculated LDA energy bands of wurtsite GaN. For the calculated spectra in Fig. 3, the conduction bands have been upshifted about 1 eV to correct for self energy effects. The symmetry labeling of the bands is discussed in Lambrecht et al. (1995). (Reprinted from Lambrecht et al. (1995), with permission from The American Physical Society.)

R from Eq. (3). The result is shown in the top panel of Fig. 3. The overall agreement of the theoretical curve with experiment is reasonable for most of the strongest features, but a certain ambiguity remains in the assignments of some transitions. The labeling is done according to the standard notations (Lambrecht et al., 1995) for the high symmetry positions in the Brillouin zone for GaN, shown in Fig. 2.

TABLE I

SUGGESTED INTERPRETATION OF THE REFLECTIVITY PEAKS FOR $\mathbf{E} \perp \mathbf{c}$ IN THE TOP PANEL IN FIG. 3 IN TERMS OF TRANSITIONS IN THE BAND STRUCTURE (FIG. 4). THE NOTATIONS REFER TO THE BRILLOUIN ZONE IN FIG. 2. FOR A DETAILED DISCUSSION, SEE LAMBRECHT ET AL. (1995)

Reflectivity Peak	Theory (eV)	Experiment (eV)	Identification
A	3.8	—	$\Gamma_6 - \Gamma_1$
B	6.5	—	$\Gamma_5 - \Gamma_3$
C	6.9	6.9	$U_4 - U_3$ + L-point
D	7.7	8.0	$M_4 - M_3$
E	9.2	9.3	$T_2 - T_2$
F	10.4–11.5	10.4–11.5	$A_{5,6} - A_{1,3}$; $L_{1,3} - L_{1,3}$
G	12.2–13.3	12.2–13.4	Many transitions
H	14.0–14.5	13.9	Many transitions
I	17.0	16.8	Many transitions
J	20.0	19.0	Many transitions
K	23.0	23.0	Many transitions

Table I summarizes the energies of the main features in the optical spectra above the bandgap. Only the most prominent peaks are included here, a more detailed account of all transitions expected is given in Lambrecht *et al.* (1995).

An alternative technique to measure the optical constants is spectroscopic ellipsometry, which directly measures both the real and the imaginary part of the dielectric function (Aspnes and Studna, 1975). Typical data are shown in Fig. 5, in the energy range 3–10 eV (Logothetidis *et al.*, 1994; Petalas *et al.*, 1995). Such data can be analyzed in an analogous way by comparison with simulated data obtained from band structure calculations.

2. THE NEAR BANDGAP REGION, EXCITON EFFECTS

In the near bandgap region additional details in the band structure need to be considered, for example, spin orbit coupling and electron-hole interaction, including the exciton effects. The near bandgap structure including spin–orbit coupling is schematically shown in Fig. 1(*b*). The near bandgap intrinsic absorption spectrum is therefore expected to be dominated by transitions from three valence bands of symmetry Γ_9, Γ_7, and Γ_7 to a single lowest Γ_7 conduction band (Fig. 1(*b*)), and the corresponding exciton states are usually denoted by A, B, and C, respectively. The optical properties of these transitions will be discussed in some detail later, covering both

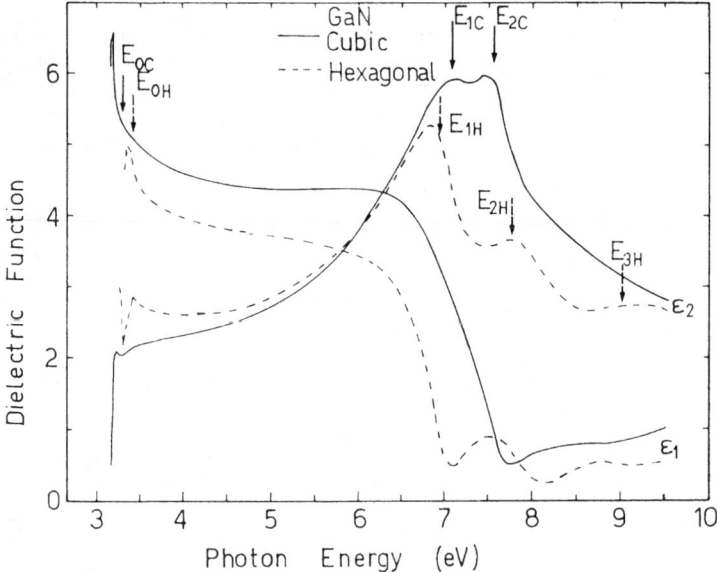

FIG. 5. Experimental data for the real (ε_1) and imaginary part (ε_2) of the dielectric function in the range 3–10 eV obtained with spectroscopic ellipsometry. Both cubic (—) and hexagonal (---) GaN was studied. (Reprinted from Logothetidis et al. (1994), with permission from The American Physical Society.)

absorption, reflectance, and luminescence data. It should be noted, that these properties are quite sensitive to built-in strain, which commonly occurs when the material is grown on a foreign substrate with heteroepitaxy, as discussed separately below.

a. Near Bandgap Absorption Data

The traditional way to measure the absorption coefficient in the near bandgap region of a semiconductor is to use straightforward optical transmission experiments. Using a set of samples with different thicknesses, a sufficiently large range of the absorption spectrum around the bandgap can be covered. The problem at present with GaN and related materials, is to prepare good quality samples thin enough (0.1–1 μm) to cover the region around the bandgap. So far only epitaxially grown films have the proper material quality to clearly exhibit the excitonic structure close to the fundamental band edge. In principle, a thin GaN layer could be grown on a transparent sapphire substrate, but the region close to the substrate

(0.1–0.2 μm) is in general unsuitable for optical studies, due a high structural defect density.

Transmission experiments using such thin epilayers were performed already in the early 1970s (Kosicki, Powell, and Burgiel, 1970; Pankove, Maruska, and Berkeyheiser, 1970; Dingle et al., 1971a; Hovel and Cuomo, 1972; Osamura, Nakajima, and Murakami, 1972), and usually show a knee in the absorption coefficient α around the excitonic edge, indicating a broadened excitonic peak. The values of α at 300 K and 3.42 eV vary between the different sets of data, in the interval $0.8 \times 10^5 - 4 \times 10^5 \mathrm{cm}^{-1}$. A high electron density may have affected the data in some of the cases above. If the electron concentration is in the high $10^{18}\,\mathrm{cm}^{-3}$ range or larger the excitons are screened, and the absorption edge becomes broadened and modified (Cingolani, Ferrara, and Lugara, 1986). Further, at high electron densities the absorption edge shifts towards higher energy, due to filling of states in the conduction band (Teisseyre et al., 1994).

As mentioned above, an alternative way to measure the absorption coefficient is to measure the entire complex dielectric function with spectroscopic ellipsometry (Aspnes and Studna, 1975). The absorption coefficient

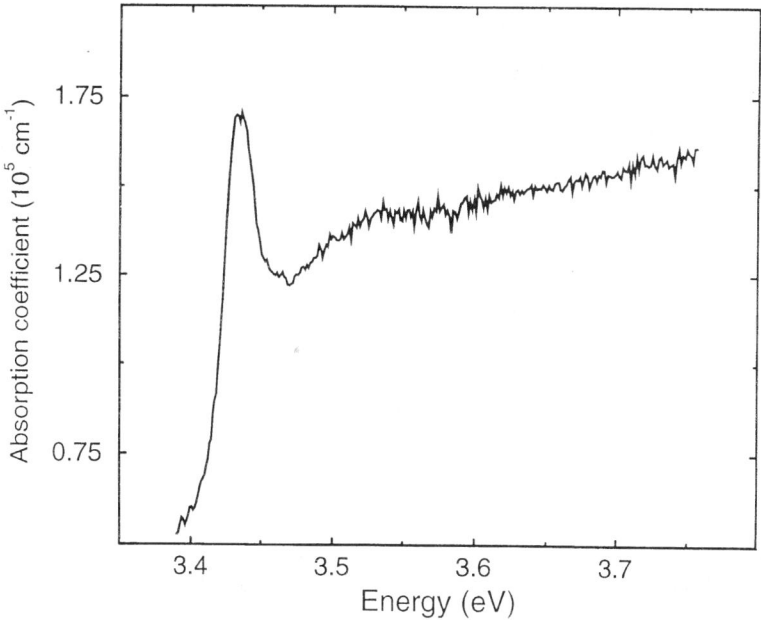

FIG. 6. Absorption coefficient at 300 K for wurtzite GaN, obtained from spectrometric ellipsometry. The exciton peak is dominated by the A exciton, but also contains the contribution from the B exciton within the linewidth shown. (H. Amano, unpublished data.)

$\alpha(\omega)$ can then be obtained directly from the imaginary part of $\varepsilon(\omega)$. Data from this technique are more recent, and probably reflect the development towards better material quality. The lowest exciton peak is typically resolved even at room temperature (Petalas *et al.*, 1995; Akasaki and Amano, 1994; Logothetidis *et al.*, 1995). An example is shown in Fig. 6. The literature values for the peak A-exciton absorption coefficient vary as $1.6 \times 10^5 \text{cm}^{-1}$ (300 K (Akasaki and Amano, 1994)), $2.2 \times 10^5 \text{cm}^{-1}$ (300 K (Logothetidis *et al.*, 1995), $3.2 \times 10^5 \text{cm}^{-1}$ (230 K (Petalas *et al.*, 1995).

The absorption coefficient α at the lowest A-exciton peak is then probably determined at least within a factor 2 as $\alpha = 2 \times 10^5 \text{cm}^{-1}$ at 300 K. This peak value is expected to be slightly higher at cryogenic temperatures, since the exciton peak then becomes much narrower. More accurate data will hopefully appear in the near future, as samples of better purity and optical quality will become available.

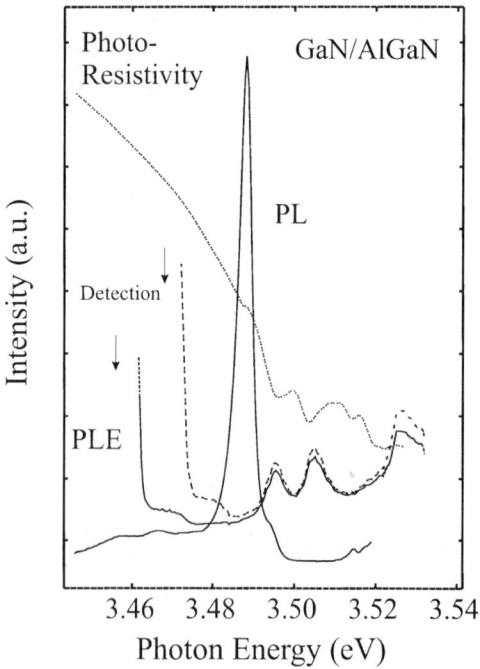

FIG. 7. Comparison of photoluminescence (PL), PL excitation (PLE), and photoresistivity measurements at 2 K for a GaN sample grown on sapphire. Note that the energies are upshifted due to the strain. The PL is dominated by the donor bound excitons, while the PLE and photoresistivity curves are dominated by the intrinsic exciton features. Two PLE curves of different detection energies are shown. (Data obtained by J. P. Bergman and A. Buyanov.)

The intrinsic bandgap is not directly observed in optical absorption experiments, in a material like GaN, where the band edge transitions are typically completely dominated by excitons even at room temperature. This is true also for photoconductivity (PC) spectra (Fig. 7), which turn out to be dominated by exciton effects. This is possibly due to electric field ionization of the excitons, which produce the carriers detected in the PC experiment. From the data available in the literature, one can conclude a similar approximate value $\alpha = 2 \times 10^5 \text{cm}^{-1}$ for the absorption strength at the lowest bandedge of GaN, for reasonably good material.

A few examples of absorption data for AlGaN (Akasaki and Amano, 1994) and InGaN (Osamura, Nakajima, and Murakami, 1972) exist in the literature. The α-values reported do not differ significantly from the GaN values (at least for low In concentrations (Osamura, Nakamima, and Murakami, 1972).

b. Intrinsic Excitonic Structure in the Near Bandgap Region

As mentioned above, the intrinsic optical properties of GaN is dominated by excitons. The three exciton states with symmetries of Γ_9 (A), Γ_7 (B), and Γ_7 (C) obey the following selection rules in optical one-photon transitions: all excitons (A, B, and C) are allowed in the σ polarization (light field vector $\mathbf{E} \perp \mathbf{c}$ axis, light propagation vector $\mathbf{k} \perp \mathbf{c}$ axis), but the C exciton is quite weak. The C exciton is strongly allowed in the π polarization ($\mathbf{E} \parallel \mathbf{c}$ axis, $\mathbf{k} \perp \mathbf{c}$), however, where the B exciton is also weakly allowed, while the A exciton is forbidden in this geometry. In the α polarization ($\mathbf{E} \perp \mathbf{c}$, $\mathbf{k} \parallel \mathbf{c}$) all three transitions are clearly observed (Dingle et al., 1971b).

Each of these fundamental exciton states is expected to have a substructure, for two main reasons. One effect expected to be of the order 1–2 meV for GaN, is the exciton polariton longitudinal-transverse splitting (Hopfield and Thomas, 1963; Ivchenko, 1982). Another effect, which could also be of the order 1 meV in GaN, is the splitting caused by the electron-hole exchange interaction (Pikus and Ivchenko, 1982). None of this substructure has so far been properly resolved for GaN, due to limitations in spectroscopic linewidth of the free excitons to $\geqslant 1$ meV in the samples so far available.

The standard technique for optical spectroscopy of the intrinsic excitons is reflectance measurements. In order to fully investigate the polarization properties and the selection rules, measurements in three different geometries are needed: the α polarization ($\mathbf{E} \perp \mathbf{c}$, $\mathbf{k} \parallel \mathbf{c}$), the σ polarization ($\mathbf{E} \perp \mathbf{c}$, $\mathbf{k} \perp \mathbf{c}$), and the π polarization ($\mathbf{E} \parallel \mathbf{c}$, $\mathbf{k} \perp \mathbf{c}$). For thin epilayers the measurement is straightforward for the α polarization (the light comes perpendicular

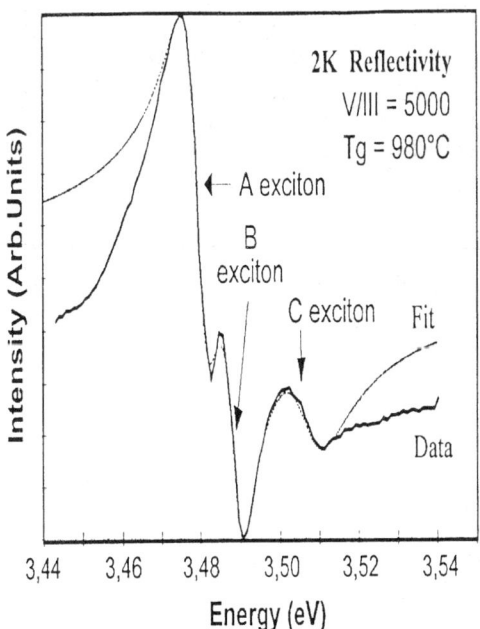

FIG. 8. Reflectance spectrum at 2 K for a 2 μm thick GaN sample grown on sapphire (—), compared with a theoretical fit with a damped oscillator model. (Reprinted from M. Tchounkeu *et al.*, 1996), with permission from The American Institute of Physics.)

to the surface of the layer). The other polarizations require the measurement of reflected light from a cleaved edge of the sample, which is often difficult for thin epilayers grown on a foreign substrate. Very often derivative techniques (like photo-reflectance) are employed, in order to improve resolution and sensitivity in the spectral measurements. An example of a reflectance spectrum obtained in the α polarization is given in Fig. 8 for a 2 μm thick GaN epilayer on sapphire. The corresponding excitonic transition energies can be evaluated with a classical model involving a damped oscillator transition (Tchounkeu *et al.*, 1996). The evaluated exciton energies and other properties are summarized in Table II.

As will be discussed in more detail later, the exact spectral positions of the excitons for an epitaxial layer is strongly dependent on the strain field present in the heteroepitaxial samples. Low-doped unstrained bulk GaN of good crystalline quality is not yet available, therefore fundamental optical studies have to be done on epitaxial layers. In order to obtain reliable values for bulk unstrained GaN, one has to study either very thick epilayers, where the strain has been relieved to a large extent via cracking, or powder type

TABLE II

FREE EXCITON PROPERTIES OF UNSTRAINED GaN AT 2 K. THE ENERGY VALUES ARE ONLY ACCURATE WITHIN ABOUT 2 meV, THE OSCILLATOR STRENGTHS (PER MOLECULE) ARE ONLY ESTIMATED TO BE ACCURATE WITHIN A FACTOR 3, THE BINDING ENERGY WITHIN 4 meV AND THE LONGITUDINAL-TRANSVERSE (LT) POLARITON SPLITTING WITHIN 1 meV

	Energy (eV)	Oscillator Strength	Binding Energy (meV)	L-T Splitting (meV)
A	3.478	1.5×10^{-3}	24	1–2
B	3.484	1.5×10^{-3}		
C	3.503	5×10^{-4}		

samples. Alternatively, one can study homoepitaxial layers, which until very recently have not been available.

There are several reports in literature on reflectance studies for thick GaN epilayers (Dingle et al., 1971b; Monemar, 1974; Gil, Briot, and Aulombard, 1995) as well as on homoepitaxial layers (Korona et al., 1996). Some recent data on samples with rather good line widths seem to give a quite consistent picture of the positions of the A, B, and C excitons at 2 K. We quote here the values at 2 K: $E_A = 3.478$ eV, $E_B = 3.484$ eV and $E_C = 3.502$ eV, with an accuracy of about 2 meV (up or down). This limited accuracy is mainly due to the possible problems of interpreting the varying reflectance line shapes for different samples, but also due to the unresolved fine structure discussed earlier. The magnitude of residual strain in epilayers has, in general, not been well characterized in samples used for optical measurements, which gives another source of error in determination of the exciton energies for unstrained bulk GaN. For the A exciton, a correlation with the spectral position of the corresponding photoluminescence (PL) line can be made, see below. For GaN grown on sapphire, a series of rather thick samples were studies by Amano, Hiramatsu, and Akasaki (1988), and an extrapolation towards zero stress gave an approximate value of 3.475 eV for the A exciton (Amano, Hiramatsu, and Akasaki, 1988). In a more recent study of many epitaxial layers by Chichibu et al. (1996b), a value of about 3.477 eV at 10 K for the A exciton was extrapolated for zero strain (Shikanai et al., 1996). For homoepitaxial samples, the values given above for the A, B, and C excitons were reported from reflectance data (Fig. 9) (Korona et al., 1996). (These homoepitaxial samples are perhaps not quite strainfree, however, due to a different lattice constant in the highly conductive GaN substrate.) The value of the A exciton energy for small crystallites of bulk GaN, supposed to be strain free, is 3.479 eV (Monemar et al., 1996). It should be noted that the broadened shape of the C-line compared to the A and B lines might indicate

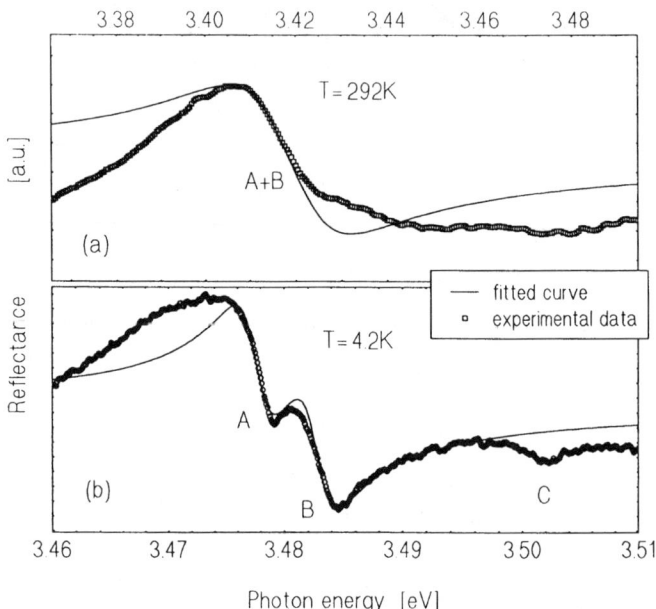

FIG. 9. Reflectance data for a homoepitaxial GaN sample at 4.2 K and 292 K, respectively, along with theoretical fits. (K. P. Korona, unpublished data, 1996.)

that the C line is actually resonant with the conduction band. It is close to, or above the lowest $\Gamma_{7c} - \Gamma_{9v}$ bandgap of GaN.

With the aid of the above exciton energies, the corresponding ones for the splittings of the three top valence bands have been estimated with the aid of the so-called quasi-cubic model (Hopfield, 1960). Values as $\Delta_{cr} \approx 20 \pm 2$ meV for the crystal field splitting and $\Delta_{so} \approx 10 \pm 2$ meV for the spin-orbit splitting are reported (Dingle et al., 1971b). The data for the A, B, and C exciton splitting are also consistent with another set of values $\Delta_{cr} \approx \Delta_{so} \approx 16 \pm 2$ meV (Edwards et al., 1996). The quasi-cubic model is quite accurate for CdS (Thomas and Hopfield, 1959), but is expected to be less reliable for GaN since the valence bands are closer in energy, and consequently, the mixing between the bands may not be negligible. So far, no direct experimental data exist for the splittings of the valence band top, comparable to the work done on the wurtzite II–VI compounds (Mang, Reimann, and Rübenacke, 1995). Higher quality samples are needed to accurately resolve the excited states of the A, B, and C excitons in GaN. When these excited states can be resolved and properly interpreted, one can

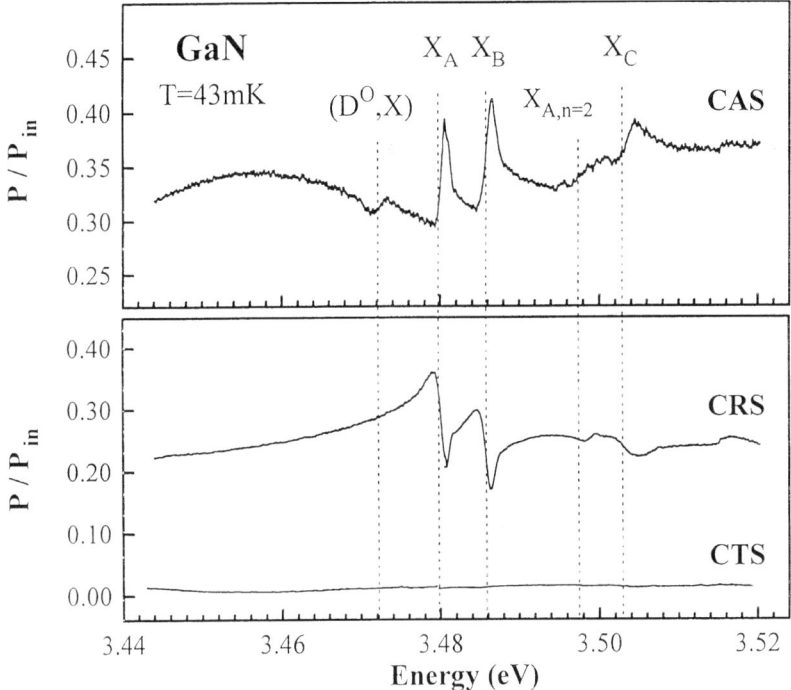

FIG. 10. Data from calorimetric absorption (CAS), reflectance (CRS) and transmission (CTS) spectra measured at very low temperatures (43 mK) on a thick (400 μm) GaN sample grown on sapphire. This sample is assumed to be nearly strain free. (Reprinted from Eckey et al., 1996a, with permission from IOP Publishing Ltd., Bristol, UK.)

obtain reliable estimates of the exciton binding energies, and consequently, the energy positions of the different valence band edges.

A different technique to measure the fundamental exciton resonances is calorimetric absorption or reflection. Such data for GaN at 43 mK were recently reported for a thick epilayer (Eckey et al., 1996a) (Fig. 10). The values for the A, B, and C excitons agree with those quoted above, apart from a small upshift of about 2 meV for the exciton energies in Eckey et al. (1996a). Such minor differences might be explained by different residual strain fields in different samples.

A very powerful technique for studying exciton structure is photoluminescence (PL). Figure 11 gives an example of a PL spectrum in the range of the fundamental excitons taken at 2 K with high spectral resolution. The

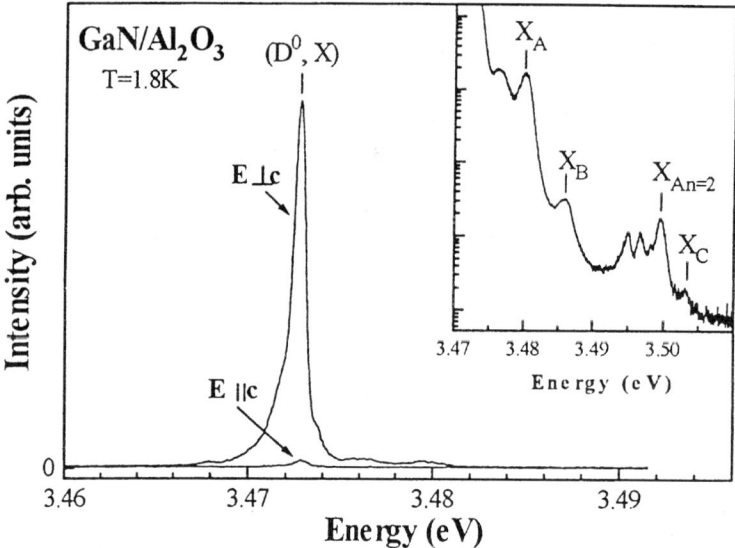

FIG. 11. Photoluminescence data for the same samples as in Fig. 10 at 1.8 K. The spectrum is dominated by the donor bound exciton peak (D^0, X), but the intrinsic exciton states are also resolved at higher energies (see inset). (Reprinted from Eckey *et al.*, 1996a, with permission from IOP Publishing Ltd., Bristol, UK.)

sample is a thick epitaxial layer on sapphire, and therefore, approximately representative for bulk unstrained GaN. The PL spectrum is dominated by the donor bound exciton (DBE) related to residual donors present at a concentration of about 10^{17} cm^{-3} (bound excitons will be further discussed later). At higher energies, the free A exciton FE_A is observed, and weaker structure at higher energy represent the FE_B state as well as excited states related to the A exciton (Fig. 11). Note that the excitation is above the bandgap, and due to efficient relaxation of the excitons to lower energies before recombination, the thermal population of the higher energy states is small, leading to lower PL intensities at higher energies.

A close examination of the excitonic spectra in homoepitaxial GaN layers shows that the FE_A transmission appears split in two components about 2 meV apart, with the lowest component at about 3.477 eV at 2 K (Fig. 12). A possible interpretation is that these two components actually correspond to recombination from the lower and upper polariton branches. More experimental data are needed to test this possibility.

At higher temperatures the higher intrinsic exciton PL peaks grow in relative intensity, as seen in Fig. 13. A reproducible weak structure is present

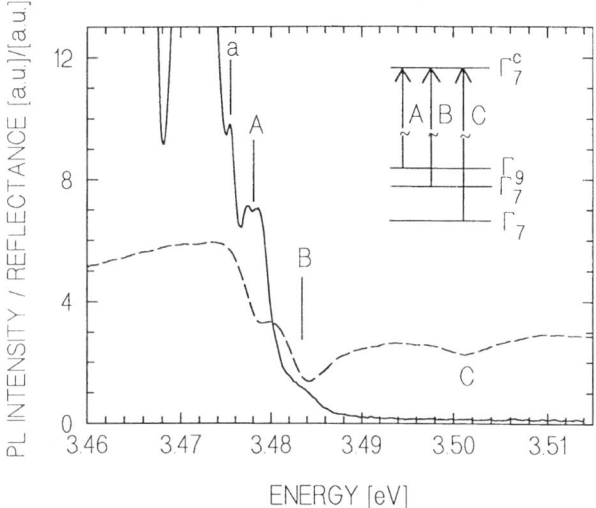

FIG. 12. Photoluminescence spectrum at 4.2 K of a homoepitaxial GaN layer, showing a doublet structure of the A exciton. Corresponding reflectance data for the same sample are also shown (---). The line *a* corresponds to a shallow bound exciton. (Reprinted from Pakula *et al.*, (1996), with permission from Elsevier Science Ltd., Kidlington, UK.)

and can be followed up to higher temperature (Fig. 13). This structure is largely due to the combination of polariton structure and exchange interaction, as described later. The spectral linewidth (of the order 1–2 meV) is however still too large to allow the spectroscopic studies needed to clarify the finer details of the electronic structure of the intrinsic excitons in GaN.

The binding energy of the excitons is an interesting quantity. It determines the energy of the bandgap of the material, but the strength of the binding is also an important factor for the thermal stability of the excitons. A theoretical estimate of the A exciton binding energy as 27.7 meV has been given (Mahler and Schröder, 1974). There is, in general, no feature at the band edge position in the optical spectra, therefore, the exciton binding energy (and the bandgap) has to be obtained in an indirect way, via an extrapolation procedure if the position of the excited states are known. Some such structure which may be attributed to the $n = 2$ FE$_A$ state is seen in Fig. 11. If this identification is correct, an extrapolation procedure gives an exciton binding energy of about 26 ± 2 meV for FE$_A$, placing the lowest Γ_{7c}–Γ_{9v} bandgap at about 3.504 eV at 2 K, in agreement with an early estimate based on data from somewhat lower quality samples (Monemar, 1974). This value for the exciton binding energy is very close to the

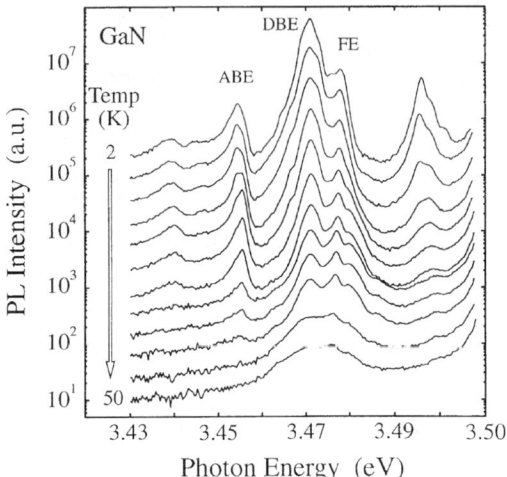

FIG. 13. Temperature dependence of photoluminescence spectra (2 K to 50 K) for a 500 μm thick GaN layer grown on sapphire. (Data obtained by J. P. Bergman.)

theoretical estimate from effective mass theory (Mahler and Schröder, 1974). Since there are several peaks around the $n = 2\, FE_A$ peak in Fig. 11, the exact identification of these peaks, however, needs to be done before extrapolation procedure mentioned above can be claimed to be accurate. Note again the close proximity estimated here between the position of the C exciton (Table II) and the bandgap. It should be mentioned, however, that in some recent work on strained epilayers, where the excited states were also resolved, much lower estimated values for the A and B binding energies (20 meV and 22 meV, respectively) were reported (Smith et al., 1995a; Reynolds et al., 1996). The binding energies of the B and C excitons are not yet accurately known. They are expected to be similar, but not necessarily identical, to the value of the A exciton. Also, due to the proximity of the three top valence bands, a variation of the free exciton binding energies with strain may also occur.

Free exciton transitions in wide bandgap materials have a characteristic coupling to LO phonons, dictated by their polariton–phonon scattering inside the crystal (Permogorov, 1982). An example of this LO-phonon coupling is demonstrated in Fig. 14 (Kovalev, 1996). As expected, the first two replicas are strongest from the theory of LO-phonon coupling for exciton polaritons (Permogorov, 1982). The 3-LO and 4-LO replicas are also clearly observable, however, in analogy with similar results for CdS (Permogorov, 1975). In fact, with other wide bandgap semiconductors, the

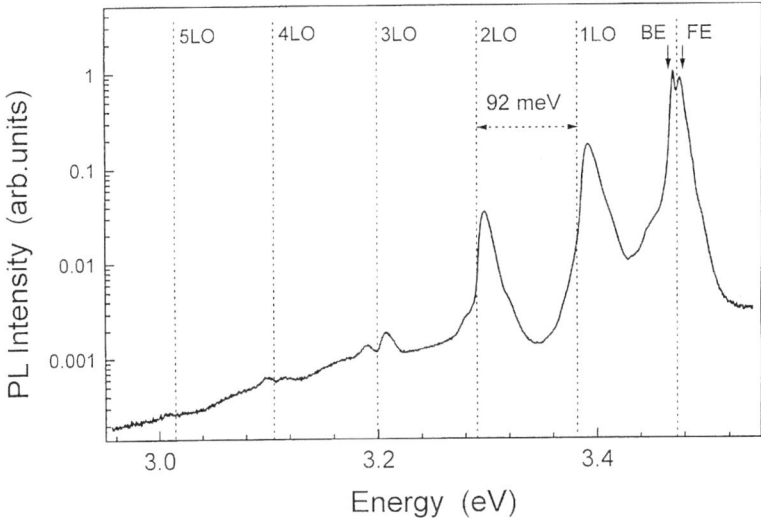

FIG. 14. Photoluminescence spectrum showing typical phonon replicas of the A exciton in GaN at 50 K. (Reprinted from Kovalev *et al.*, 1996, with permission from The American Physical Society.)

no-phonon line of the FE transition is quite weak at low temperatures, typically much weaker than the replicas corresponding to 1-LO and 2-LO phonon emission (Klochikhin, Permogorov, and Reznitsky, 1976). This is not so for GaN, the no-phonon FE^A emission line is stronger than its LO-phonon replicas at low T (see Fig. 15) (Buyanova *et al.*, 1996a, 1996b). This might be due to some details in the exciton polariton scattering for GaN, facilitating for the exciton polaritons to be scattered from the bottleneck at larger K-values to the radiative branch at the K-values, corresponding to the photon (Gross *et al.*, 1972; Wiesner and Heim, 1975; Benoit a la Guillaume, Bonnot, and Debever, 1970). We cannot rule out that impurity scattering of exciton polaritons (in particular, shallow donor scattering) plays a significant role in promoting the no-phonon line strength in PL at low temperatures in the spectra of presently available samples, where defect and impurity concentrations are much higher than in the previous studies of CdS and other wide bandgap II–VI materials. This impurity scattering of exciton polaritons has been found to be quite efficient in GaAs (Steiner *et al.*, 1986), and the impurity-polariton scattering might explain the peculiar doublet shape often seen in high quality crystals for the A exciton in PL emission (Fig. 12) (Stepniewski and Wysmolek, 1996). The doublet structure in this case might be explained as the PL from the low

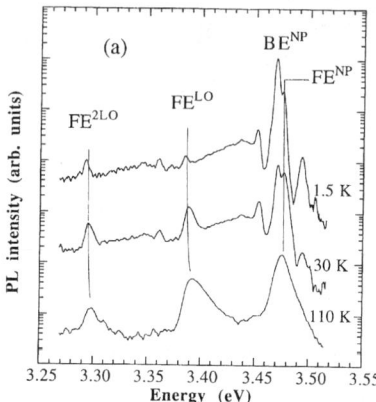

FIG. 15. Photoluminescence spectra of a 500 μm thick GaN sample on sapphire, at 1.5 K, 30 K, and 90 K, respectively. The no-phonon line of the A exciton (FE^{NP}) is stronger than its replicas at all temperatures. (Data obtained by I. A. Buyanova.)

$K(K \approx 0)$ region of the lower and upper polariton branches related to the A exciton.

The importance of excitons in the intrinsic absorption process is manifested in photoluminescence excitation (PLE) spectra. An example of such a PLE spectrum for the FE emission is shown in Fig. 16, where regular LO-phonon replicas are observed well above the bandgap energies, at

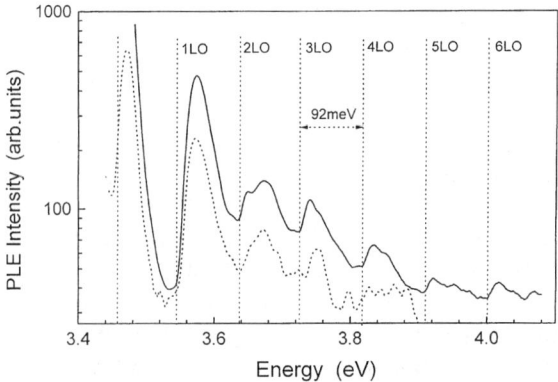

FIG. 16. Photoluminescence excitation spectra at $T = 70$ K for the free A exciton in GaN for two different detection energies, the no-phonon line (—) and the 1-LO phonon replica (---). (Reprinted from Kovalev et al., 1996, with permission from The American Physical Society.)

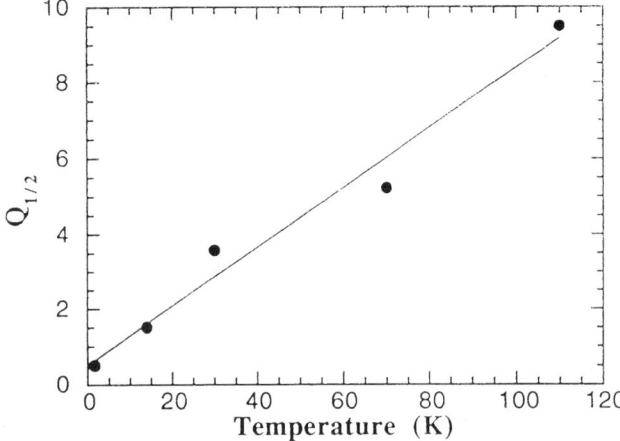

FIG. 17. Intensity ratio Q1/2 of the photoluminescence of the 1-LO replica and the 2-LO replica of the A exciton in GaN, as a function of temperature, obtained from the same sample as in Fig. 15. (Data obtained by I. A. Buyanova.)

intervals that are consistent with the LO-phonon energy (Kovalev *et al.*, 1996). This is a manifestation of the creation of indirect hot excitons in the absorption process, at energies up to 0.5 eV above the bandgap, which shows that the exciton binding is quite stable. These hot excitons then relax very fast with LO-phonon emission, towards the bottom of the exciton-polariton band. The final scattering of exciton polaritons into the emitting state ($K \approx 0$) occurs with the assistance of acoustic phonons, and probably also via impurity scattering.

The characteristic temperature dependence of the intensities of the LO phonon replicas has been investigated (Fig. 15). Theoretically, the general behavior of the exciton–polariton dispersion suggests a temperature dependence of the intensities of the phonon replicas Q_{1LO} and Q_{2LO} as $Q_{1LO}/Q_{2LO} \propto T$ (Permogorov, 1975, 1982). This behavior seems to be confirmed for the A exciton in GaN in the range $T < 100\,\text{K}$ (Fig. 17) (Buyanova *et al.*, 1996b).

c. *Dependence on Substrate, Strain Effects on Excitons*

The values for the near bandgap exciton energies were found to vary substantially in literature, between various authors. The main reasons for this is, that a heteroepitaxial layer is usually subject to a large built in strain field. A strain field is known to generally have a rather strong effect on bandgap and exciton energies for a semiconductor. For typical substrates for heteroepitaxy of III–V nitrides, the lattice mismatch is quite large. It may commonly be assumed, however, that the epilayer is essentially relaxed at the growth temperature via the development of a dense dislocation network. The layer is then more or less strain-free at the growth temperature. When the structure is cooled down after growth, a difference in the thermal expansion coefficients between substrate and epilayer may cause strain to develop, to a degree dependent on the cool down procedure (i.e., whether time is allowed for further relaxation during cool down).

The relevant linearized thermal expansion coefficients $\perp$ c axis are: $5.6\,10^{-6}\,K^{-1}$ for GaN, $4.2\,10^{-6}\,K^{-1}$ for 6H SiC, and $7.5\,10^{-6}\,K^{-1}$ for sapphire (Strite and Morkoc, 1992). At cool down, after growth on sapphire, the GaN epilayer will therefore experience a compressive biaxial strain field in the plane of the layer. In the perpendicular direction (along the c axis, for growth on (0001) oriented substrates), there will be a corresponding expansion to minimize the elastic energy. This has been confirmed by measurements of lattice parameters (Tsaregorodtsev *et al.*, 1995; Amano, 1996). This means that there will be an increase in the c/a ratio of the GaN lattice, which in turn, is expected to lead to an increase in the A, B, and C exciton splittings (Gil, Briot, and Aulombard, 1995; Edwards *et al.*, 1996; Kim, Lambrecht, and Segall, 1994; Lambrecht *et al.*, 1996). The behavior of GaN layers on sapphire substrates is therefore understood in the sense that an increased bandgap and increased ABC exciton splittings compared to the case of unstrained bulk GaN are consistently observed. Upshifts of the A and B excitons by as much as 20 meV have been observed at 2 K, the C exciton has been found to shift as much as 50 meV (Gil, Briot, and Aulombard, 1995). An example of the systematic shifts observed in reflectance data for the A, B, and C excitons is shown in Fig. 18.

A different situation is expected to occur for GaN growth on SiC substrates, since here a tensional biaxial strain in the GaN layer is expected to develop upon cool down, which in turn, would be expected to lead to a decrease in the c/a ratio, a decrease in the overall exciton energies (and the bandgap), and also a decrease on the A, B, and C splittings. Experimentally, this prediction seems to be confirmed. Most work reported on strain measurements (obtained via the lattice parameters) seems to confirm that the c-value for GaN grown on SiC without buffer layer is typically lowered by an amount $\Delta c/c \approx 0.5\,10^{-3}$ (Buyanova *et al.*, 1996a; Tsaregorodtsev *et al.*, 1995; Amano, 1996). In Fig. 19(*a*) is shown a reflectance spectrum at 2 K for a typical sample grown on 6H SiC without buffer layer, showing two excitonic resonances about 17 meV apart. A corresponding PL spectrum is also shown in Fig. 19(*b*). PL spectra at elevated temperature clearly show

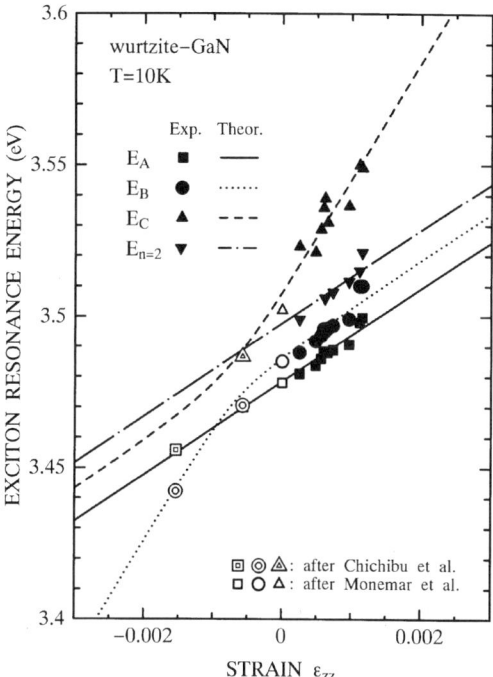

FIG. 18. Free exciton energies in GaN grown on sapphire substrates, determined experimentally from photoreflectance data, versus strain along the c axis. The strain was obtained from the measured lattice parameter values for each sample. Theoretical modeling of the strain dependence of exciton energies are also shown. (Reprinted from Shikanai et al., 1997), with permission from The American Institute of Physics.)

the same two intrinsic exciton states in PL emission (Fig. 20). A careful search for the third exciton gave a negative result, which complicates the interpretation. An obvious possibility is that the A and B excitons are close enough in energy (<5 meV) so that they are not separately resolved (Gil, Briot, and Aulombard, 1995; Buyanova et al., 1996a). In that case, the lowest energy line is the combined (A, B) exciton, and the higher energy line is the C exciton. This would be consistent with the above prediction, and is supported by measurements of lattice parameters. The energy of the lowest exciton at 2 K is also decreased in our samples, compared to bulk unstrained GaN, by up to 10 meV, for growth on SiC. An alternative explanation was recently suggested (Edwards et al., 1996; Jacobson et al., 1995), however, assigning the two reflectance peaks to the A and B excitons, respectively.

Recently, theoretical predictions have been made of the influence of the strain field in an epilayer on the exciton energies, covering the whole

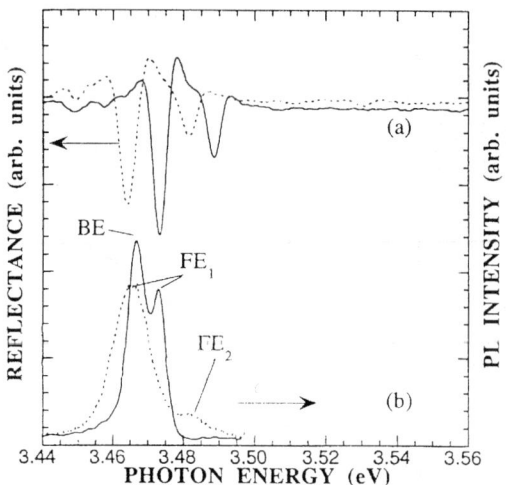

FIG. 19. (a) Derivative reflectance spectra for a GaN layer grown on a SiC substrate without buffer layer. (b) Photoluminescence spectra for the same sample. For both (a) and (b) full lines (—) are for 15 K data, while (.....) are the data at 150 K. (Data obtained by I. A. Buyanova.)

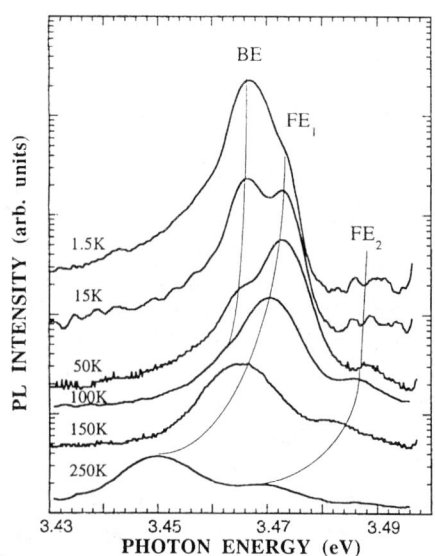

FIG. 20. Photoluminescence spectra at different temperatures, plotted in log scale, for the same sample as in Fig. 19. The same two intrinsic exciton states as seen in reflectance in Fig. 19(a) are seen in PL data at elevated temperatures. (Data obtained by I. A. Buyanova.)

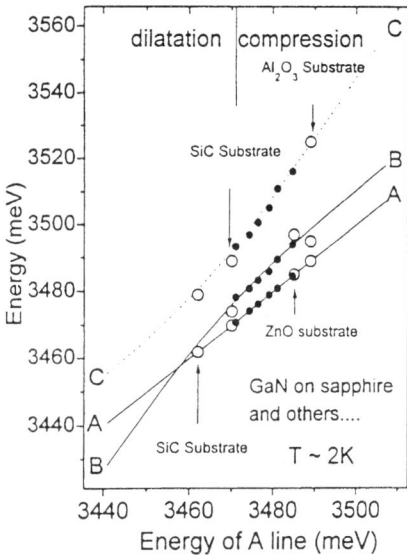

FIG. 21. Theoretical strain dependence of intrinsic exciton energies in GaN, here plotted versus the energy of the A exciton. Experimental data from literature are included to support the curves. The strain dependence is qualitatively similar as the one in Fig. 18. (B. Gil, unpublished data, 1996.)

spectrum of strain (both negative and positive) expected for growth of GaN on sapphire and SiC, respectively (Tchounkeu et al., 1996). Interestingly, it is found that the A and B excitons tend to cross at a certain negative value of the strain, relevant for layers grown on SiC substrates (Fig. 21). For the B and C excitons, on the other hand, an anticrossing behavior is expected. These data support the above interpretation of the spectra obtained on SiC substrates as being related to (A, B) and C excitons (Figs. 19 and 20).

Another interesting theoretical prediction concerns the strain dependence of the relative oscillator strengths of the A, B, and C excitons (Gil, Hamdani, and Morkoc, 1996). These properties are experimentally found to vary substantially with strain for growth of GaN on sapphire. Gil, Hamdani, and Morkoc (1996) predict a strong interaction between the B and C excitons (Fig. 22), affecting their relative oscillator strengths so that the C excitons becomes much weaker under strong axial compression (sapphire substrate), while under biaxial tension C is expected to be much stronger than B. This is also in line with the above interpretation of our data for GaN on SiC, where only two excitons labeled (A, B) and C were observed, with comparable amplitude.

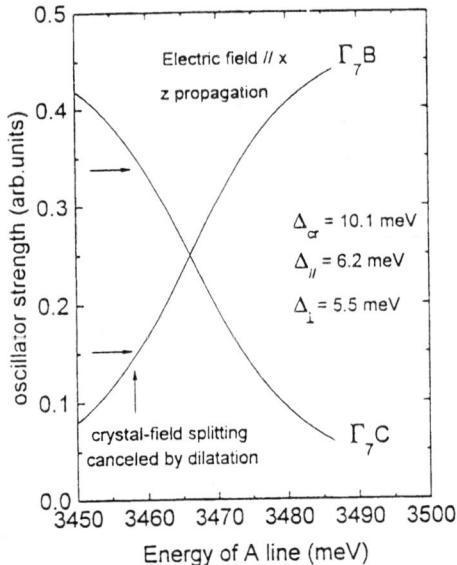

FIG. 22. Theoretical predictions of the variation of the relative oscillator strengths of the A and B excitons in GaN versus strain, here plotted against the A exciton energy. (Reprinted from Gil, Hamdani, and Morkoc, 1996, with permission from The American Physical Society.)

An interesting recent study has been made on GaN grown on SiC substrates with an AlN buffer layer. By changing the growth conditions, the authors were able to change the residual strain in the GaN epilayer over a large range, from tension to compression (Perry, 1996). The main PL peak (the donor bound exciton peak, see later) was detected, and the lattice parameters were measured at room temperature for all samples (Fig. 23). The expected donor BE position for unstrained material is indicated in the figure. Assuming a donor BE binding energy of 7 meV, a value of 3.476 eV for the A FE in unstrained GaN is evaluated from these data (Perry, 1996).

This set of data can tentatively be explained by an incomplete relaxation of the heteroepitaxial strain during growth on a SiC substrate, employing a buffer layer. Such a built-in hetero epitaxial strain would obviously be composed of two components of different origin and sign: The difference in thermal expansion coefficient will tend to develop a tensional biaxial strain in the GaN layer, while a built-in heteroepitaxial strain due to the difference in lattice constants of GaN and SiC (SiC has a smaller lattice constant) would tend to compress the GaN layer in the plane of the layer. The combination of these two effects apparently allows the tuning of the exciton energies as demonstrated by the data in Fig. 23.

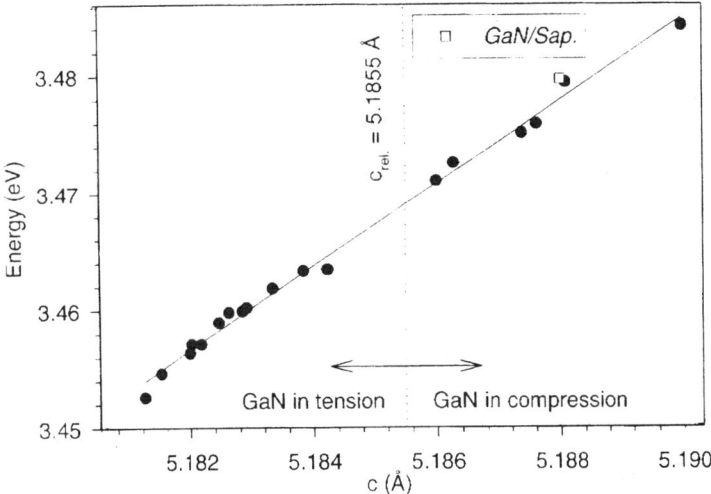

FIG. 23. Low temperature photoluminescence (donor bound exciton) peak position versus lattice parameter (measured at room temperature) for a set of samples grown on SiC with an AlN buffer layer. Since the donor bound exciton energy follows closely the A exciton energy with a shift of 6–7 meV, a value of the A exciton energy of about 3.476 eV for unstrained GaN at 2 K is suggested from these data. (Courtesy of Perry, 1996, unpublished data.)

The temperature dependence of the intrinsic exciton energies also show a considerable variation between different reports in literature (Teisseyre *et al.*, 1994; Monemar, 1996; Ilegems, Dingle, and Logan, 1972; Pankove, 1973; Matsumoto and Aoki, 1974; Shan *et al.*, 1995; Zubrilov *et al.*, 1995; Monemar *et al.*, 1995a). Quite large differences occur, which are certainly outside any possible experimental error. Clearly, epitaxial strain is an important parameter also for the temperature dependence of exciton energies in GaN. In Fig. 24, we show a comparison between the exciton energies measured between 2K and room temperature for three samples: one thick reference bulk sample, one epilayer on sapphire, and one epilayer on SiC without buffer layer (Monemar, 1995a). Assuming that the built-in elastic strain is partly released as the temperature is increased, the trend of a decreased energy difference at higher temperatures is at least partly understood.

To explain the detailed behavior of the exciton energies as a function of temperature, it is necessary to know the full strain tensor, and in addition, the corresponding deformation potentials for the induced shifts of exciton energies. The detailed temperature dependence of the thermal expansion coefficients are also needed. These may not be known accurately, since

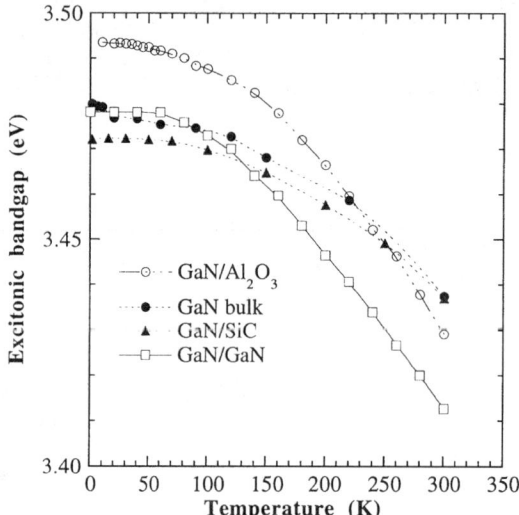

FIG. 24. Temperature dependence for the A exciton energy in samples grown on different substrates. The data are from photoluminescence (Monemar et al., 1996a), except for the homoepitaxial layer where reflectance was employed (Korona et al., 1996).

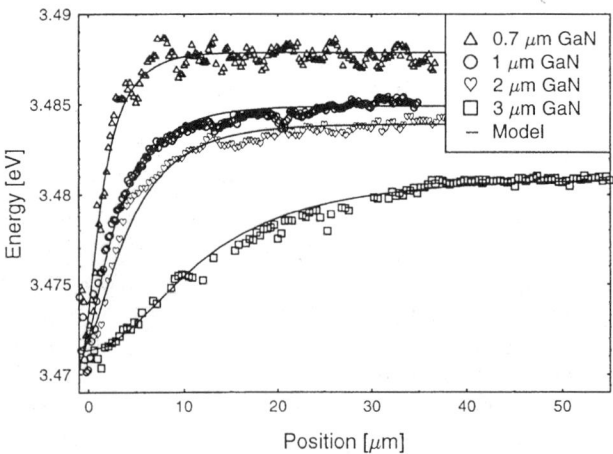

FIG. 25. Position dependence of the 2 K bound exciton photoluminescence peak energy in GaN epilayers grown on sapphire. The position is measured from the edge of the wafer, where the GaN layer is strain free, employing a set up with a very small laser excitation spot. (Reprinted from Gfrörer et al., 1996, with permission from Elsevier Science SA, Lausanne, Switzerland.)

literature values for GaN are conflicting (Strite and Morkoc, 1992; Edjer, 1974). The strain situation for a layer at a particular temperature depends on the details of the grown structure (number of layers and their thicknesses, as well as on type of substrate), the cool down procedure (i.e., whether there is sufficient time for strain relaxation during cool down), presence and thickness of buffer layer, and also to some extent the doping and point defect densities, since these influence the lattice parameters (Lagerstedt and Monemar, 1979). Obviously, a good pure bulk GaN sample is needed for reference, before data like those in Fig. 24 can be fully understood.

Recently, a more detailed study on the strain relaxation of GaN grown on sapphire was performed (Gfrörer et al., 1996). The authors studied the lattice parameters and PL exciton energies versus annealing temperature after growth. Exciton energies for unstrained material could be measured by focusing on the very edge of the sample (Fig. 25). It was concluded that under the conditions of these experiments, the strain was relaxed due to development of dislocation defects down to a temperature of about 650 K for a 3 μm layer. Below that temperature dislocations did not develop at a sufficient rate to relax the strain, and then at room temperature, the layer is strained by an amount corresponding to the elastic misfit between 650 K and room temperature. The freeze-in temperature T_f for dislocation defects was found to be higher for thinner epilayers, $T_f = 700$ K for $d = 2\,\mu$m, $T_f = 810$ K for $d = 1\,\mu$m, and $T_f = 1030$ K for $d = 0.7\,\mu$m (Fig. 26). The

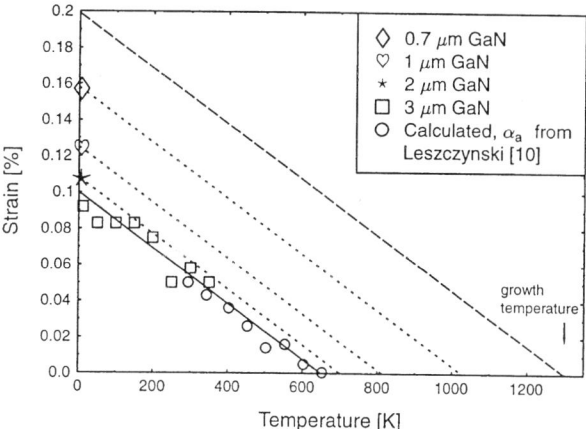

FIG. 26. Temperature dependent strain relaxation versus layer thickness for GaN grown on sapphire. (Reprinted from Gfrörer et al., 1996, with permission from Elsevier Science SA, Lausanne, Switzerland.)

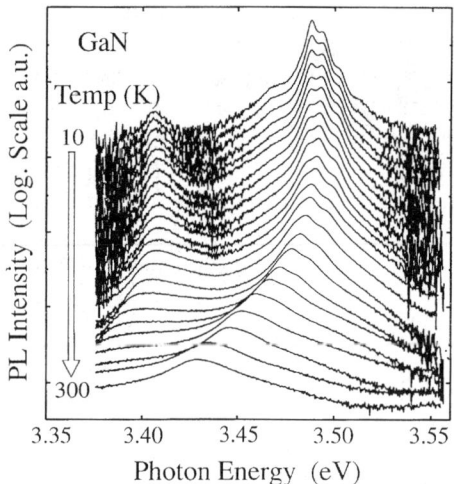

FIG. 27. Temperature dependent photoluminescence spectra for GaN grown on sapphire. The approximate values of the temperatures are as follows: 10 K, 20 K, 25 K, and with interval 5 K up to 60 K, interval 10 K in the range 60 K to 100 K, interval 20 K between 100 K and 300 K. Clearly, at 300 K, the A, B exciton spectra still dominate over the transition at the bandgap. (Data obtained by J. P. Bergman.)

authors did not perform any similar study for growth on SiC substrates. Similar results were concluded from a study made by another group (Lesczynski *et al.* (1996)).

The dominating PL process at room temperature in GaN is the free exciton recombination. This has been established by PL spectral data over a wide range of temperatures for nominally undoped samples (Monemar *et al.*, 1995b, 1996a, 1996b) (Fig. 27). It is clear from the spectra in Fig. 27, that the PL intensity at the position of the bandgap at 300 K is considerably lower than the A-exciton intensity. We may conclude that in light emitting GaN devices with not too high doping or too high carrier injection (in the 10^{18} cm^{-3} range in the active of a device), the near bandgap emission process is predicted to be of excitonic origin, at least in regions of moderate electric fields, and if the active layer structure is not too disordered.

3. EXCITON RECOMBINATION DYNAMICS

The free exciton recombination in direct wide bandgap materials is known to be fast, typically in the few ns time scale at 300 K. What is of major interest for optical applications for light emitters is the radiative recombination rates. They should ideally be fast to be able to compete efficiently with the nonradiative defect-related processes. Here we shall discuss some recent data on the recombination rates of excitons in GaN, relevant to material with not too high doping or carrier injection densities (i.e., below the high $10^{18}\,\mathrm{cm}^{-3}$ range).

Experimentally, the PL transient measurements are done with short laser pulse excitation. Laser pulses as short as 100 fs are nowadays commonly available. For the data discussed here, the pulse lengths of the order 1–5 ps (also common) are sufficient. Upconversion into UV with a nonlinear element is needed, to reach photon energies above the bandgap of GaN. For detection of transients in the range $10\,\mathrm{ps} < \tau < 2\,\mathrm{ns}$ the streak camera detection technique is useful. For longer times (up to a few μs) the conventional time correlated photon counting technique is mainly used.

PL transient data represent an excellent tool to reveal the quality of a material, since it is sensitive to both the radiative and the nonradiative processes. The measured decay time τ can be related to the radiative and nonradiative recombination processes via

$$\tau^{-1} = \tau^{-1}\,\mathrm{rad} + \tau^{-1}\,\mathrm{nonrad}. \tag{5}$$

Most GaN material produced so far has a very large defect density, causing a strong nonradiative recombination path at all temperatures, if the excitation is not sufficiently strong to saturate these processes. Usually, a narrow PL linewith is taken as a criterium of good material. This is true in the sense that the linewidth for a heteroepitaxial layer senses the inhomogeneous strain distribution in the layer, for example, a measure of structural quality. The lifetime measurement gives additional information about the density of the serious nonradiative defects in the material. Therefore, ideally, a combination of low temperature PL spectra and PL transient data should be employed for a proper optical characterization.

Typical transients are shown in Fig. 28, which were obtained at different excitation intensities for a GaN sample grown on sapphire (Harris *et al.*, 1995). Decay times at 2 K were found to vary between 60 ps and 115 ps, dependent on the excitation intensity employed, indicating a strong influence of nonradiative defects under these (rather low) excitation intensities. The exponential decay at low intensities is clearly representing a nonradiative process. At higher intensities, the radiative process of the FE_A recombination becomes important, but there is in addition, a slower background process with a decay time of about 300 ps present at longer times (Fig. 28). This slow process might be due to weak localization of the FEs in potential fluctuations induced by the inhomogeneous strain field.

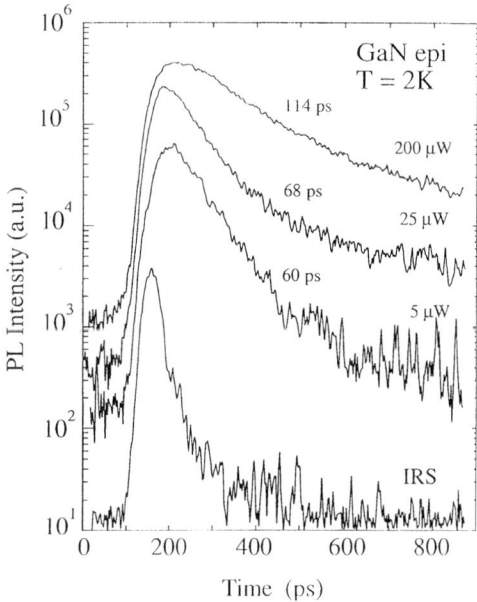

FIG. 28. Photoluminescence transients at 2 K for a GaN epilayer grown on sapphire, measured with different (rather low) average excitation power. IRS denotes the instrumental response in the experiment. The variation with excitation power is a clear indication of the importance of nonradiative processes. (Data obtained by J. P. Bergman.)

Above 10 K the decay time goes down with temperature in this sample, which is another sign of the importance of nonradiative processes. The decay time of the radiative process of the FE is expected to increase with temperature (Hooft et al., 1987). At present, it is not clear whether residual point defects or dislocations are responsible for the observed nonradiative recombination.

Figure 29 shows another set of PL transient data for a bulk sample (a 500 Å thick epilayer), with the PL spectrum given in Fig. 13. Both the donor bound exciton (DBE) and the FE_A decays are studied, since they influence each other. A decay time of about 200 ps at 2 K is obtained for FE_A, the DBE decay time is slightly longer (Monemar et al., 1995c; Bergman et al., 1996). The longer decay time in this sample (compared to the epilayer in Fig. 28) is presumably related to a lower defect density. The value 200 ps for the FE_A decay time is independent on excitation intensity, which can most naturally be taken as a confirmation that this value is the radiative lifetime of FE_A at 2K.

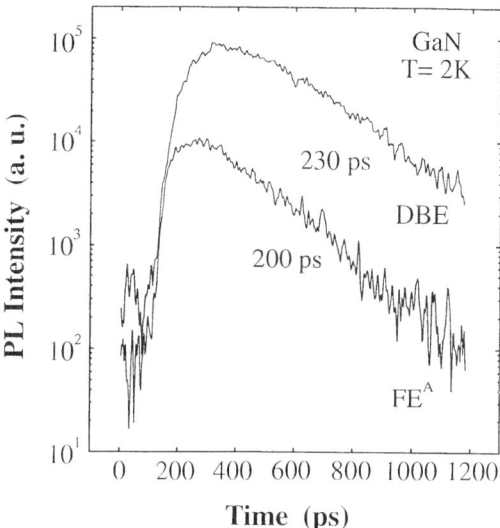

FIG. 29. Photoluminescence transients at 2 K for a 500 μm thick bulk like layer grown on sapphire. (Data obtained by J. P. Bergman.)

The problem with theoretical estimates of radiative lifetimes of free excitons seems to be that there is no good theory available for it. GaAs is an excellent example. The established oscillator strength of the FE in GaAs has been determined as $f = 7 \times 10^{-5}$ per unit cell (Hooft et al., 1987). Using the classical relation between oscillator strength and radiative lifetime (Dexter, 1958):

$$\tau_{\text{rad}} = 2\pi e_0 m_0 c^3 / n e^2 \omega^2 f \tag{6}$$

one can readily calculate an expected value of about 40 μs for the radiative lifetime of the free exciton in GaAs (Hooft et al., 1987). The experimental value of τ_{rad} at 2K in bulk GaAs is 3.3 ns, a deviation of more than four orders of magnitude from the classical theory. It has been suggested that the oscillator strength of the FE should be associated with the excitonic volume instead of one molecule for the purpose of estimating the radiative lifetime. In that case, the oscillator strength has to be multiplied with a factor $(a_x/a)^3$. Here, a_x is the exciton Bohr radius and a is the lattice constant. This rectifies the theoretical value for τ_{rad} in GaAs so it agrees with experiment within a factor 2. One might suspect a similar problem occurs in GaN. The experimentally estimated oscillator strength for the A exciton is only

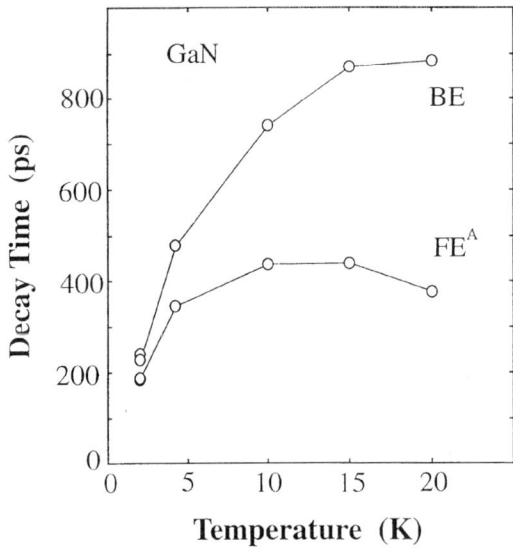

FIG. 30. Photoluminescence decay times as a function of temperature for the same sample as in Fig. 29. Nonradiative recombination dominates for the A exciton already above 15 K under the rather excitation conditions employed (about 10 μW average power). (Data obtained by J. P. Bergman.)

approximately known for GaN (Table II), but is substantially larger than for GaAs, due to a much higher absorption strength of the free exciton. A similar correction factor $(a_x/a)^3$ for GaN can be calculated as about $3.5\ 10^3$. Therefore, the observed value of about 200 ps τ_{rad} for GaN in the low temperature limit does not appear unreasonable. Some recently estimated values for τ_{rad} at 2K in GaN are very different, however, ranging from about 20 ps (Eckey et al., 1996b) to several tens of ns (Shan et al., 1996). More experiments on pure GaN samples are necessary to get an accurate value of the radiative lifetime.

The temperature dependence of the excitonic decay times has been investigated for a bulk GaN sample (Fig. 30). From theory, it is expected that τ_{rad} varies approximately with temperature as $T^{3/2}$ (Hooft et al., 1987). The experimental values go up rapidly with temperature, as expected for a radiative free exciton transition, but saturate already below 20 K, and then decrease at higher temperatures (Fig. 30) (Monemar et al., 1995b; Monemar et al., 1996b). This is a sign of an increased influence of the nonradiative processes at elevated temperatures, which could not be saturated under our present (quite weak) excitation conditions (about $100\,\mu W\,cm^{-2}$ time averaged). Unfortunately, no experimental value for the radiative lifetime at room temperature could therefore be obtained from these measurements, since the nonradiative processes made the decay time faster than our temporal resolution above 150 K. A theoretical extrapolation of the low temperature value to room temperature, employing the experimental values for the homogeneous linewidth of the free A exciton at the different temperatures, suggests a free exciton radiative lifetime of the order 2–3 ns in GaN (Monemar *et al.*, 1996b). This is nearly three orders of magnitude faster than for intrinsic (band-to-band) recombination in GaAs at room temperature (Smith *et al.*, 1990). Experimental results on carrier lifetimes in InGaN lasers in the range 2–3 ns at room temperature seem to support such a short value of the radiative lifetime (Nakamura *et al.*, 1996). The fast radiative lifetime for GaN explains why it is relatively easy to observe PL emission at room temperature from thin heteroepitaxial GaN layers, in spite of the fact that the defect density is extremely high (often of the order $10^{10}\,\text{cm}^{-2}$).

4. Near Bandgap Optical Properties at High Carrier (Exciton) Densities

The above description of optical data applies to the low intensity regime, for example, when the excited minority carrier densities are low compared to normal doping levels. This is the case in all conventional spectroscopy, unless very high intensity pulsed lasers, such as N_2 lasers or excimer lasers, are employed. Some such high intensity optical data exist in literature, in particular, luminescence data.

In Fig. 31, we see the early data on HVPE GaN by Hvam and Ejder, 1976), where the development of the PL spectrum with increasing excitation intensity at 2K is interpreted in terms of exciton-exciton interaction at the high densities. The PL spectrum upon increasing excitation intensity transforms from a donor bound exciton line to a broadening into the free exciton region at higher energy, and finally, the appearance of additional structure at lower energies at an excitation of about $2.5\,\text{MW}\,\text{cm}^{-2}$ (Fig. 31). This low energy structure at 3.453 eV and 3.448 eV respectively, was interpreted as the process, where two interacting excitons in their ground states, upon recombination, produced one exciton in its excited state in the crystal, while a photon with a correspondingly reduced photon energy was emitted. Consistent with this interpretation of the low energy lines their intensity varied quadratically with the excitation intensity. An exciton binding energy of about 28 meV would be consistent with this interpretation, very close to some recent values quoted earlier.

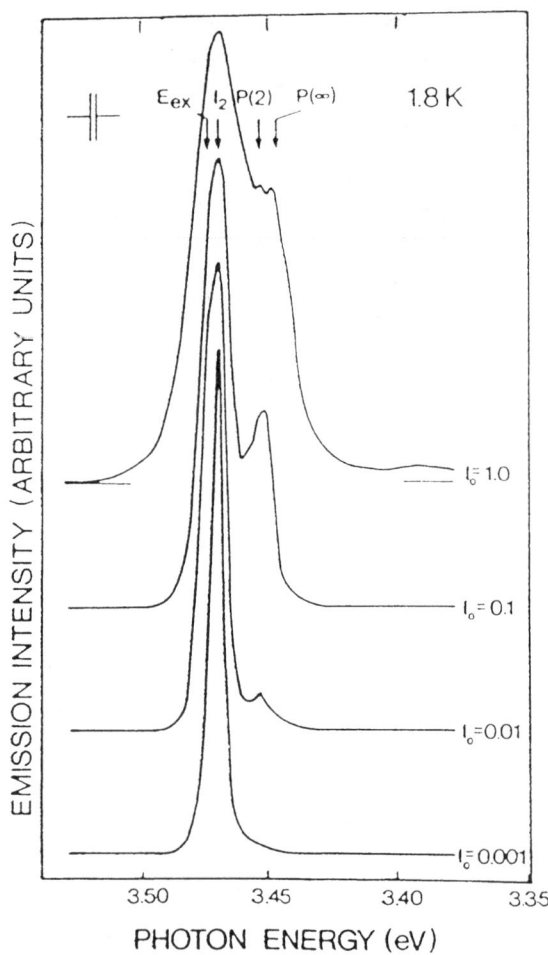

FIG. 31. Photoluminescence spectra at 1.8 K for a thick GaN layer grown on sapphire for various quite high excitation intensities with an N_2 laser. I_0 corresponds to about 2.5 MW/cm². (Reprinted from Hvam and Ejder, 1976, with permission from Elsevier Science, Amsterdam, The Netherlands.)

Recently, more studies have been made at varying excitation density, where the direct spectral observation of the biexciton in GaN epitaxial layers has been claimed (Okada et al., 1996). In Fig. 32 we see the intrinsic excitonic spectra as a function of excitation intensity, the line XX 5.3 meV below the free A exciton line E_x is interpreted as the biexciton line. It has a

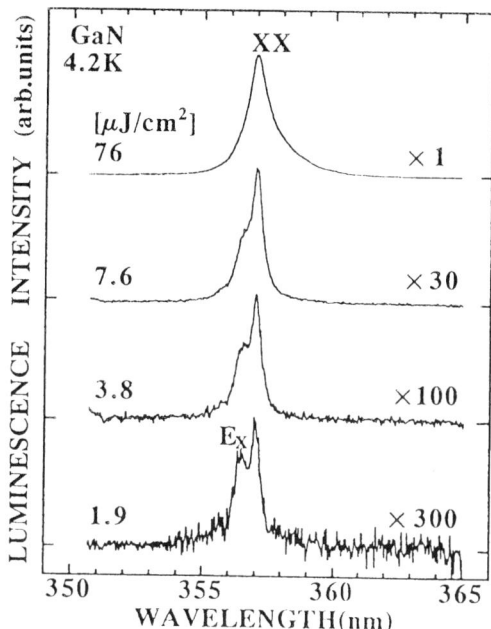

FIG. 32. Excitation intensity dependence of intrinsic exciton photoluminescence spectra for GaN, obtained at 4.2 K employing short fs excitation pulses with an average intensity as indicated. The peak XX is interpreted as the biexciton (After Okada et al., 1996). (Reprinted from Okada et al., 1996, with permission from Japanese Journal of Applied Physics, Tokyo, Japan.)

superlinear dependence on excitation intensity I, $I_{XX} \approx I^{1.6}$, which is similar to the case of the biexciton in, for example, GaAs/AlGaAs QWs (Phillips et al., 1992). In another recent work, a value of 5.7 eV for the biexciton binding energy E_{bXX} (with respect to the free A exciton energy) was deduced (Kawakami et al., 1996). This binding energy is what could be expected from a comparison with other wide bandgap semiconductors like CdS (Shionoya et al., 1973) and ZnS (Yamada et al., 1996), where a ratio of about 0.2–0.25 for E_{bXX}/E_{bX} is observed.

An earlier study was reported by Dai et al. (1982). They found that in the high density regime, the exciton interaction process was actually the dominating one at 10 K, but at even higher excitation intensities (3.5 MW cm^{-2}) another broad line peaking around 3.438 eV was most prominent. The authors interpreted this line as the evidence for an electron hole plasma at these high densities, that is, the excitons were essentially screened. The corresponding carrier (exciton) densities were not reported.

In another similar study (Cingolani, Ferrara, and Lugara, 1986), the e–h plasma region was also treated. The corresponding emission (called the X-band in Cingolani, Ferrara, and Lugara (1986)) shifted from 3.438 eV to 3.42 eV with increasing excitation intensity up to the maximum intensity of 10 MW cm^{-2}. The results were compared with calculations for the stability of the e–h plasma phase versus the interacting exciton phase. It was concluded that only at e–h densities of about $5\,10^{18}$ cm^{-3} and higher would the e–h plasma process be the dominating one. This result seems to be confirmed by more recent theoretical treatments of the recombination and gain in the interacting e–h system for GaN (Chow, Knorr, and Koch, 1995).

Recently, a large number of papers have appeared, discussing experimental data on stimulated emission and gain in both GaN bulk material and related heterostructures. The results are so far not interpreted in terms of a specific gain mechanism, but with the very large threshold observed for stimulated emission, it can be concluded that all experiments were done in the e–h plasma range. This is expected, as long as the defect density is high and consequently the threshold for stimulated emission remains high. It is not clear at the moment, whether this threshold under favorable conditions of material quality and with a standard laser structure could be reduced to the excitonic density region, for example, whether excitonic lasers based on GaN are possible. Interesting work on vertical cavity laser configurations is in progress, however, where theoretically, much reduced threshold current densities are predicted for optimized structures, in which case the excitonic region would be well within reach. Recently, Amano *et al.* (1988) have estimated the threshold carrier density for stimulated emission at room temperature as about $5\,10^{18}$ cm^{-3}, assuming the e–h plasma as the dominating medium (Amano and Akasaki, 1995).

A detailed discussion of the large amount of recent work on stimulated emission and gain in various III–V nitride structures is outside the scope of this chapter.

III. Below Bandgap Optical Properties, Refractive Index

Below the bandgap the absorption becomes much smaller, and the expression for the reflectance (for normal incidence) may be approximated by

$$R = \frac{(n-1)^2 + k^2}{(n+1)^2 + k^2} \approx \frac{(n-1)^2}{(n+1)^2} \tag{7}$$

an expression valid from the near bandgap region down to the strong ir absorption resonances associated with the TO and LO phonons, in reasonably pure samples. Reflectance data may therefore be used to conveniently obtain the value for the refractive index, and its energy dispersion. A simple alternative method is to study interference fringes in transmission of thin platelets, and derive the refractive index from the simple relation:

$$2nd = p\lambda_p \tag{8}$$

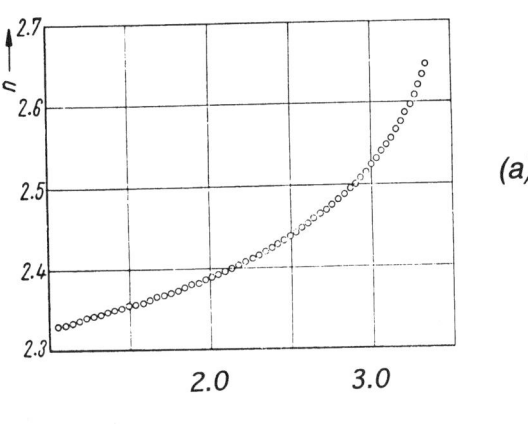

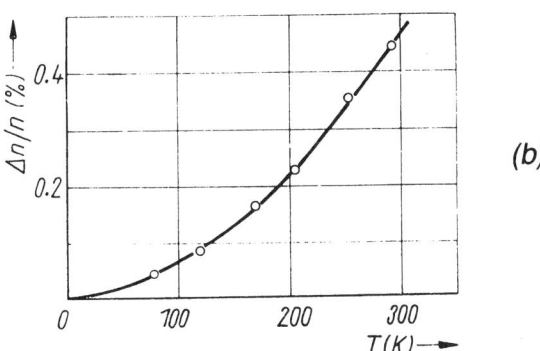

FIG. 33. (a) Refractive index ($E \perp c$) for GaN at 300 K. (b) Temperature dependence $\Delta n/n$ in %. (Reprinted from Ejder, 1971, with permission from Akademie Verlag GmbH, Berlin, Germany.)

where n is the refractive index, d is the sample thickness, p the interference order, and λ_p the wavelength for the p:th transmission maximum (Ejder, 1971). The refractive index (as well as the dielectric function itself) is anisotropic in wurtzite crystals, and should therefore be characterized by two parameters, $n_\perp$ and $n_\parallel$, for the light electric vector $\mathbf{E} \perp$ and $\parallel$ to the c axis, respectively.

In Fig. 33(a) is shown experimental data for $n_\perp$ for GaN in the energy region $1\,\text{eV} < hv < 3.4\,\text{eV}$, measured with the above interference method at room temperature. The corresponding temperature dependence between 2 K and 300 K was also measured, and is shown in Fig. 33(b). Barker and Ilegems (1973) have described this set of data with a simple expression:

$$\varepsilon_\perp = 3.6 + \frac{1.75}{1 - (h\omega/4.85)^2} \tag{9}$$

with $h\omega$ given in eV. The approximately constant value of $\varepsilon_\perp = (n_\perp)^2$ in the ir region below 1 eV is 5.35. A complete set of data for the configuration $\mathbf{E} \parallel \mathbf{c}$ was not measured, but the value of n was estimated as about 1.5% smaller in this direction at 2.5 eV (Ejder, 1971).

Additional polarized ir reflectance measurements were made by Barker and Ilegems (1973), in samples where the free electrons were removed by electron bombardment, to avoid the corresponding absorption process. The reflectance was measured over a wide range of photon energies, and a Kramers–Kronig analysis then yielded relevant data for the optical dielectric constant, in both polarizations. From these data, the corresponding static dielectric constants, as well as the effective Born lattice charge (Wallis, 1965), and the Fröhlich polaron coupling constant (Kittel, 1963) were derived. The data obtained are summarized in Table III.

TABLE III

INFRARED DIELECTRIC PROPERTIES FOR GaN AT 300 K. ε_0 DENOTES THE STATIC DIELECTRIC CONSTANT, ε'_0 THE OPTICAL VALUE IN THE IR LIMIT, n THE REFRACTIVE INDEX, ω_{TO} AND ω_{LO} THE TRANSVERSE AND LONGITUDINAL OPTICAL PHONON ENERGIES AT Γ, e^*_B DENOTES THE BORN CHARGE (WALLIS, 1965) AND α DENOTES THE FRÖHLICH POLARON COUPLING CONSTANT (KITTEL, 1963). VALUES ARE FROM BARKER AND ILEGEMS, (1973), EXCEPT FOR THE n VALUE, WHICH IS GIVEN IN EJDER, (1971).

	ε_0	ε'_0	n	ω_{TO}	ω_{LO}	e^*_B	α
$\mathbf{E} \perp \mathbf{c}$	9.5	$5.35 \pm 3\%$	2.29 ± 0.05	$560 \pm 1\%$	$746 \pm 1\%$	$2.65 \pm 2\%$	$0.44 \pm 2\%$
$\mathbf{E} \parallel \mathbf{c}$	10.4	$5.35 \pm 3\%$		$533 \pm 1\%$	$744 \pm 1\%$	$2.82 \pm 2\%$	$0.49 \pm 2\%$

These ir data also give rather accurate values for the ir active phonon modes in GaN. Phonons in III–V nitrides are described in detail in a separate chapter in this book.

IV. Optical Properties of Cubic GaN

It is only rather recently that the art of growing cubic GaN has been developed to an extent that the basic optical properties of the material can be characterized. The bandgap of the cubic phase (β-GaN) is about 0.2 eV smaller than for the wurtzite phase (α-GaN). The valence band structure of β-GaN is simpler than for α-GaN (see Fig. 1). There is no crystal field splitting of the valence band top for β-GaN, but the spin-orbit splitting can be assumed to be very similar to the one for α-GaN, and therefore, the valence band top of β-GaN is expected to be split into two components, as compared to three for α-GaN.

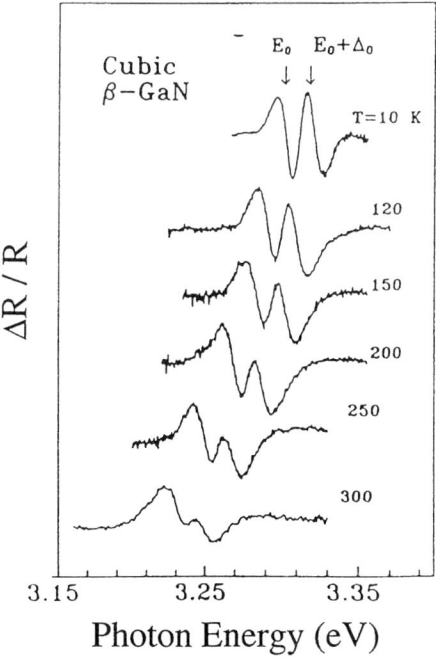

FIG. 34. Modulated photoreflectance data at different temperatures for β-GaN grown on MgO. (Reprinted from Ramirez-Flores *et al.*, 1994, with permission from The American Physical Society.)

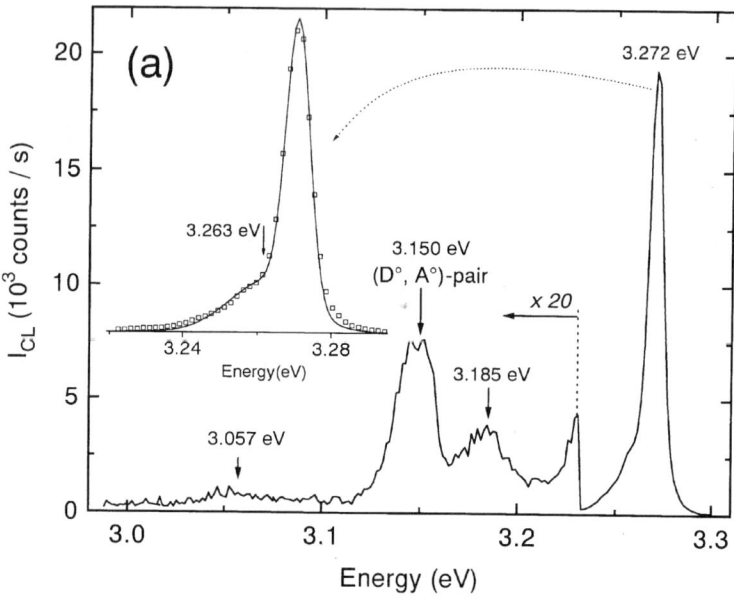

FIG. 35. Near bandgap cathodoluminescence spectra at 2 K of β-GaN. (Reprinted from Menniger et al., 1996a, with permission from The American Physical Society.)

Recently, modulated photoreflectance data for β-GaN grown on MgO substrates were presented for the near bandgap region in the temperature range 10–300 K (Ramirez-Flores et al., 1994). The electron concentration in the layers was high ($\geqslant 10^{18}\,\text{cm}^{-3}$), and the data were analyzed assuming band-to-band transitions (i.e., neglecting excitonic effects) (Fig. 34). A bandgap value of $3.302 \pm 0.004\,\text{eV}$ was reported at 10 K, the corresponding value at 300 K was estimated as 3.23 eV (Ramirez-Flores et al., 1994). The spin-orbit split bandgap occurred at 17 meV higher energies, that is, a value $\Delta_{SO} = 17 \pm 1\,\text{meV}$ was reported for β-GaN. These values were recently confirmed by another group (Ploog et al., 1995), who clearly resolved the near bandgap excitons in PL spectra for β-GaN grown on GaAs substrates (Fig. 35). The FE transition was in Ploog et al. (1995) estimated as 3.272 eV at 5K, consistent with a bandgap of 3.300 eV at 5K, assuming an exciton binding energy of 28 meV (Ploog et al., 1995). The possible influence of built-in strain on these transition energies has not been considered so far, but the above PL data (Ploog et al., 1995) were obtained from small crystallites grown on the surface of the β-GaN layer, and may therefore, represent rather strain-free material. The room temperature PL linewidth is about 100 meV, however, indicating a rather large defect density, making inhomogeneous broadening the dominating mechanism for the linewidth.

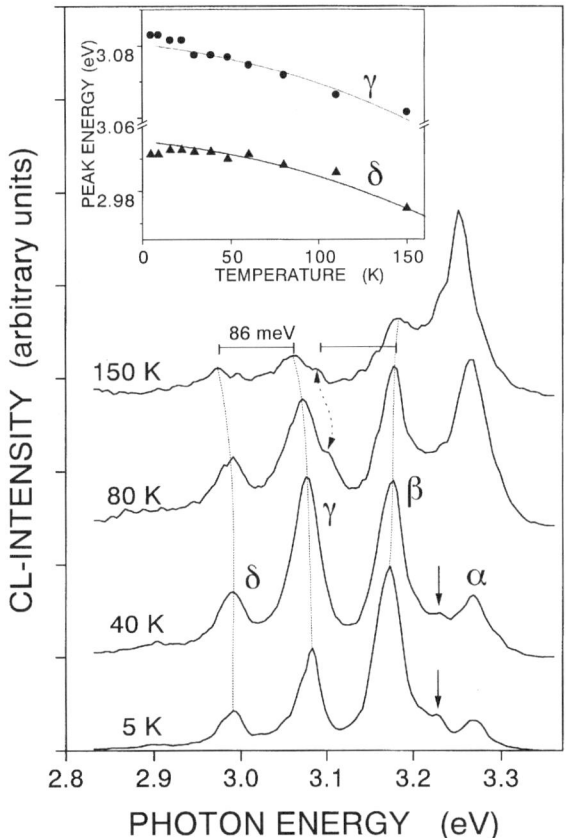

FIG. 36. Cathodoluminescence spectra of β-GaN at different temperatures. The possible origin of the lines are discussed in the text. (Reprinted from Menniger et al., 1996b, with permission from The American Institute of Physics.)

The exciton dynamics on a picosecond timescale has recently been studied in β-GaN (Klann et al., 1995). The near bandgap exciton PL peak at 3.27 eV at 5K is suggested to be composed of both free and bound exciton components. The PL intensity is linear in excitation intensity up to 10^{19} cm^{-3} carrier (exciton) density. The free excitons show a fast relaxation (15–40 ps) at short times, and a slower PL decay time (100–400 ps) at longer times is interpreted in terms of localized excitons (Klann et al., 1995).

Photoluminescence spectra below the excitonic region have also been observed recently (Menniger et al., 1996a). As shown in Fig. 36, there are a series of 3 PL peaks occurring at low temperatures in β-GaN grown on GaAs, in the photon energy range 3.2–3.0 eV. Studies of different samples reveal that the 3.15 eV peak, denoted β in Fig. 36, is not related to the lower energy peaks at 3.08 eV and 2.99 eV, respectively. These peaks have been interpreted as due to two different donor-acceptor (D-A) pair bands in β-GaN (Menniger et al., 1996c), since both peaks are rather broad and exhibit an upshift with excitation intensity at low temperatures. It appears likely that the peaks γ and δ in Fig. 36 are connected, so that, δ is an LO-phonon replica of γ. The corresponding acceptor energy would then be around 220 meV, very similar to the unidentified shallow acceptor in GaN (Lagerstedt and Monemar, 1974). The β peak, on the other hand, has a position and shape that varies slightly with sample, and it shows a continuous upshift with temperature up to 200 K, which is not expected for a DA pair transition. The appearance of this peak resembles the broad band often observed in PL spectra around 1.40–1.43 eV in α-GaN grown at low temperatures (Chen et al., 1996), in literature described as due to deep donors or shallow DA pairs. In α-GaN, this band seems to disappear with increasing structural quality of the sample. It is possible that such bands are due to carrier localization in strained regions around structural defects. Assuming a DA pair transition for the β peak, on the other hand, implies that there are quite shallow acceptors in β-GaN, of the order of 100 meV (a value of 94 meV for a shallow acceptor binding energy was recently independently suggested for β-GaN from the identification of a weak PL peak at 3.208 eV as an electron-acceptor transition (Jahn et al., 1996). A strong difference in shallow acceptor binding energies between α-GaN and β-GaN would be surprising, but cannot be ruled out. It is possible that the 3.08 peak in β-GaN corresponds to the 3.26 eV peak in α-GaN, representing the common unidentified shallow acceptor (Lagerstedt and Monemar, 1974). The presence of shallower acceptors in optical spectra should be treated with caution as long as investigations are made with the present rather poor sample quality of β-GaN.

Since doping with donors and/or acceptors has so far not been systematically studied in β-GaN, we have to conclude that there is as yet, no reliable information about energy levels of acceptors or donors.

V. Defect-Related Optical Properties

Traditionally optical characterization has been a very powerful way to detect the presence of impurities and defects in a semiconductor (Jahn et al., 1996; Pankove, 1971; Queisser and Heim, 1974; Queisser, 1976). A defect most often exhibits a characteristic optical spectrum, which can be used for detection of its presence, and often also for the characterization of other defect-related properties. This is certainly true also for the III–V nitrides. This is a very virgin field for GaN, however. While many defect-related spectra are observed, their identification is still in most cases not clear. We shall try to reflect the present situation in this subfield for GaN.

1. BOUND EXCITONS IN GaN

A bound exciton (BE) is an excited multiparticle state of a defect. The literature on studies of bound excitons in semiconductors is very extensive (Dean and Herbert, 1979a; Monemar, Lindefelt, and Chen, 1987; Monemar, 1988). In Fig. 37 is shown schematically, the electronic structure of excitons bound to neutral single donors (DBEs) and acceptors (ABEs), respectively. In these cases, the excited state has three interacting particles (two electrons and one hole for the DBE, two holes and one electron for the ABE). These two classes of BEs are by far the most important cases for direct bandgap materials. A different class of BEs are those bound to neutral "isoelectronic" defects, where the excited state has two particles (one electron and one hole). The latter BEs are less prominent in direct bandgap materials.

The recombination of BEs typically give rise to sharp spectral lines, with a photon energy characteristic for each defect. There is some substructure expected in the bound exciton lines due to electron-hole and hole-hole exchange interaction as well as crystal field effects, the splittings caused by

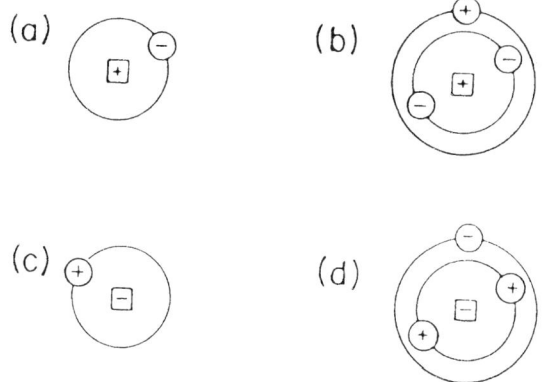

FIG. 37. Schematic picture of the neutral donor ground state (a) and neutral donor bound exciton state (b). Similarly in (c) is shown the neutral acceptor ground state, and in (d) the neutral acceptor bound exciton state.

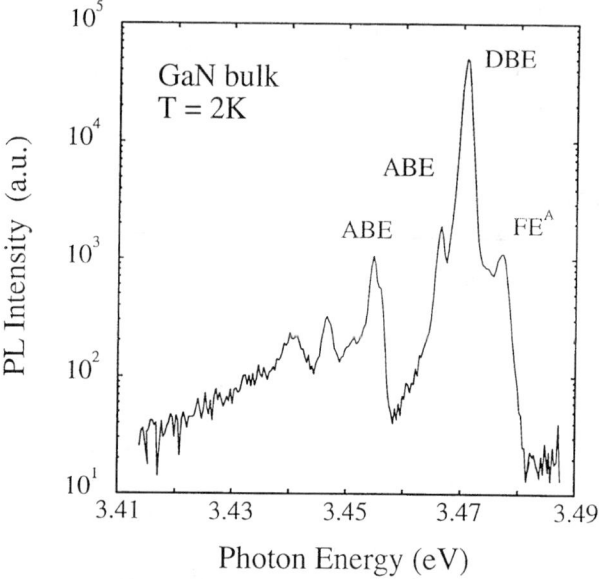

FIG. 38. Near bandgap photoluminescence spectrum at 2 K for a 500 μm thick GaN layer on sapphire, showing both donor- (DBE) and acceptor-related (ABE) bound exciton lines. (Data obtained by J. P. Bergman.)

these effects are usually of the order or below 1 meV for shallow donor- and acceptor-bound excitons. Also, for each particular BE spectrum, there is a characteristic phonon coupling, which can involve both lattice modes and defect-related vibrational modes.

Most PL spectra of GaN at low temperatures are dominated by a sharp shallow DBE line at about 3.472 eV in unstrained material (Fig. 38). This is believed to be due to an unidentified residual shallow donor in GaN (Dingle et al., 1971b; Eckey et al., 1996a, c; Matsumoto and Aoki, 1974; Lagerstedt and Monemar, 1974; Volm et al., 1995, 1996; Kaufmann et al., 1996a). The binding energy of this usually dominating shallow donor BE is about 6 meV (the binding energy for an ABE or a DBE is traditionally defined as the energy distance between the BE line and the lowest FE line). The corresponding binding energy of the dominating shallow donor electron has recently been estimated as about 35.5 meV from ir absorption data (Meyer et al., 1995). The identity of this residual shallow donor is still unknown. There are arguments that it is related to the nitrogen vacancy in GaN (Perlin et al., 1995). This defect is believed to have a 0/+ level resonant with the conduction band, thus inducing a shallow donor state just below the bottom of the conduction band (Perlin *et al.*, 1995). This conclusion has been challenged on theoretical grounds, and some authors argue that the nitrogen vacancy will not be a dominant defect in n-GaN (Neugebauer and Van de Walle, 1994; Boguslawski, Briggs, and Bernholc, 1996). Another suggestion is O on N site. O is known to be a typical contaminant in GaN, grown by all techniques, and is a good candidate for a residual shallow donor. O implanted in GaN has recently been shown to have an activation energy for electron conduction of 30–40 meV (Pearton *et al.*, 1996). Recent data from a combination of electrical and optical measurements on Si-doped GaN samples indicate that the Si donor has a binding energy of about 20 meV, while a deeper donor (35 meV) was also present, suggested to be O-related (Götz *et al.*, 1996a).

As evident from Fig. 38, the main DBE line has some substructure, due to the previously mentioned internal interaction effects. It cannot be concluded that some of this structure is due to the presence of more than one DBE (i.e., related to different donors) with slightly different binding energies. In a similar study on hydride VPE (HVPE) material another more

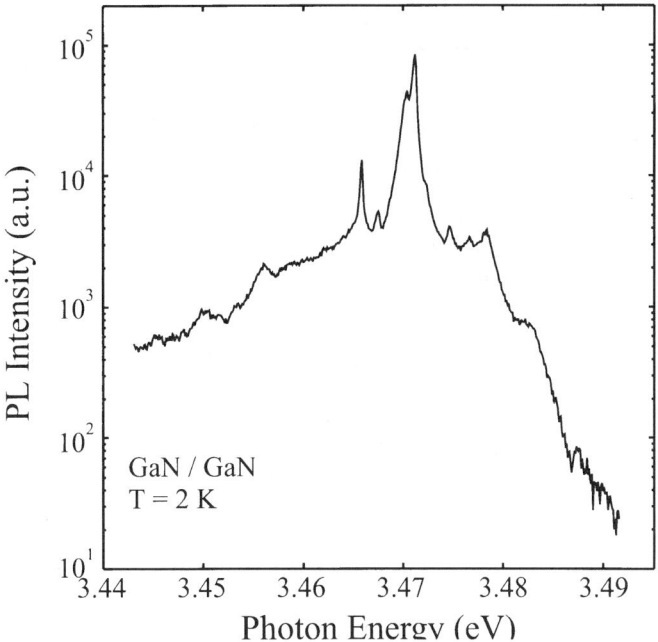

FIG. 39. Photoluminescence spectrum at 2 K of a nominally undoped n-type homoepitaxial layer of GaN. (Data obtained by J. P. Bergman, in cooperation with J. M. Baranowski.)

shallow donor BE line with a binding energy of about 3.7 meV was clearly observed, and speculated to be associated with the Si donor on Ga site (Ben-Chorin et al., 1996). Such a shoulder is also present in Fig. 39 and in Fig. 12 (the line labeled a).

Recently, PL spectra with much improved linewidths were obtained from homoepitaxial layers grown with MOCVD. In Fig. 39, we see that there are at least three well-resolved peaks in the DBE region, at about 3.4709 eV, 3.4718 eV and 3.4755 eV. This would indicate that there are actually three dominant donors in the samples, so another donor in addition to O and Si is needed. There is a possibility that the two lines at 3.4709 eV and 3.4718 eV are the exchange split components of the same BE, but in many samples one of the lines seems to be absent. It cannot be excluded that strain splitting occurs in some cases though. The shallowest line at 3.4755 eV could also be a DBE related to the B valence band. The line at 3.4664 eV dominates the acceptor related BE spectra, but there is also a weak line at about 3.468 eV, of unknown origin.

In the spectrum shown in Fig. 40, for a homoepitaxial layer grown by MBE (Teisseyre et al., 1996; Baranowski and Porowski, 1996), there are

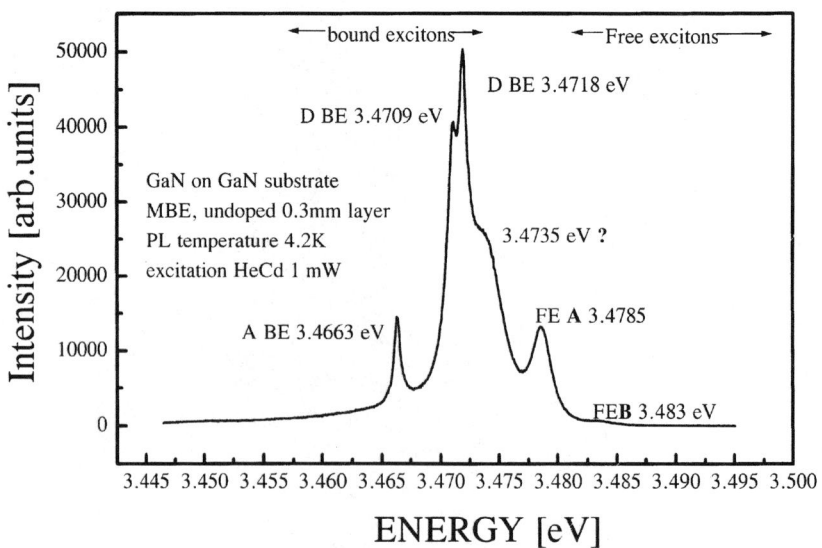

FIG. 40. Photoluminescence spectrum at 4.2 K for a nominally undoped homoepitaxial MBE grown GaN layer. Three donor bound exciton peaks are clearly resolved. (Courtesy of Teisseyre et al., 1996, unpublished data.)

clearly two sharp DBE lines at about 3.4709 eV and 3.4718 eV, respectively, while there is only one sharp ABE line at 3.4663 eV. In the case of strain splitting one would expect that both DBE and ABE lines would split. A broader more shallow DBE line occurs at 3.4735 eV.

In the energy region below the principal DBE line there is usually a rich spectrum of BEs assumed to be acceptor-related (Fig. 39). A number of rather sharp transitions are shown, assumed to be ABE transitions related to several different residual acceptors. The identity of these acceptors are not known for this sample. The peak at about 3.466 eV (Fig. 38) is usually strong in Mg-doped samples (Pakula *et al.*, 1996). The main peak at about 3.455 eV could be due to Zn, but this is only a guess so far. The broader peak at about 3.44 eV seems to be a phonon wing connected to the main 3.455 eV ABE. (The peak at about 3.447 eV might be the two-electron transition related to the main DBE line, for example, a DBE transition leaving the neutral donor in the 2s excited state. This is consistent with a 1s–2s distance of about 24 meV for the 35 meV effective mass like shallow donor.) Recently, similar studies were made in backdoped samples by Kaufmann *et al.* (1996a). They found a binding energy of 19 meV for the Mg acceptor and 34 meV for the Zn acceptor in thin epitaxial layers grown by MOCVD, where the free A exciton was considerably upshifted (Kaufmann *et al.*, 1996a). Recently, Pakula *et al.* (1996) reported the position of the main ABE in Mg-doped samples as 3.4666 eV in a homoepitaxial GaN sample (assumed to be strain free). The ABE binding energy in this case is then only about 12 meV, considerably smaller than for the strained layer. This result suggests a strong strain dependence on the binding energy for shallow BEs in GaN. Alternatively, this can be expressed as a reduced deformation potential for ABEs, compared to FEs in GaN. This is not surprising, since it has been observed for acceptors in GaAs as well (Schairer and Schmidt, 1974; Schairer *et al.*, 1976). For DBEs the reduction of the deformation potentials is much smaller, the binding energy of the DBE increases slightly from about 6 meV in unstrained GaN to about 7 meV when the A exciton has been shifted to about 3.49 eV in epilayers grown on sapphire (Chichibu *et al.*, 1996).

The decay time of the Si donor bound excitons in GaN is short, about 250 ps at 2 K (Bergman *et al.*, 1996). This value is in fact, similar to the value found for shallow donor BEs in GaAs (Finkman, Sturge, and Bhat, 1986). Decay times for excitons bound to residual acceptors (at 3.455 eV) in GaN (of unknown identity) have also been studied recently (Bergman *et al.*, 1996), and are much longer, of the order 1.2 ns. This is also analogous to GaAs, where ABEs have a decay time of about 1 ns (Finkman, Sturge, and Bhat, 1986). It is interesting to note, that the value for the ABE lifetime for the

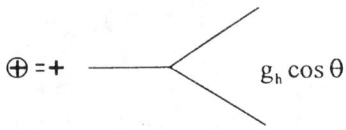

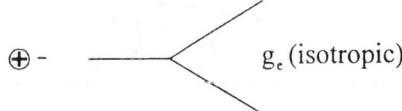

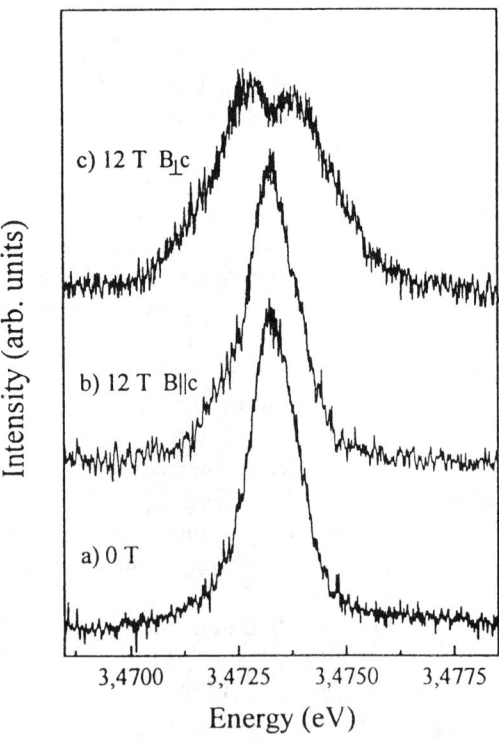

FIG. 41. (a) The expected magnetic field splitting for a neutral donor bound exciton in GaN, and the neutral donor ground state splitting. Both states are involved in the optical transition. (b) Experimental Zeeman data in photoluminescence for the donor bound exciton at 2 K. (Reprinted from Volm et al., 1995, with permission from Elsevier Science Ltd., Kidlington, UK.)

shallowest ABE at 3.463 eV has a lifetime of 800 ps (Monemar et al., 1996c), as compared to a deeper acceptor with an ABE peak at 3.455 eV, which has a lifetime of 1200 ps (Bergman et al., 1996). The corresponding oscillator strengths for the DBE is given by the general expression $f = 4.5\,10^4 \lambda^2/n\tau$ (Rashba and Gurgenischvili, 1962; Henry and Nassau, 1970), where λ is the wavelength in m. The value for the DBE is $f \approx 9$, while for the shallow ABE we get a value $f \approx 3$ and $f \approx 2$ for the deeper ABE. These values are very similar to the oscillator strengths for shallow BEs in CdS (Henry and Nassau, 1970). The radiative lifetime of the BE is expected to scale approximately as $E_b^{3/2}$, where E_b is the binding energy with respect to the FE (Steiner, Thewalt, and Bhargava, 1985). The above values support the view that the observed lifetimes are the radiative lifetimes. Nonradiative Auger processes may also be present for the BE recombination, as discussed for ZnTe (Schmidt and Dean, 1982) and in particular, for indirect bandgap materials (Dean and Herbert, 1979b). In the Auger case, the lifetime is expected to decrease drastically with increasing binding energy, due to the more localized bound particles, which is not observed for these BEs in GaN.

The influence of a magnetic field on bound exciton lines has recently been investigated (Stepniewski and Wysmolek, 1996; Volm et al., 1996). The magnetic field splittings can be analyzed by a simple model (Fig. 41(a)) where, for the DBE, the two electrons couple to a $J = 0$ state, so that the g-value is determined by the $J = 3/2$ hole, while in the ground state (the final state in the PL transition, the neutral donor) there is a donor electron with a nearly isotropic g-value $g_e \approx 1.95$ (Volm et al., 1996). The splitting is thus determined by the splitting of both the DBE state and the neutral donor. For $B \perp$ the c axis the hole g-value equals zero, and only the isotropic ground state splitting is observed (Volm et al., 1996). Knowing this, a value of $g_h = 0.65$ for the hole in the DBE is obtained from the Zeeman data in Fig. 41(b), neglecting the internal structure of the bound exciton (Volm et al., 1996). A similar conclusion can be drawn on the g-values for the 3.466 eV ABE case (Stepniewski and Wysmolek, 1996). In this case, the g-values for the electrons (in the BE state) and the hole (in the neutral acceptor ground state) are all nearly isotropic and close to 2. Again, in a proper treatment, the internal structure of the BEs should be included, which require samples with much better spectroscopic linewidths than presently available.

2. OTHER DONOR- OR ACCEPTOR-RELATED OPTICAL SPECTRA

The shallow donors are mainly seen optically via their near bandgap DBE line, as shown above. In principle donors may manifest themselves in other ways, such as free-to-bound transitions, like the donor to valence band transition (D^0h). For the dominant shallow donor with binding energy about 35 meV, the corresponding transition energy would be observed at about 3.468 eV, which is, on the low energy side of the DBE emission line. Therefore, we believe this emission is very difficult to resolve. In the literature, there are many claims of observations of optical processes of this nature, though. Some authors have recently attributed two weak PL bands at somewhat lower energies in thin GaN layers grown on sapphire to the D^0h process involving two residual donors with slightly larger binding energies (54 meV and 57 meV, respectively) (Kaufmann et al., 1996a). An early unpublished report using similar data claims that the Si donor has a binding energy of 50 meV, while the Ge donor is slightly deeper, 67 meV (Gershenzon and Wang, 1979). In MBE GaN material, there is often present a strong PL peak at about 2.42 eV at low temperatures (Smith et al., 1995b). This peak is believed to be due to a slightly deeper donor state, with a binding energy of about 90 meV. The authors suggest that oxygen might be responsible for this donor state (Smith et al., 1995b). Such data for weak or broad peaks in material with a very high defect density should so far be

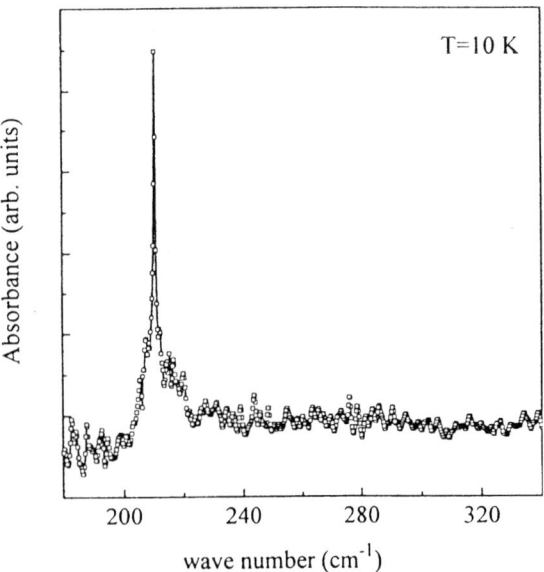

FIG. 42. Infrared absorption spectrum at 10 K for the dominant residual shallow donor in a 400 μm thick GaN epilayer grown on sapphire. (Reprinted from Meyer et al., 1995, with permission from Elsevier Science Ltd., Kidlington, UK.)

interpreted with caution, and therefore, the values in the range 50–90 meV related here may not be due to donors at all, the weak spectra may have quite different origin.

The most common residual donor in GaN has recently been observed in optical absorption by Meyer *et al.* (1995). The authors observe a sharp absorption line at 26.6 meV in a few hundred μm thick VPE grown sample, which they interpret as the allowed 1s–2p transition of the dominating shallow donor (Fig. 42). A value of the binding energy of this donor as 35.5 ± 0.5 meV is extrapolated (Meyer *et al.*, 1995). This donor has recently been argued to be O on N site (Götz *et al.*, 1996a). O is a common contaminant in GaN. The Si donor is argued to be shallower (Götz *et al.*, 1996a).

Donor-acceptor pair (DAP) spectra are also very common examples of radiative defect recombination. A very common such spectrum for GaN is shown in Fig. 43. A no-phonon peak occurs at about 3.26 eV at low T, with 2–3 well-resolved LO phonon replicas towards lower energies (Lagerstedt and Monemar, 1974; Dingle and Ilegems, 1971; Grimmeiss and Monemar,

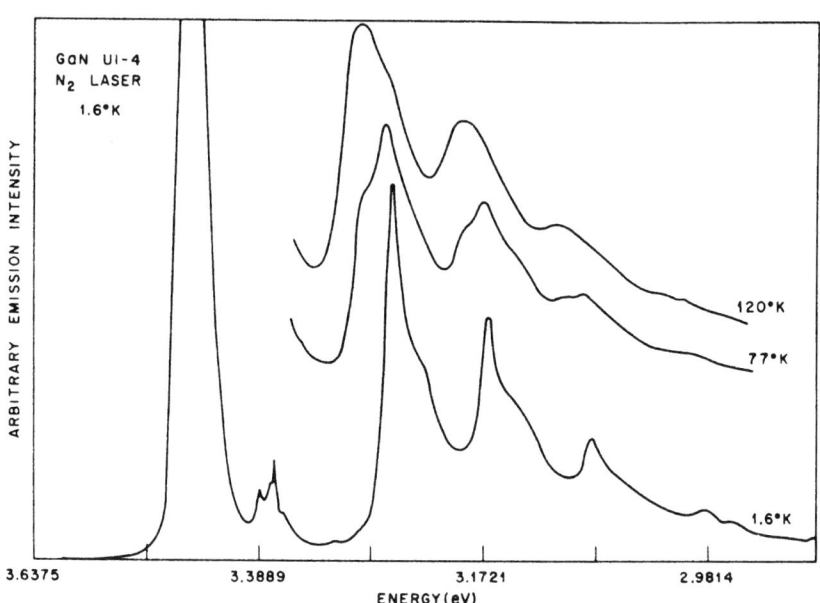

FIG. 43. Donor-acceptor pair spectrum for the dominant residual shallow acceptor in GaN, at 3 different temperatures. At 120 K, the conduction band to acceptor transition dominates. (Reprinted from Dingle and Ilegems, 1971, with permission from Elsevier Science Ltd., Kidlington, UK.)

1970; Fisher et al., 1995). The temperature dependence of the spectrum reveals the evolution from a DA pair spectrum at low T to a free-to-bound (FB) conduction band to acceptor transition at higher temperatures (120 K) (Lagerstedt and Monemar, 1974; Dingle and Ilegems, 1971). At intermediate temperatures, both processes may be resolved (Fig. 43). The acceptor binding energy has been estimated as about 230 meV. There is so far no positive identification of the identity of this acceptor, but it has recently been argued that it could be the C acceptor substitutional on N sites (Fischer et al., 1995). Another possibility is that it is related to acceptor-like native defects (Lagerstedt and Monemar 1974). It apparently occurs very frequently in GaN grown by any method, independent of doping species, arguing against any simple connection with the common impurities.

Recently, structure on the high energy tail of this emission in Mg-doped samples has tentatively been identified as due to discrete pair lines (Wysmolek et al., 1996). A type II spectrum is argued, meaning that the donor and acceptor are situated at opposite lattice sites (i.e., one on Ga site and one on N site) (Wysmolek et al., 1996). The acceptor binding energy is deduced as about 270 meV, which is then deeper than the above mentioned common shallow acceptor in GaN. This binding energy would be consistent with the Mg-related acceptor giving rise to the broad PL band centered at about 3.0 eV, however.

We have recently observed that the PLE spectrum for the 3.26 eV DAP band in the near bandgap excitonic region is very broad, in contrast to the PLE spectrum from the discrete pair lines observed from the Mg related emission (Baranowski, Bergman, and Monemar, 1996). This indicated that this 3.26 eV DAP spectrum comes from perturbed regions in the crystals, probably close to extended crystal defects, causing severe inhomogeneous broadening of the spectra. This is also consistent with the fact that no discrete pair lines were seen connected for this 3.26 eV emission.

Several reports exist in literature on PL spectra from GaN intentionally doped with the column II elements, like Li (Grimmeis and Koelmans, 1959; Pankove et al., 1973), Be (Ilegems and Dingle, 1973; Pankove and Hutchby, 1976), Mg (Ilegems and Dingle, 1973; Amano et al., 1990; Pankove, Berkeyheiser, and Miller, 1974; Nakamura, Senoh, and Mukai, 1991), Zn (Monemar, Lagerstedt, and Gislason, 1980a; Monemar, Gislason, and Lagerstedt, 1980b; Boulou, Jacob, and Bois, 1978; Bergman et al., 1987), Cd (Lagerstedt and Smith, 1974; Ejder and Grimmeiss, 1974) and Hg (Ejder and Grimmeiss, 1974). (For more details about these PL emissions we refer to Strite and Morkoc, (1992), and a large number of additional references therein.) For all these acceptors the PL spectrum has the shape of a rather broad band with a half width varying from about 0.25 eV and up for different dopants. This broad lineshape is naturally interpreted as an

TABLE IV

ACCEPTOR-RELATED PHOTOLUMINESCENCE AND BINDING ENERGIES OF THE CORRESPONDING RADIATIVE CENTERS IN GaN

Acceptor	PL Peak Position (eV at 2 K)	Binding Energy (eV)	Reference
Residual	3.26	0.23	94, 128
Li	2.2	0.75	133, 134
Be	2.2	1.1	135, 136
Mg	2.95	0.25–0.3	135, 137–139
Zn	2.87	0.34	140–143
Cd	2.72	0.5	94, 144
Hg	2.9	0.4	144
Ca	2.5	0.8	136

envelope due to rather strong phonon coupling. On the other hand, this does not allow any detailed assessment of the recombination mechanism, since the no-phonon line is not well (or most often, not at all) resolved. If it is assumed that the emissions are due to a recombination of an electron from the conduction band with a hole bound to an acceptor, approximate values for the binding energies of these acceptors can be obtained. A collection of such preliminary data is given in Table IV. The observed PL decay times for Zn and Cd related emissions (hundreds of ns) are in fact consistent with an FB type emission (Bergman *et al.*, 1987).

It is here assumed that the acceptor like defects related above (Table 4) are simple substitutional acceptors on Ga site. The group 4 elements Si and Ge have been found to be shallow donors (Koide *et al.*, 1991; Nakamura, Mukai, and Senoh, 1992). The fact that most of the group II acceptors have a rather strong phonon coupling could also indicate an alternative identification of these optical spectra as due to complex defects, such as defect pairs or relaxed acceptor configurations. In fact, in Ilegems, Dingle, and Logan (1972), it was argued that these deep acceptor related PL spectra could be explained by assuming that all acceptors were in fact shallow and responsible for the 3.26 eV DAP emission, while the deeper bands were acceptor-related complex defects. The problem with such an assumption is that all substitutional acceptors would have the same binding energy (at least within a few meV), which is not very plausible. Therefore, it is more natural to assume that the substitutional acceptors are in fact, involved in the major characteristic PL bands for each acceptor. It should be pointed out, however, that this is not proven so far. It is also possible that the characteristic radiative defects for each acceptor correspond to a different configuration than the one responsible for the major hole activation (see below for the Mg case). The Zn acceptor is clearly more complicated, since up to four progressively deeper emissions are observed in highly doped material (Boulou, Jacob and Bois, 1978; Nakamura, Mukai, and Senoh, 1992).

It should also be mentioned that bistability or multistability of acceptors is a common phenomenon in wide bandgap materials (Chadi and Park, 1995; Park and Chadi, 1995; Garcia and Northrup, 1995). In ZnSe, the Se site acceptors, except for N, are deep levels and have a relaxed low symmetry configuration. Whether there is a similar reason why so many acceptors in GaN are quite deep is a good topic for future detailed investigations. We note that the Mg related level associated with the violet emission band (binding energy $\geqslant 250$ meV) appears to be much deeper than the level associated with the electrical conductivity (about 160–170 meV) (Akasaki et al., 1991, 1993; Detchprohm et al., 1994; Götz et al., 1996b; Nakamura et al., 1992). The electrical data are obtained with very high doping levels ($10^{19} - 10^{20}$ cm^{-3}), causing potential fluctuations, which give the apparent lower activation energy. The 3.26 eV shallow DAP emission appears strongly with Mg doping (Akasaki et al., 1993; Detchprohm et al., 1994; Götz et al., 1996b), until very high doping densities are obtained, which causes confusion in literature. Recently, evidence for metastability of the Mg acceptor in GaN has been presented (Johnson et al., 1996; Li et al., 1996). Persistent photoconductivity was observed in Mg doped p-type GaN (Johnson et al., 1996), indicating metastability of the Mg acceptor. A barrier of about 130 meV for hole capture in the electrically active Mg acceptor state is reported (Li et al., 1996).

There is apparently an optically active Mg state with moderate strong lattice relaxation, giving rise to the broad-band centered at around 3.0 eV in GaN at low temperature. The corresponding Mg-related center has a quite localized wavefunction with a Bohr radius of about 3 Å as deduced from ODMR data by several authors (Glaser et al., 1995; Kunzer et al., 1994). This strong localization explains the moderately strong phonon coupling, involving a broad phonon spectrum. The metastability reported from electrical measurements on Mg-doped GaN needs further confirmation, in particular since theoretically such a metastability is not expected for Mg (Park and Chadi, 1997).

Recently, there have been doping experiments conducted with Ca as a group II acceptor. Similar rather small electrical activation energies of about 170 meV were reported (Lee et al., 1996). Interestingly, there is no report of a corresponding shallow DAP luminescence at low temperatures. Whether other group II acceptors which have deep PL bands might possess shallower configurations is not clear at present. Detailed experimental data from transport studies are still lacking for the Ca acceptor.

A much debated issue recently has been the so-called yellow emission in GaN, for example, a broad PL band centered around 2.2 eV (Glaser et al., 1995; Koschnick et al., 1995; Hofmann et al., 1995), present in most GaN material. Detailed PL investigations indicate that the emission is due to a shallow donor-deep double donor transition. Hofmann et al., 1995, although Glaser et al., 1995, suggest a different deep donor-shallow acceptor model. The nature of the deep donor has not been determined, it has been argued that it is of intrinsic origin, such as the nitrogen vacancy (Perlin et al., 1995). The intensity of the yellow PL band has been found to correlate with dislocation-related defects (Ponce et al., 1996). It has not been shown to be a serious defect in terms of determining the minority carrier lifetimes, however. The PL decay time is rather long and nonexponential (Hofmann et al., 1995), it appears to be a radiative defect (Glaser et al., 1995; Koschnick et al., 1995; Hofmann et al., 1995), with no characteristic negative ODMR signals observed on the dominant near bandgap recombination. All these facts point to the conclusion that the defect itself is harmless as a recombination center. Its presence might well correlate spatially in some sense with the presence of serious (nonradiative) recombination centers, however.

3. Optical Spectra Related to Transition Metal Centers in GaN

Transition metal impurities in semiconductors most often have deep levels in the bandgap, and are consequently important recombination centers in the material. If the concentration is sufficient, they also may control the Fermi level position. In GaN transition metals are easily introduced as contaminants during growth, unless careful precautions are taken (Monemar and Lagerstedt, 1979). In the early GaN work during the 1970s transition metal contamination was one reason why low ohmic p-type material was not obtained (Monemar and Lagerstedt, 1979). (Another reason was that H-passivation effects were not properly understood at that time).

The characteristic optical properties of transition metal impurities in semiconductors is based on the fact that a particular charge state of such a defect possesses a set of crystal field split levels (within the same charge state) in the bandgap (Ballhausen, 1962), allowing internal optical transitions (absorption or emission) between these states. A characteristic spectrum for each transition metal ion is obtained, usually showing several sharp lines and a phonon structure also characteristic for the immediate environment of the impurity (Di Bartolo, 1968). The energy position of such a characteristic spectrum for a particular metal impurity varies only weakly

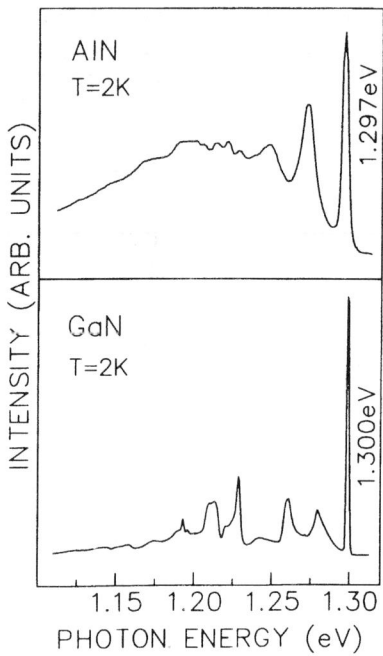

FIG. 44. Fe-related photoluminescence spectrum at 2 K for AlN and GaN. (Reprinted from Baur *et al.*, 1994, with permission from The American Institute of Physics.)

between different materials. In Figure 44 is shown a typical such PL spectrum for the same defect (Fe) in both AlN and GaN.

For thin epitaxial layers PL is the most convenient way of recording these transitions. The characteristic spectrum of Fe^{3+} has been found to be connected to a PL spectrum with a principal no-phonon line at 1.30 eV in GaN (Maier *et al.*, 1994; Bauer *et al.*, 1994a). The same emission occurs in AlN, with a no-phonon line at 1.297 eV (Baur *et al.*, 1994b, 1995a). Another characteristic line occurs at 1.193 eV in GaN. It has been assigned to Cr^{4+} or Ti^{2+} (Baur *et al.*, 1994a; Wetzel *et al.*, 1994, 1995b, 1995c), but in the latest work, it has been argued that it should be due to Ti^{2+} (Heitz *et al.*, 1995). More work is needed to settle this question. The same structure occurs at 1.201 eV in AlN (Baur *et al.*, 1995c). Other PL lines of similar nature are observed in GaN at lower energies, at 1.047 eV (Wetzel *et al.*, 1994), 0.931 eV (Baur *et al.*, 1995b, 1995c) and 0.820 eV (Kaufmann *et al.*, 1996b). The correlation with V^{3+} has been tentatively suggested in the different papers, but a firm assignment may not be possible at present.

ACKNOWLEDGMENTS

The author is very grateful to I. Akasaki, H. Amano, and K, Hiramatsu for supplying numerous samples studied in this work. J. P. Bergman, I. A. Buyanova, and A. Buyanov have collected the experimental data for a substantial part of this chapter. Many foreign colleagues have supplied articles, unpublished results and illustrations to be included in this chapter. We are particularly indebted to H. Amano, J. M. Baranowski, O. Brandt, S. Chichibu, L. Eckey, N. Edwards, E. Ejder, B. Gil, J. E. Greene, A. Hangleiter, C. I. Harris, A. Hoffmann, J. M. Hvam, I. Ivanov, U. Jahn, H. X. Jiang, U. Kaufman, K. P. Korona, W. Lambrecht, W. Li, T. Lundström, B. K. Meyer, H. Morkoc, T. D. Moustakas, K. Okada, P. Perlin, W. G. Perry, D. C. Reynolds, W. Shan, H. Teisseyre, and A. Zubrilov, in this respect.

REFERENCES

Akasaki, I., Amano, H., Kito, M., and Hiramatsu, K. (1991). *J. Lumin.* **48/49,** 666.
Akasaki, H., Amano, H., Murakami, H., Sassa, M., Kato, H., and Manabe, K. (1993). *J. Cryst. Growth* **128,** 379.
Akasaki, I., and Amano, H. (1994). *J. Electrochem. Soc.* **141,** 2266.
Amano, H., Hiramatsu, K., and Akasaki, I. (1988). *Jap. J. Appl. Phys.* **27,** 1384.
Amano, H., Kitoh, M., Hiramatsu, K., and Akasaki, I. (1990). *J. Electrochem. Soc.* **137,** 1639.
Amano, H., and Akasaki, I. (1995). *Proc. TWN 95 Nagoya, 21–23 Sept. 1995, Solid State Electronics,* in press.
Amano, H. (1996). Private communication.
Aspnes, D. E., and Studna, A. A. (1975). *Appl. Opt.* **14,** 220.
Ballhausen, C. J. (1962). "Introduction to Crystal Field Theory," McGraw-Hill, New York.
Baranowski, J. M., and Porowski, S. (1996). *Proc. 23rd Int. Conf. on the Physics of Semiconductors,* World Scientific, Berlin, 497.
Baranowski, J., Bergman, J. P., and Monemar, B. (1996). Unpublished data.
Barker, A. S., and Ilegems, M. (1973). *Phys. Rev.* **B7,** 743.
Baur, J., Maier, K., Kunzer, M., Kaufmann, U., Schneider, J., Amano, H., Akasaki, I., Detchprohm, T., and Hiramatsu, K. (1994a). *Appl. Phys. Lett.* **64,** 857.
Baur, J., Maier, K., Kunzer, M., Kaufmann, U., and Schneider, J. (1994b). *Appl. Phys. Lett.* **65,** 2211.
Baur, J., Kunzer, M., Maier, K., Kaufmann, U., and Schneider, J. (1995a). *Mat. Sci. Eng.* **B29,** 61.
Baur, J., Kaufmann, U., Kunzer, M., Schneider, J., Amano, H., Akasaki, I., Detchprohm, T., and Hiramatsu, K. (1995b). *Appl. Phys. Lett.* **67,** 1140.
Baur, J., Kaufmann, U., Kunzer, M., Schneider, J., Amano, H., Akasaki, I., Detchprohm, T., and Hiramatsu, K. (1995c). *Mat. Sci. Forum.* **197–201,** 55.
Ben-Chorin, M., Diener, J., Meyer, B. K., Dreschsler, M., Volm, D., Amano, H., Akasaki, I., Detchprohm, T., and Hiramatsu, K. (1996). Symposium "Gallium Nitride and Related Materials," Boston, Nov. 27–Dec. 1, 1995, *Mat. Res. Symp. Proc.* **395,** 467.

Benoit a la Guillaume, C., Bonnot, A., and Debever, J. M. (1970). *Phys. Rev. Lett.* **24**, 1235.
Bergman, P., Ying, G., Monemar, B., and Holtz, P. O. (1987). *J. Appl. Phys.* **61**, 4589.
Bergman, J. P., Monemar, B., Amano, H., Akasaki, I., Hiramatsu, K., Sawaki, N., and Detchprohm, T. (1996). *Inst. Phys. Conf. Ser.* **142**, 931.
Boguslawski, P., Briggs, E. L., and Bernholc, J. (1996). *Phys. Rev.* **B51**, 17255.
Boulou, M., Jacob, G., and Bois, D. (1978). *Rev. Phys. Appl.* **13**, 555.
Buyanova, I. A., Bergman, J. P., Li, W., Monemar, B., Amano, H., and Akasaki, I. (1996a). *Mater. Res. Soc. Symp. Proc.* **423**, in press.
Buyanova, I. A., Bergman, J. P., Monemar, B., Amano, H., and Akasaki, I. (1996b). *Appl. Phys. Lett.* **69**, 1255.
Chadi, D. J., and Park, C. H. (1995). *Mat. Sci. Forum* **197-201**, 285.
Chen, G. D., Smith, M., Lin, J. Y., Yiang, H. X., Salvador, A., Sverdlov, B. N., Botchkarev, A., and Morkoc, H. (1996). *J. Appl. Phys.* **79**, 2675.
Chichibu, S., Azuhata, T., Sota, T., and Nakamura, S. (1996). In *Proc Int. Symp. on Blue Laser and Light Emitting Diodes* (A. Yoshikawa *et al.*, eds.), Ohmsha Ltd., Tokyo, p. 202.
Chow, W. W., Knorr, A., and Koch, S. W. (1995). *Appl. Phys. Lett.* **67**, 754.
Cingolani, R., Ferrara, M., and Lugara, M. (1986). *Solid State Commun.* **60**, 705.
Dai, R., Zhuang, W., Bohnert, K., and Klingshirn, C. (1982). *Z. Physik B-Condensed Mat.* **46**, 189.
Dean, P. J., and Herbert, D. C. (1979a). In *Topics in Current Physics, Vol. 14, "Excitons,"* (K. Cho, ed.), Springer Verlag, Berlin, p. 55.
Dean, P. J., and Herbert, D. C. (1979b). In *Excitons* (K. Cho, ed.), Springer Verlag, Berlin, p. 165.
Detchprohm, T., Hiramatsu, K, Sawaki, N., and Akasaki, I. (1994). *J. Cryst. Growth* **137**, 170.
Dexter, D. L. (1958). In *Solid State Physics* (F. Seitz and D. Turnbull, eds.), Academic Press, New York, **6**, 353.
Di Bartolo, B. (1968). "Optical Interactions in Solids," Wiley, New York.
Dingle, R., Sell, D. D., Stokowski, S. E., Dean, P. J., and Zetterstrom, R. B. (1971a). *Phys. Rev. B* **3**, 497.
Dingle, R., Sell, D. D., Stokowski, S. E., and Ilegems, M. (1971b). *Phys. Rev. B* **4**, 1211.
Dingle, R., and Ilegems, M. (1971). *Solid State Commun.* **9**, 175.
Eckey, L., Podlowski, L., Göldner, A., Hoffmann, A., Broser, I., Meyer, B. K., Volm, D., Streibl, T., Hiramatsu, K., Detchprohm, T., Amano, H., Akasaki, I. (1996a). *Inst. Phys. Conf. Ser.* **142**, 943.
Eckey, L., Holst, J. Ch., Maxim, P., Heitz, R., Hoffmann, A., Broser, I., Meyer, B. K., Wetzel, C., Mokhov, N., and Baranov, P. G. (1996b). *Appl. Phys. Lett.* **68**, 1.
Eckey, L., Heitz, R., Hoffmann, A., Broser, I., Meyer, B. K., Hiramatsu, K., Detchprohm, T., Amano, H., and Akasaki, I. (1996c). *Proc. 6th Int. Conf. on Silicon Carbide and Related Materials, Kyoto, Sept 18-21, 1995, Inst. Phys. Conf. Ser.* **142**, 927.
Edwards, N. V., Bremser, M. D., Weeks, T. W. Jr., Kern, R. S., Liu, H., Stall, R. A., Wickenden, A. E., Doverspike, K., Gaskill, D. K., Freitas, J. A. Jr., Rossow, U., Davis, R. F., and Aspnes, D. E. (1996). Presented at the MRS Symposium "Gallium Nitride and Related Materials," Boston, Nov. 27-Dec. 1, 1995, *Mater. Res. Soc. Symp. Proc.* **395**, 405.
Ejder, E. (1971). *Phys. Stat. Solidi (a)* **6**, 445.
Ejder, E. (1974). *Phys. Stat. Sol. A* **23**, K87-88.
Ejder, E., and Grimmeiss, H. G. (1974). *Appl. Phys.* **5**, 275.
Finkman, E., Sturge, M. D., and Bhat, R. (1986). *J. Lumin.* **35**, 235.
Fischer, S., Wetzel, C., Haller, E. E., and Meyer, B. K. (1995). *Appl. Phys. Lett.* **67**, 1298.
Garcia, A., and Northrup, J. E. (1995). *Phys. Rev. Lett.* **74**, 1131.
Gershenzon, M., and Wang, C-H, D. (1979). Abstract EL 10, *Bull. Am. Phys. Soc.* **24**.

Gfrörer, O, Schlüsener, T., Härle, V., Scholz, F., and Hangleiter, A. (1996). Proc E-MRS Spring Meeting, Strasbourg, June 4–7, 1996, *Mat. Sci. Eng. B* to be published.
Gil, B., Briot, O., and Aulombard, R. L. (1995). *Phys. Rev. B* **52**, 17028.
Gil, B., Hamdani, F., and Morkoc, H. (1996). *Phys. Rev. B* **54**, 7678.
Glaser, E. R., Kennedy, T. A., Doverspike, K, Rowland, L. B., Gaskill, D. K., Freitas, Jr., J. A., Asif Khan, M., Olson, D. T., Kuznia, J. N., and Wickenden, D. K. (1995). *Phys. Rev. B* **51**, 13326.
Götz, W., Johnsson, N. M., Chen, C., Liu, H., Kuo, C., and Imler, W. (1996a). *Appl. Phys. Lett.* **68**, 3144.
Götz, W., Johnsson, N. M., Walker, J., Bour, D. P., and Street, R. A. (1996b). *Appl. Phys. Lett.* **68**, 667.
Grimmeiss, H. G., and Koelmans, H. (1959). *Z. Naturforshg* **14a**, 264.
Grimmeiss, H. G., and Monemar, B. (1970). *J. Appl. Phys.* **41**, 4054.
Gross, E., Permogorov, S., Travnikov, V., and Selkin, A. (1972). *Solid State Commun.* **16**, 1071.
Harris, C. I., Monemar, B., Amano, H., and Akasaki, I. (1995). *Appl. Phys. Lett.* **67**, 806.
Heitz, R., Thurian, P., Loa, I., Eckey, L., Hoffmann, A., Broser, I., Pressel, K., Meyer, B. K., and Mohkov, N. (1995). *Phys. Rev. B* **52**, 16508.
Henry, C. H., and Nassau, K. (1970). *Phys. Rev. B* **1**, 1628.
Hofmann, D. M., Kovalev, D., Steude, G., Meyer, B. K., Hoffmann, A., Eckey, L., Heitz, R., Detchprohm, T., Amano, H., and Akasaki, I. (1995). *Phys. Rev. B* **52**, 16702.
Hooft, G. W., van der Poel, W. A. J., Molenkamp, L. W., and Foxon, C. T. (1987). *Phys. Rev. B* **35**, 8281.
Hopfield, J. J. (1960). *J. Phys. Chem. Sol.* **15**, 97.
Hopfield, J. J., and Thomas, D. G. (1963). *Phys. Rev.* **132**, 563.
Hovel, H. J., and Cuomo, J. J. (1972). *Appl. Phys. Lett.* **20**, 71.
Hvam, J. M., and Ejder, E. (1976). *J. Lumin.* **12/13**, 611.
Ilegems, M., Dingle, R., and Logan, R. A. (1972). *J. Appl. Phys.* **43**, 3797.
Ilegems, M., and Dingle, R. (1973). *J. Appl. Phys.* **44**, 4234.
Ivchenko, E. L. (1982). In *Excitons* (E. I. Rashba and M. D. Sturge, eds.), North Holland Publishing Company, p. 141.
Jacobson, M. A., Nelson, D. K., Melnik, Yu. V., and Selkin, A. V. (1995). *Il Nuovo Cimento* **17D**, 1509.
Jahn, U., Menniger, J., Brandt, O., Yang, H., and Ploog, K. H. (1996). *Proc. 23rd Int. Conf. on the Physics of Semiconductors,* World Scientific, Berlin, p. 2857.
Johnsson, C., Lin, J. Y., Jiang, H. X., Khan, M. Asif., and Sun, C. J. (1996). *Appl. Phys. Lett.* **68**, 1808.
Kaufmann, U., Kunzer, M., Merz, C., Akasaki, I., and Amano, H. (1996a). *Presented at the MRS Symposium "Gallium Nitride and Related Materials,"* Boston, Nov. 27–Dec. 1, 1995, *Mater. Res. Soc. Symp. Proc.* **395**, 633.
Kaufmann, U., Dörnen, A., Härle, V., Bolay, H., Scholz, F., and Pensl, G. (1996b). *Appl. Phys. Lett.* **68**, 203.
Kawakami, Y., Peng, Z. G., Narukawa, Y., Fujita, S., and Fujita, S. (1996). *Appl. Phys. Lett.* **69**, 1414.
Kim, K., Lambrecht, W. R. L., and Segall, B. (1994). *Phys. Rev. B* **50**, 1502.
Kittel, C. (1963). Quantum Theory of Solids, Wiley, New York.
Klingshirn, C. F. (1995). "Semiconductor Optics," Springer Verlag, Berlin, p. 90.
Klann, R., Brandt, O., Yang, H., Grahn, H. T., and Ploog, K. (1995). *Phys. Rev. B* **52**, R11615.
Klochikhin, A. A., Permogorov, S. A., and Reznitsky, A. N. (1976). *Sov. Phys. JETP* **44**, 1176.
Koide, N., Kato, H., Sassa, M., Yamasaki, A., Manabe, K., Hashimoto, M., Amano, H., Hiramatsu, K., and Akasaki, I. (1991). *J. Cryst. Growth* **115**, 639.

Korona, K. P., Wysmolek, A., Pakula, K., Stepniewski, R., Baranowski, J. M., Grzegory, I., Lucznik, B., Wroblewski, M., and Porowski, S. (1996). Unpublished.
Koschnick, F. K., Spaeth, J. M., Glaser, E. R., Doverspike, K., Rowland, L. B., Gaskill, D. K., and Wickenden, D. K. (1995). *Mat. Sci. Forum* **197–201**, 37.
Kosicki, B. B., Powell, R. J., and Burgiel, J. C. (1970). *Phys. Rev. Lett.* **24**, 1421.
Kovalev, D., Averboukh, B., Volm, D., and Meyer, B. K. (1996). *Phys. Rev. B* **54**, 2518.
Kunzer, M., Kaufmann, U., Maier, K., Schneider, J., Herres, N., Akasaki, I., and Amano, H. (1994). *Mat. Sci. Forum* **143–147**, 87.
Lagerstedt, O., and Monemar, B. (1974). *J. Appl. Phys.* **45**, 2266.
Lagerstedt, O., and Monemar, B. (1979). *Phys. Rev. B* **19**, 3064.
Lambrecht, W. R. L., Segall, B., Rife, J., Hunter, W. R., and Wickenden, D. K. (1995). *Phys. Rev. B* **51**, 13516.
Lambrecht, W. R. L., Kim, K., Rashkeev, S. N., and Segall, B. (1996). Presented at the MRS Symposium "Gallium Nitride and Related Materials," Boston, Nov. 27–Dec. 1, 1995, *Mater. Res. Soc. Symp. Proc.* **395**, 455.
Lattice Dynamics (1965) (R. F. Wallis, ed.), Pergamon, Oxford, Sec C1.
Lee, J. W., Pearton, S. J., Zolper, J. C., and Stall, R. A. (1996). *Appl. Phys. Lett.* **68**, 2102.
Leszcynski, M., Teisseyre, H., Suski, T., Grzegory, I., Bockowski, M., Jun, J., Porowski, S., Pakula, K., Baranowski, J. M., Foxon, C. T., and Cheng, T. S. (1996). *Appl. Phys. Lett.* **69**, 73.
Li, J. Z., Lin, J. Y., Jiang, H. X., Salvador, A., Boytchkarev, A., and Morkoc, H. (1996). *Appl. Phys. Lett.* **69**, 1474.
Logothetidis, S., Petalas, J., Cardona, M., and Moustakas, T. D. (1994). *Phys. Rev. B* **50**, 18017.
Logothetidis, S., Petalas, J., Cardona, M., and Moustakas, T. D. (1995). *Mat. Sci. and Eng. B* **29**, 65.
Mahler, G., and Schröder, U. (1974). *Phys. Stat. Sol (b)* **61**, 629.
Maier, K., Kunzer, M., Kaufmann, U., Schneider, J., Monemar, B., Akasaki, I., and Amano, H. (1994). *Mat. Sci. Forum* **143–147**, 93.
Mang, A., Reimann, K., and Rübenacke, St. (1995). *Solid State Commun.* **94**, 251.
Matsumoto, T., and Aoki, M. (1974). *Jap. J. Appl. Phys.* **13**, 1804.
Menniger, J., Jahn, U., Brandt, O., Yang, H., and Ploog, K. (1996a). *Phys. Rev. B* **53**, 1881.
Menniger, J., Jahn, U., Brandt, O., Yang, H., and Ploog, K. (1996b). *Appl. Phys. Lett.* **69**, 836.
Meyer, B. K., Volm, D., Graber, A., Alt, H. C., Detchprohm, T., Amano, H., and Akasaki, I. (1995). *Solid State Commun.* **95**, 597.
Monemar, B. (1974). *Phys. Rev. B* **10**, 676.
Monemar, B., and Lagerstedt, O. (1979). *J. Appl. Phys.* **50**, 6480.
Monemar, B., Lagerstedt, O., and Gislason, H. P. (1980a). *J. Appl. Phys.* **51**, 625.
Monemar, B., Gislason, H. P., and Lagerstedt, O. (1980b). *J. Appl. Phys.* **51**, 640.
Monemar, B., Lindefelt, U., and Chen, W. M. (1987). *Phys. B* **146**, 256.
Monemar, B. (1988). CRC Critical Reviews in Solid State and Materials Sciences **15**, p. 111.
Monemar, B., Bergman, J. P., Buyanova, I. A., Amano, H., Akasaki, I., Detchprohm, T., Hiramatsu, K., and Sawaki, N. (1995a). *Proc. TWN 95 Nagoya, 21–23 Sept.* Solid State Electronics, **41**, 239.
Monemar, B., Bergman, J. P., Lundström, T., Harris, C. I., Amano, H., Akasaki, I., Detchprohm, T., Hiramatsu, K., and Sawaki, N. (1995b). *Proc. TWN 95 Nagoya, 21–23 Sept.* Solid State Electronics, **41**, 181.
Monemar, B. et al. (1996). Unpublished data.
Monemar, B., Bergman, J. P., Buyanova, I. A., Li, W., Amano, H., and Akasaki, I. (1996a). *MRS Internet Journal Nitride Semiconductor Research,* **1**, Article 2.
Monemar, B., Bergman, J. P., Amano, H., Akasaki, I., Detchprohm, T., Hiramatsu, K., Sawaki,

N. (1996b). In *Proc. Int. Symp. on Blue Laser and Light Emitting Diodes* (A. Yoshikawa *et al.*, eds.), Ohmsha Ltd., Tokyo, 135–140.
Nakamura, S., Senoh, M., and Mukai, T. (1991). *Jpn. J. Appl. Phys.* **30**, L1708.
Nakamura, S., Mukai, T., and Senoh, M. (1992). *Jpn. J. Appl. Phys.* **31**, 195.
Nakamura, S., Iwasa, N., Senoh, M., and Mukai, T. (1992). *Jpn. J. Appl. Phys.* **31**, 1258.
Nakamura, S., Senoh, M., Nagahama, S., Iwasa, N., Yamada, T., Matsushita, T., Sugimoto, Y., and Kiyoko, H. (1996). *Appl. Phys. Lett.* **69**, 1568.
Neugebauer, J., and Van de Walle, C. G. (1994). *Phys. Rev. B* **50**, 8067.
Okada, K., Yamada, Y., Taguchi, T., Sasaki, F., Kobayashi, S., Tani, T., Nakamura, S., and Shinomiya, G. (1996). *Jpn. J. Appl. Phys.* **35**, L787.
Osamura, K., Nakajima, K., and Murakami, Y. (1972). *Solid State Commun.* **11**, 617.
Pakula, K., Wysmolek, A., Korona, K. P., Baranowski, J. M., Stepniewski, R., Grzegory, I., Bockowski, M., Jun, J., Krukowski, S., Wroblewski, M., and Porowski, S. (1996). *Solid State Commun.* **97**, 919.
Pankove, H. P., Maruska, H. P., and Berkeyheiser, J. E. (1970). *Appl. Phys. Lett.* **17**, 197.
Pankove, J. I. (1971). "Optical Processes in Semiconductors," Prentice Hall, Englewood Cliffs, New Jersey.
Pankove, J. I. (1973). *J. Lumin.* **7**, 114.
Pankove, J. I., Duffy, M. T., Miller, E. A., and Berkeyheiser, J. E. (1973). *J. Lumin.* **8**, 89.
Pankove, J. I., Berkeyheiser, J. E., and Miller, E. A. (1974). *J. Appl. Phys.* **45**, 1280.
Pankove, J. I., and Hutchby, J. A. (1976). *J. Appl. Phys.* **47**, 5387.
Park, C. H., and Chadi, D. J. (1995). *Phys. Rev. Lett.* **75**, 1134.
Park, C. H., and Chadi, D. J. (1997). *Phys. Rev. B* **55**, 12995.
Pearton, S. J., Abernathy, C. R., Lee, J. W., Vartuli, C. B., Mackenzie, J. D., Ren, F., Wilson, R. G., Zavada, J. M., Shul, R. J., and Zolper, J. C. (1996). *Mater. Res. Soc. Symp. Proc.* **423**, in press.
Perlin, P., Suski, T., Teisseyre, H., Leszczynski, M., Grzegory, I., Jun, J., Porowski, S., Boguslawski, P., Bernholc, J., Chervin, J. C., Polian, A., and Moustakas, T. D. (1995). *Phys. Rev. Lett.* **75**, 296.
Permogorov, S. (1975). *Phys. Stat. Sol.* **68**, 9.
Permogorov, S. (1982). In *Excitons* (E. I. Rashba and M. D. Sturge, eds.), North Holland Publishing Company, p. 177.
Perry, W. G. (1996). Unpublishied data.
Petalas, J., Logothetidis, S., Boultadakis, S., Alouani, M., and Willis, J. M. (1995). *Phys. Rev. B* **52**, 8082.
Phillips, R. T., Lovering, D. J., Denton, G. J., and Smith, G. W. (1992). *Phys. Rev. B* **45**, 4308.
Pikus, G. E., and Ivchenko, E. L. (1982). In *Excitons* (E. I. Rashba and M. D. Sturge, eds.), North Holland Publishing Company, p. 205.
Ploog, K. H., Brandt, O., Yang, H., Menniger, J., and Klann, R. (1995). *Proc. TWN 95 Nagoya, 21–23* Solid State Electronics, in press.
Ponce, F. A., Bour, D. P., Götz, W., and Wright, P. J. (1996). *Appl. Phys. Lett.* **68**, 57.
Queisser, H. J., and Heim, U. (1974). *Ann. Rev. Mater. Sci.* **4**, 125.
Queisser, H. J. (1976). *Appl. Phys.* **10**, 275.
Ramirez-Flores, G., Navarro-Contreras, H., Lastras-Martinez, A., Powell, R. C., and Greene, J. E. (1994). *Phys. Rev. B* **50**, 8433.
Rashba, E. I., and Gurgenischvili, G. E. (1962). *Fiz. Tverdogo Tela 4 (1029),* Soviet Phys-Solid State **4**, 759.
Reynolds, D. C., Look, D. C., Kim, W., Ösgür, A., Botchkarev, A., Salvador, A., Morkoc, H., Talwar, D. N. (1996). *J. Appl. Phys.* **80**, 594.
Schmid, W., and Dean, P. J. (1982). *Phys. Stat. Sol. B* **110**, 591.

Schairer, W., and Schmidt, M. (1974). *Phys. Rev. B* **10**, 2501.
Schairer, W., Bimberg, D., Kottler, W., Cho, K., and Schmidt, M. (1976). *Phys. Rev. B* **13**, 3452.
Shan, W., Schmidt, T. J., Yang, X. H., Hwang, S. J., Song, J. J., and Goldenberg, B. (1995). *Appl. Phys. Lett.* **66**, 985.
Shan, W., Schmidt, T., Yang, X. H., Song, J. J., and Goldenberg, B. (1996). *J. Appl. Phys.* **79**, 3691.
Shikanai, A., Azuhata, T., Sota, T., Chichibu, S., Kuramata, A., Horino, K., and Nakamura, S. (1997). *J. Appl. Phys.* **81**, in press; Chichibu, S., Shikanai, A., Azuhata, T., Sota, T., Kuramata, A., Horino, K., and Nakamura, S. (1996). *Appl. Phys. Lett.* **68**, 3766.
Shionoya, S., Saito, H., Hanamura, E., and Akimoto, O. (1973). *Solid State Commun.* **12**, 223.
Smith, L. M., Wolford, D. J., Venkatasubramanian, R., and Ghandi, S. K. (1990). *Mater. Res. Soc. Symp. Proc.* **163**, 95.
Smith, M., Chen, G. D., Li, J. Z., Lin, J. Y., Jiang, H. X., Salvador, A., Kim, W. K., Aktas, O., Botchkarev, A., Morkoc, H. (1995a). *Appl. Phys. Lett.* **67**, 3387–3389.
Smith, M., Chen, G. D., Lin, J. Y., Jiang, H. X., Salvador, A., Sverdlov, B. N., Botchkarev, A., and Morkoc, H. (1995b). *Appl. Phys. Lett.* **66**, 3474.
Steiner, T., Thewalt, M. L. W., and Bhargava, R. N. (1985). *Solid State Commun.* **56**, 933.
Steiner, T., Thewalt, M. L. W., Koteles, E. S., and Salerno, J. P. (1986). *Phys. Rev. B* **34**, 1006.
Stepniewski, R., and Wysmolek, A. (1996). In press.
Strite, S., and Morkoc, H. (1992). *J. Vac. Sci. Technol.* **B10**, 1237.
Tchounkeu, M., Briot, O., Gil, B., Alexis, J. P., and Aulombard, R. L. (1996). *J. Appl. Phys.* **80**, 5352.
Teisseyre, H., Perlin, P., Suski, T., Grzegory, I., Porowski, S., Jun, J., Pietrazko, A., and Moustakas, T. D. (1994). *J. Appl. Phys.* **76**, 2429.
Teisseyre, H., Nowak, G., Leszcynski, M., Grzegory, I., Bockowski, M., Krukowski, S., Porowski, S. (1996). In press.
Thomas, D. G., and Hopfield, J. J. (1959). *Phys. Rev.* **116**, 573.
Tsaregorodtsev, A. M., Nikitina, I. P., Scheglov, M. P., and Zubrilov, A. S. (1995). Private communication.
Volm, D., Streibl, T., Meyer, B. K., Detchprohm, T., Amano, H., and Akasaki, I. (1995). *Solid State Commun.* **96**, 53.
Volm, D., Oettinger, K., Streibl, T., Kovalev, D., Ben-Chorin, M., Diener, J., Meyer, B. K., Majewski, J., Eckey, L., Hoffman, A., Amano, H., Akasaki, I., Hiramatsu, K., Detchprohm, T. (1996). *Phys. Rev. B* **53**, 16543.
Wallis, R. F. (ed.) (1965). *Lattice Dynamics*, Pergamon, Oxford.
Wetzel, C., Volm, D., Meyer, B. K., Pressel, K., Nilsson, S., Mohkov, E. N., and Baranov, P. G. (1994). *Appl. Phys. Lett.* **65**, 1033.
Wiesner, P., and Heim, U. (1975). *Phys. Rev. B* **11**, 3071.
Wysmolek, A., Baranowski, J. M., Pakula, K., Korona, K. P., Grzegory, I., Wroblewski, M., and Porowski, S. (1996). In *Proc. Int. Symp. on Blue Laser and Light Emitting Diodes* (A. Yoshikawa et al., eds.), Ohmsha Ltd, Tokyo, p. 492.
Yamada, Y., Yamamoto, T., Nakamura, S., Taguchi, T., Sasaki, F., Kobayashi, S., and Tani, T. (1996). *Appl. Phys. Lett.* **69**, 88.
Yu, P. Y., and Cardona, M. (1996). "Fundamentals of Semiconductors," Springer, p. 253.
Zubrilov, A. S., Nilolaev, V. I., Tsvetkov, D. V., Dmitriev, V. A., Irvine, K. G., Edmond, J. A., and Carter, Jr. C. H. (1995). *Appl. Phys. Lett.* **67**, 533.

CHAPTER 12

Band Structure of the Group III Nitrides

W. R. L. Lambrecht

DEPARTMENT OF PHYSICS
CASE WESTERN RESERVE UNIVERSITY
CLEVELAND, OHIO

I. INTRODUCTION 369
II. OVERVIEW OF CALCULATIONS 370
   1. *Early Semi-Empirical Studies* 370
   2. *Local Density Functional Calculations* 372
   3. Beyond LDA 378
III. RELATIONS BETWEEN BRILLOUIN ZONES OF WURTZITE AND ZINC-BLENDE 379
IV. TRENDS IN BAND STRUCTURE 385
V. EXPERIMENTAL PROBES 391
   1. *Photoemission* 391
   2. *UV Optics* 394
   3. *X-Ray Absorption* 397
   4. *Other Nitrides* 398
VI. DETAILS NEAR THE BAND EDGES 399
VII. OUTLOOK FOR FUTURE WORK 404
   References 405

I. Introduction

In this review, we present the current state of knowledge of the electronic band structure of group III nitrides as well as some historic perspective. The number of papers dealing with the electronic band structure of the nitrides and the variety of methods applied may seem bewildering to the nonexpert. It turns out that many of these are only "slight variations on a theme" and the good news is that most recent calculations agree pretty well with each other on the basic aspects. Thus, rather than attempting to be complete, we try here to highlight the key aspects and the most significant contributions.

In Section II, we give a brief historic overview of the most significant calculations, along with a short description of the various methods that have been employed.

In Section III, we turn to a discussion of the relation between the band structures in the two crystal structures of most practical interest, namely the zinc-blende and the wurtzite. As is well known, the III nitrides (except for BN which is more similar to diamond in many aspects, and is therefore mostly excluded from this review) naturally occur in the wurtzite structure, but the zinc-blende phase can be stabilized by epitaxial growth techniques. For GaN, this is done rather commonly and successfully, while for InN and AlN there have been only a few reports. We note that there is a third crystal structure of interest, namely the rocksalt structure, which occurs under high-pressure conditions. For electronic applications, however, the zinc-blende and wurtzite phases are far more important, and hence, we confine our discussion of the band structures to those two phases.

In Section IV, we discuss some of the trends in the band structures and related properties in terms of ionicity and cation.

In Section V, we discuss experimental probes of the band structures, such as X-ray and UV photoemission spectroscopy (XPS), UV reflectivity and spectroscopic ellipsometry (SE) and X-ray absorption near-edge spectroscopy (XANES).

Up to this point, we consider the band structures mostly at the scale of several eV. For most of the intended applications of the nitrides, the most important part of the band structure, however, is the region right next to the band edges, at a scale of meV. This pertains to questions on the effective masses, exciton splittings under strain, etc. Recently, several papers have appeared dealing with these aspects. A discussion of this work is given in Section VI.

Finally, in Section VII, we present an outlook for the future. What are still unresolved questions or where should we expect most progress in the coming years?

Overall, our presentation will focus on the qualitative aspects, and the issues that have been under discussion over the years, and will somewhat have the nature of a comment. While we include quantitative figures and tables where appropriate, the paper is not intended as a reference work for finding the most accurate, complete, or recent set of numerical values, and is hence not organized material by material. For that, we refer the reader to the original publications, and to the EMIS Datareview on Group III Nitrides (Lambrecht and Segall, 1994).

II. Overview of Calculations

1. EARLY SEMI-EMPIRICAL STUDIES

The first studies of the band structure of the nitrides appeared in the late 1960's to early 1970's. The work of Bloom (1971), Bloom *et al.* (1974), and Jones and Lettington (1969) used the semi-empirical pseudopotential

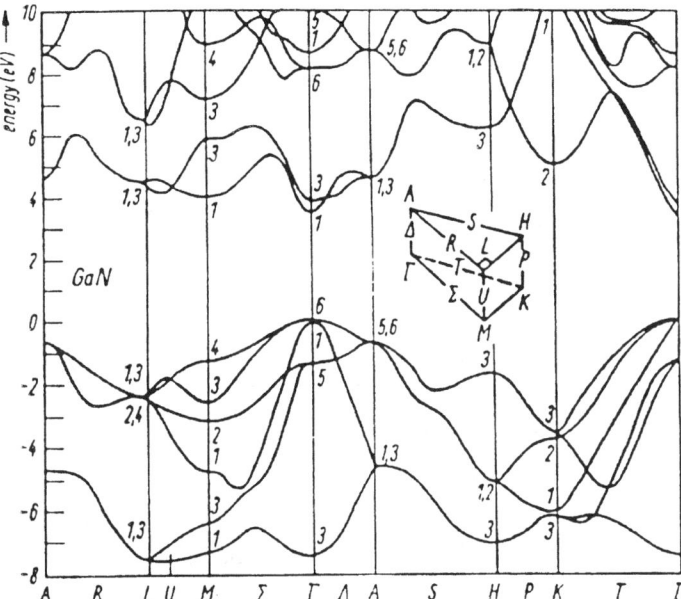

FIG. 1. Semiempirical pseudopotential band structure of wurtzite GaN by Bloom et al. (1974).

approach, in conjunction with optical reflectivity measurements. In this method, the Fourier components of the pseudopotential are adjusted in an iterative procedure, so as to adjust a few interband transitions to experimentally observed peak positions in reflectivity. The disadvantage of this approach, is that the assignment of experimental features to specific transitions is not always straightforward and not unique. Usually, similarity to other better known semiconductors is exploited to arrive at a consistent assignment.

The band structures obtained in this manner agree reasonably well with more recent calculations for the valence bands, but they differ significantly from the more recent calculations in the conduction bands. This is evident by comparing the bands of GaN in Fig. 1 with a recent calculation in Fig. 5 given below in Section III. Of particular importance, is that the energy difference between the lowest conduction band minimum at Γ and local minima at other k-points is much smaller in the older calculation. Also, the first Γ_6 band is much lower, and the second Γ_3 band below it is missing. Nevertheless, these band structures gave a fairly good account of the then available reflectivity spectra up to about 10 eV.

Other early calculations for GaN by Bourne and Jacobs (1972) using a muffin-tin potential differ substantially from the now accepted band structure. Even more recent semi-empirical pseudopotential calculations by Grinajew, Malachow, and Czaldyszew (1986) obtained an indirect gap incorrectly.

For AlN, there were even earlier Orthogonalized Plane Wave (OPW) calculations (Hejda and Hauptmanová, 1969), an approach which is a direct precursor of the pseudopotential methods.

For InN, the first experimental studies by Tyagai *et al.* (1977) suggested an indirect gap, while a later semi-empirical pseudopotential calculation by Foley and Tansley (1986) indicated a direct gap. These difficulties were in large part due to the problems in obtaining adequate samples.

In summary, the intrinsic ambiguities in the semi-empirical pseudopotential method left much uncertainty on even elementary questions on the band structures, such as direct or indirect nature of the gap. Results by different groups led to contradictory results, but nevertheless, provided important guidance to the optical studies of the time.

2. Local Density Functional Calculations

A boom in band structure calculations of the nitrides occurred in the late 1980's and early 1990's. During the intervening decade, the practice of band-structure theory had evolved considerably. The "standard" approach until this date is density functional theory (DFT) in the local density approximation (LDA) (Hohenberg and Kohn, 1965). Before discussing the application of this method to the nitrides, we briefly review some basics of the method.

a. Introduction to DFT, LDA, and GW Methods

In the present context, DFT can be thought of as a self-consistent field method for obtaining the crystal potential. The emphasis of band-structure theorists, however, has also mostly changed from electronic structure *per se* to using it as an intermediate step in calculating total binding energies of solids and from it, properties such as lattice constant, bulk modulus, elastic constants, phonon frequencies, etc. These are the properties, density functional theory is, strictly speaking, designed for. The electronc charge density is the basic quantity in this theory, and is determined by minimizing the total electronic energy of the solid. The expression for the total energy as functional of the density includes exchange and correlation effects in an average "electron-gas"-like way, which is referred to as the "local density approximation." In its most commonly used form, the charge density is expressed in terms of occupied one-electron wave functions (the so-called Kohn-Sham states) whose corresponding eigenvalues constitute the band structure. The Kohn-Sham equation reads

$$\left\{-\frac{\hbar^2}{2m}\Delta + v_H[n(\mathbf{r})] + v_{xc}[n(\mathbf{r})]\right\}\psi_{nk}(\mathbf{r}) = \varepsilon_{nk}\psi_{nk}(\mathbf{r}), \tag{1}$$

with $v_H[n(\mathbf{r})]$ and $v_{xc}[n(\mathbf{r})] = \delta E_{xc}/\delta n$ the Hartree and exchange-correlation potentials, which are both known functionals of the density. More precisely,

$$v_H[n(\mathbf{r})] = \int\int \frac{n(\mathbf{r})n(\mathbf{r}')}{|\mathbf{r}-\mathbf{r}'|}\, d^3r d^3r' \tag{2}$$

is known exactly, and is the classic Coulomb interaction and $v_{xc}[n(\mathbf{r})]$ is approximated by its corresponding expression from the homogeneous electron gas, where n is a constant, but now applied at each point in space. Hence, the name, *local* density approximation. The density itself must be determined self-consistently from

$$n(\mathbf{r}) = \sum_{nk}^{occ} |\psi_{nk}(\mathbf{r})|^2 \tag{3}$$

In a purist point of view, the band structures ε_{nk}, appearing in this theory have nothing to do with the single-particle excitations (or quasiparticles) of the system. In other words, they do not correspond to the energies for adding or removing a single electron from the system. They correspond even less to the two-particle excitations involved in the optical spectra. Nevertheless, it is common practice to view the Kohn-Sham band structure as a first approximation to the true band structure. The reason why this is reasonable, is that the true quasiparticle band structure can be obtained from the closely related Dyson equation,

$$\left[-\frac{\hbar^2}{2m}\Delta + v_H(\mathbf{r})\right]\Psi_{nk}(\mathbf{r}) + \int \Sigma_{xc}(\mathbf{r},\mathbf{r}',E_{nk})\Psi_{nk}(\mathbf{r}')d^3r' = E_{nk}\Psi_{nk}(\mathbf{r}), \tag{4}$$

the only difference being that the exchange-correlation potential is here replaced by a more complicated energy-dependent and non-local self-energy operator (Hedin and Lundqvist, 1969; Sham and Kohn, 1966). One can indeed view the exchange-correlation potential as a local approximation to the self-energy operator

$$\Sigma_{xc}(\mathbf{r}, \mathbf{r}', E) \approx \delta(\mathbf{r} - \mathbf{r}')v_{xc}[n(\mathbf{r})]. \tag{5}$$

The current state-of-the-art for calculating self-energies is the so-called "GW" approach, in which the letters stand for the symbols used in Hedin's original paper (Hedin and Lundqvist, 1969) on this method for respectively the one-electron Green's function and the screened Coulomb interaction. The GW approximation is the first term in a perturbation expansion of the self-energy, or, alternatively, can be viewed as a "screened" Hartree-Fock theory. The method was computationally developed by Hybertsen and Louie (1986), Godby, Schlüter, and Sham (1988), and more recently, in an alternative basis set formulation by Aryasetiawan and Gunnarsson (1994).

In semiconductors, the "GW" band structures differ from the LDA band structures primarily by a shift of the conduction band relative to the valence band. In other words, LDA Kohn-Sham band structures underestimate band gaps in semiconductors, which is a well-known problem (Perdew and Levy, 1983; Sham and Schlüter, 1983). We emphasize that it is not a fundamental problem of LDFT, but rather a question of "using the wrong equation." We immediately clarify that the gap underestimate is not the only change, or that the shifts are strictly k-independent. In fact, thay are not. Nevertheless, the LDA band structures + constant shift form a convenient starting point to analyze the true band structure.

Most band structure calculations for the nitrides have been of the LDA type, and are discussed below. Recent calculations beyond LDA are discussed in Subsection 3.

b. Introduction to Band-Structure Methods

A variety of band structure methods have been used within this general framework. We distinguish here between pseudopotential and all-electron methods. In the first of these, the atomic potential is replaced by a smooth (norm-conserving non-local) pseudopotential (Hamann, Schlüter, and Chiang, 1979; Kerker, 1980) whose valence-eigenstates agree with those of the true system outside the core of the atom, but differ from them in the core region which is unimportant for chemical bonding. Usually, the wave functions are expanded in plane waves although localized atom-centered basis sets such as Gaussian orbitals or mixed basis sets have also been used. There is some flexibility in how to construct the potential: one refers to hard or soft pseudopotentials depending on the core radius applied. In principle, the softer one makes the potential, the less it becomes transferable from one situation to another, although new approaches to overcome these difficulties have been introduced (Vanderbilt, 1990; Blöchl, 1994). At the same time, the softer the potential, the easier it becomes to do the calculation because fewer plane waves are required to obtain convergence of the wave functions. Thus, a soft pseudopotential may be more accurate than a hard pseudopotential when used with the same plane wave energy cut-off. These variations and the computationally affordable cut-offs explain the differences between the results of the earlier calculations on the nitrides using this approach.

Most all-electron methods treat the inner part of the atoms in another way, known as augmentation. That is, the actual solutions of the spherical part of the potential within a sphere around the atom and their energy derivatives are used to expand the wave functions and matched to envelope functions outside the spheres. These methods include the linear muffin-tin orbital (LMTO) and linearized augmented plane-wave (LAPW) methods (Andersen, 1975). While nowadays, these methods can be applied without shape approximations and are designated as full-potential (FP) methods (Methfessel, 1988), the earlier applications of LMTO made the so-called atomic sphere approximation (ASA). In the ASA, the crystal potential is approximated by a superposition of slightly overlapping spherical potentials centered on atomic sites in such a way that the spheres are filling space. For open structures, one usually also includes spheres centered on interstitial sites. It turns out that this potential is rather close to the true full potential (Andersen, Postnivok, and Savrasov, 1992), and much better than the original muffin-tin potentials which are constant in the interstitial region and spherical within non-overlapping spheres. In practice, it is usually combined with a spherical approximation to the charge density in the same spheres. While this is quite satisfactory for most aspects of the band structure to a precision of $\sim 0.1\,\text{eV}$ or better, it is not entirely satisfactory when addressing very delicate aspects of the band structure, which crucially depend on local symmetry breaking, or for total energy variations such as those involved in frozen phonon calculations. In particular, the very small total energy differences between the zincblende and wurtzite are not accurately obtained within ASA-LMTO. Another quantity that is very subtle is the crystal field splitting of the valence-band maximum. Thus, the early calculations of this type have the order of the doublet and singlet of wurtzite inverted for GaN.

c. Calculations for the Nitrides

With these considerations in mind, we now examine some of the calculations that have been performed for the nitrides. Gorczyca et al. (1992) and Christensen and Gorczyca (1994) applied the ASA-LMTO to the III-nitrides in a series of papers, that was mostly focused on the study of the high-pressure transition to rocksalt. A FP-LMTO study of AlN structural aspects under pressure was also performed by these authors (Christensen and Gorczyca, 1993) as well as a study of some of the important phonon modes (Gorczyca et al., 1995). Lambrecht and Segall (1992) applied ASA-LMTO to study the relations between zinc-blende and wurtzite band structures, discussed direct versus indirectness of the gaps and trends in the materials (Lambrecht, 1994). They went on to study photoemission (Lambrecht et al., 1994) and UV optical properties (Lambrecht et al., 1995). Kim, Lambrecht, and Segall (1994, 1996) subsequently used FP-LMTO calculations to study elastic constants and strain effects on the band structures. Fiorentini, Methfessel, and Scheffler (1993) also applied FP-LMTO to zincblende GaN and focused on the effects of Ga3d electrons. The latter were also discussed by Lambrecht et al. (1994). Muñoz and Kunc (1993) presented a FP-LMTO study for InN. These various all-electron calculations all included Ga3d and In4d band dispersion and hybridization with the rest of the bands.

The early norm-conserving pseudopotential calculations by Van Camp, Van Doren, and Devreese (1991), Muñoz and Kunc (1991), and Palummo et al. (1993) did not include the effects of the d-electrons. Reaching convergence in plane wave expansions proved difficult, because of the deep N pseudopotential and was continuously found to require higher cut-offs from one calculation to the next. Min, Chan, and Ho (1992) used a mixed basis set to overcome this difficulty, but obtained the somewhat unexpected result that zinc-blende GaN would be lower in energy than wurtzite, which was not bourne out by later well-converged plane-wave calculations (Yeh et al., 1992; Wright and Nelson, 1995a) and seems to contradict the experimental fact that natural bulk crystals of GaN have the wurtzite structure. Yeh et al. (1992) focused on precisely this question for a number of semiconductors including the nitrides. Their work is also the first to fully relax the structural parameters for wurtzite.

An issue that was discussed repeatedly between pseudopotential and all-electron practitioners is the role of the d-bands. The reason for this is that the semi-core Ga3d and In4d states overlap in energy with the deep N2s states. They thus play a non-insignificant role in the bonding. While this is also known to be the case in GaAs, it is more strongly pronounced here because of the close proximity of the Ga atoms, the lattice constants of these materials being rather small. The various ways in which d-electrons can be treated: as core orbitals with or without nonlinear core corrections (i.e., including or excluding core-valence exchange), or as pseudized valence states in pseudopotential methods; frozen core or relaxed core or valence with single or two-panel approaches in linear methods, lead to some confusions, which were discussed rather extensively by Fiorentini, Methfessel, and Scheffler (1993) and Lambrecht et al. (1994). In the pseudopotential framework, accurate calculations treating the Ga3d and In4d explicitly as valence states were performed by Wright and Nelson (1995a). These authors also pushed the plane wave cut-offs to 240 Ry, to insure convergence. An interesting aspect is that while d-electron hybridization is important in bonding, they nevertheless appear as separate states, when considering quasiparticle excitations in photoemission (Lambrecht et al., 1994).

Another aspect which influences the accuracy of the calculations is whether or not full structural relaxation is taken into account. One is faced with the dilemma of taking either the experimental crystal structure parameters or the theoretical equilibrium values obtained self-consistently from the density functional calculation. Typically, the local density functional calculations underestimate the lattice constants by a percent or so. Since the bandgap deformation potentials in these materials are substantial, that does affect the band structure considerably. For wurtzite, the structure is not entirely fixed by symmetry: the c/a ratio and the internal parameter u (defining the bond length along the c-axis by $d = cu$) need to be determined by total energy minimization. While c/a is well known from experiment, the internal parameter is more difficult to obtain accurately experimentally, and substantial variation exists among the reported experimental values. Hence, the most accurate calculations are those in which u is determined theoretically. Of course, accuracy should not be our only criterion. Understanding the trends is what will be emphasized in the later sections.

Other groups have presented pseudopotential calculations mainly as a test to go on to defect calculations and surface calculations (Neugebauer and Van de Walle, 1996; Boguslawski, Briggs, and Bernholc, 1995) dielectric constant and nonlinear susceptibility (Chen, Levine, and Wilkins, 1995), or phonon calculations (Miwa and Fukumoto, 1993; Karch et al., 1996). Other calculations worth mentioning are those of Corkill, Rubio, and Cohen, (1994), who addressed the role of d-bands on the gaps and Yeh, Wei, and Zunger (1994) addressing the relation of bandgaps in wurtzite and zincblende. Some other methods were also applied: LCAO with gaussian type orbitals by Huang and Ching (1985), Ching and Harmon (1986), Xu and Ching (1993), and tight-binding by Kobayashi et al. (1983) and Jenkins and Dow (1989).

Although slight variations persist among the results of the different methods, the origin of which is hard to trace back, the well-converged LDA calculations agree well on basic predictions of the topology of the bands, bandwidths, size of the bandgaps (to within a few 0.1 eV), lattice constants, bulk moduli and other total energy properties.

3. BEYOND LDA

A few calculations for the nitrides have been performed in the Hartree-Fock approximation rather than the local density functional theory. These include a calculation for GaN by Pandey, Jaffe, and Harrison (1993) and for AlN by Ruiz, Alvarez, and Alemany (1994). In Hartree-Fock, exchange is treated exactly, but correlations are completely neglected. Contrary to LDA, Hartree-Fock calculations overestimate the gaps of semiconductors.

As mentioned in the introduction of the previous subsection, a considerable better approach is the GW approximation. GW calculations were performed for wurtzite and zinc-blende GaN and AlN by Rubio et al. (1993), for zinc-blende GaN by Palummo et al. (1994, 1995). These calculations agree fairly well among each other in their basic predictions.

One finds that the valence bands shift down with respect to the LDA results, and the conduction bands shift up. The downward shift of the valence bands increases with increasing energy below the valence band maximum, and in particular, with the more localized character of these

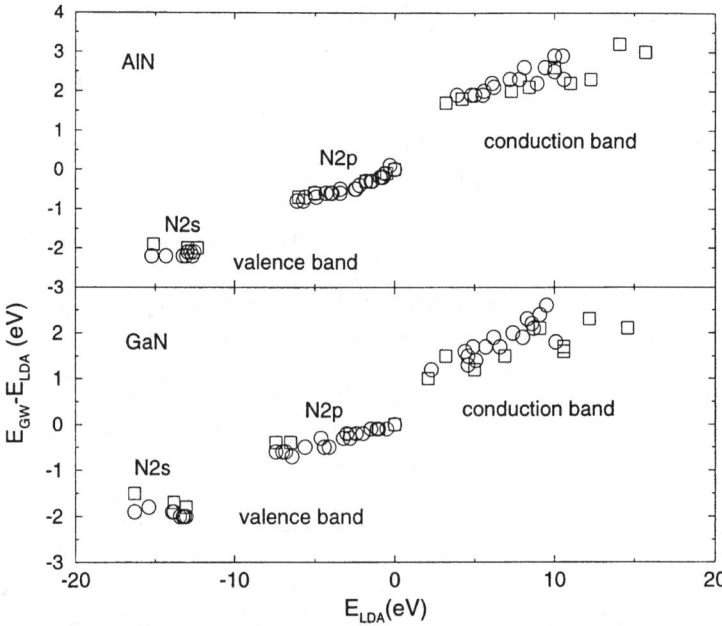

FIG. 2. Self energy corrections to the LDA band structures, that is, GW eigenvalues minus corresponding LDA eigenvalue as function of LDA eigenvalues for GaN and AlN from tabulations in Rubio et al. (1993). Circles correspond to wurtzite and squares to zinc-blende.

states. Thus, the N2s states shift by about 1.5–2 eV more than the valence band maximum. The bottom of the N2p valence band which is of mixed N2p cation-s character shift by about 0.5 eV more than the maximum. This can be seen clearly in Fig. 2, which is based on tabulated results by Rubio et al. (1993).

The absolute shift of the valence-band maximum is a somewhat more problematic quantity in current GW theories. Even for well studied materials such as Si (Godby, Schlüter, and Sham, 1988), the various calculations agree less upon this than for the gap corrections. In fact, it appears to depend sensitively on which parameterization is used for the LDA starting point of the calculation. The GW changes the gap by about 1 eV in GaN and about 2 eV in AlN, thus giving excellent agreement with experiment. The conduction band corrections are found to vary by a few 0.1 eV among different k-points and specific states, and appear to increase as one goes up in energy (see Fig. 2). However, strict tests of this cannot be done until these states have been determined more accurately experimentally, which is the subject of Section V.

III. Relations Between Brillouin Zones of Wurtzite and Zinc-blende

The relation between the band structures in zinc-blende (zb) and wurtzite (wz) can largely be understood in terms of band structure folding effects. While there are not exact relations, they are quite helpful in comparing the band structures. The relation is based on the fact that the wurtzite basal planes are essentially the same as the zinc-blende {111} planes, but with a different relative stacking, ABC in cubic and AB in wurtzite. Thus, the first step to obtain a common framework to compare the band structures is to use a set of cartesian coordinates for cubic materials which is more closely related to the one used for the wurtzite. We can do this by choosing a z' axis along [111] of cubic. In the (111) plane of zinc-blende and the (0001) plane of wurtzite, the crystal structure consists of buckled hexagonal rings of alternating nitrogen and cation. Thus, by aligning these hexagons, we specify completely the relation between the new set of coordinate axes in zinc-blende and those in wurtzite. Specifically, one has the rotation matrix:

$$\begin{pmatrix} x' \\ y' \\ z' \end{pmatrix} = \begin{pmatrix} 1/\sqrt{2} & -1/\sqrt{2} & 0 \\ 1/\sqrt{6} & 1/\sqrt{6} & -2/\sqrt{6} \\ 1/\sqrt{3} & 1/\sqrt{3} & 1/\sqrt{3} \end{pmatrix} \begin{pmatrix} x \\ y \\ z \end{pmatrix} \quad (6)$$

In other words, the following directions in zinc-blende and wurtzite corre-

TABLE I

Equivalence Between Wurtzite and Zinc-blende Symmetry k-Points, Assuming the Ideal c/a Ratio for Wurtzite

Zinc-blende	Wurtzite
$L_{\parallel}$	Γ
$L_{\perp}$	U at 2/3 of $M - L$
X	U at 2/3 of $M - L$
W	T at 3/4 of $\Gamma - K$
U, K	Σ at 3/4 of $\Gamma - M$

spond to each other:

$$\text{wurtzite } [0001] \parallel \text{zincblende } [111]$$
$$\text{wurtzite } [11\bar{2}0] \parallel \text{zincblende } [10\bar{1}]$$
$$\text{wurtzite } [1\bar{1}00] \parallel \text{zincblende } [1\bar{2}1]$$

The above transformation can now be applied to any k-point of the reciprocal space to find equivalences between cubic BZ high-symmetry points and their location in the hexagonal Brillouin zone. These are summarized in Table I. The relations are further clarified in Fig. 3.

In Figs. 4 and 5, we compare the zinc-blende and wurtzite band structures of AlN and GaN both shown in the same wurtzite BZ. These were obtained in the LDA using the FP-LMTO method at the experimental lattice constants and with optimized u values. Although deformation potential results obtained from these band structures were given elsewhere (Kim, Lambrecht, and Segall, 1996), these full and accurate band structure figures were not published before. One may now appreciate the similarities and the true differences brought about by the different structure without being confused by the conventional way of plotting these bands in their own and different BZs. Figures of the zinc-blende band structures in the conventional zinc-blende BZ can be found in Lambrecht and Segall (1994).

It should be understood, that when we plot the zinc-blende bands in a wurtzite Brillouin zone, we plot the bands of a unit cell twice the size of the primitive unit cell of ZB, and consisting of two {111} layers. Even so, the reciprocal lattice of this supercell does not correspond to the wurtzite one. We merely show the bands along the high-symmetry lines of wurtzite in k-space, which we can always do, even if this is not a proper BZ. This implies among others that zinc-blende bands do not have the full symmetry

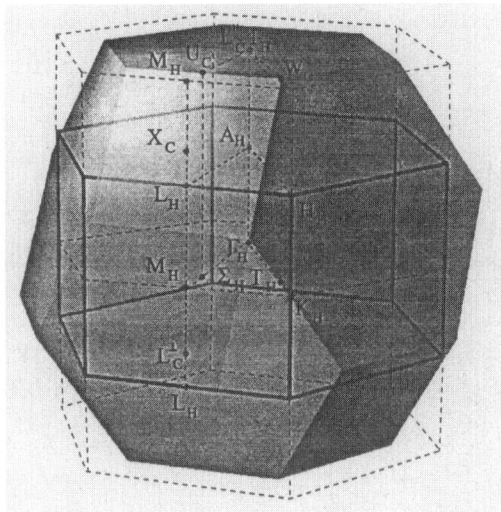

FIG. 3. Relation between zinc-blende and wurtzite Brillouin zone. Half of the adjacent wurtzite Brillouin zones along the c-axis are shown in addition to the central one. This way, one can see that both the X_C (center of the square face and $L_C^\perp$ point (center of one of the hexagonal sides) lie at 2/3 of the hexagonal ML_H line. The $L_C^\parallel$ point on the top and bottom hexagonal faces correspond to the center of the hexagonal BZ, for example, fold on Γ_H. One may also see that the cubic LW_C line will be folded onto the hexagonal $\Gamma - K_H$ line.

required of a hexagonal Brillouin zone. Thus, some of the eigenvalues at Γ_H are true Γ_C eigenvalues of zb and some are folded $L_C^\parallel$ points. Specifically, the Γ_3 states are essentially folded L_C states. Similarly, the bands at the point 2/3 of ML_H correspond to both the X_C and $L_C^\perp$ points of zb. While in the true hexagonal crystal, some of the hidden cubic symmetries are broken, there are also new symmetries related to the presence of a hexagonal axis not present in cubic. The wurtzite structure has point group C_{6v}. This is not a subgroup of the tetrahedral group T_d of zinc-blende. However, both have a common subgroup C_{3v}. In spite of these differences resulting from the different symmetry, the corresponding states are closely related in physical character. We now discuss some of these in more detail.

We first focus on the states near the gap at the Γ-point. The gaps obtained from our FP-LMTO band-structure calculations are given in Table II. One finds that the direct gap at Γ is slightly larger in wurtzite than in zinc-blende in all cases. The splitting of the Γ_{15}^v valence band maximum into Γ_6^v and Γ_1^v leads to a slightly repulsive interaction of the Γ_1^v state with the conduction band Γ_1^c, and hence, an opening of the gap. Since the Γ_1^v however, is mostly p_z-like, and the Γ_1^c state is mostly s-like, this effect is rather small.

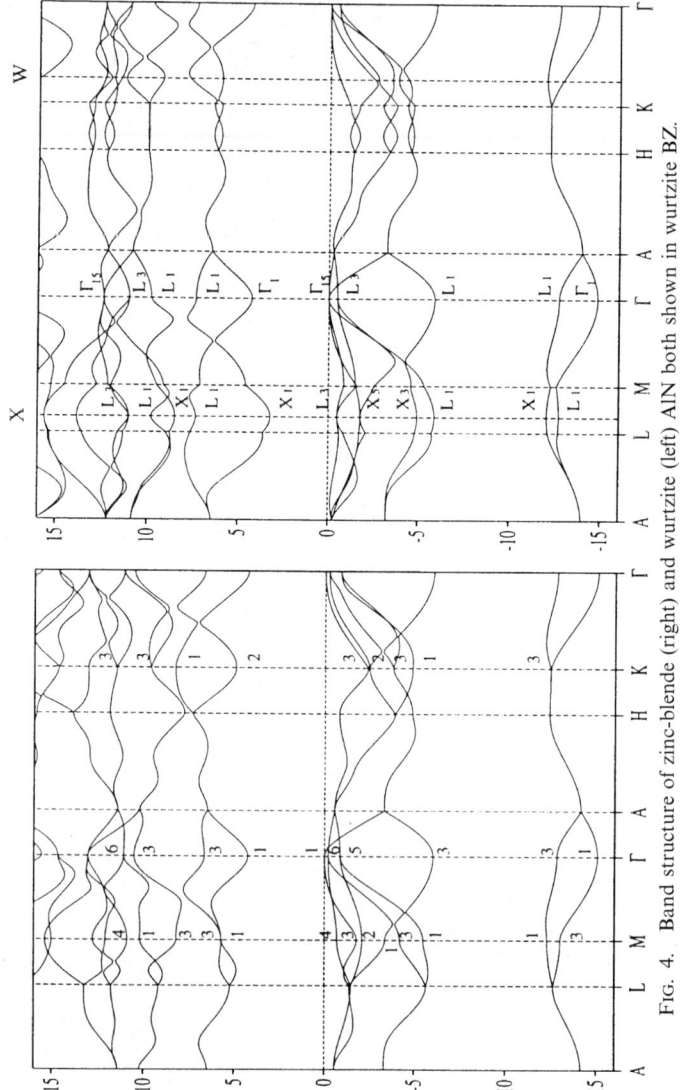

FIG. 4. Band structure of zinc-blende (right) and wurtzite (left) AlN both shown in wurtzite BZ.

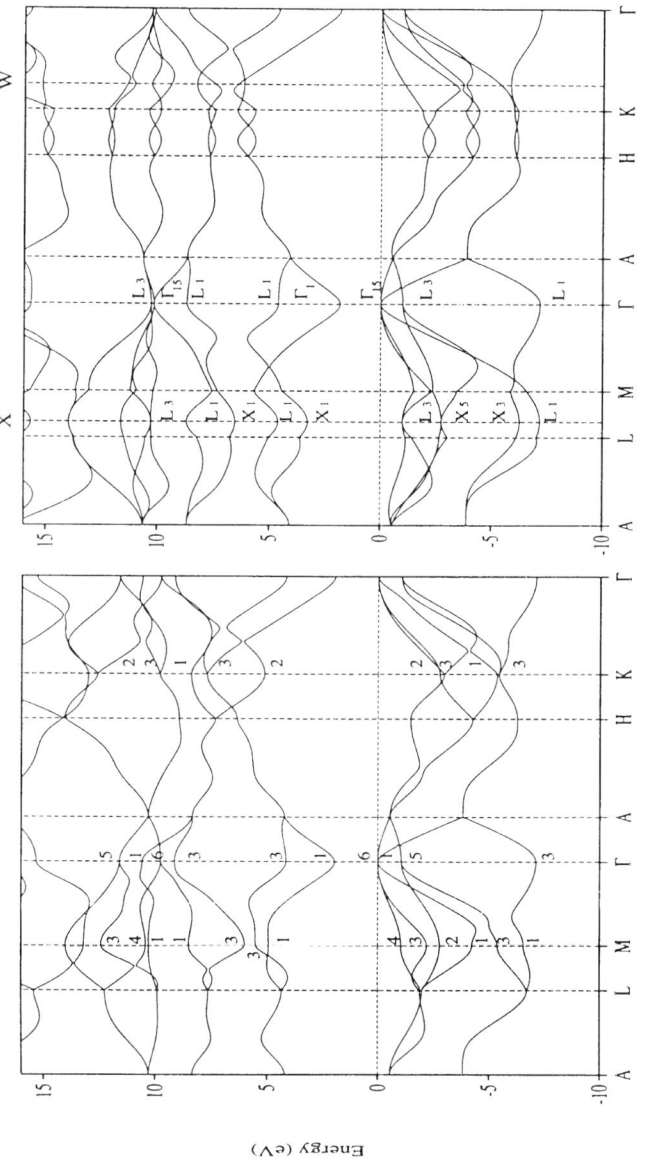

FIG. 5. Band structure of zinc-blende (right) and wurtzite (left) GaN both shown in wurtzite BZ. The lower N2s and Ga3d bands are not shown here.

TABLE II

LDA BANDGAPS IN WURTZITE AND ZINC-BLENDE (IN eV) FROM LDA FP-LMTO CALCULATIONS

	Zinc-blende	Wurtzite	Difference
AlN	4.236	4.255	0.019
GaN	1.763	1.935	0.172
InN	−0.399	0.064	0.463

Furthermore, it is partially cancelled by the fact that the minimum gap is actually the $\Gamma_6^v - \Gamma_1^c$ gap. Also, this ignores interactions with the higher lying Γ_{15}^c or rather its derived Γ_1^c conduction band. The latter lies farther away in the nitrides than the valence band maximum, and furthermore, has generally smaller interaction matrix elements with the state of interest. In earlier ASA-LMTO calculations, we obtained a value of about 0.3 eV for the gap difference in GaN (Lambrecht and Segall, 1994). Pseudopotential calculations (Palummo et al., 1993; Wright and Nelson, 1995a; Rubio et al., 1993) give values 0.1–0.2 eV. Experimentally, the most accurate values for the exciton gap in wurtzite GaN appears to be 3.48 eV (Eckey et al., 1996) and for zinc-blende 3.27 eV (Okumura et al., 1996), and 3.30 eV (Ramírez-Flores et al., 1994), giving a difference of 0.2 eV. In AlN, the c/a ratio is significantly lower than the ideal value. This leads to an inversion of the Γ_1^v and Γ_6^v states. This upward shift of the Γ_1^v tends to lower the gap.

Other notable differences are the lowering of the bands (at 2/3 ML_H equivalent to X when going to the cubic structure) and the lowering of the K_2 state in wurtzite. These effects are purely structurally and symmetry-lowering related. The same kind of shifts occur in other materials (e.g., diamond, SiC).

For example, X_C is a high symmetry point in cubic but not in wurtzite. Because of the higher symmetry of this state in the cubic material, fewer states between conduction band and valence band can interact. In fact, the minimum has X_1 symmetry and only interacts with the N2s like states about 12 eV below in the zinc-blende structure. In the wurtzite structure, the band at this same k-point have symmetry U_3 and hence can interact with N2p-like states in the valence band much closer by, which also have this same symmetry. Of course, we should also consider interactions with states higher up in the conduction band, which would have the opposite effect. However, in a $\mathbf{k} \cdot \mathbf{p}$ framework, these generally have smaller matrix elements. Hence, one expects these states to be pushed up in wurtzite because of their lower symmetry.

Similarly, we can see that the zinc-blende states at the K_{wz} point in the valence band are considerably more spread out, the highest one of them lying only 2 eV below the valence-band maximum instead of 3 eV in wz. One can think of this as being a result of increasing interactions between these bands. This in turn affects the interactions with the conduction band, pushing up the conduction band minimum at K in zb versus that of the K_2 state in wz. In wz, the latter has purely p_x, p_y symmetry with x and y in the c-plane. In zb, this symmetry is broken and additional interactions push up the state. We note in passing, that the lowest conduction band state at K appears to be rather sensitive to nonspherical corrections to the potential. It is significantly lowered in FP versus ASA calculations. The reason for this is not fully understood, but may be related to the pure p-like and hence strongly asymetric character of this state.

There is a slightly different way of viewing the zb bandstructure which may further clarify these relations. Consider the cubic structure in a supercell containing three cubic unit cells in ABC stacking along [111]. Now, consider a slight distortion of the lattice along this axis (as for example, produced by a uniaxial pressure). This structure obviously still exhibits the 3-fold symmetry C_{3v} and belongs to the rhombohedral (or trigonal) Bravais lattice. We can hence call it the 3R structure, using a nomenclature of polytypes often used in the SiC literature. The Brillouin zone corresponding to this supercell has the hexagonal shape of the wurtzite BZ, but its height is only 2/3 of that of the wurtzite BZ. The primitive unit cell of the rhombohedral lattice is actually three times smaller and corresponds in this case to the conventional fcc unit cell with a slight distortion along [111]. Hence, its BZ (the conventional fcc BZ) is folded three times into the smaller BZ corresponding to the 3-layer supercell along the [111] stacking axis. The X-point of zinc-blende according to Table 1 corresponds to a point 2/3 of the ML axis in wz. This point would, in fact, be folded down onto the M point in the 3R supercell BZ.

IV. Trends in Band Structure

By inspection of the band structures of these and related wide bandgap materials, it soon becomes clear that the valence band maximum in all these materials is at Γ while for the conduction band, there are a few k-points "in competition" for the minimum: the Γ point, the X point of zinc-blende and the K-point of wurtzite. Figure 6 shows the variation of these eigenvalues with cation.

This figure allows us to explain the directness or indirectness of the gaps. In BN, the conduction band minimum is at K in wurtzite and at X in

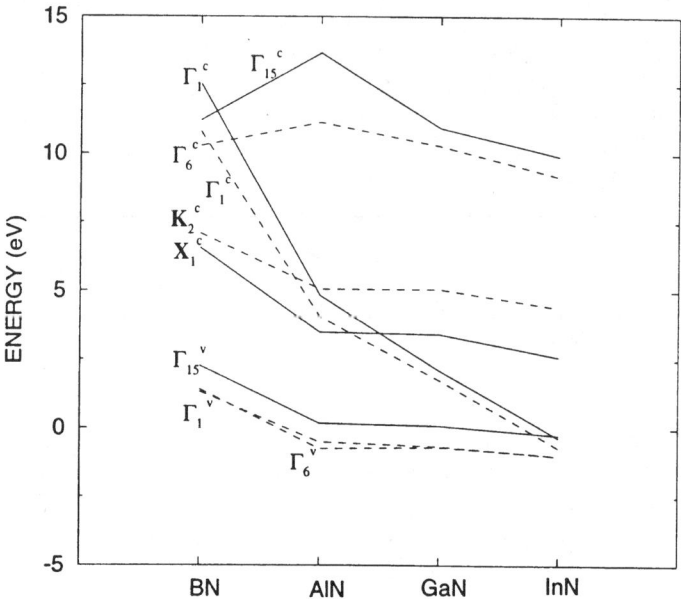

FIG. 6. Trend of band-edge eigenvalues of nitrides with cation. Solid line: zb eigenvalues, dashed line: wz.

zinc-blende. In AlN, the minimum is still at X in zinc-blende (but quite close to Γ), but switched to Γ in wurtzite. In GaN and InN, the minimum is at Γ in both cases. The reason for this trend is that the Γ_1^c states have a strong cation s-character. As one moves down in the periodic table, these states are lowered in energy with respect to corresponding p-like states. In other words, the s-p splitting is increasing. This is because s-states are nonzero at the nucleus and hence feel a stronger and stronger attractive potential as the atomic number Z increases. The states at X, on the other hand have mixed s-p character and also a strong component in the interstitial region. The relevant state at K has purely p-character. For second row elements such as C, B, and N, the p and s-states are particularly close in spatial and energetic extent. This is responsible for the k-point location of the minimum in BN. In fact, one can see in Fig. 6 that also the p-like conduction band states Γ_{15}^c in zb and Γ_6^c in wz are below the s-like Γ_1^c state in BN. The conduction-band minima locations in BN is the same as in SiC, and nearly the same as in diamond (because of the higher symmetry, in diamond, the conduction band moves slightly away from X leading to the well-known camel-back structure). The closeness between s and p-states in first row is also what leads to competition with π bonding and hence layered graphitic structures of carbon and BN.

The question discussed in the previous paragraph is not entirely straightforward. In fact, the problem is the choice of reference level. In a periodic crystal, there is no well-defined absolute reference level for the electrostatic potential (Kleinman, 1981). So, strictly speaking, the question of whether the gaps decrease from AlN to InN because the valence band goes up or the conduction band goes down cannot be answered meaningfully. One can only discuss such questions quantitatively when considering the two systems in contact, which leads to the band-offset problem at heterojunctions. That question has been discussed elsewhere (Albanesi and Lambrecht, 1994; Lambrecht et al., 1996a). Here, we are only interested in a qualitative insight and some choice of reference level can be made. The choice adopted in Fig. 6 is the ASA reference level, which is the average electrostatic potential of the net charges per sphere and naturally suited for this purpose. One can think of this reference level as being closely related to the average potential in the interstitial region. On the other hand, Corkill, Rubio, and Cohen (1994) have argued that the gap reduction is related to the effect of the semicore d-electrons pushing up the valence band maximum. We have shown elsewhere (Lambrecht et al., 1994) that this leads to only small direct effects near the valence band maximum once their effect on the bonding and hence lattice constant is properly taken into account. Also, under the superposition that the main effect comes from the d-states, it is difficult to understand where the difference between InN and GaN comes from.

It may be asked why in AlN, where the X minimum and Γ minimum are close in zb, the Γ point nevertheless wins out in wz. In fact, we have seen in the previous section that the Γ gap in wz is slightly larger than the zb gap (although by a very small amount). So, why does not the folded X state at $U = (2/3)ML$ end up below Γ in wz? Yeh, Wei, and Zunger (1994) have noted that in several compounds, the wz minimum at U appears to be the average of the folded X_c and L_c states. Since the folded L state is well above Γ and X, this then explains why the U state in wz is pushed up. However, there is no fundamental reason why this "average rule" should hold. We have argued above that the symmetry breaking is responsible for the X-folded state being pushed up in wz. We have also shown above that the symmetry breaking at Γ results only in a very small increase in gap because the states still are essentially of different angular momentum character.

The next question, one might ask is why this variation of gaps with cation is so much stronger in the nitrides series than for example in the As or P series. The reason is the high ionicity of the nitrides. As a result, the Γ_1^c

conduction band minimum has a more pure cation s-like character with less covalent admixing of the anion s-state. Hence, there is a greater sensitivity to the cation s-level trend with cation.

The ionic character of the nitrides deserves some further discussion. Several ionicity scales are in use, some of which are given in Table III for the nitrides and for comparison also for GaAs and ZnSe, a typical III–V and II–VI compound. In the popular Phillips scale of ionicity (noted below as f_i) (Phillips, 1973), the nitrides are less ionic than II–VI materials such as ZnSe, and slightly more ionic than other III–Vs such as arsenides or phosphides. The same is true in Harrison's sp^3 hybrid polarity scale (Harrison, 1989). In both of these scales, polarity* is determined by the ratio of the heteropolar part of the gap C to the total average bonding anti-bonding gap $\sqrt{E_h^2 + C^2}$, with E_h the homopolar part of the gap (corresponding to the group IV element of the appropriate row of the periodic table). The difference is that Phillips's scale is based on optical properties of semiconductors while Harrison's is based on atomic energies. In Harrison's definition, the heteropolar component of the gap is the difference in sp^3 hybrid energy levels of cation and anion and the covalent gap is based on the covalent interaction V between sp^3 orbitals pointing toward each other. Specifically,

$$\alpha_P = (E_c - E_a)/\sqrt{(E_c - E_a)^2 + 4V^2}. \tag{7}$$

One can obtain a similarly defined polarity scale based on ASA-LMTO calculations by making a transformation to a minimal basis set (Lambrecht and Segall, 1990). Unlike Harrison's polarity, this definition uses an internal reference level rather than the vacuum level for the orbital energies. More importantly, it is based on orbitals assigned to equal size atomic spheres rather than atomic orbitals. Garcia and Cohen (1993) introduced an ionicity scale based on the asymmetry in the charge distribution along the bonds, which is denoted as g in Table 3.

Both the ASA-LMTO polarity and Garcia and Cohen's ionicity scale assign a significantly more ionic character to the nitrides than to the II–VI compound ZnSe. In this view, the reason for the high ionicity of the nitrides is that the N2s and N2p levels are very deep compared to corresponding P3s and P3p valence states in phosphides and As4s and As4p states in arsenides. This in turn is due to the lack of a lower p-like core state. This reflects itself in the charge distribution when considering the latter in an unbiased way. That is the case in Garcia and Cohen's definition and in the ASA-LMTO based scale because the latter uses equal sphere sizes on anion

---

*Phillips ionicity is formally the square of what is here defined as polarity.

TABLE III

IONICITY AND RELATED PROPERTIES. ASA-LMTO BASED POLARITY α_p^{ASA}, GARCIA IONICITY g, HARRISON POLARITY α_p^H, SQUARE ROOT OF PHILLIPS IONICITY f_i, EFFECTIVE CHARGE PARAMETER S, WURTZITE STABILIZATION ENERGY $E_{wz} - E_{zb}$, TRANSITION PRESSURES TO ROCKSALT, AND BAND SPLITTINGS AT X

	α_p^{ASA}	g	α_p^H	$\sqrt{f_i}$	S^a	$E_{wz} - E_{zb}^{\,b}$ (mev/atom)	p_t^c (GPa)	$X_3^v - X_1^v$ (eV)	$X_3^c - X_1^{c\,d}$ (eV)
BN	0.475	0.484	0.41	0.506	0.86	20	850	5.6	4.9
AlN	0.807	0.794	0.59	0.670	1.55	−18.41	16.6	7.2	5.4
GaN	0.771	0.780	0.62	0.707	1.25	−9.88	51.8	6.4	3.5
InN	0.792	0.853	0.64	0.760	1.98	−11.44	21.6	7.0	3.3
GaAs	0.404	0.316	0.50	0.557	0.44	12.02		4.0	0.4
ZnSe	0.663	0.597	0.72	0.794	0.69	5.3		5.8	0.6

[a] From Kim, Lambrecht, and Segall (1996) for nitrides, and Martin (1970) for other.
[b] From Yeh et al. (1992) except for BN which is from Lam, Wentzcovitch, and Cohen (1990).
[c] From Christensen and Gorczyca (1994).
[d] From Lambrecht and Segall (1994).

and cation. This special character of second row elements of the periodic table is not captured well by the Phillips and Harrison scales.

Now, we discuss the relative ionicity among the nitrides. All scales agree that BN is the least ionic. From the bond-orbital point of view, this is because B also has quite deep levels being from the same row of the periodic table as N. The remaining nitrides have rather close ionicities with all models placing GaN in the middle. They differ however, in predicting either Al or In to be the most ionic. In any case, the difference in ionic character between these two appears to be quite small.

To judge ionicity scales, one needs to look at their predictive qualities for directly measurable quantities. One such quantity which is intuitively associated with ionicity is the dynamic effective charge Z^* and the associated parameter S, defined by Martin (1970),

$$S = Z^{*2}/\varepsilon = (\Omega/4\pi e^2)\mu(\omega_l^2 - \omega_t^2), \tag{8}$$

where Ω is the unit cell volume, μ the reduced mass, ε the electronic dielectric constant (i.e., at high frequency with respect to phonons but low frequency with respect to optical transitions), and ω_l and ω_t the longitudinal and transverse optic phonon frequencies at Γ. One can see in Table III, that S is significantly lower for GaAs and ZnSe than for the nitrides (in agreement with ASA-LMTO and Garcia scales) and follows exactly the same order as Garcia's scale for the nitrides.

High ionicity is also closely related to the preference for the hexagonal stacking of wurtzite. While this is a rather subtle problem requiring a delicate balance of band structure and electrostatic terms in the total energy, Cheng, Needs, and Heine (1987) showed that ionicity favors wurtzite from the point of view of the Madelung energy. A more elaborate relation to ionicity scales and atomic orbital radii was discussed by Yeh et al. (1992). Insofar as one considers the energy difference between wz and zb as a measure of ionicity, their calculations also predict the order of increasing ionicity to be B–Ga–In–Al. Also, the considerably higher ionicity of the nitrides than ZnSe is consistent with that materials preference for zincblende.

It is well known that high ionicity favors the rocksalt structure. At high pressure bonds tend to become even more ionic. Thus, one expects that the transition pressures from tetrahedrally bonded to rocksalt structure would decrease with ionicity. Table III shows that the same non-monotonic order of ionicities with cation is followed by the transition pressures as by the ASA-LMTO scale.

The previously discussed quantities related to ionicities reflect in some sense average properties of the electronic structure. Returning now to the band structure itself, there are some band splittings that can be viewed as direct measurements of the ionic character, because they are zero in the purely covalent semiconductors by symmetry, and therefore, are related to the asymmetric part of the potential. This is, for example, the case for the $X_3 - X_1$ splittings in zinc-blende. As Table 3 shows, the $X_3^v - X_1^v$ splitting, which is close to the ionicity gap between N2s and N2p like bands, follows exactly the same order as the ASA-LMTO polarity scale. The $X_3^c - X_1^c$ splitting on the other hand is smaller in GaN and InN than in BN. Of course, there is no reason why the band splittings of specific k-states should reflect exactly the same polarity as the average overall ionicity in the bonding that is embodied in the above ionicity scales. Nevertheless, the large values of the $X_3^c - X_1^c$ splittings compared to GaAs and ZnSe still reflect the high ionicity of the nitrides.

V. Experimental Probes

In this section, we discuss the experimental information on the band structures. Unlike in a naive single particle picture, we emphasize that there is no such thing as "the experimental band structure." While intuitively, it is clear that optical measurements probe differences between eigenstates of the system and hence should provide us with information on the band structure, such as the various gaps at different k-points, it is important to keep in mind, that we are in reality, probing a many-body system of interacting particles, which is only approximately described by the one-electron theory.

We have already outlined the general theoretical framework above. The experimental band structure could be defined as the spectrum of quasiparticle excitations from the ground state. In other words, it is the energy required to extract a particle or to add a particle to the system. In other words, they are differences of total energies of the N and $N - 1$ or N and $N + 1$ particle system. This spectrum is essentially obtained by photoemission and inverse photoemission spectroscopies respectively.

1. PHOTOEMISSION

The many-body problem consists in the fact that when we extract an electron, the remaining electrons respond to this change. For example, if we extract an electron from a strongly localized state, such as a core-state, the remaining electrons will screen the hole left behind in a timeframe short with

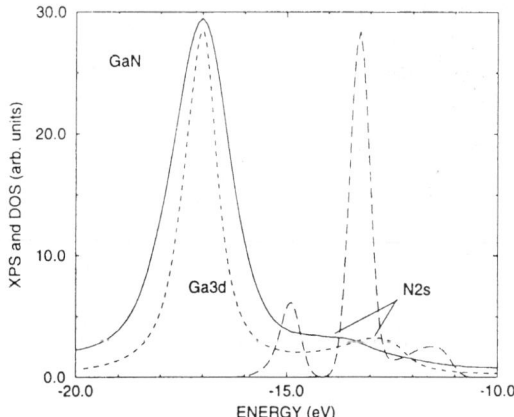

FIG. 7. XPS spectrum of w-GaN compared to calculated density of states. Solid line: experiment from Lambrecht *et al.* (1994); long-dashed line: LDA band theory with Ga3d band at -13 eV and forming bonding and antibonding states with N2s at -15 eV and -11.5 eV; short-dashed line: theory with shifted Ga3d band center by calculated self-energy shift of 4 eV. All energies are referenced to the valence band maximum.

respect to the time it takes the hole to delocalize over the solid. In other words, the translational symmetry of the solid is broken. Hence, one can think of this almost as of an impurity problem (the impurity being the atom which misses a core electron) and the binding energy of that core electron can thus be calculated as the energy of formation of such an impurity. This approach, which is called the ΔSCF approach, applies well, even to the semicore Ga3d and In4d electrons, and explains why these states appear about 4 eV lower than predicted by the straightforward LDA band theory. This is shown in Fig. 7. Further details can be found in Lambrecht *et al.* (1994).

For delocalized states, the above effects become in-principle negligibly small and even LDA theory should provide an adequate first approximation to the excitation spectrum, except for the gap correction. By comparing the density of states calculated from the Kohn-Sham LDA band structure with photoemission spectra, we can obtain information on the expectation value of the difference between the true self-energy operator and its LDA approximation. Figure 7 shows that the N2s peak from LDA (even after correcting for the Ga3d) still is shifted by about 1.2 ± 0.2 eV from its experimental value.

Figure 8 shows this comparison for AlN, where the situation is clearer because there is no semicore d-band. One can see that when the valence-

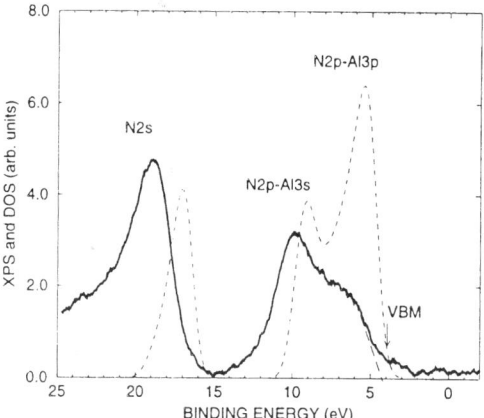

FIG. 8. X-ray photoemission spectrum and Density of States in AlN from King *et al.* (1996).

band maximum is aligned, the states near the bottom of the first band are shifted down in the experiment by about 0.5 eV and the N2s-like band is 1.9 eV underbinding in the LDA. Thus LDA underestimates bandwidths. These shifts are in quite satisfactory agreement with the GW calculations of Rubio, discussed in subsection 3. The N2s peak turns out to be near the X_1^v (in zb) or the deep M_1^v eigenvalues (in wz) for which they find GW corrections compared to LDA of 1.9 ± 0.1 eV in GaN and 2.0 ± 0.2 eV in AlN. Nevertheless, the GW calculations appear to slightly overestimate these shifts. This may be due to the fact that GW calculations are not entirely self-consistent but evaluated essentially by perturbation theory from LDA results.

We have here focused on recent XPS data that we could compare directly to the theory. Similar spectra have been obtained by UPS for GaN by Hunt *et al.* (1993), by XPS for GaN by Hedman and Martensson (1980), by XPS and UPS for AlN by Gautier, Duraud, and Le Gressus (1987), and for all three nitrides by Martin *et al.* (1996).

Very recently, a photoemission study of the band dispersion in zb GaN was carried out using synchrotron radiation by Ding *et al.* (1996). They used an angular resolved normal emission with varying photon energy in the range 30–80 eV. By assuming a parabolic final state band, one can then deduce the dispersion of the initial state band from the dispersion of the peaks with photon energy. They found the dispersion of the Δ_1 branch to coincide closely with LDA theory, placing the X_3^v band at -6.1 eV below the valence-band maximum, in close agreement with the FP-LMTO calculations of Fiorentini, Methfessel, and Scheffler (1993), and our own group (Fig. 5). The GW calculations of Rubio et al. (1993) predict a 0.4 eV shift from the LDA values for this state and furthermore, their LDA value appears to be too low resulting in a value of -6.9 eV for this state. For the X_5^v state, the experiments find a higher value than the FP-LMTO calculations by about 0.8 eV. Again, GW calculations would further shift this band down by 0.2 eV. This discrepancy, however, as pointed out in Ding et al. (1996) may be due to experimental problems in resolving the peak position near the valence-band maximum. For the N2s–Ga3d region, they confirm the results of the earlier study presented above, that the Ga3d band lies well below the N2s band at respectively -17.7 eV and -14.2 eV. They find, however, a signficantly smaller dispersion of the N2s band than our theoretical predictions, only 0.65 eV compared to about 3 eV in the theory. We note, however, that the N2s band is only clearly resolved as a peak near the X_1^v state. Near the Γ_1^v state, the N2s peak only appears as a shoulder on the dominating Ga3d peak. This can also be seen from the errorbars on their measurements.

From the above, it appears that the occupied valence states and even semi-core states are quite well understood. Unfortunately, the situation is not as clear for conduction band states. At this time, there are to our knowledge, no inverse photoemission data available for the nitrides. Hence, we must obtain our information more indirectly from optical measurements. This has the added complication that one really measures two-particle excitations: in other words, the electron and hole interact with each other and produce excitonic effects. While these are clearly important near the band edges, where they lead to well defined bound states, they might be hoped to be somewhat less important within the continuum. We will see that even this is not quite the case in GaN!

2. UV Optics

Since the most reliable data are available on GaN, we focus first on GaN. UV reflectivity measurements for GaN for a range of 10–30 eV have been available since 1980 (Olson, Lynch, and Zehe, 1981). Recently, there has been renewed activity. In Lambrecht et al. (1995), a detailed analysis is presented of UV-reflectivity measurements in the range 0–35 eV using synchrotron radiation for wurtzite GaN. The experimental data were in good agreement with the older data, but had previously not been analyzed in similar detail. The analysis is based on calculated optical response functions using an ASA muffin-tin orbital basis set and using the LDA except for a constant shift of the conduction band states. Spectroscopic ellipsometry measurements were presented for both zinc-blende and wurtzite GaN by Logothetidis et al. (1994), Petalas et al. (1995), and Janowitz et al. (1994). Petalas et al. (1995) includes comparisons to FP-LMTO calculations of the optical response function. Christensen and Gorczyca (1994) used essentially the same computational approach as Lambrecht et al. (1995) for optical response functions of the whole series of III-nitrides. As expected, the calculated response functions agree very well between these two sets of calculations. There are nevertheless, some differences in assignments of peaks to particular parts of the BZ, reflecting the fact that this interpretational part of the work is not an entirely automatic and straightforward task. From both our (Lambrecht et al., 1995) work and the work of Christensen and Gorczyca (1994), it is clear that the peaks in optical response functions do not, generally speaking, correspond to well-defined transitions at high-symmetry k-points, but rather to extended regions of k-space. They are dominated by joint density of state effects, that is, extended regions of nearly parallel bands. The peaks in the measurements often correspond to superpositions of different contributions. Thus, the assignments to high-symmetry k-point transitions on the basis of energy coincidence and visual inspection of the bands as given in Logothetidis et al. (1994) is clearly a huge oversimplification.

At present, the various papers present somewhat conflicting interpretations. Both Petalas et al. (1995) and Janowitz et al. (1994) compare their measured results directly to LDA, calculated response functions, in the latter, to the calculated results of Christensen and Gorczyca (1994). In that case, in zinc-blende, the peaks around 7.0–7.5 eV (labeled E_{1C} and E_{2C}) agree well with theory, while higher peaks and the minimum gap are underestimated. In wurtzite, the experimental E_{1H}, E_{2H} peaks resolved in the spectroscopic ellipsometry are aligned with the peaks labeled D and E in Lambrecht et al. (1995), while if we shift up the theory by 1 eV, they align with C and D. These are reproduced in Fig. 9. The (rather weak in theory) peak C is also present in the FP-LMTO calculated spectra of Petalas et al. (1995), but is somchow ignored in this work, presumably under the assumption that it would be obscured by the broadening.

In our view, this interpretation is not very satisfactory. For wurtzite, we obtain consistent results for the minimum gap and for transitions up to 15 eV when including a 1 eV upward shift of the gap. Also, the Ga 3d transitions to the conduction band (in the range 20–30 eV) are clearly identified, and fall in the right place when including the 1 eV shift of the conduction band and the 4 eV downward shift of the corehole consistent with XPS. As mentioned earlier, a gap correction by 1 eV is required by GW calculations. Unlike the GW calculations discussed in Section 3, however,

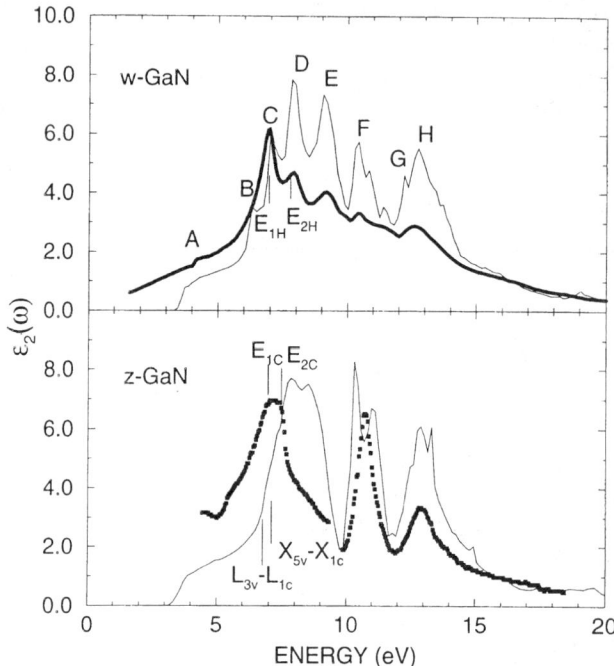

FIG. 9. UV optical response functions of wurtzite and zincblende GaN. Thick lines: experiment from Lambrecht *et al.* (1995) and Janowitz *et al.* (1994) for wz and zb respectively. For label explanation, see text.

one finds no clear trend for an increasing "correction" with increasing energy into the conduction band.

For zinc-blende GaN, there is, however a clear discrepancy in that the E_{1C}, E_{2C} transitions appear almost 1 eV below their prediction from the LDA + constant (1 eV) shift model. This can be seen in Fig. 9 in the lower panel. Clearly, this cannot be attributed to non-constant GW corrections because the GW calculations would predict it to shift even higher. Our proposed explanation is that this reflects the importance of two-particle response function effects. In other words, excitonic effects (electron-hole interaction) play a significant role even in the continuum. This is equivalent to the statement that our present calculation of linear response which is at the RPA level does not include exchange or electron-hole coupling effects. This is actually well known to affect the detailed shapes of the spectra in other semiconductors (del Castillo-Mussot and Sham, 1985; Hanke and Sham, 1980), but appears here to take more the appearance of a peak shift rather than a peak amplitude change. This is related to the fact that the excitonically shifted peaks appear in a region of low spectral density. This in turn, is related to the shape of the bands near the Γ_1 minimum. The effective mass is rather low and the next minima occur at significantly higher energy (several eV). This leads to a rather low spectral density between the minimum gap and the next transitions, with strong spectral weight in the E_{1C}, E_{2C} region. Furthermore, the energetic proximity of E_{1C} and E_{2C} in this material, the fact that they both form part of an extended region in k-space of very nearly parallel bands, and the small screening in GaN (due to the high ionicity) all help to increase excitonic effects. Further detail is provided in Lambrecht *et al.* (1996a). We note that similar effects probably play a role in wurtzite. In that case, however, they rather appear as discrepancies in peak amplitude rather than shifts. They correspond to the significant increase in spectral strength of peak C in the experiment compared to the theory. This may appear as a less dramatic discrepancy than for zinc-blende, because intensities are always more difficult to describe and are also affected by surface roughness (Lambrecht *et al.*, 1995), but is really a manifestation of the same underlying physics of shifting of spectral weight by the excitonic effects. Including these properly will also require to include local-field corrections.

3. X-Ray Absorption

There remains the question of whether the conduction band shifts are really constant up to about 10 eV in the conduction band or are increasing as predicted by GW, and then possibly recorrected by excitonic effects. To that end, it is useful to consider a more direct probe of conduction band states, namely X-ray absorption near edge fine structure, known as NEXAFS or XANES. The N K-edge spectrum in GaN was recently measured by several groups (Katsikini *et al.*, 1996; Lambrecht *et al.*, 1996b). This spectrum essentially probes the N-p partial density of states in the conduction band because it corresponds to excitation of a N2s state to the N p-like empty states.

Figure 10 shows the absorption spectrum extracted from glancing angle X-ray reflectivity compared to our theoretical prediction based on the band structures using our usual LDA + constant shift approach and a shift of the core-level extracted from XPS (Martin *et al.*, 1994). The two measurements are for s-polarization and p-polarization respectively, which as explained elsewhere correspond to $\mathbf{E} \perp \mathbf{c}$ and $\mathbf{E} \parallel \mathbf{c}$ respectively. The calculation is polarization averaged. Although, there are intensity discrepancies, which are mostly due to the polarization effects, one can clearly see that the

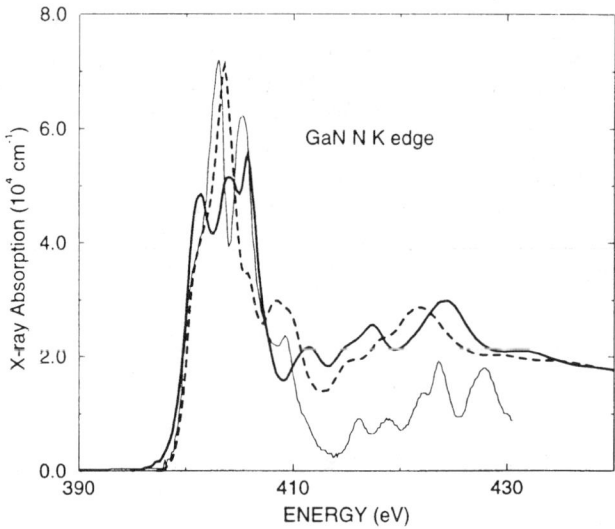

FIG. 10. Nitrogen K-edge, X-ray absorption spectrum of GaN, extracted from glancing angle reflectivity. Thick solid line, experiment for **E** ∥ **c**, dashed line, **E** ⊥ **c**, thin solid line: polarization averaged theory.

peaks agree in position up to at least 10 eV above the conduction band minimum. This confirms the UV reflectivity results. If core-hole to conduction band excitonic effects play a role here, it is hard to see why they would equal the ones in UV and would similarly affect the higher peaks so much more than the states near the minimum gap. Hence, the most plausible explanation is that the actual quasiparticle energy shifts from the LDA Kohn Sham eigenvalues in the conduction band are indeed almost constant. The increasing trend found in the GW calculations thus is not bourne out by the measurements. This may be consistent with the fact that also the valence band state shifts appear to be a bit overestimated by current GW as was discussed above in connection with the photoemission results.

4. OTHER NITRIDES

For the other nitrides, the experimental situation is much more incomplete. For AlN, there have been UV-reflectivity measurements by Loughin et al. (1993). They were analyzed in terms of an LCAO model. A comparison with our LMTO calculations is given in Lambrecht et al. (1996a) and reveals significant discrepancies. This may be related to oxygen contamination of the samples. Our calculations agree well with those of Christensen and Gorczyca (1994). For InN, a comparison with experimental data by Guo et al. (1992) is presented in Christensen and Gorczyca (1994).

VI. Details Near the Band Edges

While in the remainder of the paper, we have discussed overall aspects of the band structure, such as bandwidths and transitions to states up to 10 eV or so, into the conduction band, the most important region of the band structure for practical applications in optoelectronics and transport is clearly the region in the immediate vicinity of the gap. For w-GaN, the reflectivity spectra of Dingle et al. (1971), Shan et al. (1995, 1996), Gil, Briot, and Aulombard (1995), photoluminescence excitation spectra by Monemar (1974), and photoluminescence spectra as well as calorimetric absorption and reflection spectroscopy by Eckey et al. (1996) show three well separated excitons labeled A, B, and C. Temperature dependent reflectivity data for zincblende GaN in the spectral region of the gap were presented by Ramírez-Flores et al. (1994), and reveal two excitons. To deal with these aspects of the band structure and the associated questions of effective masses, related to the curvature of the bands near the minima, we need meV precision. At this level, it is of crucial importance to include strain effects and spin-orbit coupling.

The appropriate theoretical framework for discussing the states near the band edges is the theory of invariants which provides the most general allowed form by group theory of the Hamiltonian up to terms of second order in $\mathbf{k}$. In addition, $\mathbf{k} \cdot \mathbf{p}$ theory provides perturbation theoretical expressions for the parameters in this Hamiltonian in terms of momentum matrix elements and energy band differences. The application of these theories to the zinc-blende and wurtzite have been described by Luttinger and Kohn (1955), Bir and Pikus (1974), Cho (1976), Sirenko et al. (1996). Following Sirenko et al. (1996), who traced the historic origins of this work in the Russian literature, we refer to the wurtzite Hamiltonian as the Rashba-Sheka-Pikus (RSP) Hamiltonian, while the zinc-blende Hamiltonian will be referred to as the Kohn-Luttinger Hamiltonian. Traditionally, the parameters in these models have been determined by experiment. For the nitrides, this information is still rather incomplete. Only the conduction band mass has been obtained by ODCR (Drechsler et al., 1995). Attempts were made to extract crystal field splitting and spin-orbit coupling parameters from the exciton splittings. However, this has been hindered by the incomplete knowledge of the strain state of the materials. A few groups have attempted to determine the parameters in these models by fitting these model band structures to first-principle calculations (Suzuki, Uenoyama, and Yanase, 1995; Kim, Lambrecht, and Segall, unpublished; Wei and Zunger, unpublished).

We first discuss the valence-band splittings at Γ. In the absence of spin-orbit coupling, the hexagonal crystal field splits the zincblende Γ_{15} state into a doublet Γ_6 and a singlet Γ_1 state (not counting spin degeneracy), whose splitting we shall denote by Δ_c. With the inclusion of the spin-orbit coupling, the Γ_6 states splits into a Γ_9 and Γ_7 state, while the Γ_1 becomes a Γ_7 state in the notation of the double group.

On the basis of symmetry, there are two spin-orbit coupling terms in the effective Hamiltonian for the valence-band manifold,

$$H_{SO} = \Delta_2 L_z \sigma_z + \sqrt{2}\Delta_3(L_+\sigma_- + L_-\sigma_+), \tag{9}$$

in which L_z, $L_\pm$ are the spherical tensor components of the $L=1$ angular momentum operator and σ is the Pauli spin vector, also given in spherical tensor components. However, the spin-orbit coupling derives mainly from the inner part of the atom, and hence, is rather insensitive to the crystal structure and strain and one thus finds $\Delta_2 = \Delta_3 = \Delta_0/3$ with the Δ_0 the spin-orbit splitting in zinc-blende. For the same reasons, it can be confidently calculated within ASA-LMTO. Using this quasi-cubic approximation (Hopfield, 1960; Bir and Pikus, 1974), one finds for the $\Gamma_9 - \Gamma_7$ splittings, which correspond to the A-B, A-C exciton splittings,

$$E_{\Gamma_9} - E_{\Gamma_7} = \frac{\Delta_c + \Delta_0}{2} \pm \frac{1}{2}\sqrt{(\Delta_c + \Delta_0)^2 - \frac{8}{3}\Delta_0\Delta_c}. \tag{10}$$

The ASA-LMTO calculated spin-orbit parameters for the III-nitrides are given in Table IV. For GaN, the value is in good agreement with the data for zinc-blende by Ramīrez-Flores et al. (1994). We also compare with the values obtained with LAPW calculations by Wei and Zunger (1996a). The very small value for InN is of interest. At first sight, it is counterintuitive that for the heavier element, spin-orbit coupling is smaller. The reason is that the effective spin-orbit splitting contains a negative contribution from the semicore d-states. The latter is the strongest in InN and almost completely cancels the otherwise dominant N2p contribution. This d-band effect was first noted by Cardona (1963) for Cu-halides and explained by Shindo, Morita, and Kamimura (1965). Further details are given in Lambrecht et al. (1996a).

TABLE IV

SPIN-ORBIT SPLITTING Δ_0 (IN MEV) IN ZINC-BLENDE NITRIDES

BN	22			
AlN	19			19
GaN	19	21	17 ± 1	15
InN	3	19		6

[a]Ramírez-Flores, et al. (1994).
[b]Wei, S. H. and Zunger, A. (1996a).

Unlike the spin-orbit coupling, the crystal field splitting Δ_c is very sensitive to strain. Gil, Briot, and Aulombard (1995) analyzed the data available on the exciton splittings and showed that they are correlated with the thickness of the films grown in sapphire. From the bandgap reduction (overall shift of the excitons) and knowing the bandgap deformation potential, one can then extract the hydrostatic component of the strain. Assuming that the crystals are biaxially strained, one obtains a further relation $\varepsilon_{zz} = -2C_{13}/C_{33}\varepsilon_{xx}$ between the two independent strain components (normal to the basal plane ε_{zz} and in-plane $\varepsilon_{xx} = \varepsilon_{yy}$). Here, C_{13} and C_{33} are elastic constants which can be found in Kim, Lambrecht, and Segall (1994). From the thickest films ($\sim 500\,\mu$m) (Eckey et al., 1996) one can obtain the zero strain limit (Δ_1) and from the slope of the splitting with strain, one obtains a uniaxial deformation potential, known as D_3 in Bir and Pikus (1974) notation. Explicitly, one has

$$\Delta_c = E_{\Gamma_6} - E_{\Gamma_1} = \Delta_1 + (3/2)D_3\varepsilon_{zz}. \tag{11}$$

The analysis of Gil, Briot, and Aulombard (1995) gives $\Delta_1 = 10\,\text{meV}$. Their value for D_3, 3.71 eV, is somewhat inaccurate due to the use of old and inaccurate data on the elastic constants. We have recently reanalyzed these data including also newer data on GaN on SiC substrates and found $D_3 = 5.7\,\text{eV}$ obtained directly from our first-principles calculations is in good agreement with the data. This can be seen in Fig. 11.

The calculated values for the zero-strain value of the crystal field splitting both by us (19 meV at $u = 0.379$ and 36 meV at the more accurate $u = 0.377$) and Wei and Zunger (1996a) (42 meV) appear too high. This happens even though both calculations include carefully the internal parameter u relaxation and use accurate experimental lattice constants a and c/a. Clearly, we are struggling with the accuracy limitations of current band structure methods at this point. If one uses the ideal value of $u = 3/8$ instead, one obtains a value that is even larger, about 73 meV as in Suzuki, Uenoyama, and Yanase (1995). The remaining discrepancy is related to the LDA.

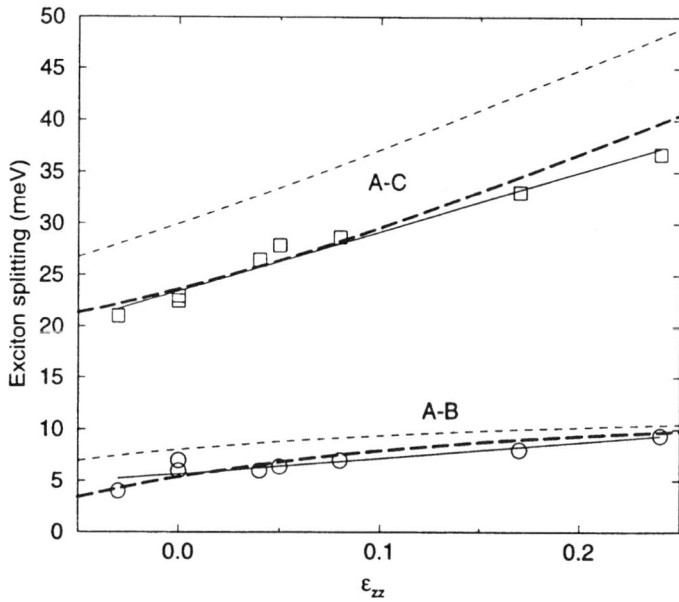

FIG. 11. Exciton splittings as function of uniaxial strain component ε_{zz}. The circles and squares indicate experimental values for AB and AC splitting respectively. Full lines are linear fits. The dashed lines are our calculated results using $\Delta_0 = 19$ meV, $D_3 = -5.7$ meV and $\Delta_1 = 19$ and 10 meV respectively, for the short-thin and long-thick dashes.

For AlN, both Wei and Zunger (1996a), Suzuki, Uenoyama, and Yanase (1995), and Kim, et al. (1997) find that the Γ_1 state lies above the Γ_6 by respectively 217 meV, 58 meV, and 215 meV. Again, the discrepancy in Suzuki, Uenoyama, and Yanase (1995) is related to the use of the ideal u value. No experimental data on these splittings are available to our knowledge. The strain dependence is again given in terms of the D_3 deformation potential for which we find a value of 9.6 eV (Kim, et al., 1997). For InN, Wei and Zunger (1996a) give a value of 41 meV for Δ_1, close to that of GaN. In the case of InN, however, one should be strongly concerned about correction effects beyond LDA, because the LDA bandgap turns out to be negative or very close to zero, depending on slight details of the calculations.

We now turn to the inverse mass-like parameters of the effective Hamiltonians. In wurtzite, the independent parameters are essentially the effective masses of light, heavy, and split-off hole bands in the two inequivalent crystal directions: in-plane and perpendicular to the basal plane. The values

TABLE V

KOHN-LUTTINGER PARAMETERS (IN UNITS $\hbar^2/2m_0$) FOR ZINC-BLENDE AlN AND GaN

	AlN	GaN
A	−2.81	−5.05
B	−0.69	−1.17
C	−3.51	−5.85

given by Kim, *et al.* (1997) for GaN and AlN generally agree well with those by Suzuki, Uenoyama, and Yanase (1995). The largest discrepancies occur for the in-plane heavy hole mass, which is so large in AlN, that it is difficult to extract from parabolic fits to the bands. The results also agree closely with the predictions of the quasicubic model (Bir and Pikus, 1974). We can hence describe both wurtzite and zinc-blende valence bands fairly well, with the three usual Kohn-Luttinger parameters given in Table V. In terms of these, one can obtain the RSP parameters and the effective masses in WZ using the relations given in Table VI, where numerical values are also given. We give both the values fitted directly to wurtzite bands, and the ones obtained from the quasicubic model using fits to the zinc-blende bands. For InN, these data are still lacking because of the more substantial problems of LDA for this material.

TABLE VI

WURTZITE RASHBA-SHEKA-PIKUS PARAMETERS IN TERMS OF ZINC-BLENDE KOHN-LUTTINGER PARAMETERS AND WURTZITE EFFECTIVE MASSES AND THEIR VALUES IN AlN AND GaN OBTAINED FROM DIRECT CALCULATION AND IN QUASICUBIC APPROXIMATION (IN PARENTHESES)*

		AlN	GaN
A_1	$(A + 2B + 2C)/3$	−3.9 (−3.7)	−6.4 (−6.4)
A_2	$(A + 2B − C)/3$	−0.3 (−0.2)	−0.5 (−0.5)
A_3	$-C$	3.6 (3.5)	5.9 (5.9)
A_4	$C/2$	−1.3 (−1.7)	−2.5 (−2.9)
A_5	$-(A - B + 2C)/6$	1.5 (1.5)	2.6 (2.6)
A_6	$-(2A - 2B + C)/3\sqrt{2}$	1.6 (1.8)	3.1 (3.2)
$m_{hh}^{\parallel} = m_{lh}^{\parallel}$	$-(A_1 + A_3)^{-1}$	3.5 (4.4)	2.0 (2.0)
$m_{\text{split}}^{\parallel}$	$-A_1^{-1}$	0.3 (0.3)	0.2 (0.2)
$m_{hh}^{\perp}$	$-(A_2 + A_4 + A_5)^{-1}$	11.1 (2.2)	2.0 (1.2)
$m_{lh}^{\perp}$	$-(A_2 + A_4 - A_5)^{-1}$	0.4 (0.3)	0.2 (0.2)
$m_{\text{split}}^{\perp}$	$-A_2^{-1}$	4.0 (4.5)	2.0 (2.0)

*Units are $\hbar^2/2m_0$ for A_i and m_0 for masses.

VII. Outlook for Future Work

From the previous sections, it becomes clear that the basic features of the band structures of the pure nitrides, their trends, and relations between zinc-blende and wurtzite are by now rather well understood. To be sure, there are some aspects that still require refinements, for example, inclusion of spin-orbit coupling effects on the higher conduction band states, determining the corrections beyond LDA with greater accuracy, understanding the relations between the fitted KL and RSP model Hamiltonian parameters and their expressions in terms of the critical point energy gaps and the matrix elements in $\mathbf{k}\cdot\mathbf{p}$ theory.

Mostly, however, at this point, there is a lack of experimental data to test the theoretical data base. Angular resolved photoemission and inverse photoemission spectroscopy are highly desirable to challenge the accuracy of the present band structure theories, in particular, to study the deviations from LDA and to test the accuracy of GW on the k-dependence and state dependence of the self-energy corrections as well as the lifetime effects on those states. In the area of UV optical spectra, many-body effects need to be included. It should be added, that even for much better understood semiconductors, this statement is equally valid. Also, at this moment, only for GaN, there have been extensive studies in the UV range for both zinc-blende and wurtzite. Even in GaN, the polarization dependence has not been studied experimentally. For AlN, InN, and BN, synchrotron radiation studies of UV reflectivity and/or ellipsometry have not been performed, presumably for lack of adequate samples. Electroreflectivity of all these materials may further help to identify critical point transitions.

The present review was limited to the pure nitrides. The other area where one can expect much progress in the near future is the band structure of mixed materials. There has already been some body of work on alloys (Jenkins and Dow, 1989; Wright and Nelson, 1995b; Albanesi, Lambrecht, and Segall, 1993; Lambrecht, 1997; Kim, Limpijumnong, Lambrecht, and Segall; Wei and Zunger, 1996b; Neugebauer and Van de Walle, 1995), but it has been mostly limited to zinc-blende alloys. Virtual crystal theories will clearly be inadequate because of the large size mismatches between the cations (especially those containing In) and strong bandgap variation among the nitrides. The effects of short-range and long-range ordering in the wurtzite structure based alloys are not understood at all. Finally, the electronic structure of superlattices and other layered structures of reduced dimensionality will clearly attract great interest. These are particularly interesting because of the polar nature of the crystals. One may expect that just like (111) oriented superlattices in other III–Vs, the wurtzite (0001) superlattices will exhibit interesting spontaneous electric field and piezoelectric effects. While some studies of the band line-ups have already appeared

(Lambrecht et al., 1996a; Albanesi and Lambrecht, 1994), the study of the optical properties and band structure of these systems is clearly still in its initial stages.

Acknowledgments

It is a pleasure to thank my colleague Professor Segall, and the students (Kwiseon Kim, Sukit Limpijumnong) and research associates (Sergey Rashkeev, Eduardo Albanesi) of our research group who collaborated on various aspects of the study of nitrides over the past few years. I would also like to thank various people for sharing their experimental data with us, H. Morkoç and G. Martin, J. C. Rife and D. K. Wickenden, T. Suski and K. Jablonska, R. F. Davis and S. W. King, C. Janowitz, S. Loughlin. Finally, this work would not have been possible without the support of the National Science Foundation and the Office of Naval Research.

Note Added in Proof

After submission of this manuscript, other studies relevant to section V.1, and V.3 appeared: C. B. Stagarescu, L. C. Duda, K. E. Smith, J. H. Guo, J. Nordgren, R. Singh, and T. D. Moustakas, (1996) *Phys. Rev. B* **54**, R17335 and K. E. Smith, S. S. Dhesi, L. C. Duda, C. B. Stagarescu, J. H. Guo, J. Nordgren, R. Singh, and T. D. Moustakas, (1997) in *III–V Nitrides*, F. A. Ponce, T. D. Moustakas, I. Akasaki, and B. A. Monemar, Mater. Res. Soc. Symp. Proc. MRS Pittsburgh, **449**, 787–92.

References

Albanesi, E. A., Lambrecht, W. R. L., and Segall, B. (1993). *Phys. Rev. B* **48**, 17841.
Albanesi, E. A., Lambrecht, W. R. L. (1994). In *Diamond, SiC and Nitride Wide Bandgap Semiconductors* (C. H. Carter, Jr., G. Gildenblat, S. Nakamura, and R. J. Nemanich, eds.), *Mater. Res. Soc. Symp. Proc.* MRS, Pittsburgh **339**, 607–612.
Andersen, O. K. (1975). *Phys. Rev. B* **12**, 3060.
Andersen, O. K., Postnikov, A. V., and Savrasov, Yu. S. (1992). In *Applications of Multiple Scattering Theory to Materials Science*, (W. H. Butler, P. H. Dederichs, A. Gonis, and R. L. Weaver, eds.), *Mater. Res. Soc. Symp. Proc. MRS, Pittsburgh*, **253**, 37.
Aryasetiawan, F., and Gunnarsson, O. (1994). *Phys. Rev. B* **49**, 7219; *Phys. Rev. B* **49**, 16214.
Bloom, S. (1971). *J. Phys. Chem. Sol.* **32**, 2027.
Bloom, S., Harbeke, G., Meier, E., and Ortenburger, I. B. (1974). *Phys. Stat. Sol. B* **66**, 161.
Bir, G. L., and Pikus, G. E. (1974). *Symmetry and Strain-Induced Effects in Semiconductors*, John Wiley & Sons, New York.
Blöchl, P. E. (1994). *Phys. Rev. B* **50**, 17953.
Boguslawski, P., Briggs, E. L., and Bernholc, J. (1995). *Phys. Rev. B* **51**, 17255.
Bourne, J., and Jacobs, R. L. (1972). *J. Phys. C: Solid State Physics* **5**, 3462.

Cardona, M. (1963). *Phys. Rev.* **129,** 69.
Cheng, C. Needs, R. J., and Heine, V. (1987). *Europhys. Lett.* **3,** 475.
Chen, J. Levine, Z. H., and Wilkins, J. W. (1995). *Appl. Phys. Lett.* **66,** 1129.
Ching, W. Y., and Harmon, B. N. (1986). *Phys. Rev. B* **34,** 5305.
Cho, K. (1976). *Phys. Rev. B* **14,** 4463.
Christensen, N. E., and Gorczyca, I. (1993). *Phys. Rev. B* **47,** 4307.
Christensen, N. E., and Gorczyca, I. (1994). *Phys. Rev. B* **50,** 4397.
Corkill, J. L., Rubio, A., and Cohen, M. L. (1994). *J. Phys. Condens. Mat.* **6,** 961.
del Castillo-Mussot, M., and Sham, L. J. (1985). *Phys. Rev. B* **31,** 2092.
Ding, S. A., Neuhold, G., Weaver, J. H., Häberle, P., Horn, K., Brandt, O., Yang, O., and Ploog, K. (1996). *J. Vac. Sci. Technol. A* **14,** 819.
Dingle, R., Sell, D. D., Stokowski, S. E., and Ilegems, M. (1971). *Phys. Rev. B* **4,** 1211.
Dreschsler, M., Meyer, B. K., Hoffmann, D. M., Detchprohm, D., Amano, H., and Akasaki, I. (1995). *Jpn. J. Appl. Phys.* **34,** L1178.
Eckey, L., Podlowskii, L., Göldner, A., Hoffman, A., Broser, I., Meyer, B. K., Volm, D., Streibl, T., Detchprohm, T., Amano, H., and Akasaki, I. (1996). In *Silicon Carbide and Related Materials 1995*, (S. Nakashima, H. Matsunami, S. Yoshida, and H. Harima, eds.), Institute of Physics Conference Series No. 142, IOP, London, p. 943.
Fiorentini, V., Methfessel, M., and Scheffler, M. (1993). *Phys. Rev. B* **47,** 13353.
Foley, C. P., and Tansley, T. L. (1986). *Phys. Rev. B* **33,** 1430.
Garcia, A., and Cohen, M. L. (1993). *Phys. Rev. B* **47,** 4215.
Gautier, M., Duraud, J. P., and Le Gressus, C. (1986). *Surf. Sci.* **178,** 201; (1987). *J. Appl. Phys.* **61,** 574.
Gil, B., Briot, O., and Aulombard, R.-L. (1995). *Phys. Rev. B* **52,** R17028.
Godby, R. W., Schlüter, M., and Sham, L. J. (1988). *Phys. Rev. B* **37,** 10159.
Gorczyca, I., Christensen, N. E., Perlin, P., Grzegory, I., Jun, J., and Bockowski, M. (1991). *Solid State Commun.* **79,** 1033; Gorczyca, I., and Christensen, N. E. (1991). *Solid State Commun.* **80,** 335; Gorczyca, I., and Christensen, N. E. (1993). *Phys. B* **185,** 410; Perlin, P., Gorczyca, I., Porowski, S., Suski, T., Christensen, N. E., and Polian, A. (1993). *Jpn. J. Appl. Phys.* **32,** Suppl. 32-1, 334; Perlin, P., Gorczyca, I., Christensen, N. E., Grzegory, I., Teisseyre, H., and Suski, T. (1992). *Phys. Rev. B* **45,** 13307.
Gorczyca, I., Christensen, N. E., Pelzer y Blanca, E. L., and Rodriguez, C. O. (1995). *Phys. Rev. B* **51,** 11936.
Grinajew, S. H., Malachow, W. J., and Czaldyszew, W. A. (1986). *Izv. Vuz. Fiz. (USSR)* **29,** 69.
Guo, Q., Kato, O., Fujisawa, M., and Yoshida, A. (1992). *Solid State Commun.* **83,** 721.
Hamann, D. R., Schlüter, M., and Chiang, C. (1979). *Phys. Rev. Lett.* **43,** 1494.
Hanke, W., and Sham, L. J. (1975). *Phys. Rev. B* **12,** 4501; (1980). *Phys. Rev. B* **21,** 4656.
Harrison, W. A. (1989). *Electronic Structure and Properties of Solids*, Dover, New York.
Hedin, L., and Lundqvist, S. (1969). In *Solid State Physics* **23,** 1.
Hedman, J., and Martensson, N. (1980). *Physica Scripta* **22,** 176.
Hejda, B., and Hauptmanová, K. (1969). *Phys. Status Solidi* **36,** K95.
Hohenberg, P., and Kohn, W. (1964). *Phys. Rev.* **136,** B864; Kohn, W., and Sham, L. J. (1965). *ibid.* **140,** A1133.
Hopfield, J. J. (1960). *J. Phys. Chem. Solids* **15,** 97.
Huang, M. Z., and Ching, W. Y. (1985). *J. Chem. Phys. Sol.* **46,** 977.
Hunt, R. W., Vanzetti, L., Castro, T., Chen, K. M., Sorba, L., Cohen, P. I., Gladfelder, W., Van Hove, J. M., Kuznia, J. N., Asif Khan, M., and Franciosi, A. (1993). *Physica B* **185,** 415.
Hybertsen, M., and Louie, S. G. (1986). *Phys. Rev. B* **34,** 5390.
Janowitz, C., Cardona, M., Johnson, R. L., Cheng, T., Foxon, T., Günther, O., and Jungk, G. (1994). *BESSY Jahresbericht*, p. 230.
Jenkins, D. W., and Dow, J. D. (1989). *Phys. Rev. B* **39,** 3317.
Jones, D., and Lettington, A. H. (1969). *Solid State Commun.* **11,** 701; (1969). *ibid.* **7,** 1319.

Karch, K., Portish, G., Bechstedt, F., Pavone, P., and Strauch, D. (1996). In *Silicon Carbide and Related Materials 1995* (S. Nakashima, H. Matsunami, S. Yoshida, and H. Harima, eds.), Institute of Physics Conference Series No. 142, IOP, London, p. 967.

Katsikini, M., Paloura, E. C., Kalomiros, J., Bressler, P., and Moustakas, T. (1996). In *Proceedings of the 23rd International Conference on the Physics of Semiconductors* (M. Scheffler and R. Zimmermann, eds.), World Scientific, Singapore, p. 573.

Kerker, G. (1980). *J. Phys. C: Solid State Physics* **13**, L189.

Kim, K., Lambrecht, W. R. L., Segall, B., and van Sdrilfgarde M. (1997). *Phys. Rev. B* **56**, (Sept 15) in press.

Kim, K., Lambrecht, W. R. L., and Segall, B. (1994). *Phys. Rev. B* **50**, 1502.

Kim, K., Lambrecht, W. R. L., and Segall, B. (1996). *Phys. Rev. B* **53**, 16310.

Kim, K., Limpijumnong, S., Lambrecht, W. R. L., and Segall, B. (1997). In *III–V Nitrides* (F. A. Ponce, T. D. Moustakas, I. Akasaki, B. A. Monemar) Mater. Res. Soc. Symp. Proc. MRS Pittsburgh, **449**, 929–934.

King, S. W., Benjamin, M. C., Nemanich, R. J., Davis, R. F., and Lambrecht, W. R. L. (1996). In *Gallium Nitride and Related Materials* (F. Ponce, R. D. Dupuis, S. Nakamura, and J. A. Edmond, eds.), *Mater. Res. Soc. Symp. Proc. MRS, Pittsburgh*, **395**, 375.

Klienman, L. (1981). *Phys. Rev. B* **24**, 7412.

Kobayashi, A., Sankey, O. F., Volz, S. M., and Dow, J. D. (1983). *Phys. Rev. B* **28**, 935.

Lambrecht, W. R. L., and Segall, B. (1990). *Phys. Rev. B* **41**, 2832.

Lambrecht, W. R. L., and Segall, B. (1992). In *Wide Band Gap Semiconductors* (T. D. Moustakas, J. I. Pankove, and Y. Hamakawa, eds.), *Mater. Res. Soc. Symp. Proc.* **242**, 367.

Lambrecht, W. R. L., and Segall, B. (1994). In *Properties of Group III Nitrides* (J. H. Edgar, ed.), Electronic Materials Information Service (EMIS) Datareviews Series (Institution of Electrical Engineers, London, Chapter 5.

Lambrecht, W. R. L. (1994). In *Diamond, Silicon Carbide and Nitride Wide Bandgap Semiconductors* (C. H. Carter, Jr., G. Gildenblat, S. Nakamura, and R. J. Nemanich, eds.), *Mater. Res. Soc. Symp. Proc.* **339**, 565–582.

Lambrecht, W. R. L. (1997). *Solid State Electronics*, **41**, 195–199.

Lambrecht, W. R. L., Segall, B., Strite, S., Martin, G., Agarwal, A., Morkoç, H., and Rockett, A. (1994). *Phys. Rev. B* **50**, 14155.

Lambrecht, W. R. L., Segall, B., Rife, J., Hunter, W. R., and Wickenden, D. K. (1995). *Phys. Rev. B* **51**, 13516.

Lambrecht, W. R. L., Kim, K., Rashkeev, S. N., and Segall, B. (1996a). In *Gallium Nitride and Related Materials* (F. Ponce, R. D. Dupuis, S. Nakamura, and J. A. Edmond, eds.), *Mater. Res. Soc. Symp. Proc. MRS, Pittsburgh* **395**, 455.

Lambrecht, W. R. L., Rashkeev, S. N., Segall, B., Lawniczak-Jablonska, K., Suski, T., Gullikson, E. M., Underwood, J. H., Perera, R. C. C., Rife, J. C., Grzegory, I., Porowski, S., and Wickenden, D. K. (1997). *Phys. Rev. B*, **55**, 2612.

Lam, P. K., Wentzcovitch, R. M., and Cohen, M. L. (1990). In *Materials Science Forum*, Trans Tech Publications, Switzerland, **54/55**, 165–192.

Landolt-Börnstein: *Numerical Data and Functional Relationships in Science and Technology* (O. Madelung, ed.), Group III, Vol. 22a, Springer, Berlin.

Logothetidis, S., Petalas, J., Cardona, M., and Moustakas, T. D. (1994). *Phys. Rev. B* **50**, 18017.

Loughlin, S., French, R. H., Ching, W. Y., Xu, Y. N., and Slack, G. A. (1993). *Appl. Phys. Lett.* **63**, 1182.

Luttinger, J. M. (1956). *Phys. Rev.* **102**, 1030; Luttinger, J. M., and Kohn, W. (1955). *Phys. Rev.* **97**, 869.

Martin, G., Strite, S., Botchkarev, A. Agarwal, A., Rockett, A., Morkoç, H., Lambrecht, W. R. L., and Segall, B. (1994). *Appl. Phys. Lett.* **65**, 610.

Martin, G., Botchkarev, A., Rockett, A., and Morkoç, H. (1996). *Appl. Phys. Lett.* **68**, 2541.

Martin, R. (1970). In *Phys. Rev. B* **1**, 4005.

Methfessel, M. (1988). *Phys. Rev. B* **38**, 1537.
Min, B. J., Chan, C. T., and Ho, K. M. (1992). *Phys. Rev. B* **45**, 1159.
Miwa, K., and Fukumoto, A. (1993). *Phys. Rev. B* **48**, 7897.
Monemar, B. A. (1974). *Phys. Rev. B* **10**, 676.
Muñoz, A., and Kunc, K. (1993). *Physica B* **185**, 422; (1991). *Phys. Rev. B* **44**, 10372.
Muñoz, A., and Kunc, K. (1993). *J. Phys. Condens. Matter* **5**, 6015.
Neugebauer, J., and Van de Walle, C. G. (1994). *Phys. Rev. B* **50**, 8067; Northrup, J. E., and Neugebauer, J. (1996). *Phys. Rev. B* **53**, R10477.
Neugebauer, J., and Van de Walle, C. G. (1995). *Phys. Rev. B* **51**, 10568.
Okumura, H., Ohta, K., Ando, K., Rühle, W. W., Nagamoto, T., and Yoshida, S. (1996). In *Silicon Carbide and Related Materials 1995* (S. Nakashima, H. Matsunami, S. Yoshida, and H. Harima, eds.), Institute of Physics Conference Series No. 142, IOP, London, p. 939.
Olson, C. G., Lynch, D. W., and Zehe, A. (1981). *Phys. Rev. B* **24**, 4629.
Palummo, M., Bertoni, C. M., Reining, L., and Finocchi, F. (1993). *Physica B* **185**, 404.
Palummo, M., Reining, L., Godby, R. W., Bertoni, C. M., and Börnsen, N. (1994). *Europhysics Lett.* **26**, 607.
Palummo, M., Del Sole, R., Reining, L., Bechstedt, F., and Cappellini, G. (1995). *Solid State Commun.* **95**, 393.
Pandey, R., Jaffe, J. E., and Harrison, N. M. (1993). *J. Mater. Res.* **8**, 1922.
Perdew, J. P., and Levy, M. (1983). *Phys. Rev. Lett.* **51**, 1884.
Petalas, J., Logothetidis, S., Boutadakis, S., Alouani, M., and Wills, J. M. (1995). *Phys. Rev. B* **52**, 8082.
Phillips, J. C. (1973). *Bonds and Bands in Semiconductors*, Academic, New York.
Ramírez-Flores, G., Navarro-Contreras, H., Lastras-Martínez, A., Powell, R. C., and Greene, J. E. (1994). *Phys. Rev. B* **50**, 8433.
Rubio, A., Corkill, J. L., Cohen, M. L, Shirley, E. L., and Louie, S. G. (1993). *Phys. Rev. B* **48**, 11810.
Ruiz, E., Alvarez, S., and Alemany, P. (1994). *Phys. Rev. B* **49**, 7115.
Sham, L. J., and Kohn, W. (1966). *Phys. Rev.* **145**, A561.
Sham, L. J., and Schlüter, M. (1983). *Phys. Rev. Lett.* **51**, 1888.
Shan, W., Schmidt, T. J., Yang, X. H., Hwang, S. J., Song, J. J., and Goldenberg, B. (1995). *Appl. Phys. Lett.* **66**, 985.
Shan, W., Fisher, A. J., Song, J. J., Bulman, G. E., Kong, H. S., Leonard, M. T., Perry, W. G., Bremser, M. D., and Davis, R. F. (1996). *Appl. Phys. Lett.*, **69**, 740.
Shindo, K., Morita, A., and Kamimura, H. (1965). *J. Phys. Soc. Jpn.* **20**, 2054.
Sirenko, Yu. M., Jeon, J.-B., Kim, K. W., Littlejohn, M. A., and Stroscio, M. A. (1996). *Phys. Rev. B* **53**, 1997.
Suzuki, M., Uenoyama, T., and Yanase, A. (1995). *Phys. Rev. B* **52**, 8132.
Tyagai, V. A., Evtigneev, A. M., Krasiko, A. N., Andreeva, A. F., and Malakov, V. Ya. (1977). *Fiz. Tekh. Poluprovodn.* **11**, 2142, English translation in (1977). *Sov. Phys. Semicond.* **11**, 1257.
Van Camp, P. E., Van Doren, V. E., and Devreese, J. T. (1989). *Solid State Commun.* **71**, 1055; (1992). *ibid.* **81**, 23; (1991). *Phys. Rev. B* **44**, 9056.
Vanderbilt, D. (1990). *Phys. Rev. B* **41**, 7892.
Wei, S. H., and Zunger, A. (1996c). *Appl. Phys. Lett.* **69**, 2719.
Wei, S. H., and Zunger, A. (1996b). *Phys. Rev. Lett.* **76**, 664.
Wright, A. F., and Nelson, J. S. (1994). *Phys. Rev. B* **50**, 2159; (1995a). *ibid.* **51**, 7866.
Wright, A. F., and Nelson, J. S. (1995). *Appl. Phys. Lett.* **66**, 3051; (1995a). *ibid.* **66**, 3465.
Xu, Y.-N., and Ching, W. Y. (1993). *Phys. Rev. B* **48**, 4335.
Yeh, C.-Y., Lu, Z. W., Froyen, S., and Zunger, A. (1992). *Phys. Rev. B* **46**, 10086.
Yeh, C.-Y., Wei, S.-H., and Zunger, A. (1994). *Phys. Rev. B* **50**, 2715.

CHAPTER 13

Phonons and Phase Transitions in GaN

N. E. Christensen

INSTITUTE OF PHYSICS AND ASTRONOMY
UNIVERSITY OF AARHUS
DK-8000 AARHUS
DENMARK

P. Perlin

HIGH PRESSURE RESEARCH CENTER
POLISH ACADEMY OF SCIENCES
SOKOLOWSKA 29/37
PL-01-142 WARSAW
POLAND

I.	INTRODUCTION	409
II.	LATTICE STABILITY OF GaN	410
	1. *Stability of the Wurtzite Phase*	410
	2. *Determination of the Compressibility of GaN*	413
III.	PHONONS IN GaN	415
	1. *Zone-Center Modes at Zero Pressure*	415
	2. *Pressure-Dependence of the Phonon Frequencies*	420
	3. *Temperature Dependence of the Phonon Frequencies*	421
	4. *Phonons as Probe of Internal Stress in the Crystal*	423
	5. *Two-Phonon Raman Spectra*	424
	6. *Local Vibrational Modes*	425
IV.	SUMMARY AND CONCLUSIONS	426
	References	427

I. Introduction

Among the III–V nitrides GaN is technologically the most interesting compound, in particular because of its optical properties making it a promising material (Mohammad, and Morkoç, 1996) for optoelectronic devices operating in the blue regime. The understanding of the electronic, optical, mechanical and vibrational properties is based on a variety of experimental and theoretical studies. These also include the examinations of how the material behaves under applied pressure since these yield information about structural stability and bonding properties. The measurements of phonon frequencies as well as their pressure dependence also contribute to this. Such studies will be the subject of the present review. We shall discuss experimental as well as theoretical results in this connection. The theory will be based on *ab initio* electronic structure calculations, that is, methods which in more detail are described by Lambrecht (1997) in this book as well as in (Christensen, 1997) and references given therein.

In Section II, we discuss experimental observations of the pressure-induced transition from the wurtzite to the rocksalt structure and compare these to theoretical predictions. Also, the bulk modulus as well as its pressure coefficient are discussed. Section III discusses the zone-center phonons in wurtzite GaN, in particular those which are Raman active because Raman spectroscopy is a very accurate experimental technique. In this context it is shown that good agreement between the measured and calculated frequencies as well as their pressure coefficients can be obtained. In addition local modes associated with Mg_{Ga}-H complexes are briefly described. Section IV contains the summary and conclusion.

II. Lattice Stability of GaN

Free standing crystalline samples of gallium nitride always have the hexagonal wurtzite structure. The cubic, zinc-blende-type, GaN can be grown only as a heteroepitaxial layer on the cubic substrates like GaAs (Strite, *et al.*, 1991), β-SiC (Paisley, *et al.*, 1989), or Si (Lei, *et al.*, 1991). Cubic crystals contain almost always some amount of the hexagonal phase (Lei, Ludwig, and Moustakas, 1993). Recently it was shown, that in high quality hexagonal GaN, the presence of the cubic phase can be observed (Strauss, *et al.*, 1996). This cubic phase can be related rather to the large number of stacking faults than to the simple inclusions of cubic material.

1. STABILITY OF THE WURTZITE PHASE

It was predicted by Van Vechten (1969) that the pressure-induced phase transition of GaN from the hexagonal wurtzite structure into an idealized metal phase should occur at a pressure as high as 100 GPa. Probably the high value of the transition pressure predicted by Van Vechten, and also the difficulty in obtaining crystalline samples, caused the experimental studies of structure stability of this compound to be made rather late. To our knowledge, the first experimental measurement of the structural transformation in GaN was performed in 1991 by Perlin et al. (1991). X-ray absorption spectroscopy (XAS), was used to study the equation of state and to determine the transition pressure. However, in this experiment it was not possible to identify the symmetry of the high pressure phase. XAS measurements provide mostly the information about the local arrangements of the atoms, and are not an appropriate tool to define the symmetry of the lattice. At the same time the theoretical first-principles calculations of the total energy of GaN in various crystal structures indicated that the cubic rocksalt structure could be a likely candidate for the high pressure phase of GaN (Muñoz, and Kunc, 1991; Gorczyca, and Christensen, 1991; Van Camp, Van Doren, and Devreese, 1992). As is typical to highly ionic crystals the rocksalt structure appeared to have much lower energy than the metallic β-tin structure (see Fig. 1). These theoretical predictions soon found experimental confirmation. Xia, et al. (1993) and Ueno, et al. (1994) performed X-ray diffraction studies of GaN under high pressure. Both groups observed the disappearance of the wurtzite peaks and appearance of new peaks which

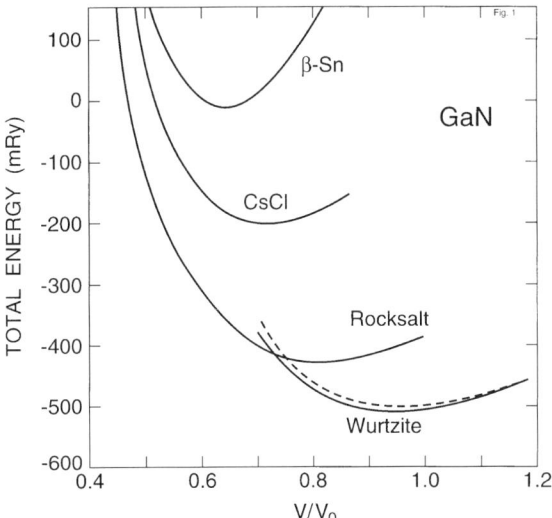

FIG. 1. Total energies as calculated (Christensen, and Gorczyca, 1994) from first principles for GaN in various structures as functions of volume. V_0 is the experimental equilibrium volume. The dashed curve shows the calculations for the zinc-blende structure.

TABLE I

EXPERIMENTAL DATA FOR THE PRESSURE-INDUCED TRANSITION FROM WURTZITE TO ROCKSALT STRUCTURE

Authors	Critical Pressure	Type of Crystal	Structure of High Pressure Phase	Technique
Perlin, et al. (1991)	47 GPa	Single crystal	Not identified	XAS
Xia, et al. (1993)	37	Powder	Rocksalt	X-ray diffraction
Ueno, et al. (1994)	52.2	Powder	Rocksalt	X-ray diffraction

could be attributed unambiguously to the cubic rocksalt structure. The transition pressures, P_t, obtained in all three experiments are given in Table I.

The existing experimental data show that the cubic rocksalt structure is the high pressure phase of GaN. The large scatter in the values of P_t can be related to two factors, (i) sensitivity of the method, and (ii) the type of sample used. Concerning the first point, the method of Ueno, et al. (1994) is probably the less sensitive. As for the second point it is important to note that different results are obtained when the sample used in the experiment is a monocrystal and when a powdered technique is applied. If the grains of the powder sample are small, the surface energy can start playing a role, and cause the phase transition to occur at a lower pressure. Thus, the experimental value of critical pressure does not only depend on the sensitivity of the measurement method but also on the form of the sample. Further, the pressure medium applied in the experiment may influence the measurements substantially. Large pressure gradients in the pressure cell can initiate the phase transition at lower average pressures than those in a medium with a completely uniform pressure distribution.

The transition pressure derived from *ab initio* theroetical calculations now seems to agree well with experiments. The density-functional calculations, of (Muñoz, and Kunc, 1991) and Christensen, and Gorzcyca, 1994) yield $P_t = 55$ and $51.8\,\text{GPa}$, respectively. These two calculations both use the local approximation for exchange and correlation (LDA), but they differ with respect to the choice of method of solution of the one-particle equations. The former (Muñoz, and Kunc, 1991) uses norm-conserving pseudopotentials whereas the Linear Muffin-Tin Orbital (LMTO) method was used in (Christensen, and Gorczyca, 1994). These two calculations also find essentially the same volume changes during the wurtzite → rocksalt transition. Letting V_1 and V_2 denote the volume per formula unit of the wurtzite and rocksalt phases at $P = P_t$, and V_0 the (experimental) equilibrium volume (Muñoz, and Kunc, 1991) finds $V_1/V_0 = 0.82$ and $V_2/V_0 = 0.71$. The same quantities as calculated in (Christensen, and Gorczyca, 1994) are 0.81 and 0.69, respectively. The transition pressure has been calculated by other groups and by means of other methods. Comparisons to these are made in (Christensen, 1997).

2. DETERMINATION OF THE COMPRESSIBILITY OF GaN

High-pressure X-ray diffraction studies of GaN also provide an opportunity to determine the bulk modulus (or compressibility) of the crystal. Usually these experiments can give the dependence of the lattice constant (or more detailed structural data) on the applied hydrostatic pressure. The results are most frequently interpreted by fitting the Murnaghan (Murnaghan, 1944) equation of state:

$$a = a_0 \left(1 + \frac{B'_0 P}{B_0}\right)^{-1/(3B'_0)}. \qquad (1)$$

Here a is the lattice constant, B_0 is the bulk modulus and B'_0 is the pressure derivative of the bulk modulus. Accurate determination of the bulk modulus from this kind of data is difficult because it depends sensitively on value of B'_0. This is not easily obtained because it requires a large pressures (P) range, which in turn are always limited by the phase transition pressure. This explains the scatter in the obtained empirical results which is mainly related to inter-dependency between two parameters, B_0 and B'_0 in the Murnaghan equation. One should also notice that theoretical predictions (Majewski, and Vogl, 1987) indicate that the value of B'_0 should be similar for compounds having similar chemical bonds, and this value should be very close to 4.5 for GaN. Table II shows experimental data. One should also notice that bulk modulus of GaN was also determined in the very accurate way by ambient pressure Brillouin scattering (Polian, Grimsditch, and Grzegory, 1996). In the light of the present results the value of the bulk modulus of GaN is equal to 210 ± 10 GPa.

The values of B_0 obtained from theoretical calculations range from 176 to 240 GPa as discussed in (Christensen, 1997). The pseudopotential calculation of (Muñoz, and Kunc, 1991) is rather low, 179 GPa, and close to the similar kind of theoretical study (Van Camp, Van Doren, and Devreese, 1992) which yielded $B_0 = 176$ GPa. The LMTO calculation (Christensen, and Gorczyca, 1994) gave 200 GPa. The most recent (Kim, Lambrecht, and Segall, 1996) calculation yields $B_0 = 207$ GPa. The cal-

TABLE II

EXPERIMENTAL VALUES OF THE BULK MODULUS OF GaN

Authors	B_0 (GPa)	B'_0	Technique	Comment
Perlin, et al. (1991)	245	4.5 (imposed)	XAS	Low precision
Xia, et al. (1993)	188	3.2	X-ray diffraction	Very low B'_0
Ueno, et al. (1994)	237 ± 31	4.3 ± 2	X-ray diffraction	High pressure
Leszczynski, et al. (1995)	207 ± 3	Not determined	X-ray diffraction	High precision, low pressure
Polian, et al. (1996)	210 ± 10	Not determined	Brillouin scattering	Ambient pressure technique

culated value of B'_0 is 3.8 (Christensen, and Gorczyca, 1994). This is lower than the value of 4.5 suggested above, but it should be noted that the theoretical value is obtained directly from the second derivative of the pressure-volume relation and refers to its value at the equilibrium volume. Thus, it is *not* derived as a parameter optimizing the fitted Murnaghan equation over a wide pressure range.

The Brillouin scattering experiments of (Polian, Grimsditch, and Grzegory, 1996) also allowed accurate experimental determination of the shear elastic constants which otherwise were only known as estimates obtained from atomic displacements measured by X-ray diffraction (Savastenko, and Sheleg, 1978). The experimental values obtained in (Polian, Grimsditch, and Grzegory, 1996) are (wurtzite GaN): $C_{11} = 390 \pm 15$, $C_{33} = 398 \pm 20$, $C_{44} = 105 \pm 10$, $C_{66} = 123 \pm 10$, $C_{12} = 145 \pm 20$, and $C_{13} = 106 \pm 20$, all in GPa. These experimental values may be compared to those obtained from theoretical calculations by Kim, et al. (1994); 396, 476, 91, 126, 144, and 94 GPa, respectively. The theoretical values were not derived directly from total-energy calculations for wurtzite GaN under strain but by means of the transformation method suggested by Martin (1972). This relates the elastic constants of the hexagonal structure to those of the cubic (zincblende) structure. Kim, et al. (1994) calculated the shear constants of zincblende-GaN from first principles, and this also included the determination of the internal strain parameter, ζ, as defined by Kleinman, (1962). The value of ζ found in (Kim, Labrecht, and Segall, 1994) is 0.5, which is close to the values of the internal-strain parameters of Si, Ge, GaAs and SiC, but smaller than those calculated for some II-VI compounds ($\zeta \approx 0.66$ in ZnS and ZnSe), see (Christensen, 1997) and references given therein. It is seen from the numerical values above that the calculated elastic constants agree well with experiments, except in the cases of C_{33} and C_{13}.

The method of using the transformation of the elastic tensor is not sufficiently accurate in these cases. This was shown by a later total energy calculation directly for the wurtzite structure, also by (Kim, *et al.* (1996) who in that way obtained $C_{13} = 100\,\text{GPa}$, and $C_{33} = 392\,\text{GPa}$. These are in perfect agreement with experiment, taking the error bars into account. In analogy to the presence of the internal strain quantified by η in the B3 structure which determines the relative sublattice displacement under trigonal strain, the wurtzite structure is characterized by a displacement parameter, $\xi = du/d(c/a)$, where $uc = d$ is the bond length along the *c*-axis. The value of ξ found in (Kim, *et al.*, 1996) was -0.11 for GaN.

III. Phonons in GaN

Raman spectroscopy is an accurate and convenient technique for measuring zone-center phonons. Complications may arise, however, when the phonon modes couple to plasmons, and GaN samples often have electron concentrations which are so high that this occurs. This section describes this, and we present experimental as well as theoretical data for Γ phonon modes and their pressure dependence in GaN. Subsections 2–6 further discuss the effects of internal stress, two-phonon processes and local modes of defect complexes.

1. Zone-Center Modes at Zero Pressure

Gallium nitride crystallizes in the hexagonal wurtzite structure with four atoms in the unit cell and belongs to the C_{6v}^4 ($C6_3mc$) space group. Nine optical modes are predicted at the zone center (Γ) by the group theory: two A_1, two E_1, two E_2 and two B_1 modes. The A_1 and B_1 modes are Raman and infrared active, and the two E_2 modes are Raman active. The modes are illustrated in Fig. 2. The B_2 modes are silent (Arguello, Rousseau, and Porto, 1969). Because of the similarity between the zincblende and the wurtzite structures, it is sometimes useful to consider the phonon branches in wurtzite crystals as zincblende branches folded along the (111) direction. The first data on the phonon frequencies from the center of the Brillouin zone were obtained in the early seventies by means of Raman scattering experiments (Manchon, *et al.*, 1970; Burns, *et al.*, 1973; and Lemos, Arguello, and Leite, 1972). While the frequencies of both E_2 modes as well as of A_1 (TO) and E_1 (TO) were determined quite accurately, the identification of

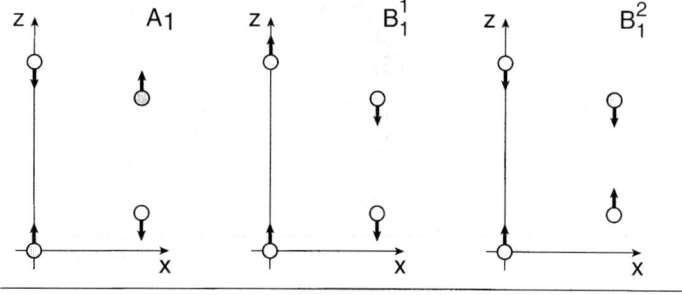
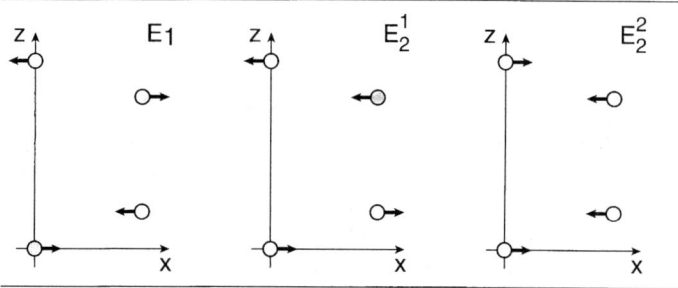

FIG. 2. Optical zone-center phonon modes in the wurtzite structure.

TABLE III

FREQUENCIES (cm^{-1}) OF ZONE CENTER PHONONS IN WURTZITE GaN

Phonon Mode	E_2 (low)	A_1 (TO)	E_1 (TO)	E_2 (high)	A_1 (LO)	E_1 (LO)
Expmt. (cm^{-1})	144 ± 1[c]	531 ± 2[c]	557 ± 1[c]	567.3 ± 0.1[d]	735 ± 1[d,e,f]	742 ± 3[e,f]
Theory[a]	150	537	555	558		
Theory[b]	146	534	556	560		

[a]Gorczyca, et al. (1997).
[b]Miwa, and Fukomoto (1993).
[c]Perlin, et al. (1992).
[d]Perlin, et al. (1995).
[e]Siegle, et al. (1995).
[f]Hung-Red, et al. (1995).

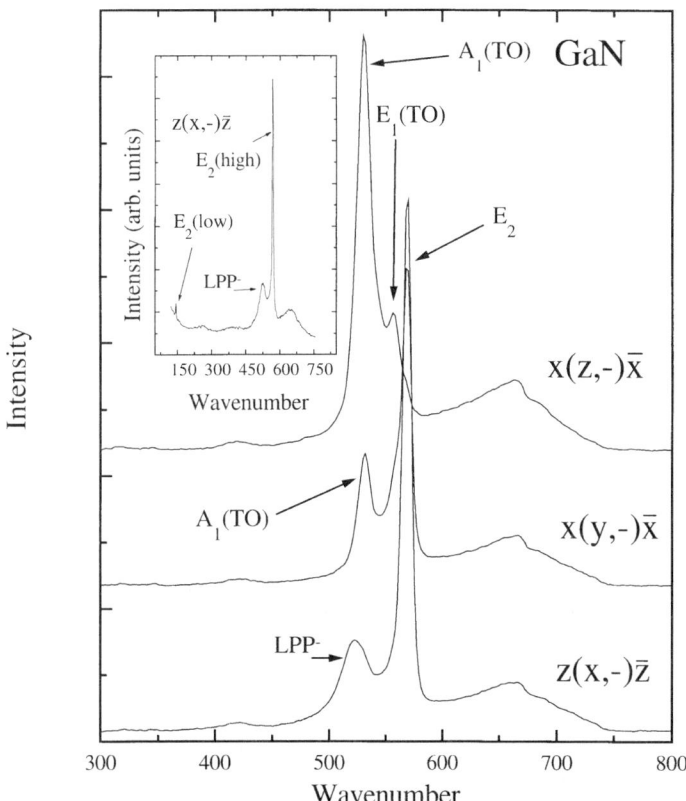

FIG. 3. Polarized Ramans spectra of bulk GaN crystals. Insert shows the spectrum measured in the broader energy range, showing low energy E_2 mode. LPP$^-$ denotes low energy LO phonon–plasmon coupled mode. Electron concentration in crystal -5×10^{19} cm^{-3}.

the LO modes was much more difficult. This problem was caused by the coupling between the LO modes in the highly polar GaN crystals and the conduction electron "gas". In that situation the first measurements of the LO frequencies were done using infrared reflectivity techniques (Barker, and Illegems, 1972). Only recently, when improved crystal growth techniques have made it possible to obtain high-resistivity GaN crystals (electron concentrations below 10^{18} cm^{-3}), have the A_1 and E_1 modes become easily detectable by Raman scattering spectroscopy (Kozawa, et al., 1994). First principles theoretical calculations of (Murnaghan, 1944) phonon frequencies in GaN give very good agreement with experimental data. Table III shows

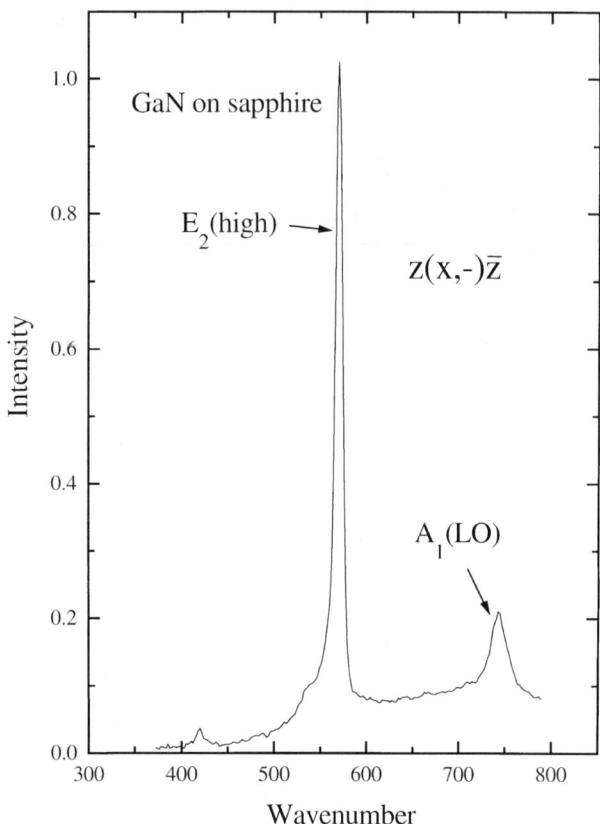

FIG. 4. Raman spectrum of epitaxial GaN layer on sapphire. Electron concentration around 3×10^{17} cm^{-3}.

the data on the Brillouin zone center phonons. We have mainly selected data obtained from bulk GaN crystals in order to avoid the influence of strain on the phonon frequencies (this problem will be discussed later in this chapter). The reader should note that the energies of the two LO modes in GaN depend on the carrier concentration via plasmon-phonon coupling and thus are more sample dependent.

Figure 3 shows the polarized Raman spectra of bulk (highly conducting) GaN crystals. One should notice the appearance of the coupled LO phonon–plasmon mode in the $z(x, -)\bar{z}$ polarization (Perlin, et al., 1995). The insert in Fig. 3 shows the Raman spectrum of GaN measured in a broader energy range, which makes it possible to observe the E_2 (low)

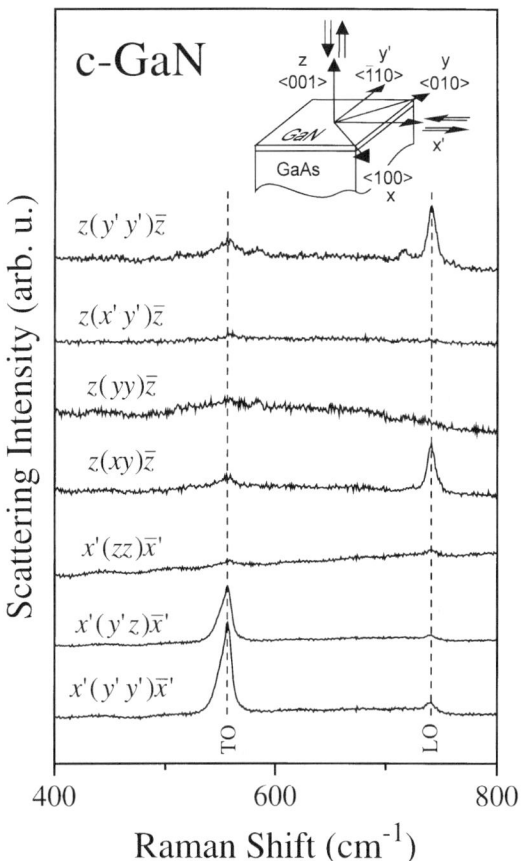

FIG. 5. Room temperature Raman spectra taken from the cubic sample in various configurations. The inset shows the corresponding scattering geometries. (From Siegle, et al., 1995).

phonon mode, close to 140 cm^{-1}. The features of the Raman spectrum related to the plasmon–phonon coupling disappear if the experiment is performed on samples with low electron concentration. Figure 4 shows the Raman spectrum measured on the GaN epitaxial layer on sapphire. In the $z(x, -)\bar{z}$ only E_2 and A_1 (LO) modes are allowed. As it was mentioned at the beginning of the chapter, GaN can be forced to grow in the cubic zincblende structure, if only the substrate is properly chosen. In the cubic structure only two optical modes exist: LO and TO. Figure 5 shows the polarized spectrum of cubic GaN grown on (001) surface of GaAs (Siegle, *et al.*, 1995). The frequencies of the TO and LO phonons in the cubic structure are 555 cm^{-1} and 740 cm^{-1}, respectively (Siegle, *et al.*, 1995). The calculated TO frequency is 551 cm^{-1} in (Gorczyca, *et al.*, 1995) and 558 in (Miwa, and Fukomoto, 1993).

2. Pressure-Dependence of the Phonon Frequencies

Application of the hydrostatic pressure generally increases the frequency of zone center phonons because of the increasing stiffening of the interatomic interactions. The pressure dependence of different phonon modes are shown in Fig. 6. Measurements are performed until the phase transition

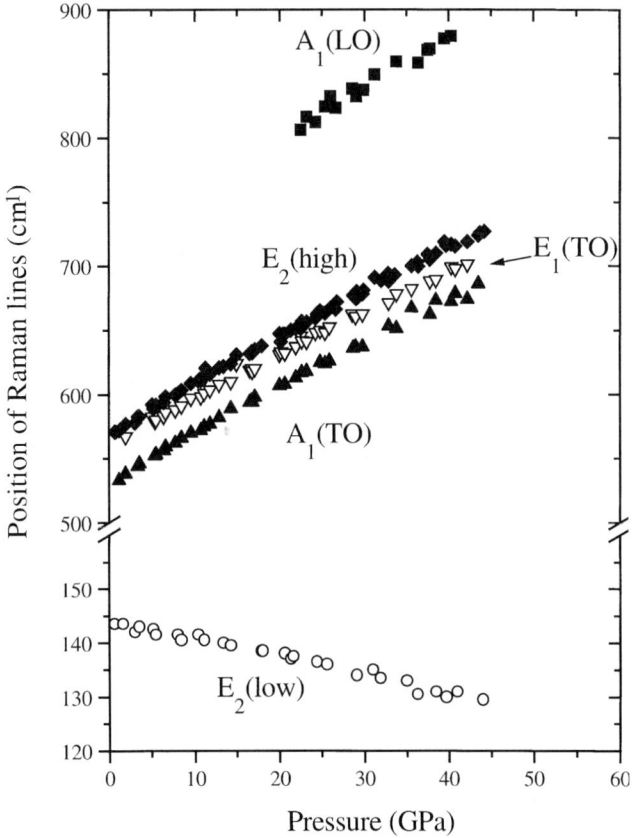

FIG. 6. Position of several Raman active phonons as a function of hydrostatic pressure.

TABLE IV

PRESSURE PROPERTIES OF PHONON MODES IN GaN

Phonon Mode	E_2 (low)	A_1 (TO)	E_1 (TO)	E_2 (high)	A_1 (LO)	E_1 (LO)
Grüneisen Parameter	−0.46 −0.20	1.50 1.52	1.24 1.48	1.33 1.60	1.09	——
Pressure coefficient (cm^{-1}/GPa)	−0.32 −0.15	3.8 4.08	3.3 4.10	3.6 4.46	3.8	---

Upper numbers give experimental values, lower numbers the calculations (Gorczyca, et al. 1995).

to the rocksalt structure occurs at around 47 GPa (Perlin, et al., 1991). The pressure coefficient and the Grüneisen parameters are listed in Table IV. The mode Grüneisen parameter is defined by:

$$\gamma = \frac{B_0}{\omega} \frac{d\omega}{dP} \qquad (2)$$

where B_0 is the bulk modulus equal to 210 GPa, and ω is the frequency (energy) of the mode. An interesting property of the wurtzite crystals is the fact that the E_2 (low) modes have negative pressure coefficients, i.e., they are so-called "soft" modes. This, for zone-center modes unusual feature, is related to the correspondence between E_2 (low) mode in wurtzite and the TA mode at the L point of the Brillouin zone of zincblende structure. Weinstein (1977) suggested the existence of a correlation between the Grüneisen parameter of this mode and the pressure for the structural phase transition. Similar analysis can be qualitatively performed in wurtzite crystals.

3. TEMPERATURE DEPENDENCE OF THE PHONON FREQUENCIES

The influence of temperature on the phonon frequencies is relatively small and mostly related to the temperature-induced change of the lattice constants (thermal expansion). Thus, with decreasing temperature the frequencies of GaN phonons increase with the exception of the "soft" E_2 (low) mode for which the energy decreases. There is a lack of systematical data of temperature dependences of phonon frequencies in GaN. The existing data are listed in Table V.

TABLE V

TEMPERATURE SHIFT OF PHONON FREQUENCIES IN GaN

Phonon Mode	E_2 (low)	A_1 (TO)	E_1 (TO)	E_2 (high)	A_1 (LO)	E_1 (LO)
$\omega (T = 20\,\text{K})^-$ $\omega (T = 300\,\text{K})\ (\text{cm}^{-1})$	-1	$+1$	$+1$	$+1$	—	---

As grown GaN is always of *n*-type. The free electron concentration ranges typically from 10^{21} up to 10^{22} cm^{-3} for MOCVD, or MBE epitaxial layers, and between 10^{23} and 10^{24} for bulk crystals. The predominantly *n*-type character of GaN is most probably due to oxygen (Perlin, *et al.*, 1995) impurities, although the possibility that nitrogen vacancies (Perlin, *et al.*,

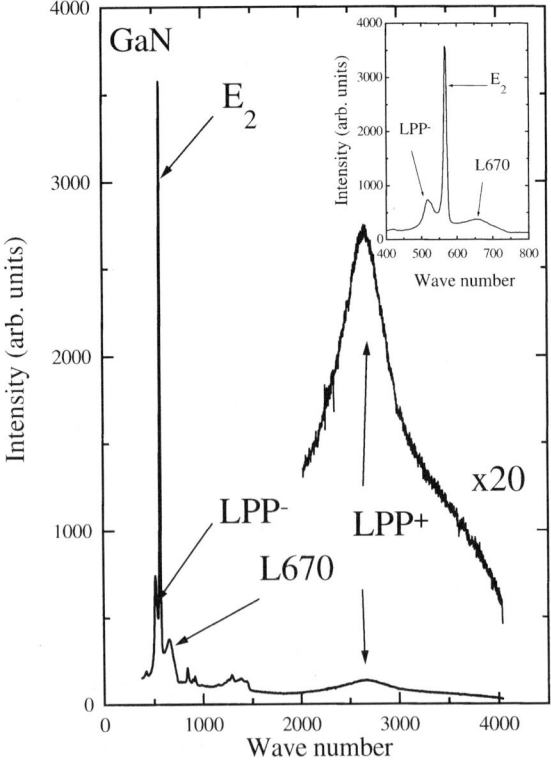

FIG. 7. Coupled phonon–plasmon modes in bulk GaN. Electron concentration close to 5×10^{19} cm^{-3}.

1996) are responsible for the high background electron concentration cannot be completely ruled out. As mentioned earlier, the existence of the electron plasma in highly ionic crystals like GaN leads to coupling between LO modes and plasmons. The plasmon phonon coupling leads to the formation of two branches of mixed excitations. The energies of these branches can be expressed approximately by the equation:

$$\omega^{\pm} = \tfrac{1}{2}\{\omega_L^2 + \omega_p^2 \pm ((\omega_L^2 + \omega_T^2)^2 - 4\omega_p^2\omega_T^2)^{1/2}\} \qquad (3)$$

where ω_L and ω_T are LO and TO phonon frequencies and ω_p is a plasmon frequency. In highly doped samples with electron concentration above 10^{19} cm^{-3}, the upper branch has almost purely plasmon character, while the lower one resembles a TO phonon. Both lower and upper branches (denoted as LPP$^-$ and LPP$^+$) were observed in bulk GaN crystals (Perlin, 1995). Figure 7 shows Raman spectrum with clearly visible LPP$^+$, and LPP$^-$ modes (see also Fig. 1). In epitaxial GaN only the upper branch of the coupled modes had been observed (Kozawa, et al., 1994) and only if electron concentration is lower than 10^{18} cm^{-3}. For higher electron concentrations the modes become damped and disappear in the background. It seems that the reason for the difference between epitaxial and bulk GaN is that the latter has much higher optical mobility than epilayers with the same electron concentration. From the line shape of the coupled mode one can obtain the damping parameters and the plasma frequency (Kozawa, et al., 1994).

4. PHONONS AS PROBE OF INTERNAL STRESS IN THE CRYSTAL

Epitaxial layers of GaN grown on the foreign substrates like SiC, Al$_2$O$_3$, Si, GaAs are usually strained, mostly because of lattice mismatch and the difference between thermal expansion coefficients of the substrate and the overlayer materials. Raman scattering is widely used as a method of monitoring the internal stress (and also disorder) in the sample by measuring the frequency (and polarization properties) of Raman active phonons (Suski, et al., 1996; Camassel, et al., 1995; and Suski, et al., 1996). Figure 8 shows the frequency of E_2 (high) phonons as a function of the position of bound exciton peak. Squares indicate the measurements on almost unstrained homoepitaxial layers. It is clear that, depending on the growth conditions, compressive or tensile strains can exist in the sample. Compressive stresses can reach 1 GPa.

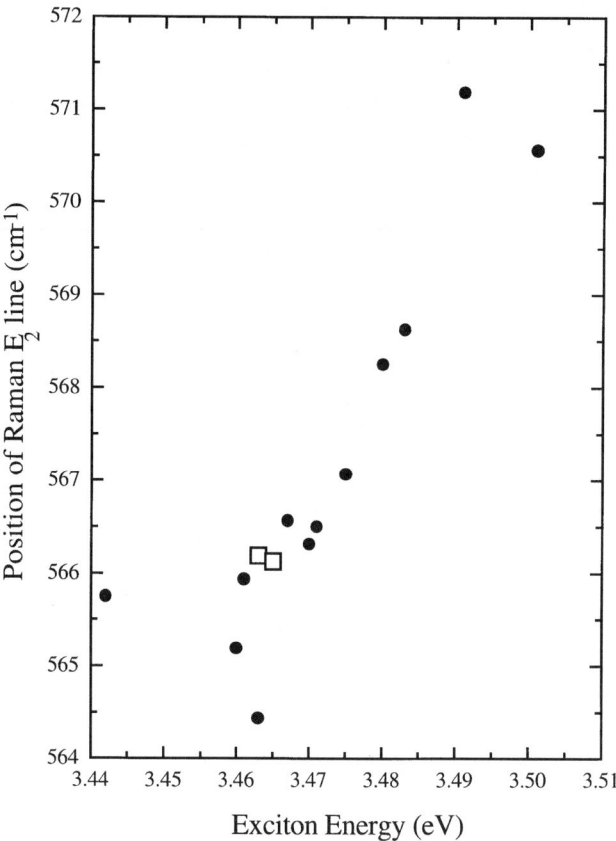

FIG. 8. Position of the E_2 (high) Raman peak as a function of energy of recombination of bound excitons. Internal stress is as high as 1 GPa for the compressive strain and 0.5 GPa for the tensile stress. (From Suski, *et al.*, 1996.)

5. Two-Phonon Raman Spectra

In contrast to the first order Raman scattering, where only phonons with zero wavenumber can be involved, two-phonon spectroscopy can provide potentially much more detailed information about the phonon dispersion relations. However, the experimental data on two phonon scattering in GaN are not abundant. Murugkar, *et al.*, 1995) have demonstrated (Fig. 9) that the second-order spectra of GaN are dominated by longitudinal phonons. The Raman band is centered around 1460 cm^{-1} which is very close to twice

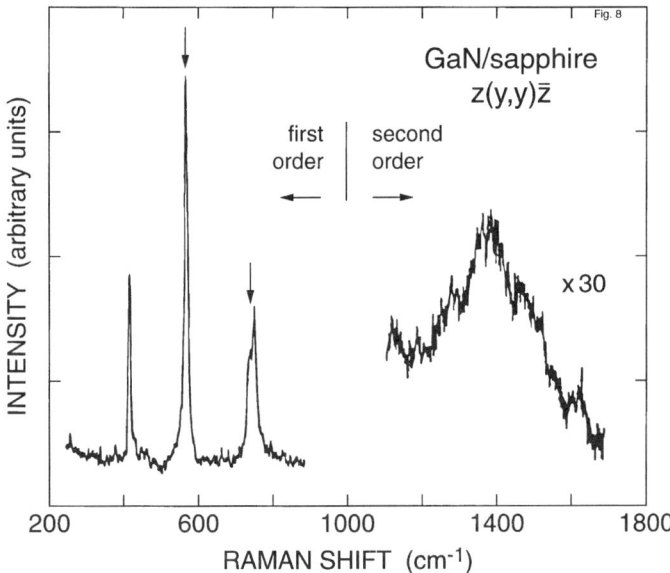

FIG. 9. First and second order Raman spectra of GaN. Arrows indicate positions of zone-center optical phonons. (From Murugkar, *et al.*, 1995.)

the energy of the $A_1(LO)$ (Γ) phonon. The contributions from the E_2 phonons were much weaker.

6. LOCAL VIBRATIONAL MODES

During the last years the search for local vibrational modes was related to the development of a model of the Mg_{Ga}-H complex in GaN. Magnesium is the most important dopant for the *p*-type conductivity and therefore the study of such impurities is of considerable technological importance. The Raman scattering study of Brandt, *et al.*, (1994) suggested that lines observed at 2168 cm^{-1} and 2219 cm^{-1} are related to the vibrations of the Mg-H complex. The first peak would be assigned to the A' vibration of the Mg-H complex oriented in the *c*-plane, and the 2219 cm^{-1} line would be associated with A_1 vibration of Mg-H parallel to the *c*-direction. Very different results appeared from the IR absorption experiments by Götz, *et al.*, (1996) who observed a line at frequency 3125 cm^{-1}. The assignment of this line to the Mg-H complex was further supported by observation of isotopic shift (after deuteration), and a decrease of line intensity after thermal activation of the sample. Theoretical studies of this important defect complex were made by Neugebauer and Van de Walle (1996), who in particular examined the role of hydrogen in enhancing the *p*-type doping.

IV. Summary and Conclusions

Experimental and theoretical studies of the structural and vibrational properties of GaN have been presented, and in particular we described the effects of applying external pressure. The hexagonal wurtzite structure is assumed by GaN at ambient conditions, but application of a pressure of ~ 50 GPa causes a transition to the cubic rocksalt structure. Experiments and theoretical calculations have shown (Christensen, and Gorczyca, 1994) that a similar transition in AlN occurs at a considerably lower pressure, 12–15 GPa. This is somewhat surprising because AlN and GaN have very similar equilibrium bond lengths and ionicities. An analysis (Christensen, and Gorczyca, 1994) of orbital partial pressures (Christensen, and Heine, 1985) has shown the difference in transition pressure to be due to the (semi)core 3*d*-states on the cation of GaN which have no counterpart in AlN. The present results for GaN show that excellent agreement between observed and calculated pressure-volume relations can be obtained. Similar agreement was found (Christensen, and Gorczyca, 1994; and Christensen, 1997) for other III-nitrides, but especially AlN has internal structural parameters (c/a and u) which vary with pressure (Christensen, and Gorzcyca, 1993).

Phonon modes have been examined by means of Raman spectroscopy, a technique that earlier was difficult to apply to GaN due to the lack of sample with low electron concentrations. Theoretical calculations support the interpretation of the experimental spectra and they further reproduce the pressure dependences well (mode Grüneisen parameters).

Local vibration modes of defect complexes were briefly mentioned, and Mg-H in GaN was taken as an example. The reason is that the use of Mg as a dopant has essentially solved the difficult problem of obtaining *p*-type GaN. Despite the progress in growth methods, several questions, mainly related to native defects in GaN (and the other III-nitrides) still have not been answered. These are essential to the optoelectronic applications of the nitrides and they are therefore studied extensively, (Neugebauer, and Van de Walle, 1996; and Mattila, Seitsonen, and Niemenen, 1996). Also the study of defects, including the local structure (relaxation), under pressure (Gorczyca, Svane, and Christensen, 1997) gives information about material parameters important to the device applications.

ACKNOWLEDGMENTS

Much of the work reviewed here results from collaboration and discussions with many of our colleagues. In particular we wish to thank I. Gorczyca, I. Grzegory, A. Polian, A. Svane, and T. Suski, A. Heiring is thanked for her expert assistance in composing the manuscript.

REFERENCES

Arguello, C. A., Rousseau, D. L., and Porto, S. P. S. (1969). "First-Order Raman Effect in Wurtzite-Type Crystals". *Phys. Rev.* **181**, 1351.
Barker, A. S., and Illegems, M. (1972). "Infrared Lattice Vibrations and Free-Electron Dispersion in GaN". *Phys. Rev.* **B7**, 743.
Brand, M. S. T., Ager, III, J. W., Götz, W., Johnson, N. M., Harris, J. S. Jr., Molnar R. J., and Moustakas, T. D. (1994). "Local Vibrational Modes in Mg-Doped Gallium Nitride". *Phys. Rev.* **B49**, 14758.
Burns, G., Dacol, F., Marinace, J. C., and Scott, B. A. (1973). "Raman Scattering in Thin-Film Waveguides". *Appl. Phys. Lett.* **22**, 356.
Camassel, J., Beamont, B., Talercio, T., Malzac, J. P., Shwedler, R., Gobart, P., and Perlin, P. (1995). "Raman and Micro-Raman Spectroscopy of GaN Layers Deposited on Sapphire". *Inst. Phys. Conf. Ser.* **142**, Ch. 5, p. 959.
Van Camp, P. E., Van Doren, V. E., and Devreese, J. T. (1992). "High Pressure Structural Phase Transformation in Gallium Nitride". *Solid State Commun.* **81**, 23.
Christensen, N. E., and Heine, V. (1985). "Analysis of the Electronic Pressure in Transition and Noble Metals", *Phys. Rev.* **B32**, 6145.
Christensen, N. E., and Gorzcyca, I. (1993). "Calculated Structural Transitions of Aluminum Nitride Under Pressure", *Phys. Rev.* **B47**, 4307.
Christensen, N. E., and Gorczyca, I. (1994). "Optical and Structural Properties of III-V Nitrides Under Pressure". *Phys. Rev.* **B50**, 4397.
Christensen, N. E. (1997). "Electronic Structures of Semiconductors Under Pressure". In "High Pressure in Semiconductor Physics", (Suski, T., and Paul, W., eds.), Academic Press.
Gorczyca, I., and Christensen, N. E. (1991). "Band Structure and High-Pressure Phase Transition in GaN". *Solid State Commun.* **80**, 335.
Gorczyca, I., Christensen, N. E., Pelzer y Blanca, E. L., and Rodriguez, C. O. (1995). "Optical Phonon Modes in GaN and AlN". *Phys. Rev.* **B51**, 11936.
Gorczyca, I., Svane, A., and Christensen, N. E. (1997). "Calculated Defect Levels in GaN and AlN and their Pressure Coefficients". *Solid State Commun.* (in print).
Götz, W., Johnson, N. M., Bour, D. P., McCluskey M. D., and Haller, E. E. (1996). "Local Vibrational Modes of the Mg-H Acceptor Complex in GaN". *Appl. Phys. Lett.* **69**, 3725.
Hung-Red, K., Ming-Shiann, F., Jeng-Dah, G., and Ming-Chih, L. (1995). "Raman Scattering of Se-Doped Gallium Nitride Films". *J. Appl. Phys.* **34**, 5628.

Kim, K., Lambrecht, W. R. L., and Segall, B. (1994). "Electronic Structure of GaN with Strain and Phonon Distortions". *Phys. Rev.* **B50**, 1502.

Kim, K., Lambrecht, W. R. L., and Segall, B. (1996). "Elastic Constants and Related Properties of the Group III-Nitrides". *Mat. Res. Symp. Proc.* Vol. 395, 399; and *Phys. Rev.* **B53**, 16310.

Kleinman, L. (1962). "Deformation Potentials in Silicon. I. Uniaxial Strain". *Phys. Rev.* **128**, 2614.

Kozawa, T., Kachi, T., Kano, H., Taga, Y., and Hashimoto, M. (1994). "Raman Scattering from LO Phonon-Plasmon Coupled Modes in Gallium Nitride". *J. Appl. Phys.* **75**, 1098.

Lambrecht, W. R. L. (1997). "Band Structure of the Group-III Nitrides".

Lei, T., Fanciulli, M., Molnar, R. J., Moustakas, T. D., Graham, R. J., and Scanlon, J. (1991). "Epitaxial Growth of Zincblende and Wurtzitic Gallium Nitride Thin Films on (001) Silicon". *Appl. Phys. Lett.* **59**, 944.

Lei, T., Ludwig, K. F., and Moustakas, T. D. (1993). "Heteroepitaxy, Polymorphism, and Faulting in GaN Thin Films on Silicon and Sapphire Substrates". *J. Appl. Phys.* **74**, 4430.

Lemos, V., Arguello, C. A., and Leite, R. C. C. (1972). "Resonant Raman Scattering of $TO(A_1)$, $TO(E_1)$ and E_2 Optical Phonons in GaN". *Solid State Commun.* **11**, 1351.

Leszczynski, M., Suski, T., Perlin, P., Teisseyre, H., Grzegory, I., Bockowski, M., Jun, J., Porowski, S., and Major, J. (1995). "Lattice Constants, Thermal Expansion and Compressibility of Gallium Nitride". *J. Phys.* **D28**, A149.

Majewski, J. A., and Vogl, P. (1987). "Simple Model for Structural Properties and Crystal Stability of sp-Bonded Solids". *Phys. Rev.* **B35**, 9666.

Manchon, D. D., Barker, A. S., Dean, P. J., and Zetterstrom R. B. (1970). "Optical Studies of the Phonons and Electrons in Gallium Nitride". *Solid State Commun.* **8**, 1227.

Martin, R. M. (1972). "Relation Between Elastic Tensors of Wurtzite and Zinc-Blende Structure Materials". *Phys. Rev.* **B6**, 4546.

Mattila, T., Seitsonen, A. P., and Niemenen, R. M. (1996). "Large Atomic Displacements Associated with Nitrogen Antisite in GaN". *Phys. Rev.* **B54**, 1454.

Miwa, K., and Fukomoto, A. (1993). "First-Principles Calculation of the Structural, Electronic, and Vibrational Properties of Gallium Nitride and Aluminum Nitride". *Phys. Rev.* **B48**, 7897.

Mohammad, S. N., and Morkoç, H. (1996). "Progress and Prospects of Group-III Nitride Semiconductors". *Progr. Quant. Electr.* **20**, 361.

Muñoz, A. and Kunc, K. (1991). "High-Pressure Phase of Gallium Nitride". *Phys. Rev.* **B44**, 10372.

Murnaghan, F. D. (1944). "The Compressibility of Media Under Extreme Pressures". *Proc. Natl. Acad. Sci. USA* **30**, 24.

Murugkar, S., Merlin, R., Botchakar???, A., Salvador, A., and Morkoç, H. (1995). "Second-Order Raman Spectroscopy of the wurtzite form of GaN". *J. Appl. Phys.* **77**, 6042.

Neugebauer, J., and Van de Walle, C. G. (1996). "Theory of Point Defects and Complexes in GaN". *Mat. Res. Symp. Proc.* **395**, 645.

Paisley, M. J., Sitar, Z., Posthill, J. B., and Davis, R. F. (1989). "Growth of Cubic Phase Gallium Nitride by Modified Molecular Beam Epitaxy". *J. Vac. Sci. Technol.* **A7**, 701.

Perlin, P., Camassel, J., Knap, W., Taliercio, T., Chervin, J. C., Suski, T., Grzegory, I., and Porowski, S. (1995). "Investigation of Longitudinal-Optical Phonon–Plasmon Coupled Modes in Highly Conducting Bulk GaN". *Appl. Phys. Lett.* **67**, 2524.

Perlin, P., Jauberthie-Carillon, C., Itie, J. P., San Miguel, A., Grzegory, I., and Polian, A. (1991). "High Pressure Phase Transition in Gallium Nitride". *High Pressure Research* 7–8, 96; Perlin, P., Jauberthie-Carillon, C., Itie, J. P., San Miguel, A., Grzegory I., and Polian, A. (1992). "Raman Scattering and X-ray Absorption Spectroscopy in Gallium Nitride Under High Pressure". *Phys. Rev.* **B45**, 83.

Perlin, P., Suski, T., Teisseyre, H., Leszczynski, M., Grzegory, I., Jun, J., Porowski, S., Boguslawski, P., Bernholc, J., Chervin, J. C., Polian, A., and Moustakas, T. D. (1995).

Perlin, P., Suski, T., Polian, A., Chervin, J. C., Knap, W., Camassel, J., Grzegory, I., Porowski, S., and Erickson, J. W. (1996). "Spatial Distribution of Electron Concentration and Strain in Bulk GaN Single Crystals Relation to Growth Mechanism". In Proceedings of Fall MRS Meeting Boston, December.

Polian, A., Grimsditch, M., and Grzegory, I. (1996). "Elastic Constants of Gallium Nitride". *J. Appl. Phys.* **79**, 3343.

Powell, R. C., Tomasch, G. A., Kim, Y. W., Thorton, J. A., and Greene, J. E. (1990). "Growth of High-Resistivity Wurtzite and Zincblende Structure Single Crystal GaN by Reactive-Ion Molecular Beam Epitaxy". *Mater. Res. Soc. Symp. Proc.* **162**, 525.

Savastenko, V. A., and Sheleg, A. U. (1978). "Study of the Elastic Properties of Gallium Nitride". *Phys. Stat. Solidi* **A48**, K135.

Siegle, H., Eckey, L., Hoffmann, A., Thomsen, C., Meyer, B. K., Schikora, D., Hankeln, M., and Lischka, K. (1995). "Quantative Determination of Hexagonal Minority Phase in Cubic GaN Using Raman Spectroscopy". *Solid State Commun.* **96**, 943.

Strauss, U., Tews, H., Riechert, H., Averbeck, R., Schienle, M., Jobst, B., Volm, D., Streibl, T., Meyer, B. K., and Rohle, W. W. (1996). "Identification of a Cubic Phase in Epitaxial Layers of Predominantly Hexagonal GaN". *MRS Internet J. Nitride Semicond. Res.* **1**, 44.

Strite, S., Ruan, J., Li, Z., Manning, N., Salvador, A., Chen, H., Smith, D. J., Choyke, W. J., and Morkoç, H. (1991). "An Investigation of the Properties of Cubic GaN Grown on GaAs by Plasma-Assisted Molecular-Beam Epitaxy". *J. Vac. Sci. Technol.* **B9**, 1924.

Suski, T., Krueger, J., Kisielowski, C., Phatak, P., Leung, M. S. H., Lilienthal-Weber, Z., Gassman, A., Newman, N., Rubin, M. D., Weber, E. R., Grzegory, I., Jun, J., Bockowski, M. Porowski, S., and Helava, H. I. (1996). "Properties of Homoepitaxially MBE-Grown GaN". In Proceedings of MRS Spring Meeting, San Francisco.

Suski, T., Ruimov, S., Krueger, J., Conti, G., Weber, E. R., Bremser, N. D., Davis, R. F., and Kuo, C. (1996). "Internal Stress Effects in Si-Doped, Ge-Doped and Undoped Gallium Nitride Epitaxial Films". In Proceedings of MRS Fall Meeting, Boston, December.

Ueno, M., Yoshida, M., Onodera, A., Shimomura, O., and Takemura, K. (1994). "Stability of Wurtzite-Type Structure Under High Pressure: GaN, InN". *Phys. Rev.* **B49**, 14.

Van Vechten, J. A. (1969). "Quantum Dielectric Theory of Electronegativity in Covalent Systems. I. Electronic Dielectronic Constant." *Phys. Rev.* **182**, 891.

Weinstein, B. A. (1977). "Phonon Dispersion of Zinc Chalcogenides Under Extreme Pressure and the Metallic Transformation". *Solid State Commun.* **24**, 595.

Xia, H., Xia, Q., and Ruoff, A. L. (1993). "High Pressure Structure of Gallium Nitride. Wurtzite to Rocksalt Phase Transition". *Phys. Rev.* **B47**, 12925.

CHAPTER 14

Applications of LEDs and LDs

S. Nakamura

R&D Department
Nichia Chemical Industries, Ltd.
491 Oka, Kaminaka, Anan, Tokushima 774, Japan

I.   INTRODUCTION 431
II. InGaN/AlGaN DOUBLE-HETEROSTRUCTURE (DH) LEDs 434
III. InGaN SINGLE-QUANTUM-WELL (SQW) STRUCTURE LEDs 439
IV. EMISSION MECHANISM OF SINGLE-QUANTUM-WELL LEDs 445
V.   InGaN MULTI-QUANTUM-WELL (MQW) STRUCTURE LDs 448
VI. SUMMARY 456
     References 456

I. Introduction

Much research has been done on high-brightness blue light-emitting diodes (LEDs) and laser diodes (LDs) for use in full-color displays, full-color indicators and light sources for lamps, with the characteristics of high efficiency, high reliability and high speed. For these purposes, II–VI materials such as ZnSe (Xie et al., 1992; Eason et al., 1995), SiC (Edmond et al., 1994) and III–V nitride semiconductors such as GaN (Pankove et al., 1971) have been investigated intensively for a long time. However, it was impossible to obtain high-brightness blue LEDs with a brightness over 1 cd and reliable LDs. Much progress has been achieved recently on green LEDs and LDs using II–VI based materials (Okuyama and Ishibashi, 1994). The short lifetimes prevent II–VI based devices from commercialization at present. It is considered that the short lifetime of these II–VI based devices is caused by the crystal defects at a density of $10^4/cm^2$ because one crystal defect would cause the propagation of other defects leading to a failure of the devices. SiC is another wide bandgap material for blue LEDs. The brightness of SiC blue LEDs is only between 10 mcd and 20 mcd because of the indirect bandgap of this material. Despite this poor performance, 6H-SiC blue LEDs have been commercialized for a long time because there has been no competition for blue light emitting devices (Koga and Yamaguchi, 1991).

On green devices, the external quantum efficiency of coventional green GaP LEDs is only 0.1% due to the indirect bandgap of this material and the peak wavelength is 555 nm (yellowish green) (Craford, 1992). As another material for green emission devices, AlInGaP has been used. The present performance of green AlInGaP LEDs is an emission wavelength of 570 nm (yellowish green) and maximum external quantum efficiency of 1% (Craford, 1992; Sugawara et al., 1994). When the emission wavelength is reduced to the green region, the external quantum efficiency drops sharply, because the band structure of AlInGaP becomes nearly indirect. Therefore, high-brightness pure green LEDs, which have a high efficiency above 1% at the peak wavelength between 510–530 nm with a narrow full-width at half maximum (FWHM), have not been commercialized yet.

GaN and related materials such as AlGaInN are III–V nitride semiconductors with the wurtzite crystal structure and a direct energy band structure which is suitable for light emitting devices. The bandgap energy of AlGaInN varies between 6.2 and 1.95 eV depending on its composition at room temperature (Fig. 1). Therefore, these III–V nitride semiconductors are useful for light emitting devices especially in the short wavelength regions. The spinel ($MgAl_2O_4$) (9.5%) and SiC substrates (3.5%) which have

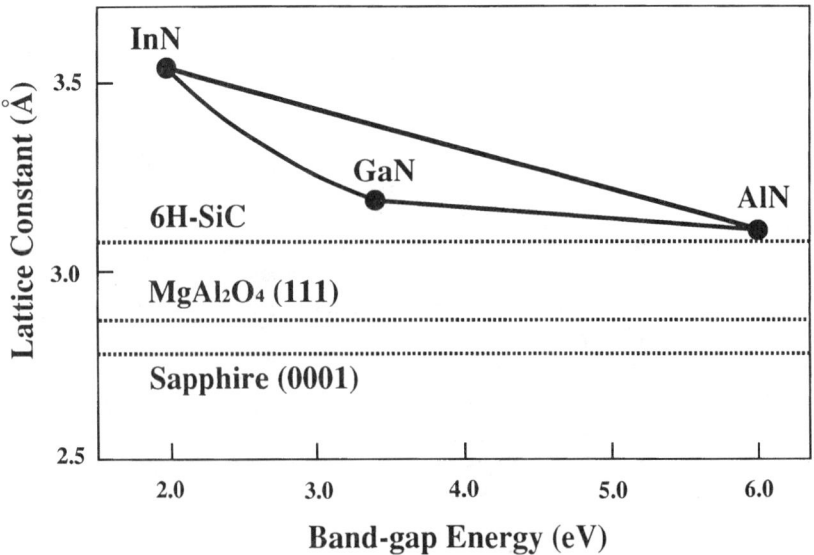

FIG. 1. Lattice constant of III–V nitride compounds as a function of their bandgap energy.

a small lattice mismatch in comparison with that (13%) between GaN and sapphire are also shown in Fig. 1. These substrates have been used for the growth of the III–V nitride based blue LED and LD structures. Recent research on III–V nitrides has paved the way for the realization of high-quality crystals of GaN, AlGaN, and GaInN, and of p-type conduction in GaN and AlGaN (Amano *et al.*, 1989; Nakamura, 1991; Strite and Morkoc, 1992; Morkoc *et al.*, 1994). The mechanism of the acceptor-compensation which prevents obtaining low-resistivity p-type GaN and AlGaN has been elucidated (Nakamura *et al.*, 1992; Rubin *et al.*, 1994; Brandt *et al.*, 1994; Zavada *et al.*, 1994). In Mg-doped p-type GaN, Mg acceptors are deactivated by atomic hydrogen which is produced from NH_3 gas used as the N source during GaN growth. High-brightness blue LEDs have been fabricated on the basis of these results, and luminous intensities over 1 cd have been achieved (Nakamura, 1994a, b). These LEDs are now commercially available. Also, high brightness single-quantum-well structure (SQW) blue, green and yellow InGaN LEDs with a luminous intensity above 10 cd have been achieved and commercialized (Nakamura *et al.*, 1995a, b, c). By combining these high-power and high-brightness blue In-GaN SQW LED, green InGaN SQW LED and red AlInGaP LED, many kinds of applications, such as LED full-color displays and LED white lamps for use in place of light bulbs or fluorescent lamps, are now possible with characteristics of high reliability, high durability and low energy consumption.

At present, the main focus of III–V nitride research is the realization of a current-injected laser diode which can be operated under a continuous-wave (CW) at room temperature. Recent developments have yielded an optically pumped stimulated emission from GaN films (Amano *et al.*, 1990; Zubrilov *et al.*, 1995), InGaN films (Khan *et al.*, 1994; Kim *et al.*, 1995), AlGaN/InGaN double heterostructures (Amano *et al.*, 1994) and GaN/AlGaN double heterostructures (Aggarwal *et al.*, 1996; Schmidt *et al.*, 1996). However, stimulated emission had been observed only with optical pumping, not current injection. The first current-injection III–V nitride-based LDs were fabricated by the present authors using the InGaN multi-quantum-well (MQW) structure as an active layer (Nakamura *et al.*, 1996a). The laser emission wavelength (417 nm) was the shortest one ever generated by a semiconductor LD. The mirror facet for the laser cavity was formed by etching of III–V nitride films due to the difficulty in cleaving the (0001) C-face sapphire substrate. The etched facet surface was relatively rough (approximately 500 Å). Also, sapphire substrate with (11$\bar{2}$0) orientation (A-face) was used to fabricate the InGaN MQW LDs because A-face sapphire could be cleaved along (1$\bar{1}$02) (R-face) (Nakamura *et al.*, 1996b). The InGaN MQW LD structures were also fabricated on spinel ($MgAl_2O_4$)

substrates, which has a small lattice mismatch (9.5%) in comparison with that (13%) between GaN and sapphire (Nakamura et al., 1996c, d). These LDs emitted coherent light at 390–440 nm from an InGaN based MQW structure under pulsed current injection at room temperature. Here, present status of III–V nitride based light emitting devices are described.

II. InGaN/AlGaN Double-Heterostructure (DH) LEDs

Figure 2 shows the structure of the InGaN/AlGaN DH LEDs (Nakamura, 1994a, b, c, 1995). The active layer is InGaN codoped with Si and Zn to enhance the blue emission. The blue emission intensity is a maximum at an electron carrier concentration of 1×10^{19} cm^{-3}. The need for codoping suggests that the high-efficiency of this InGaN/AlGaN DH LED is the result of impurity-assisted, for example, free-carrier-acceptor (FA), recombination. A p-type GaN layer was used as the contact layer for the p-type electrode in order to improve the ohmic contact. After the growth, N_2 ambient thermal annealing at a temperature of 700°C was performed to obtain highly p-type GaN and AlGaN layers. Fabrication of LED chips was accomplished as follows: the surface of the p-type GaN layer was partially etched until the n-type GaN layer was exposed. Next, a Ni/Au contact was evaporated onto the p-type GaN layer and a Ti/Al contact onto

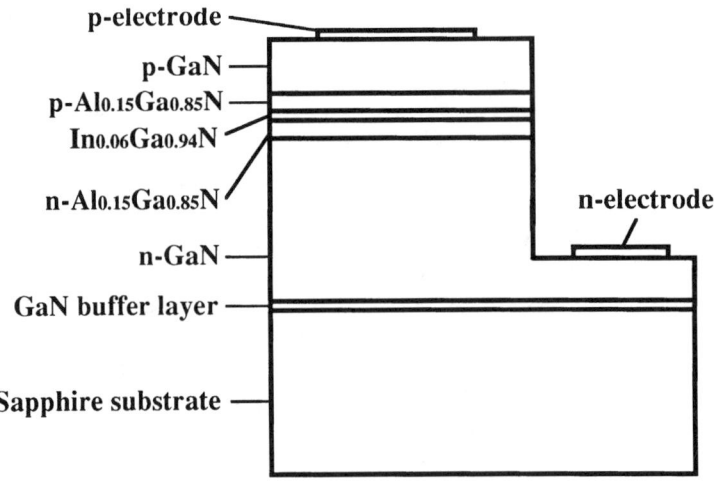

FIG. 2. The structure of the InGaN/AlGaN DH LEDs.

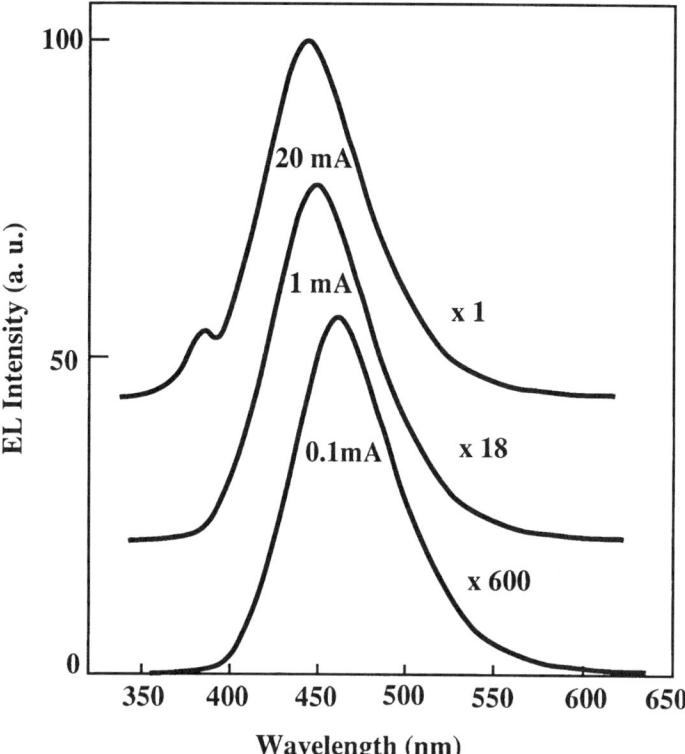

FIG. 3. Electroluminescence spectra of the InGaN/AlGaN DH blue LEDs under different forward currents.

the n-type GaN layer. The wafer was cut into a rectangular shape (350 μm × 350 μm). These chips were set on the lead frame, and were then molded. The characteristics of LEDs were measured under DC-biased conditions at room temperature.

Figure 3 shows the electroluminescence (EL) spectra of the InGaN/ AlGaN DH blue LEDs at forward currents of 0.1 mA, 1 mA and 20 mA. The carrier concentration of the InGaN active layer in this LED was 1×10^{19} cm^{-3}. Typical peak wavelengths and values of FWHM of the EL at 20 mA were 450 nm and 70 nm, respectively. The peak wavelength shifts to shorter wavelengths with increasing forward current. The peak wavelength is 460 nm at 0.1 mA, 449 nm at 1 mA and 447 nm at 20 mA. At 20 mA, a narrower, higher-energy peak emerges at around 385 nm, as shown in Fig. 3. This peak, due to band-to-band recombination in the InGaN active layer, becomes resolved at injection levels where the impurity-related

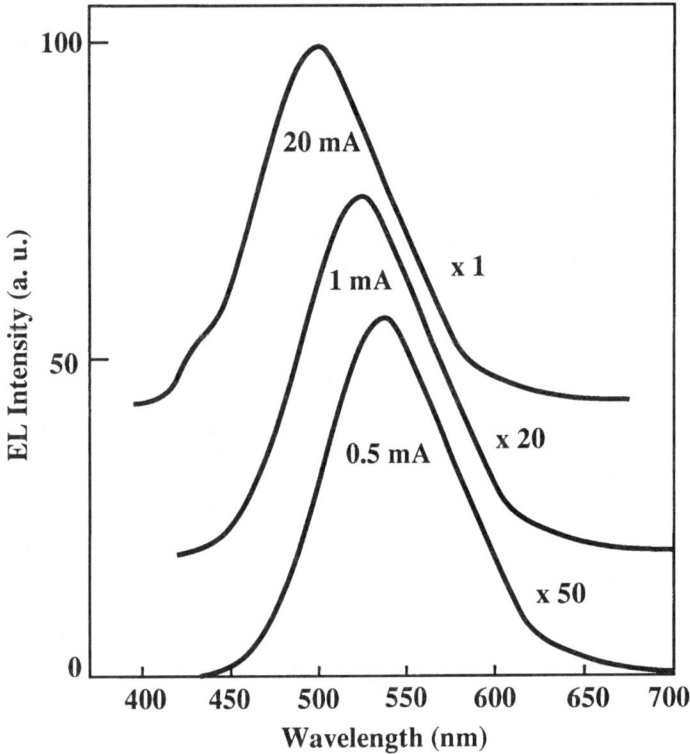

FIG. 4. Electroluminescence spectra of the InGaN/AlGaN DH blue-green LEDs under different forward currents.

recombination is saturated. The output power of the InGaN/AlGaN DH blue LEDs is 1.5 mW at 10 mA, 3 mW at 20 mA and 4.8 mW at 40 mA. The external quantum efficiency is 5.4% at 20 mA. The typical on-axis luminous intensity of InGaN/AlGaN LEDs with a 15° conical viewing angle is 2.5 cd at 20 mA. The forward voltage is 3.6 V at 20 mA. High-brightness blue LEDs with a luminous intensity over 1 cd will pave the way forward realization of full-color LED displays, especially for outdoor use.

Blue-green LEDs were fabricated for application to traffic lights by increasing the indium mole fraction of the InGaN active layer from 0.06 to 0.19 (Nakamura *et al.*, 1994b). Figure 4 shows the EL spectra of these InGaN/AlGaN DH LEDs at forward currents of 0.5 mA, 1 mA and 20 mA. Typical peak wavelengths and values of FWHM of the EL at 20 mA were 500 nm and 80 nm. The peak wavelength shifts from 537 nm at 0.5 mA to

14 Applications of LEDs and LDs

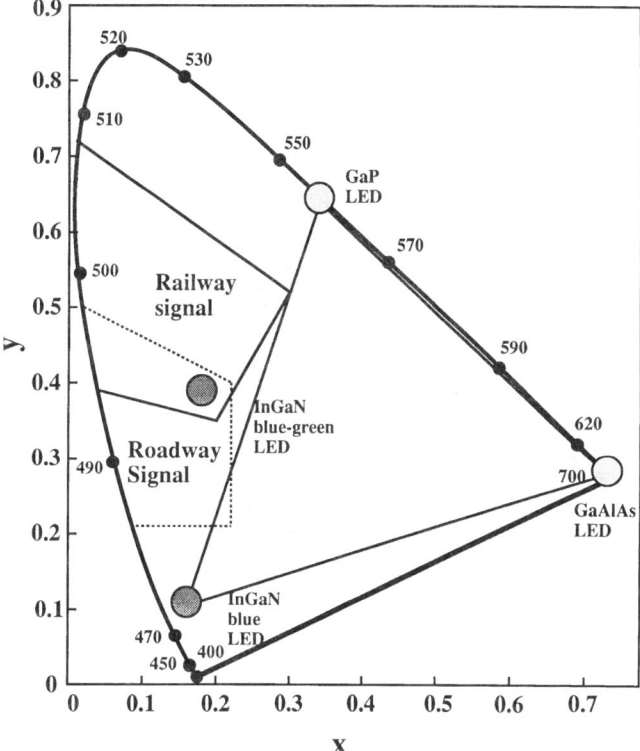

FIG. 5. Chromaticity diagram where blue InGaN/AlGaN LEDs, blue-green InGaN/AlGaN LEDs, green GaP LEDs, and red GaAlAs LEDs are shown.

525 nm at 1 mA, and 500 nm at 20 mA. The output power of the InGaN/AlGaN DH blue-green LEDs is 1.0 mW at 20 mA, where the external quantum efficiency is 2.1%. A typical on-axis luminous intensity with a 15° conical viewing angle is 2 cd at 20 mA. This luminous intensity is sufficiently bright for outdoor application, such as traffic lights and displays. The forward voltage was 3.5 V at 20 mA.

Figure 5 shows a chromaticity diagram where blue and blue-green InGaN/AlGaN LEDs are shown (Nakamura, 1995). Commercially available green GaP LEDs and red GaAlAs LEDs are also shown. From this figure, only the blue-green InGaN/AlGaN LEDs are within the applicable regions for roadway signals and railway signals. Therefore, InGaN/AlGaN blue-green LEDs can be used for those applications from the viewpoint of color. Traffic lights may prove to be a great application for the blue-green

Fig. 6. The first actual LED traffic light, which was set in Japan in 1994, using InGaN/AlGaN blue-green, AlInGaP yellow, and GaAlAs red LEDs.

LEDs. Total power consumption by traffic lights reaches the gigawatt range in Japan. InGaN/AlGaN blue-green LED traffic lights, with an electrical power consumption of only 12% that of present incandescent bulb-traffic lights, promise to save vast amounts of energy. With its extremely long lifetime of several tens of thousands of hours, the replacement of burned-out traffic light bulbs will be dramatically reduced. Using these high-brightness blue-green LEDs, safe and energy-efficient roadway and railway signals can be achieved. Figure 6 shows the first actual LED traffic light which was set in Japan in 1994 using InGaN/AlGaN blue-green, AlInGaP yellow and AlInGaP red LEDs.

Using High-brightness InGaN/AlGaN blue, GaP green and GaAlAs red LEDs, full-color LED displays, especially for outdoor use can be fabricated at present. Figure 7 shows the actual LED full-color display (10 m × 10 m) which was set in Japan in 1994 for the first time. The color range of light emitted by a full-color LED lamp in the chromaticity diagram is shown as the region inside a triangle which is drawn by connecting the positions of the three primary color LED lamps, in Fig. 5. The color range is not so wide

14 APPLICATIONS OF LEDs AND LDs 439

FIG. 7. The actual LED full-color display, which was set in Japan in 1994 for the first time. The blue InGaN/AlGaN LEDs, green GaP LEDs and red GaAlAs LEDs are used as three primary color LEDs.

because the color of the green GaP LEDs is yellowish green (555 nm), not pure green. This means that this LED full-color display cannot express an actual nature-color, expecially in the green region. For pure green LEDs, emission wavelengths between 510 nm and 530 nm are required. Also, the luminous intensity of the green GaP LEDs (0.1 cd) is too low in comparison with that of blue InGaN/AlGaN DH LEDs (2 cd) and red GaAlAs LEDs (2 cd) to fabricate high brightness LED full color displays. In order to obtain high-brightness pure blue and green LEDs, the InGaN SQW LEDs have been developed as below.

III. InGaN Single-Quantum-Well (SQW) Structure LEDs

High-brightness blue and blue-green InGaN/AlGaN DH LEDs with a luminous intensity of 2 cd have been fabricated and are now commercially available, as mentioned above (Nakamura, 1994a, 1995). In order to obtain blue and blue-green emission centers in these InGaN/AlGaN DH LEDs, Zn doping into the InGaN active layer was performed. Although these InGaN/AlGaN DH LEDs produced high-power light output in the blue and blue-green regions with a broad emission spectrum (FWHM = 70 nm), green or yellow LEDs with peak wavelengths longer than 500 nm have not been fabricated. The longest peak wavelength of the EL of InGaN/AlGaN DH LEDs achieved thus far is 500 nm (blue-green) because the crystal quality of the InGaN active layer of DH LEDs becomes poor when the indium mole fraction is increased in order to obtain a green band-edge emission.

In conventional green GaP LEDs, the external quantum efficiency is only 0.1% due to the indirect transition bandgap material, and the peak wavelength is 555 nm (yellowish green) (Craford, 1992). As another material for green emission devices, AlInGaP has been used. The present green AlInGaP LEDs have an emission wavelength of 564 nm (yellowish green) and an external quantum efficiency of 0.6% (Kish et al., 1994). When the emission wavelength is reduced to the green region, the external quantum efficiency drops sharply because the band structure of AlInGaP approaches the indirect region. Therefore, high-brightness pure green LEDs, having a high efficiency of above 1% and a peak wavelength of between 510–530 nm with a narrow FWHM, have not been commercialized yet.

Violet InGaN/AlGaN DH LEDs with a narrow spectrum (FWHM = 10 nm) at a peak wavelength of 400 nm originating from the band-to-band

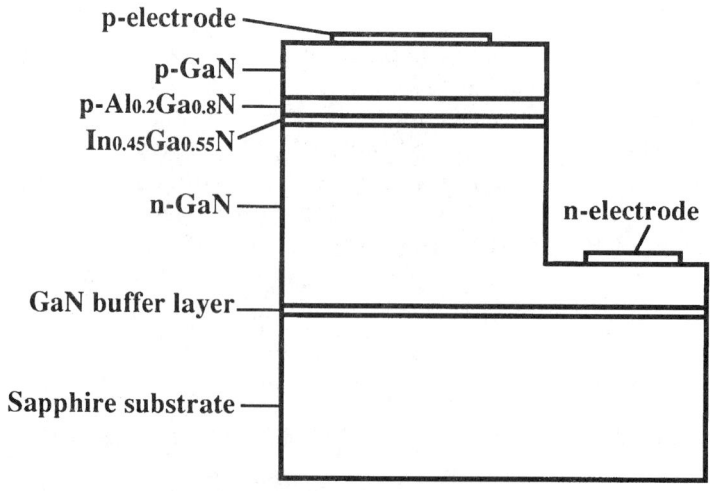

FIG. 8. The structure of green SQW LED.

emission of InGaN were fabricated (Nakamura, 1994b). However, the output power and the external quantum efficiency of these violet LEDs were only 1 mW and 1.6%, respectively, probably due to the formation of misfit dislocation in the thick InGaN active layer (about 1000 Å) caused by the stress introduced into the InGaN active layer due to lattice mismatch, and the difference in thermal expansion coefficients between the InGaN active layer and AlGaN cladding layers. When the InGaN active layer becomes thin, it is expected that the elastic strain is not relieved by the formation of misfit dislocations and that the crystal quality of the InGaN active layer improves. High-quality InGaN multi-quantum-well structures (MQW) with 30 Å well and 30 Å barrier layers were grown (Nakamura et al., 1993). Here, the quantum-well structure (QW) LEDs which have a thin InGaN active layer (about 30 Å) in order to obtain high-power emission in the region from blue to yellow with a narrow emission spectrum is described (Nakamura et al., 1995a, b, c).

The green InGaN SQW LED device structures (Fig. 8) consist of a 300 Å GaN buffer layer grown at a low temperature (550°C), a 4-μm-thick layer of n-type GaN:Si, a 30-Å-thick active layer of undoped $In_{0.45}Ga_{0.55}N$, a 1000-Å-thick layer of p-type $Al_{0.2}Ga_{0.8}N$:Mg, and a 0.5-μm-thick layer of p-type GaN:Mg. The active region is a SQW structure consisting of a 30-Å-thick $In_{0.45}Ga_{0.55}N$ well layer sandwiched by 4-μm-thick n-type GaN and 1000-Å-thick p-type $Al_{0.2}Ga_{0.8}N$ barrier layers.

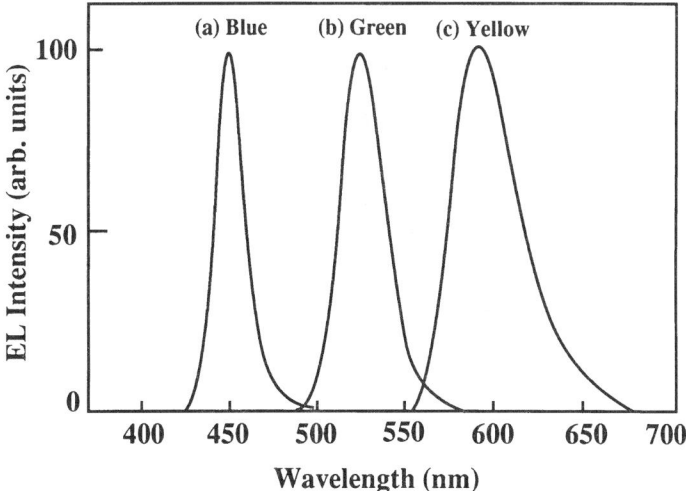

FIG. 9. Electroluminescence of (a) blue, (b) green, and (c) yellow SQW LEDs at a forward current of 20 mA.

Figure 9 shows the typical EL of the blue, green and yellow SQW LEDs with different indium mole fractions of the InGaN well layer at a forward current of 20 mA (Nakamura et al., 1995a, b, c). The peak wavelength and the FWHM of the typical blue SQW LEDs are 450 nm and 20 nm, respectively, and those of the green SQW LEDs are 525 nm and 30 nm, and those of yellow are 600 nm and 50 nm, respectively. When the peak wavelength becomes longer, the FWHM of the EL spectra increases, probably due to the inhomogeneities of InGaN layer or the strain between well and barrier layers of the SQW, which is caused by mismatch of the lattice and the thermal expansion coefficients between well and barrier layers. At 20 mA, the output power and the external quantum efficiency of the blue SQW LEDs are 5 mW and 9.1%, respectively. Those of the green SQW LEDs are 3 mW and 6.3%, respectively. A typical on-axis luminous

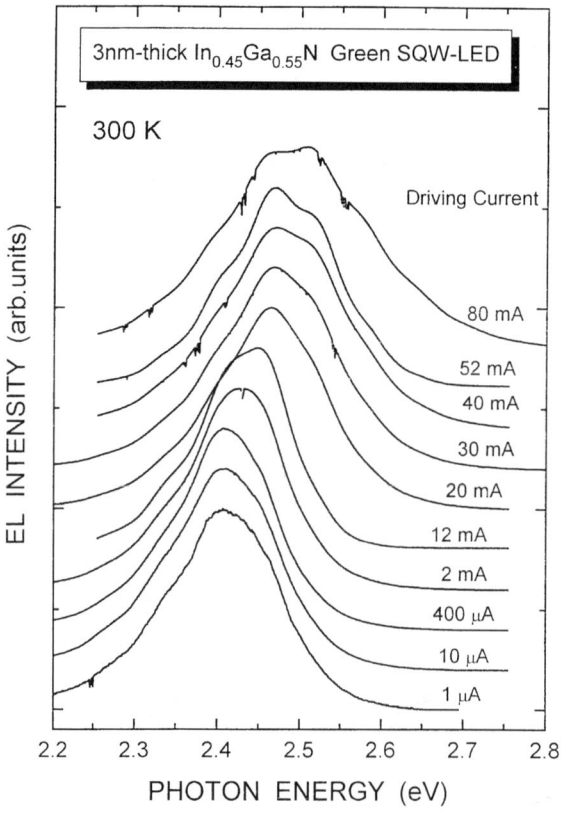

FIG. 10. Room-temperature EL of typical green SQW LEDs with different driving currents.

intensity of the green SQW LEDs with a 10° cone viewing angle is 10 cd at 20 mA. These values of output power, external quantum efficiency and luminous intensity of blue/green SQW LEDs are the highest ever reported for blue/green LEDs. By combining these high-power and high-brightness blue InGaN SQW, green InGaN SQW and red GaAlAs LEDs, many kinds of applications, such as LED full-color displays and LED white lamps for use in place of light bulbs or fluorescent lamps, are now possible with characteristics of high reliability, high durability and low energy consumption.

Figure 10 shows the typical EL of green SQW LEDs. The EL peak energy of the green SQW LEDs shows blue shifts by about 100 meV with increasing the driving current from 1 μA to 80 mA. Similar blue shift is found in both of blue and yellow SQW LEDs.

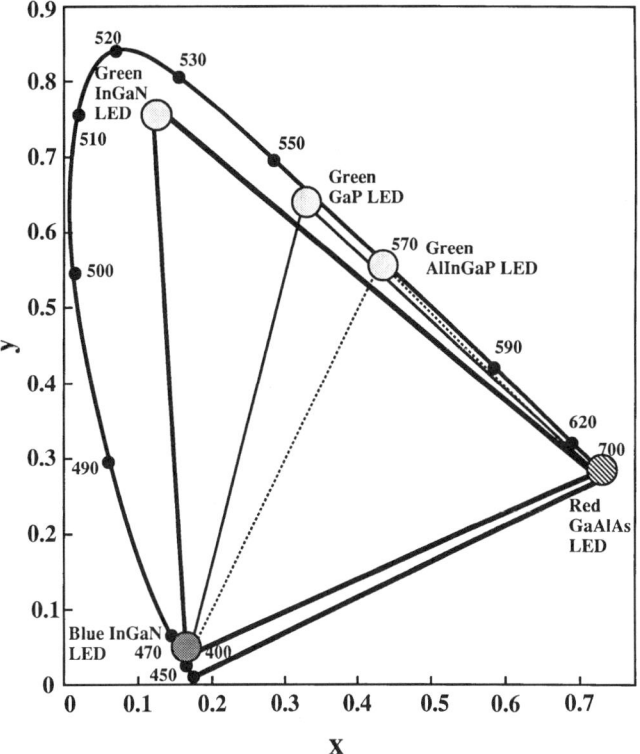

FIG. 11. Chromaticity diagram in which blue InGaN SQW LED, green InGaN SQW LED, green GaP LED, green AlInGaP LED and red GaAlAs LED are shown.

Figure 11 is a chromaticity diagram in which the blue and green InGaN SQW LEDs are shown. Commercially available green GaP LEDs, green AlInGaP LEDs and red GaAlAs LEDs are also shown. The color range of light emitted by a full-color LED lamp in the chromaticity diagram is shown as the region inside each triangle which is drawn by connecting the positions of three primary color LED lamps. Three color ranges (triangles) are shown for differences only in the green LED (green InGaN, green GaP and green AlInGaP LEDs). In this figure, the color range of lamps composed of a blue InGaN SQW LED, a green InGaN SQW LED and a red GaAlAs LED, is the widest. This means that the InGaN blue and green SQW LEDs show much better color and color purity in comparison with other blue and green LEDs. Using these blue and green SQW LEDs, much more beautiful LED full color display have been set in Japan, 1996, as shown in Figure 12. In this display, blue and green LEDs are InGaN SQW LEDs, and only red LEDs are GaAlAs LEDs. Also, these green InGaN SQW LEDs have been used for the application of traffic lights especially in the USA and Europe, as shown in Figure 13.

FIG. 12. The actual LED full-color display which was set in Japan in 1996. The blue InGaN SQW LEDs, green InGaN SQW LEDs and red GaAlAs LEDs are used as three primary color LEDs.

Fig. 13. The actual LED traffic light which was set in Berlin, Germany in 1996, using InGaN SQW green, AlInGaP yellow and AlInGaP red LEDs.

IV. Emission Mechanism of Single-Quantum-Well LEDs

Figure 14 shows the room-temperature photoluminescence (PL) spectra of the green $In_{0.45}Ga_{0.55}N$ SQW structure as a function of external bial applied to the device (Chichibu *et al.*, 1996a). The PL was excited by the 457.9 nm (2.71 eV) line of a CW Ar^+ laser (50 mW) to observe the field effect on the PL luminescence of the quantum well. The light source excites

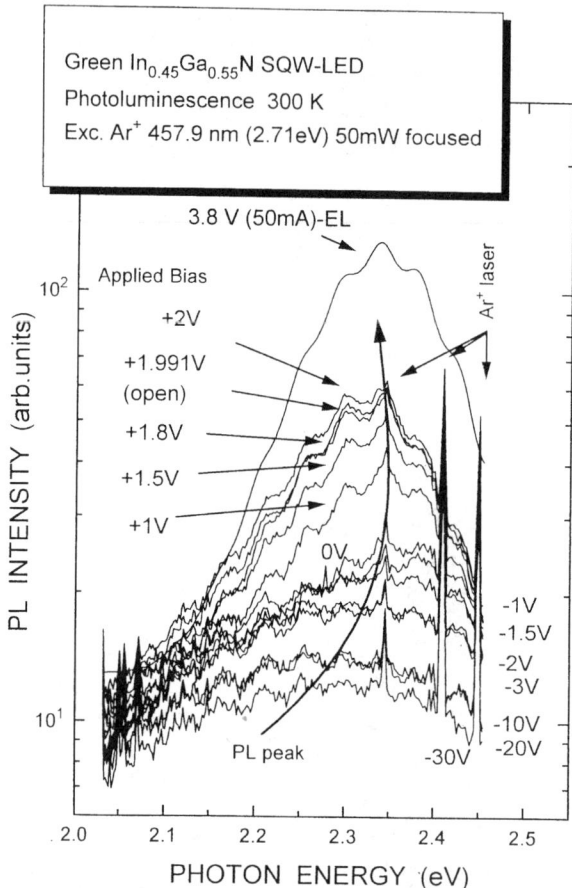

FIG. 14. Room-temperature PL spectra of green InGaN SQW structure as a function of external applied bias. The PL was excited by the 457.9 nm (2.71 eV) line of Ar+ laser (50 mW) to excite only the SQW. Variation of the PL peak position is shown by the arrow as an eye-guide.

carriers only in the SQW. The EL spectrum is also shown for the comparison. The PL peak intensity decreases with increasing reverse bias against the open-circuit condition. For the reverse bias of -2 V (the field strength is 1.2×10^6 V/cm), the PL intensity becomes one third of that for the $+2$ V bias (the field strength is 5.7×10^4 V/cm). The emission almost vanishes for the reverse bias of -10 V, where the field strength is 2.1×10^6 V/cm. Taking the mass anisotropy into account, we can calculate the exciton binding energy in the 3 nm-thick GaN quantum well with infinite barrier height. The obtained value is almost 1.5 times larger than that in the bulk

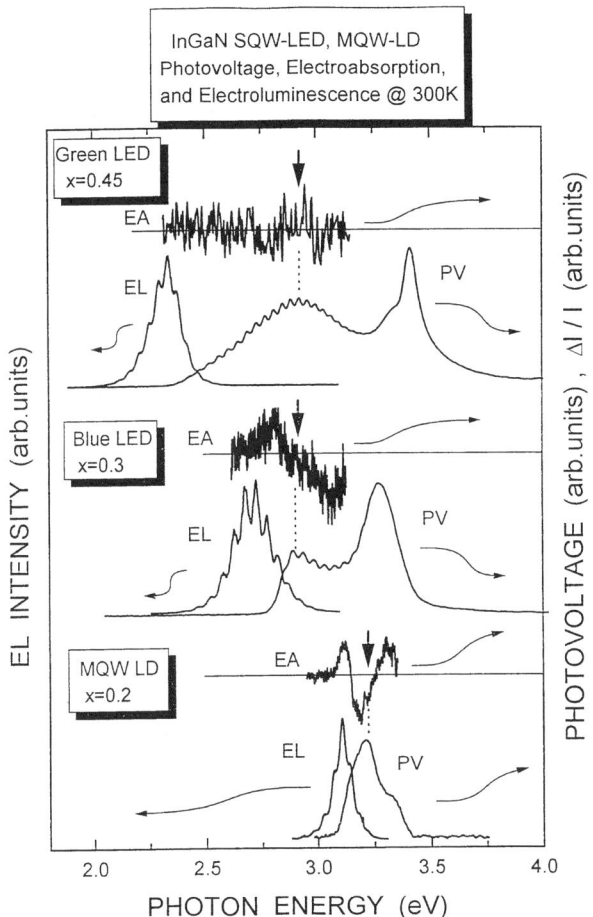

FIG. 15. Electroluminescence, PV, and EA spectra for InGaN green (510 nm) and blue (450 nm) SQW LED structure and MQW laser structure, whose lasing wavelength is 406 nm. The EL spectrum of the MQW structure was measured below the threshold current. The In composition in the $In_xGa_{(1-x)}N$ quantum well for green, blue, and MQW LED is 0.45, 0.3, and 0.2, respectively. The structure in the EA spectra correspond to the free exciton resonances.

(28 meV × 1.5 = 42 meV). Thus the electric field to dissociate free excitons in the 3 nm-thick InGaN quantum well is estimated to be as high as 1.5×10^5 V/cm. Therefore, the quenching of the PL intensity with increasing the reverse bias is considered to be due to the quantum-confinement Stark effects (Mendez et al., 1982). The emission can still be observed for higher electric field (1.2×10^6 V/cm) than the field required for the free exciton

quenching because these PL emissions are due to recombination of localized excitons as mentioned below.

Static (dc) EL, photovoltage (PV) and modulated-electroabsorption (EA) spectra were measured on the above mentioned SQW LEDs and MQW LDs (Chichibu et al., 1996a). Figure 15 summarizes room-temperature EL, PV, and EA spectra of green, blue SQW LED and MQW LD structure. The EL spectrum of the MQW LD structure was measured below the threshold current density. The lasing emission of this MQW LD appeared at 3.052 eV (406 nm) at the threshold current density of 11.3 kA/cm^2. In general, the low-field EA monitors exciton resonance rather than band to band transition even at RT in widegap semiconductor such as GaN (Chichibu et al., 1996b) under certain condition that the modulation field is smaller than that to dissociate excitons (1.5×10^5 V/cm). The EA spectra were measured using the rectangular modulation bias of -2 V to $+1.95$ V, corresponding to the field of 1.2×10^6 V/cm and 1.1×10^5 V/cm, respectively, to maintain the field strength of the upper level smaller than 1.5×10^5 V/cm. Therefore, the structure observed in the EA spectra are due to room-temperature free exciton resonances in the quantum wells.

The PV spectra were taken using a monochromated light, and the open-circuit voltage of the device was measured spectroscopically. The PV peak at 3.21, 2.91, and 2.93 eV for MQW, blue and green SQW structures correspond to exiton absorption in the quantum well, because the energies agree with those in the EA spectra. It is recognized that FWHM of the PV peak increases with increasing x. The PV peak energy decreases from 3.21 to 2.91 eV with increasing x from 0.2 to 0.45. However, the peak energy is almost unchanged for $x = 0.3$ and $x = 0.45$. This implies that InGaN does not form perfect alloys (Osamura et al., 1975), but form compositional tailing especially for larger x. Such a compositional tailing in the quantum well plane can produce two-dimensional potential minima.

The EL peak energy is smaller by 100, 215, and 570 meV than the free exciton resonance energy from MQW LD, blue and green SQW LED, respectively. All EL peaks are located at the low energy tail of the free exciton resonance. Such low energy tails of the exciton structure reflect a presence of certain potential minima in the quantum well plane. These EL emissions are considered as recombination of localized excitons in the quantum well.

V. InGaN Multi-Quantum-Well (MQW) Structure LDs

High-brightness blue and green LEDs have been fabricated using III–V nitride materials and are now commercially available as mentioned above. At present, the main focus of III–V nitride research is the realization of a

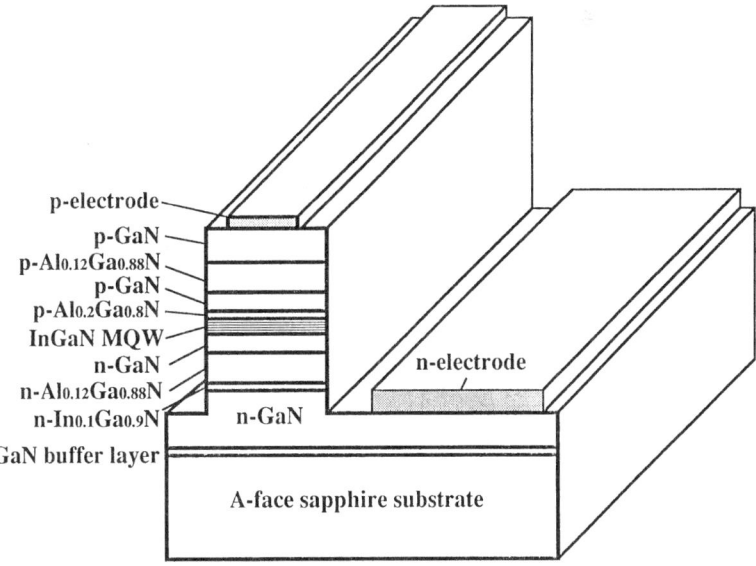

FIG. 16. The structure of the InGaN MQW LD.

current-injected laser diode which is expected to be the shortest-wavelength semiconductor laser diode ever demonstrated. However, stimulated emission has been observed only by optical pumping, and not by current injection. Here, first LDs fabricated using wide-bandgap III–V nitride materials are described (Nakamura *et al.*, 1996a, b, c, d). These LDs emitted coherent light at 390–440 nm from InGaN-based MQW structure under pulsed current injection at room temperature.

The InGaN MQW LD structure is shown in Fig. 16. The active layer is an $In_{0.2}Ga_{0.8}N/In_{0.05}Ga_{0.95}N$ MQW structure consisting of four 30-Å-thick undoped $In_{0.2}Ga_{0.8}N$ well layers forming the gain medium, separated by 60-Å-thick undoped $In_{0.05}Ga_{0.95}N$ barrier layers. The 0.1-μm-thick n-type and p-type GaN layers were light-guiding layers. The 0.5-μm-thick n-type and p-type $Al_{0.12}Ga_{0.88}N$ layers were cladding layers for confinement of the carriers and the light emitted from the active region of the InGaN MQW structure. In the stripe-geometry LDs, the stripe width was 10 μm and, the width of p-electrode was 6 μm. The cavity length of the LD was 600 μm. The electrical characteristics of LDs were measured under pulsed current-biased conditions at RT. The output power from one facet was measured using a Si photodetector.

Figure 17 shows the typical voltage-current (V-I) characteristics and the light output power per coated facet of the LDs as a function of the pulsed

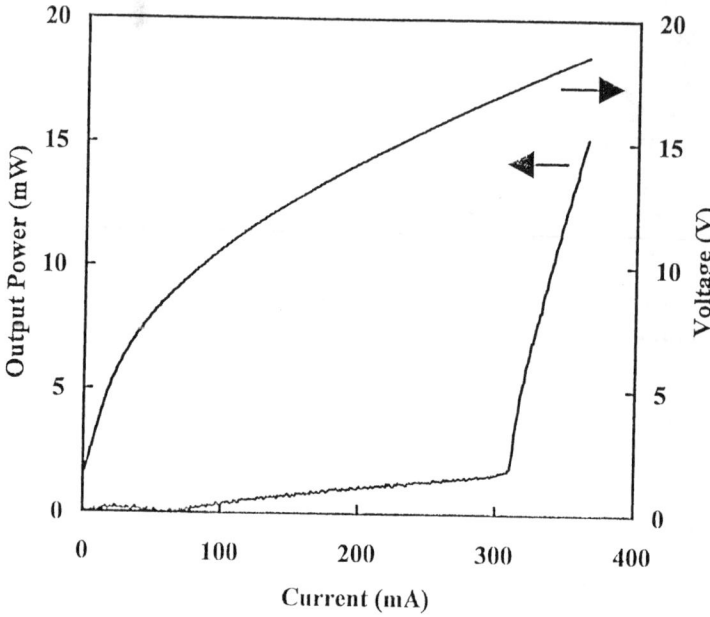

FIG. 17. The L-I and V-I characteristics of the InGaN MQW LD.

forward current (L-I). No stimulated emission of the LDs was observed up to the threshold current of 300 mA, which corresponded to a threshold current density of 8 kA/cm², as shown in Fig. 17. The differential quantum efficiency of 10% per facet and pulsed output power of 15 mW per facet were obtained at a current of 370 mA. The operating voltage of the device at the threshold current was around 17 V. Also, polarization measurement showed that the transverse electric (TE)-polarized light output intensity increased to a much larger value than that of the transverse magnetic (TM)-polarized light, above the threshold.

Figure 18 shows typical optical spectra of the InGaN MQW LDs under pulsed current injection at room temperature. These spectra were measured using the Optical Spectrum Analyzer which had a resolution of 0.01 nm. At injection currents around the threshold currents, many sharp peaks appeared with a peak separation of 0.05 nm, as shown in Figure 18(a). If these peaks arise from the longitudinal modes, the mode separation $\Delta\lambda$ is given by $\Delta\lambda = \lambda_0^2/2/L/n$, where n is the refractive index and λ_0 is the emission wavelength (404.3 nm). L was 0.06 cm. A value of 2.54 was used for the refractive index. Thus, $\Delta\lambda$ is calculated as 0.05 nm. Therefore, the observed peak separation is the longitudinal mode separation. This is the first

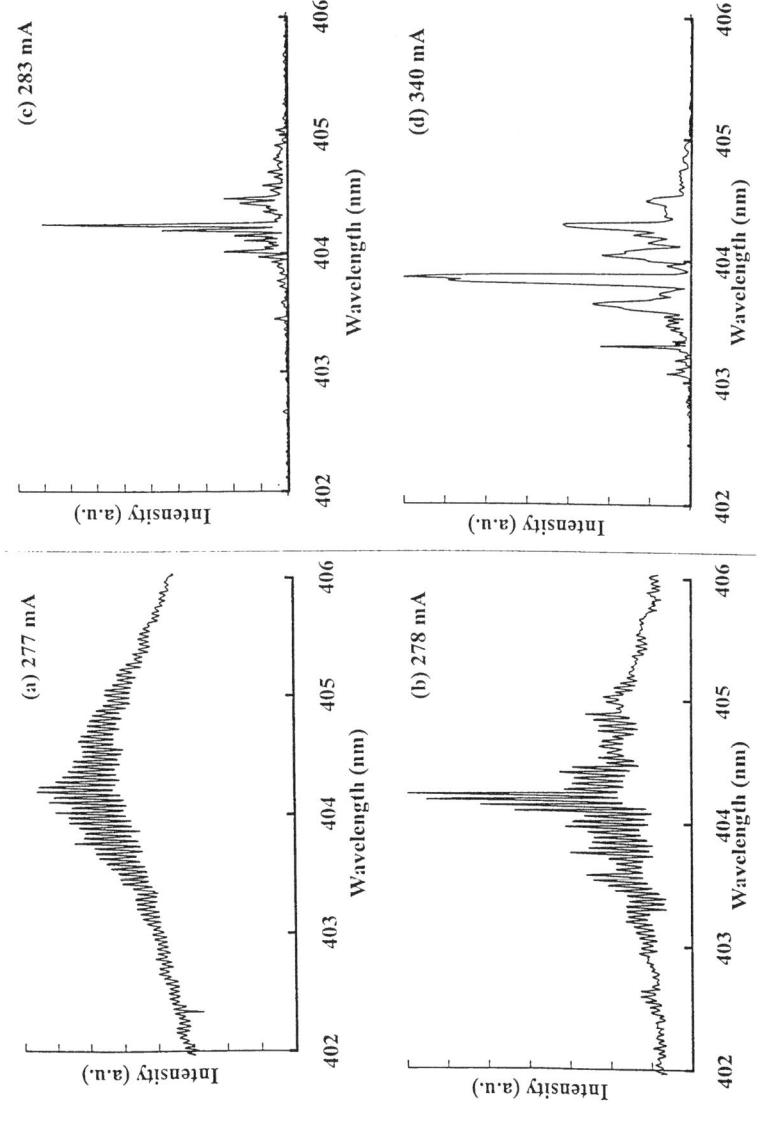

FIG. 18. The optical spectra for the InGaN MQW LD. (a) at a current of 277 mA; (b) at a current of 278 mA; (c) at a current of 283 mA; at a current of 340 mA. Intensity scales for these four spectra are in arbitrary units, and are different.

observation of the longitudinal mode of the III–V nitride based LDs. When the forward current was increased, the main peak become dominant at the wavelength of 404.3 nm, as shown in Fig. 18(b). At a current above 283 mA, several peaks which had a different peak separation from the longitudinal mode were observed, as shown in Figs. 18(c) and 18(d). The origin of these subband emissions is not clear at present. These spectral measurements were performed under different pulse widths between 0.01–1 μs in order to ensure that these spectral changes with increase of the forward current were not due to heat generation in the junction. The spectra were almost the same as those described above. This means that these spectral changes were caused by the increase in current, not by the change in temperature in the junction of the LDs. There is a possibility that inhomogeneities in the film thickness and in the layer compositions led to deviations from the ideal behavior and hence to asymmetry in the spectrum. Temperature and electron densities alter the refractive index and displace the resonance wavelength of the modes. The near-field pattern (NFP) of the stimulated emission of the LDs was measured using a microscope to check whether the stimulated emission originated from one spot or many spots in the junction with increase of the current. The emission was generated from one spot just below the p-electrode in the junction. This means that these subband emissions were not caused by the inhomogeneities of the junction. Many small peaks with a peak separation of 0.5–0.7 nm were also observed by other groups in the stimulated emission of GaN obtained by optical pumping (Zubrilov *et al.*, 1995; Aggarwal *et al.*, 1996). Zubrilov *et al.*, 1995 proposed that the short cavity mirrors formed by cracks with a width of 20–50 μm in the GaN layer caused these peaks. However, we did not observe any cracks in our LDs. It is clear, however, that these spectra are not due to simple Fabry-Perot modes. Another possible explanation for the origin of these emissions is a subband transition of quantum energy levels caused by quantum confinement of electrons and holes.

We measured the surface morphology of the InGaN well layers using atomic force microscopy (AFM), as shown in Fig. 19. The roughness of the InGaN well layer was about 10–20 Å with periods of approximately 1000 Å horizontally in the plane parallel to the junction, as shown in Fig. 19. This island-like structures is not found for thicker (>40 Å) layers. Origin of this structure is not clear, however, it might reflect the condensation of In-rich phase considering the fact that InGaN scarcely forms the perfect alloy in nature (Osamura *et al.*, 1975). These In-rich regions in InGaN well layers can form a potential minima to confine carriers. This means that electrons and holes can be confined in the plane parallel to the junction with 1000 Å periods. Therefore, the three-dimensional confinement of the electrons and holes can be considered, such as a quantum dot (box). These subband

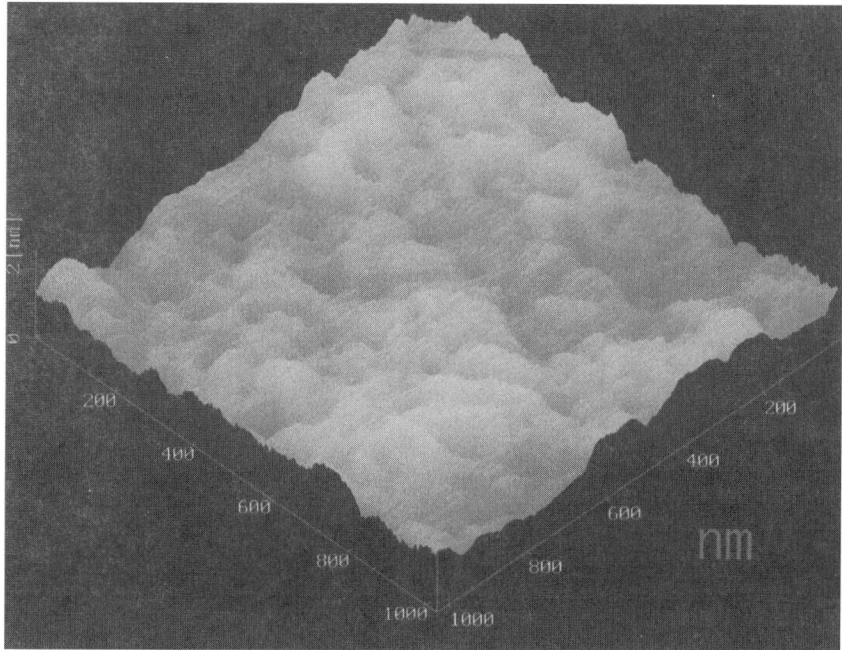

Fig. 19. AFM image of the 30 Å thick $In_{0.45}Ga_{0.55}N$ layer on n-GaN/sapphire.

emissions seem to be a result of the subband transition of quantum energy levels caused by three-dimensional quantum confinement of electrons and holes.

The EL, PV, and EA spectra of InGaN MQW LDs are shown in Fig. 20 (Chichibu *et al.*, 1996a). A distinct structure at 3.210 eV in EA and PV spectra corresponds to free exciton resonance in the MQW. This assignment is based on the facts that the spectrum was measured under low modulation field to alive excitons, as mentioned on InGaN SQW LEDs in Fig. 15. The EL peak at 3.109 eV is located in lower energy tail of the exciton structure, showing Stokes shift of 100 meV. The quantum-confinement Stark effects were already observed on the $In_{0.45}Ga_{0.55}N$ SQW structure. Therefore, the EL peak at 3.109 eV is due to the recombination of excitons localized at certain potential minima in the quantum wells.

The lifetime of the spontaneous emission of the MQW LEDs was typically 3 ns. The lasing emission appears at 3.052 eV for the current density $J = J_{th}$, where $J_{th} = 11.3 \, kA/cm^2$. The lasing emission lifetime is also 3 ns, and the internal quantum efficiency is estimated to be 90%. Thus the radiative recombination lifetime is about 3.3 ns. The threshold injected

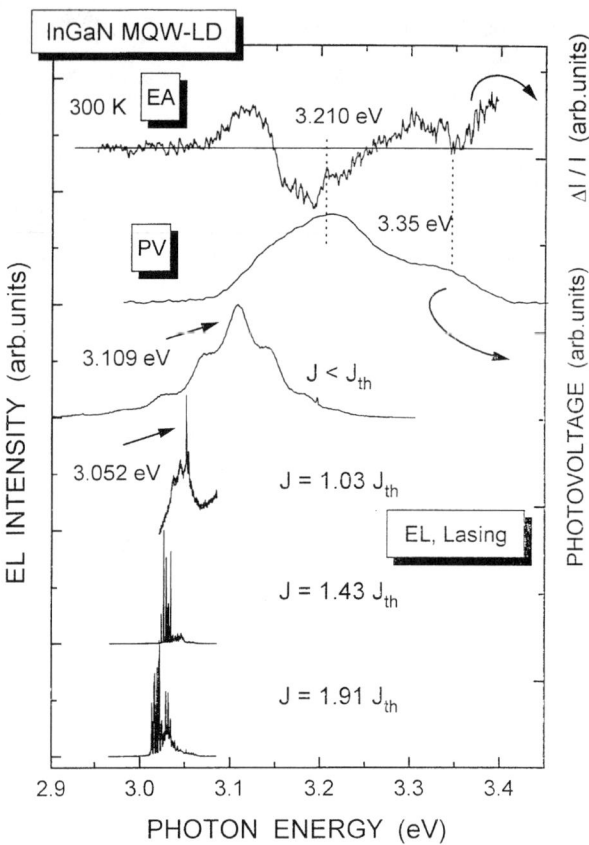

FIG. 20. EA, PV, and EL spectra of the InGaN MQW structure at RT. The EL spectra are shown as a function of relative current density. J_{th} is a threshold current density.

carrier density in the MQW is thus calculated to be about $2.1 \times 10^{19}\,cm^{-3}$. For bulk GaN, free excitons can no longer survive under this large carrier density. Note that the charge density to screen excitons at RT is about $1 \times 10^{18}\,cm^{-3}$ using the values of exciton binding energy (28 meV) and dielectric constant (Chichibu et al., 1996b).

For laser emission, various process have been proposed for the exciton-related lasing mechanisms (Sugawara, 1996). Many-body Coulomb effect in the electron-hole plasma (EHP) has also been proposed (Chow and Koch, 1995). There is a possibility for the localized biexciton-localized exciton population inversion of our MOW LD (Sugawara, 1996) because the

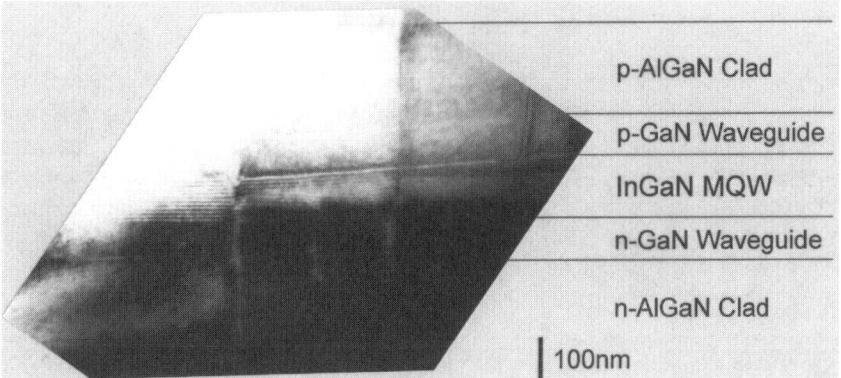

FIG. 21. TEM image of the InGaN MQW structure. The structure of the MQW layer is distorted by high-density (10^{10} cm^{-2}) threading dislocations.

emission from localized excitons is still observed for the injected carrier density of at least 5×10^{18} cm^{-3} in the $In_{0.45}Ga_{0.55}N$ SQW LED, which is due to its deep localized potential (570 meV), as shown in Fig. 10. For the MQW structure, the Stokes shift is 100 meV, which can prevent thermalization of excitons even at RT (Sugawara, 1996). The localized biexcitons in quantum wells are equivalent to biexcitons in quantum dots. The EL peak below the threshold current is originated from the recombination of excitons localized at certain potential minima in the quantum wells. Therefore, there is a possibility of the exciton-related lasing, such as the localized biexciton-localized exciton population inversion due to the large localized potential of the excitons. Also, recently, the emissions caused by biexcitons in GaN layers were observed (Okada et al., 1996).

The transmission electron micrograph (TEM) of the MQW structure is shown in Fig. 21 (Chichibu et al., 1996a). The MQW structure is clearly observed homogeneously in small area. However, the MQW structure is distorted and/or inclined caused by high density (10^{10} cm^{-2}) threading dislocations and/or other reasons, for example, each column-like homogeneous zone is distributed inhomogeneously. The column size is about 100–400 nm. It should be noted that the MQW LD device including inhomogeneous column-like structures shows lasing action. This indicates that the oscillator strength and gain at the local potential minima are very large. The several peaks which had a different peak separation from the longitudinal mode were observed in Figs. 18c, d. There is a possibility that these emissions are due to the localized potential caused by these inhomogeneous column-like structures and inhomogeneous InGaN layers.

VI. Summary

Superbright InGaN blue and green SQW LEDs were fabricated. By combining a high-power and high-brightness blue InGaN SQW LED, a green InGaN SQW LED and a red AlInGaP LED, many kinds of applications, such as LED full-color displays and LED white lamps for use in place of light bulbs or fluorescent lamps, are now possible with characteristics of high reliability, high durability and low energy consumption. Also, very recently, III–V nitride based LDs were fabricated for the first time. These LDs emitted coherent light at 390–440 nm from an InGaN based MQW structure under pulsed current injection at room temperature. These results indicate a possibility that the short wavelength LDs from green to UV will be realized in the near future using III–V nitride material. As the emission mechanism of these InGaN based quantum well devices, the recombination of excitons localized at certain potential minima in the quantum well was proposed. The laser emissions seem to be a result of the subband transition of quantum energy levels caused by three-dimensional quantum confinement of electrons and holes localized at certain potential minima.

REFERENCES

Aggarwal, R. L., Maki, P. A., Molnar, R. J., Liau, Z. L., and Melngailis, I. (1996). *J. Appl. Phys.* **79,** 2148.

Amano, H., Kito, M., Hiramatsu, K., and Akasaki, I. (1989). *Jpn. J. Appl. Phys.* **28,** L2112.

Amano, H., Asahi, T., and Akasaki, I. (1990). *Jpn. J. Appl. Phys.* **29,** L205.

Amano, H., Tanaka, T., Kunii, Y., Kato, K., Kim, S. T., and Akasaki, I. (1994). *Appl. Phys. Lett.* **64,** 1377.

Brandt, M. S., Johnson, N. M., Molnar, R. J., Singh, R., and Moustakas, T. D. (1994). *Appl. Phys. Lett.* **64,** 2264.

Chichibu, S., Azuhata, T., Sota, T., and Nakamura, S. (1996a). 38th Electronic Material Conference W10, Santa Barbara, USA, June 28.

Chichibu, S., Azuhata, T., Sota, T., and Nakamura, S. (1996b). *J. Appl. Phys.* **79,** 2784.

Chow, W. W., and Koch, S. W. (1995). *Appl. Phys. Lett.* **66,** 3000.

Craford, M. G. (1992). Circuits & Devices. September 24.

Eason, D. E., Yu, Z., Hughes, W. C., Roland, W. H., Boney, C., Cook, J. W., Jr., Schetzina, J. F., Cantwell, G., and Harasch, W. C. (1995). *Appl. Phys. Lett.* **66,** 115.

Edmond, J., Kong, H., and Dmitriev, V. (1994). *Inst. Phys. Conf. Ser.* **137,** 515.

Khan, M. A., Krishnankutty, S., Skogman, R. A., Kuznia, J. N., and Olson, D. T. (1994). *Appl. Phys. Lett.* **65,** 520.

Kim, S. T., Amano, H., and Akasaki, I. (1995). *Appl. Phys. Lett.* **67,** 267.

Kish, F. A., Steranka, F. M., DeFevere, D. C., Vanderwater, D. A., Park, K. G., Kuo, C. P., Osentowski, T. D., Peanasky, M. J., Yu, J. G., Fletcher, R. M., Steigerwald, D. A., Craford, M. G., and Robbins, V. M. (1994). *Appl. Phys. Lett.* **64,** 2839.

Koga, K., and Yamaguchi, T. (1991). *Prog. Crystal Growth and Charact.* **23**, 127.
Mendez, E. E., Bastard, G., Chang, L. L., Esaki, L., Morkoç, H., and Fisher, R. (1982). *Phys. Rev.* **B26**, 7101.
Morkoç, H., Strite, S., Gao, G. B., Lin, M. E., Sverdlov, B., and Burns, M. (1994). *J. Appl. Phys.* **76**, 1363.
Nakamura, S., Iwasa, N., Senoh, M., and Mukai, T. (1992). *Jpn. J. Appl. Phys.* **31**, 1258.
Nakamura, S., Mukai, T., Senoh, M., Nagahama, S., and Iwasa, N. (1993). *J. Appl. Phys.* **74**, 3911.
Nakamura, S., Mukai, T., and Senoh, M. (1994a). *Appl. Phys. Lett.* **64**, 1687.
Nakamura, S., Mukai, T., and Senoh, M. (1994b). *J. Appl. Phys.* **76**, 8189.
Nakamura, S., Senoh, M., Iwasa, N., and Nagahama, S. (1995a). *Jpn. J. Appl. Phys.* **34**, L797.
Nakamura, S., Senoh, M., Iwasa, N., and Nagahama, S. (1995b). *Appl. Phys. Lett.* **67**, 1868.
Nakamura, S., Senoh, M., Iwasa, N., Nagahama, S., Yamada, T., and Mukai, T. (1995c). *Jpn. J. Appl. Phys.* **34**, L1332.
Nakamura, S., Senoh, M., Nagahama, S., Iwasa, N., Yamada, T., Matsushita, T., Kiyoku, H., and Sugimoto, Y. (1996a). *Jpn. J. Appl. Phys.* **35**, L74.
Nakamura, S., Senoh, M., Nagahama, S., Iwasa, N., Yamada, T., Matsushita, T., Kiyoku, H., and Sugimoto, Y. (1996b). *Jpn. J. Appl. Phys.* **35**, L217.
Nakamura, S., Senoh, M., Nagahama, S., Iwasa, N., Yamada, T., Matsushita, T., Kiyoku, H., and Sugimoto, Y. (1996c). *Appl. Phys. Lett.* **68**, 2105.
Nakamura, S., Senoh, M., Nagahama, S., Iwasa, N., Yamada, T., Matsushita, T., Kiyoku, H., and Sugimoto, Y. (1996d). *Appl. Phys. Lett.* **68**, 3269.
Nakamura, S. (1991). *Jpn. J. Appl. Phys.* **30**, L1705.
Nakamura, S. (1994a). *Nikkei Electronics Asia* **6**, 65.
Nakamura, S. (1994b). *Microelec. J.* **25**, 651.
Nakamura, S. (1994c). *J. Cryst. Growth* **145**, 911.
Nakamura, S. (1995). *J. Vac. Sci. Technol.* **A13**, 705.
Okada, K., Yamada, T., Taguchi, T., Sasaki, F., Kobayashi, S., Tani, T., Nakamura, S., and Shinomiya, G. (1996). *Jpn. J. Appl. Phys.* **35**, L787.
Okuyama, H., and Ishibashi, A. (1994). *Microelec. J.* **25**, 643.
Osamura, K., Naka, S., and Murakami, Y. (1975). *J. Appl. Phys.* **46**, 3432.
Pankove, J. I., Miller, E. A., and Berkeyheiser, J. E. (1971). *RCA Review* **32**, 283.
Rubin, M., Newman, N., Chen, J. S., Fu, T. C., and Ross, J. T. (1994). *Appl. Phys. Lett.* **64**, 64.
Schmidt, T. J., Yang, X. H., Shan, W., Song, J. J., Salvador, A., Kim, W., Aktas, O., Botchkarev, A., and Morkoç, H. (1996). *Appl. Phys. Lett.* **68**, 1820.
Strite, S., and Morkoç, H. (1992). *J. Vac. Sci. Technol.* **B10**, 1237.
Sugawara, H., Itaya, K., and Hatakoshi, G. (1994). *Jpn. J. Appl. Phys.* **33**, 5784.
Sugawara, M. (1996). *Jpn. J. Appl. Phys.* **35**, 124.
Xie, W., Grillo, D. C., Gunshor, R. L., Kobayashi, M., Jeon, H., Ding, J., Nurmikko, A. V., Hua, G. C., and Otsuka, N. (1992). *Appl. Phys. Lett.* **60**, 1999.
Zavada, J. M., Wilson, R. G., Abernathy, C. R., and Pearton, S. J. (1994). *Appl. Phys. Lett.* **64**, 2724.
Zubrilov, A. S., Nikolaev, V. I., Tsvetkov, D. V., Dmitriev, V. A., Irvine, K. G., Edmond, J. A., and Carter, C. H. (1995). *Appl. Phys. Lett.* **67**, 533.

CHAPTER 15

Lasers

I. Akasaki and H. Amano

DEPARTMENT OF ELECTRICAL AND ELECTRONIC ENGINEERING
MEIJO UNIVERSITY
1-501 SHIOGAMAGUCHI
TEMPAKU-KU
NAGOYA 468, JAPAN

I. INTRODUCTION	. . . .	459
II. BASIC STRUCTURE	. . . .	462
III. CRITICAL LAYER THICKNESS	. . . .	463
IV. CONTROL OF CONDUCTIVITY	. . . .	464
V. THRESHOLD CURRENT DENSITY	. . . .	466
1. *Carrier Confinement and Optical Confinement*	. . . .	466
2. *Transparency Carrier Density*	. . . .	469
3. *Low Threshold Structure*	. . . .	470
VI. SUMMARY	. . . .	470
References	. . . .	471

I. Introduction

Compared to other III–V compound semiconductors such as GaAs and InP that are zinc-blende, GaN has usually the wurtzite structure. Moreover, it has a large density of states for both conduction band and valence band, and very small lattice constants. All of which is unfavorable for the laser diode. Plenty of pioneering work has been done for early GaN research (Maruska and Tietjen, 1969; Pankove et al., 1971, 1973; Manasevit, Erdman, and Simpson, 1971), much of which focused on the application of GaN to short wavelength light emitters. The first demonstration of stimulated emission from GaN was reported by Dingle et al. (1971), who succeeded in optical pump lasing at 2 K. Since then, it has taken approximately 25 years to realize lasing by current injection. Fundamental issues in addition to the above mentioned pioneering work for the realization of current injection lasing are, (1) growth of high quality epitaxial layer, (2) control of electrical conductivity, (3) growth of high quality ternary alloys, and (4) fabrication of high quality heterostructure and quantum wells.

The breakthrough for the growth of high quality films was first achieved by Amano et al. (1986), who succeeded in growing specular epitaxial films free of cracks on the sapphire substrate by organometallic vapor phase epitaxy (OMVPE) using a low temperature deposited buffer layer. They also succeeded in controlling conductivity of n-type nitrides by doping with Si (Amano and Akasaki, 1990), and realized the first p-type conduction in nitrides by doping with Mg, followed by activation of Mg by low energy electron beam irradiation (Amano et al., 1989).

Koide et al. (1988) grew single crystalline $Al_xGa_{1-x}N$ with x up to 0.4 by using the low temperature deposited buffer layer. Nagatomo et al. (1989) and Yoshimoto et al. (1991) grew single crystalline GaInN by OMVPE on sapphire substrates.

In order to realize carrier and optical confinement, the heterostructure is essential. This was critical for nitride, because nitride system such as GaN, AlGaN, and GaInN are lattice mismatched to each other. Issues of critical layer thickness of AlGaN and GaInN on GaN will be discussed later. For the realization of low threshold lasing, quantum well (QW) structure is inevitable. Khan et al. (1990) and Itoh et al. (1991) first fabricated AlGaN/GaN multi-quantum well (MQW) structures. The AlGaN/GaN MQW structures with the GaN well thickness as thin as 5 nm were fabricated (Itoh et al., 1991). GaN/GaInN layered structures were first fabricated by Nakamura et al. (1993).

The current injection lasing was based on the above mentioned results.

Amano, Asahi, and Akasaki (1990) observed room temperature stimulated emission by optical pumping from a high quality GaN epitaxial layer grown using the low temperature deposited buffer layer. Khan et al. (1991) reported the first surface mode stimulated emission by optical pumping using low temperature deposited buffer layer. Kim et al. (1994) first measured optical gain in AlGaN/GaN double heterostructure by optical pumping at room temperature. In 1995, a threshold power for stimulated emission by optical pumping of $27\,kW/cm^2$ was reported, which corresponds to $7.4\,kA/cm^2$ for current injection (Akasaki and Amano, 1995; Akasaki, Amano, and Suemune, 1995). Figure 1 shows the change of threshold power (P_{th}) for stimulated emission by optical pumping from group III nitrides. For the first time Akasaki et al. (1995) observed the onset of stimulated emission by current injection. Nakamura et al. (1996a,b,c) fabricated the first nitride based laser diode (LD). The active layer is a GaInN based MQW having a large number of well layers. GaN and AlGaN were used as the waveguide and cladding layers, respectively. The duty cycle was 0.1%. The threshold current density was $4.0\,kA/cm^2$. The operating

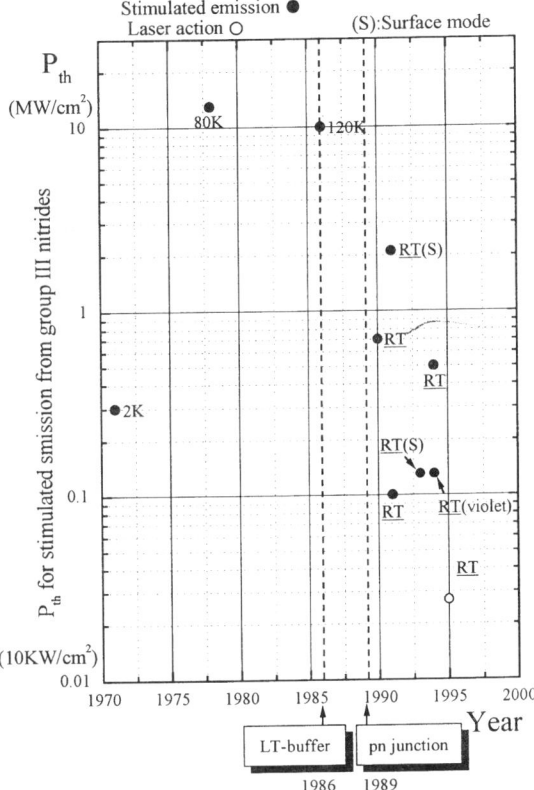

FIG. 1. Change of the threshold power for stimulated emission from nitrides as a function of calendar year.

voltage at the threshold was as high as 34 V, which was mainly caused by the high resistivity of the p-type electrode.

Akasaki et al. (1996) used a GaInN single quantum well (SQW) structure as the active layer and succeeded in fabricating the shortest wavelength semiconductor LD to date with a wavelength of 376.0 nm at room temperature. The threshold current density was $3.0\,kA/cm^2$ and the operating voltage at the threshold current density was 16 V.

Itaya et al. (1996) also succeeded in fabricating LD operated by pulsed current injection using cleaved edge mirrors.

Finally, Nakamura et al. (1996d) succeeded in continuous wave (CW) lasing at room temperature using ridge waveguide structure. The lifetime of the first CW-LD was a few seconds and lengthened quite rapidly. By the end of 1996, it reached a maximum of 35 h operation with 1.5 mW output.

In this chapter, we review the characteristics of nitrides as the laser materials.

II. Basic Structure

Figure 2 schematically shows the basic structure of edge emitting nitride based LD. The structure was that of separate confinement heterostructure (SCH). Substrate is either sapphire (Nakamura *et al.*, 1996a,b; Akasaki *et al.*, 1996) or spinel (Nakamura *et al.*, 1996c). In both cases, the low temperature deposited buffer layer was used. Thick n-GaN highly doped with Si was grown on the buffer layer. Since the substrate is insulating, thick and low resistive n-GaN is inevitable in order to reduce series resistance of the device. The cladding layer is usually AlGaN, the waveguide layer is GaN, and the active layer is GaInN based QW. For the contact, p-GaN layer highly doped with Mg was used to reduce contact resistance. Reactive ion etching (RIE) was used to form electrode for n-GaN. RIE was also used for the fabrication of cavity mirrors (Nakamura *et al.*, 1996a) or ridge waveguide structure (Nakamura *et al.*, 1996). The electrode metal for n-GaN was Ti/Al and for p-GaN it was Ni/Au or Pd/Au.

In the following sections, details of these issues will be discussed.

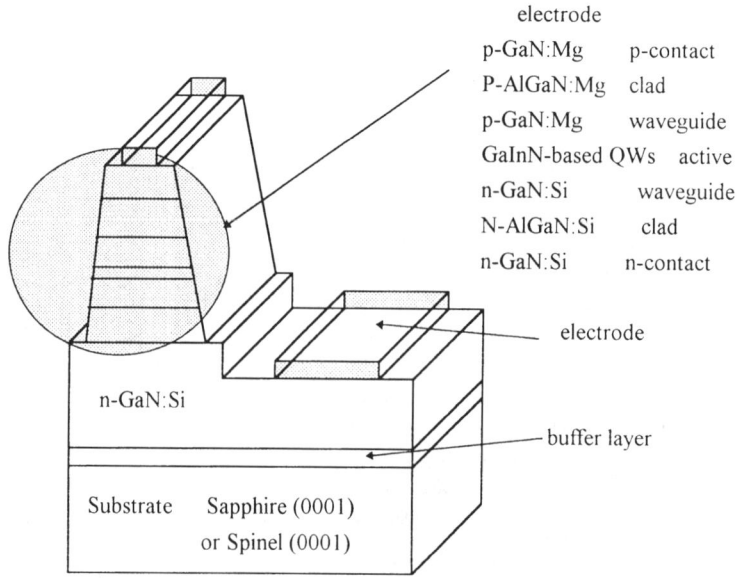

FIG. 2. Typical structure of nitride based laser diode.

III. Critical Layer Thickness

A heterostructure free of defects is essential in order to avoid the effect of excess recombination centers at the interface. Since ternary alloys (AlGaN and GaInN) and GaN are mismatched to each other, coherency or full strain of the growth is the guide for high quality interface. Figures 3(a) and (b) show the dependence of coherency of the growth of ternary alloys on thick GaN as a function of alloy composition. In both figures, solid lines are estimated from the equilibrium theory (Matthews and Blakeslee, 1974;

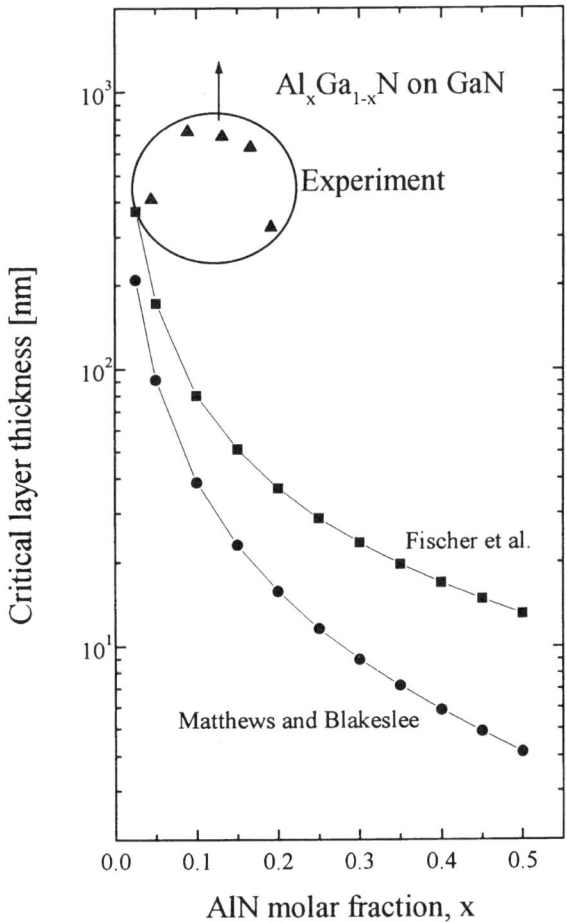

FIG. 3. (a) and (b) (*overleaf*) Coherency of the growth of ternary alloys, AlGaN and GaInN grown on thick GaN. Critical layer thickness estimated from equilibrium theory are also shown for comparison.

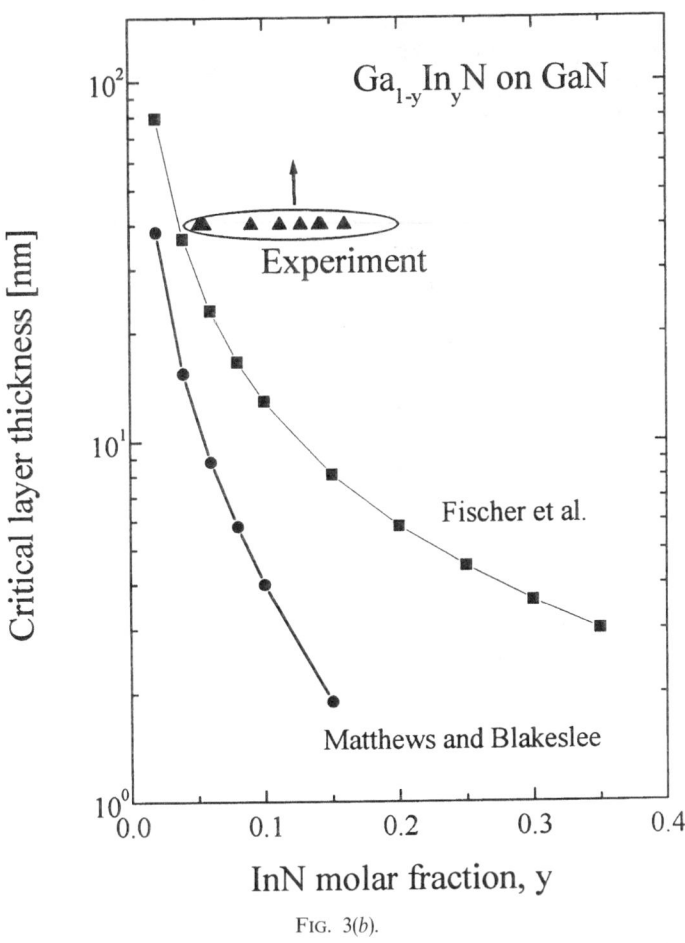

FIG. 3(b).

Fischer, Kuhne, and Richter, 1994). As shown in the figure, experimental results indicate that even though ternary layers exceeds calculated critical layer thickness, it shows coherent growth. This result shows widening of the feasibility for the design of nitride based heterostructure.

IV. Control of Conductivity

In order to inject sufficient carriers in active layer and to reduce leakage current through waveguide or cladding layer, a cladding layer having high

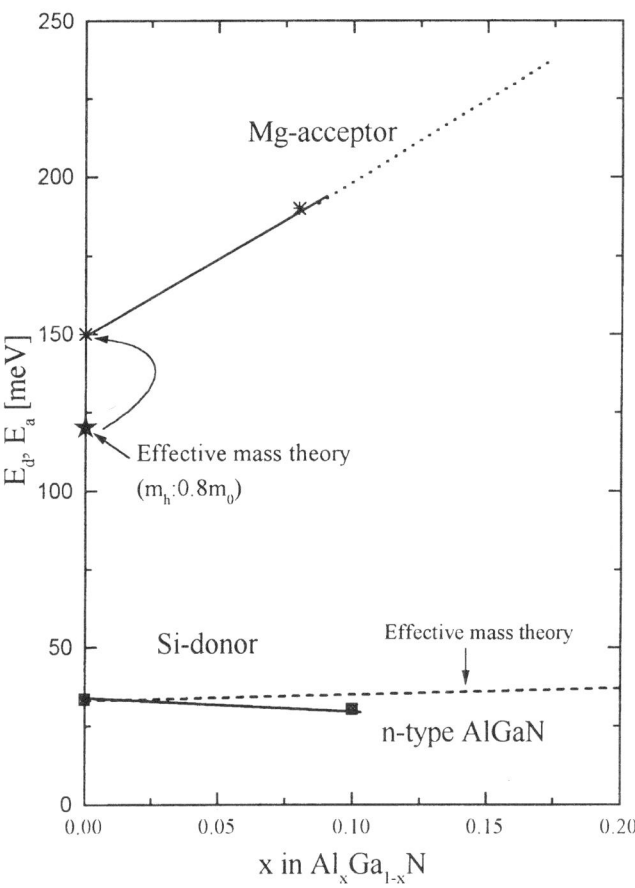

FIG. 4. Compositional dependencies of the activation energies of Mg and Si in AlGaN.

AlN content and high density of free carrier is essential, although these two demands are a tradeoff especially in p-type AlGaN. As concerns the n-AlGaN, high electron concentration of more than 10^{19} cm^{-3} at AlN molar fraction of 0.3 was achieved by doping with Si. In case of p-type, hole concentration of about 2×10^{18} cm^{-3} was achieved in GaN by low energy electron beam irradiation (LEEBI) or thermal treatment of OMVPE grown GaN highly doped with Mg. In case of molecular beam epitaxy (MBE), hole concentration up to 2×10^{19} cm^{-3} was achieved without any special treatment (Moustakas, 1993). However, the maximum hole concentration in $Al_{0.2}Ga_{0.8}N$ was as low as 2×10^{17} cm^{-3} for OMVPE grown ternary AlGaN. Activation energies of Si and Mg was characterized by Hall effect measurement with various temperatures from 70 K to 800 K (Tanaka et al.,

1994). The results were summarized in Fig. 4. As shown in the figure, by increasing the molar fraction of AlN, activation energy of Mg increases, and this causes the increase of inactive Mg at room temperature.

V. Threshold Current Density

1. Carrier Confinement and Optical Confinement

Figures 5(a) and (b) show the compositional dependence of the bandgap of AlGaN (Fig. 5(a)) and compositional dependence of the photolumines-

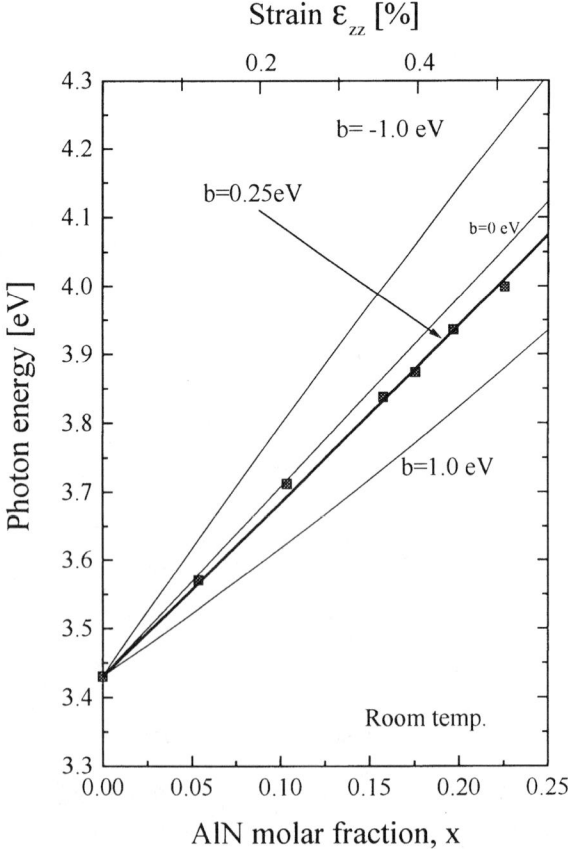

FIG. 5. (a) and (b) Compositional dependence of the bandgap of AlGaN and PL peak energy from GaInN.

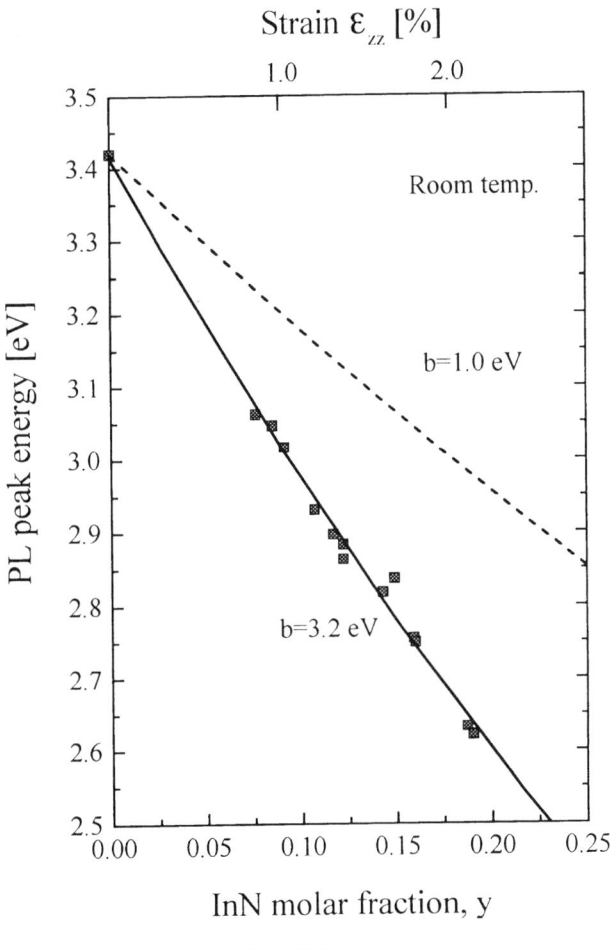

FIG. 5(b).

cence (PL) peak energy of GaInN (Fig. 5(b)), both of which were coherently grown on thick GaN. The bowing parameter of fully strained AlGaN is approximately 0.25 eV, which is close to that of unstrained AlGaN (Wickenden et al., 1994). However, the reported bandgap of GaInN is scattered from group to group. We should remember nitride is usually grown on (0001) plane that does not have inversion symmetry but has large piezoelectric constant (Bykhovski et al., 1996). Consequently, transition energy of fully strained GaInN is affected by the internal electric field. The red shift of transition energy due to the Stark effect in thick GaInN is estimated to be small because the correlation length of excited electrons and holes in thick

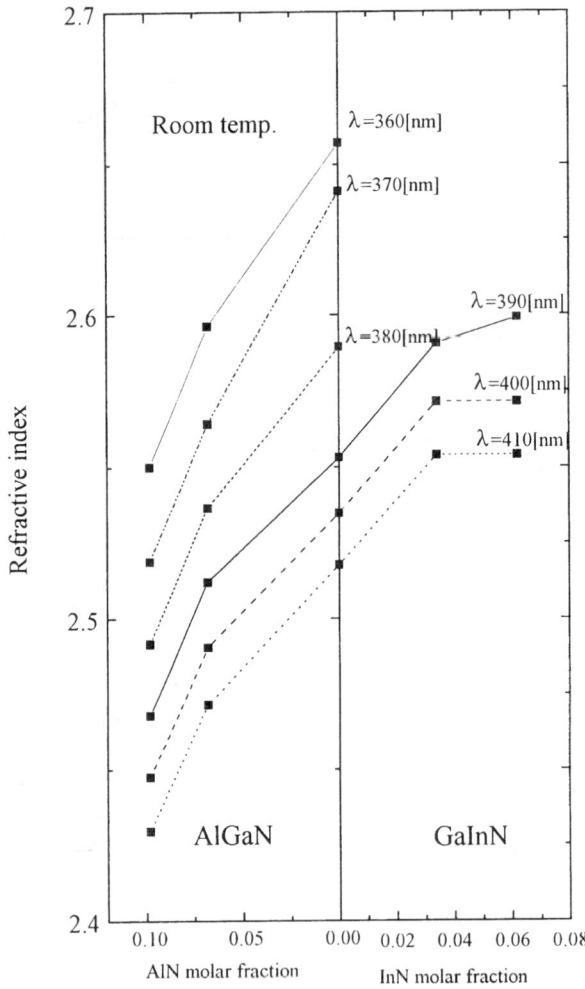

FIG. 6. Compositional dependence of refractive index at various wavelengths.

GaInN is short, therefore the situation of transition energy in GaInN QWs changes drastically. The transition energy in GaInN QWs are largely affected by the internal electric field caused by piezoelectricity because electrons and holes are confined in GaInN QWs, therefore the correlation length becomes large. These phenomena are the well known quantum confined Stark effect.

For the design of optical confinement structure and fabrication of waveguide structure, refractive index at each layer should be clarified.

Figure 6 summarizes compositional dependence of the refractive index with wave vector of the light perpendicular to c axis. As shown, with increase of bandgap, refractive index decreases, suggesting that both carrier and optical confinements can be achieved with this system.

2. TRANSPARENCY CARRIER DENSITY

It is well known that the density of states for both conduction band and valence band of GaN is very large. Accordingly, transparency carrier density is expected to be large. Conduction band effective mass of GaN perpendicular to c-axis with polaron correction was experimentally determined to be $0.22 \pm 0.02 m_0$ (Drechsler et al., 1995), while valence bands hole masses of

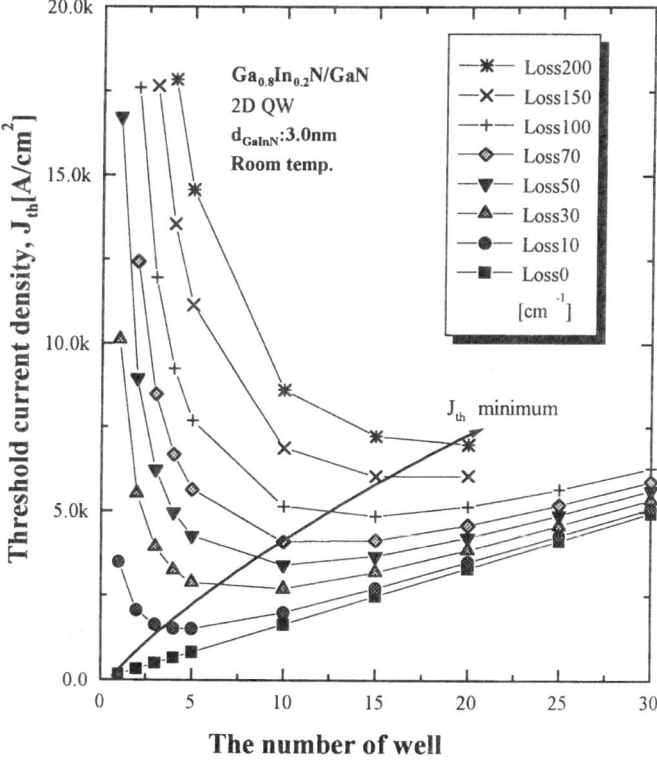

FIG. 7. Dependence of the threshold current density of SCH MQW LD on the number of well layers in active layer as a function of total loss.

GaN was calculated to be 1.76, 1.76, and $0.16m_0$ for effective masses parallel to c-axis, and 0.16, 1.61, and $0.14m_0$ for those perpendicular to c-axis for Γ_{9v}, Γ_{7v}, and Γ_{7v} bands, respectively (Suzuki et al., 1996). The corresponding transparency carrier density is 1.0×10^{19} cm^{-3} at room temperature, which is approximately 4 and 2 times higher, respectively, than those of GaAs and ZnSe.

3. Low Threshold Structure

In order to realize low threshold current density lasing, analysis of the optical loss, that is, internal propagation loss, reflection loss, and scattering loss at the mirror edge are also important. Figure 7 summarizes the dependence of threshold current density on the number of well layers as a function of total loss. InN molar fraction and thickness of the well layer are fixed at 0.2 and 3.0 nm, respectively. This calculation suggests that if the total loss is negligibly small, the number of well layers should be as small as possible for low threshold lasing. However, if the total loss is large, there is an optimum number of well layers for lowest threshold lasing. Large loss in this system was ascertained experimentally (Nakamura, 1996).

VI. Summary

In this chapter, development of the GaN based laser diode has been reviewed and key issues specific to nitride were discussed. A GaN based laser diode is applicable not only for the light source of optical storage, but also for high density and high speed printing, displays, excitation source for phosphors, medical engineering, and many more uses.

Acknowledgments

The authors are grateful to J. I. Pankove and T. D. Moustakas for their critical reading of the manuscript. The authors are also greatly indebted to the following collaborators who have made major contributions to this work: Y. Koide, T. Tanaka, B. Monemar, K. Manabe, and M. Koike. This work was partly supported by the Ministry of Education, Science, Sports, and Culture of Japan (High-Tech Research Center, Contract Nos. 06452114, 07505012, and 07650025), and Japan Society for the Promotion of Science (Research for the Future).

References

Akasaki, I., and Amano, H. (1995). *Inst. Phys. Conf. Ser.* **145**, 19.
Akasaki, I., Amano, H., and Suemune, I. (1995). *Proceedings of International Conference on Silicon Carbide and Related Mater*, Kyoto, Japan, 5.
Akasaki, I., Amano, H., Sota, S., Saki, H., Tanaka, T., and Koike, M. (1995). *Jpn. J. Appl. Phys.* **34**, L1517.
Akasaki, I., Sota, S., Sakai, H., Tanaka, T., Koike, M., and Amano, H. (1996). *Electron. Lett.* **32**, 1105.
Amano, H., Sawaki, N., Akasaki, I., and Toyoda, Y. (1986). *Appl. Phys. Lett.* **48**, 353.
Amano, H., Kito, M., Hiramatsu, K., Sawaki, N., and Akasaki, I. (1989). *Jpn. J. Appl. Phys.* **28**, L2112.
Amano, H., and Akasaki, I. (1990). *Mat. Res. Soc. Ext. Abst.* **EA-21**, 165.
Amano, H., Asahi, T., and Akasaki, I. (1990). *Jpn. J. Appl. Phys.* **29**, L205.
Bykhovski, A. D., Kaminski, V. V., Shur, M. S., Chen, Q. C., and Khan, M. A. (1996). *Appl. Phys. Lett.* **68**, 818.
Dingle, R., Shaklee, K. L., Leheny, R. F., and Zetterstrom, R. B. (1971). *Appl. Phys. Lett.* **19**, 5.
Drechsler, M., Hoffman, D. M., Meyer, B. K., Detchprohm, T., Amano, H., and Akasaki, I. (1995). *Jpn. J. Appl. Phys.* **34**, L1778.
Fischer, A., Kuhne, H., and Richter, H. (1994). *Phys. Rev. Lett.* **73**, 2712.
Itaya, K., Onomura, M., Nishio, J., Sugiura, L., Saito, S., Suzuki, M., Rennie, J., Nunoue, S., Yamamoto, M., Fujimoto, H., Kokubun, H., Ohba, Y., Hatakoshi, G., and Ishikawa, M. (1996). *Jpn. J. Appl. Phys.* **35**, L1315.
Itoh, K., Kawamoto, T., Amano, H., Hiramatsu, K., and Akasaki, I. (1991). *Jpn. J. Appl. Phys.* **30**, 1924.
Khan, M. A., Skogman, R. A., van Hove, J. M., Krishnankutty, S., and Kolbas, R. M. (1990). *Appl. Phys. Lett.* **56**, 1257.
Khan, M. A., Olson, D. T., van Hove, J. M., and Kuznia, J. N. (1991). *Appl. Phys. Lett.* **58**, 1515.
Kim, S. T., Amano, H., Akasaki, I., and Koide, N. (1994). *Appl. Phys. Lett.* **64**, 1535.
Koide, Y., Itoh, N., Itoh, K., Sawaki, N., and Akasaki, I. (1988). *Jpn. J. Appl. Phys.* **27**, 1156.
Manasevit, H. M., Erdman, F. M., and Simpson, W. I. (1971). *J. Electrochem. Soc.* **118**, 1864.
Maruska, H. P., and Tietjen, J. J. (1969). *Appl. Phys. Lett.* **15**, 327.
Matthews, J. W., and Blakeslee, A. E. (1974). *J. Crystal Growth* **27**, 118.
Moustakas, T. D. (1993). *The 183rd Electrochemical Society 93-1 Meeting*, Honolulu, 958.
Nagatomo, T., Kuboyama, T., Minamino, H., and Omoto, O. (1989). *Jpn. J. Appl. Phys.* **28**, L1334.
Nakamura, S., Mukai, T., Senoh, M., Nagahama, S., and Iwasa, N. (1993). *J. Appl. Phys.* **74**, 3911.
Nakamura, S. (1996). Abstracts of Mat. Res. Soc., Fall Meeting, Boston, Session N, 308.
Nakamura, S., Senoh, M., Nagahama, S., Iwasa, N., Yamada, T., Matsushita, T., Kiyoku, H., and Sugimoto, Y. (1996a). *Jpn. J. Appl. Phys.* **35**, L74.
Nakamura, S., Senoh, M., Nagahama, S., Iwasa, N., Yamada, T., Matsushita, T., Kiyoku, H., and Sugimoto, Y. (1996b). *Jpn. J. Appl. Phys.* **35**, L217.
Nakamura, S., Senoh, M., Nagahama, S., Iwasa, N., Yamada, T., Matsushita, T., Kiyoku, H., and Sugimoto, Y. (1996c). *Appl. Phys. Lett.* **68**, 2105.
Pankove, J. I., Miller, E. A., Richman, D., and Berkeyheiser, J. E. (1971). *J. Lumin* **4**, 63.
Pankove, J. I. (1973). *RCA Review* **34**, 336.

Suzuki, M., and Uenoyama, T. (1996). *J. Appl. Phys.* **80**, 6868.
Tanaka, T., Watanabe, A., Amano, H., Kobayashi, Y., Akasaki, I., Yamazaki, S., and Koike, M. (1994). *Appl. Phys. Lett.* **65**, 593.
Wickenden, D. K., Bargeron, C. B., Bryden, W. A., Miragliova, J., and Kistenmacher, T. J. (1994). *Appl. Phys. Lett.* **65**, 2024.
Yoshimoto, N., Matsuoka, T., Sasaki, T., and Katsui, A. (1991). *Appl. Phys. Lett.* **59**, 2251.

CHAPTER 16

Nonvolatile Random Access Memories in Wide Bandgap Semiconductors

James A. Cooper, Jr.

SCHOOL OF ELECTRICAL AND COMPUTER ENGINEERING
PURDUE UNIVERSITY
WEST LAFAYETTE, IN

I. INTRODUCTION	473
II. DYNAMIC MEMORIES IN GALLIUM ARSENIDE	475
1. *Basic Storage Capacitor Designs*	475
2. *FET-Accessed Memory Cells*	477
3. *Bipolar-Accessed Memory Cells*	478
III. NONVOLATILE MEMORIES IN SILICON CARBIDE	481
1. *Generation Mechanisms*	481
2. *Memory Cell Design*	484
3. *Monolithic NVRAM Demonstration Chips*	486
IV. POTENTIAL FOR NONVOLATILE MEMORIES IN THE AlGaN SYSTEM	486
V. CONCLUSIONS	489
References	490

I. Introduction

The ideal semiconductor memory can be read or written in zero time, retains data forever (even with no power applied), dissipates zero power, and occupies zero area. The ideal memory is also very inexpensive to produce, costing zero cents per bit. Unfortunately, real semiconductor memories can only hope to approach the ideal.

One can classify existing semiconductor memories into three general categories: random access memory (RAM), read-only memory (ROM), and electrically-alterable read-only memory (EAROM). The EAROM family includes such variants as the flash memory. All these memories are "random access", in the sense that data can be read in a random order in a relatively short time. The memories differ, however, in the ease with which data may be *written*. RAM's allow data to be read and written very rapidly, in a time on the order of 100 ns or less. Read-only memory, or ROM, cannot be written at all, and is typically used to store basic machine operating code or tables of data which will never change. Data in EAROM can be electrically altered during operation, but the "reprogramming" requires orders of magnitude longer than the read access time, typically a few ms. Thus, EAROM is not used for the main read/write memory in computer systems. Instead, the main working memory is implemented exclusively with RAM.

Random access memory, or RAM, can be further divided into static RAM and dynamic RAM. The memory cells in a static RAM, or SRAM, are small circuits consisting of up to six transistors each, configured to form a static latch. Data is retained as long as power is applied to the circuit. In a dynamic RAM, or DRAM, the cell consists of a single access transistor connected to a storage capacitor. Data is written by charging the storage capacitor to a known potential through the access transistor. The access transistor is then turned off, isolating the storage capacitor. The potential on the storage capacitor will only be retained for a short time, typically between 1 and 100 sec. at room temperature, because leakage due to thermal generation in the semiconductor gradually discharges the storage capacitor. Thus, to store data for an extended period of time it is necessary for the operating system to periodically read and refresh each cell in a DRAM. This refresh process idles the memory and, more importantly, accounts for up to 85% of the power dissipation in some computer systems. This becomes especially important in notebook and laptop computers, where use of a non-refreshing memory could extend battery life considerably.

The recent emergence of wide bandgap semiconductors such as silicon carbide (SiC) and the III–V nitrides (GaN, AlN, and AlGaN) has opened the possibility for a *new class* of semiconductor memory called nonvolatile RAM, or NVRAM. NVRAM cells are similar to DRAM cells in that they consist of an access transistor and a storage capacitor. Like DRAM's, NVRAM cells also tend to lose data because of thermal generation in the material. However, thermal generation rates in wide bandgap semiconductors are many orders of magnitude smaller than in silicon. For example, in 6H-SiC with a bandgap energy of 3.0 eV, the generation rate is about a factor of 10^{16} lower than silicon at room temperature (this factor arises from the ratio of intrinsic carrier concentrations of the two materials). This large factor effectively means that the SiC memory will not need to be refreshed in our lifetime. Moreover, if the access transistor is properly designed, the stored data can be retained even if power is removed from the chip. In this case, the memory is said to be *nonvolatile*. This principle has been demonstrated experimentally at Purdue University, where SiC memory cells have retained data continuously without power for over two years with less than 2% charge loss.

NVRAM represents a totally new class of semiconductor memory, on the same level with RAM, ROM, and EAROM. NVRAM can be written as quickly as RAM, but retains data indefinitely without refresh, much like an EAROM. In principle, semiconductor NVRAM could constitute the entire memory system of future computers, replacing both the high speed main memory and the rotating disk mass storage. In the following sections we will review NVRAM development in several wide bandgap semiconductors, beginning with GaAs ($E_G = 1.42\,\text{eV}$), proceeding to SiC ($E_G = 3.0\,\text{eV}$), and projecting to memories in the III–V nitrides ($E_G \geqslant 3.4\,\text{eV}$).

II. Dynamic Memories in Gallium Arsenide

1. Basic Storage Capacitor Designs

Early attempts to develop semiconductor dynamic memories in GaAs centered on what are termed *leakage-limited* storage cells (Cooper, Qian, and Melloch, 1986, 1989; Kleine, *et al.*, 1989; Qian, Melloch, and Cooper, 1989). In a leakage-limited cell, one places an *excess* of electrons (or holes) on the storage capacitor and hopes to prevent them from leaking off. An example might be a cell where electrons are confined behind a heterojunction potential barrier or within a quantum well. Thermionic emission of electrons over the potential barrier gradually discharges the memory cell, returning it to equilibrium. To estimate the storage time, note that the probability of an electron acquiring thermal energy ΔE from a crystal at absolute temperature T goes as $\exp(-\Delta E/kT)$, where k is Boltzmann's constant. Thus the storage time is exponentially related to the barrier height ΔE for charge exchange, i.e. $\tau_S \sim \exp(\Delta E/kT)$. This exponential relationship makes the storage time very sensitive to the confining barrier height. For example, a 0.06 eV decrease in barrier height translates into an order-of-magnitude decrease in storage time at room temperature. Because of this exponential relationship, leakage-limited GaAs memories do not exhibit useful storage time at room temperature, since the barrier heights arising from heterojunction discontinuities in the GaAs/AlGaAs system are relatively small. The interested reader is referred to the literature for details (Cooper, 1993).

The first successful dynamic memories in GaAs were *generation-limited* cells, in which a *deficit* of electrons is created in a semiconductor region and the missing electrons are gradually supplied by thermal generation (Cooper, 1993; Dungan, Cooper, and Melloch, 1987). The highly successful silicon DRAM is also an example of a generation-limited memory. In GaAs, the

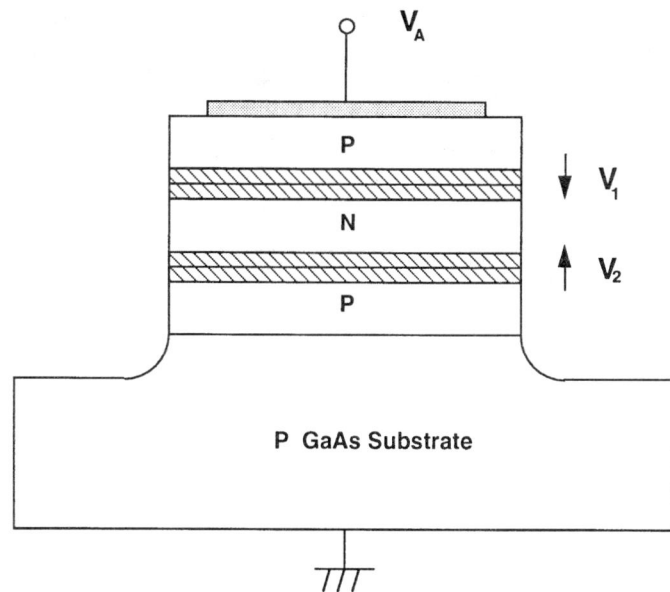

FIG. 1. Cross section of a p-n-p storage capacitor. The middle n-layer is floating, and the potential of this layer represents the stored information.

most common form of generation-limited memory is based on the pnp storage capacitor of Fig. 1. In this structure, electrons are removed from the floating n-region by applying a pulse of either polarity across the pnp stack. The pulse forward-biases one junction and reverse-biases the other. Electrons flow across the forward-biased junction until the potential across that junction is reduced to zero, at which time the current stops, leaving the entire voltage across the reverse-biased junction. The floating n-region is now at the same potential as the positive terminal of the structure. When the terminal voltage returns to ground, the floating n-layer remains at a positive potential, since electrons were removed during the pulse. This places both p-n junctions under reverse bias. Thermal generation in the depletion regions gradually supplies electrons to the n-layer, restoring the structure to equilibrium. Assuming Shockley-Read-Hall generation through midgap traps (the most efficient generation process), the effective barrier height ΔE for charge exchange is half the zero-temperature semiconductor bandgap (Cooper, 1993). In GaAs, this is 0.76 eV, much higher than the largest available conduction band discontinuities in the GaAs/AlGaAs system. This higher effective barrier height permits storage times of up to 10 h in GaAs memory cells at room temperature (Cooper, 1993). Although these storage times are not long enough to be considered nonvolatile, they are much longer than the storage times of silicon DRAM's. (Silicon DRAM cells are based on an MOS storage capacitor that is driven into deep depletion. Thermal generation in the depletion region supplies electrons to the inversion layer, returning the cell to equilibrium. The effective barrier height is half the silicon bandgap, or about 0.58 eV.)

2. FET-Accessed Memory Cells

A major consideration in the design of memory cells is the type of access transistor to be used for coupling the potential from the bit line to the storage capacitor (writing) or from the storage capacitor to the bit line (reading). In the GaAs system, several FET-type memory cells have been studied, including cells based on JFET, MESFET, and MODFET access transistors (Neudeck, *et al.*, 1989; Dungan, *et al.*, 1990; Kleine, Cooper, and Melloch, 1992). The complete memory cell can be visualized as a field-effect transistor with a floating drain, as shown in Fig. 2. The floating drain is designed to have a large capacitance to ground in order to store the desired charge. However, these FET-accessed cells have a major drawback: they all exhibit *subthreshold leakage* in the off state. In principle, subthreshold current from the source can be reduced to negligible levels by applying sufficient reverse voltage to the gate. However, since the JFET, MESFET, and MODFET do not have a true insulator between the gate and channel, increasing the reverse bias to the gate inevitably increases the gate leakage. These unavoidable leakage currents within the access transistor limit the storage times of FET-accessed cells in GaAs to well below the theoretical storage time of the capacitor itself. Even if gate leakage could be eliminated,

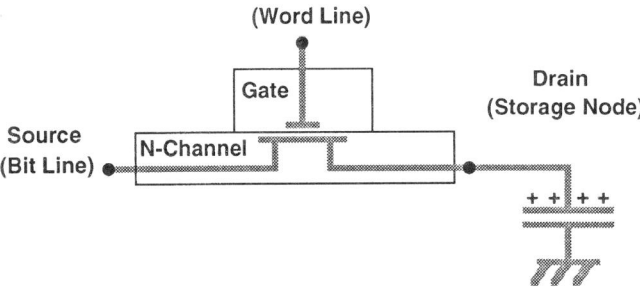

Fig. 2. Illustration of a typical FET-accessed memory cell. The device can be viewed as a field-effect access transistor with a floating drain, with charge stored on the drain capacitance.

these cells would still require negative gate bias to suppress subthreshold leakage, and applying a bias is inconsistent with the notion of a *nonvolatile* memory, in which information is stored with no power applied. All these shortcomings of a FET-access memory can be overcome with the use of a *bipolar* access transistor.

3. BIPOLAR-ACCESSED MEMORY CELLS

The GaAs bipolar memory cell is illustrated in Fig. 3 (Stellwag, Cooper, and Melloch, 1992). In this cell, an npn bipolar access transistor is vertically integrated over a pn junction storage capacitor. The emitter is connected to a bit line of the memory array and the base is connected to a word line. Operation is illustrated in Fig. 4. Figure 4(*a*) shows the equilibrium state ("store zero") of the cell, with emitter, base, and floating collector all at ground potential. To read the cell or write a logic "one", electrons are removed from the floating collector by taking both emitter and base to a positive potential, Fig. 4(*b*). This causes electrons to be injected from the floating collector into the base until the potential across the collector-base junction is reduced to zero (or, if the base is more positive than the emitter, until the collector-to-emitter voltage is reduced to zero). To store the logic "one", the base and emitter are returned to ground, Fig. 4(*c*), leaving the collector floating at a positive potential. In this situation, the collector-base and collector-substrate junctions are both reverse biased, and thermal

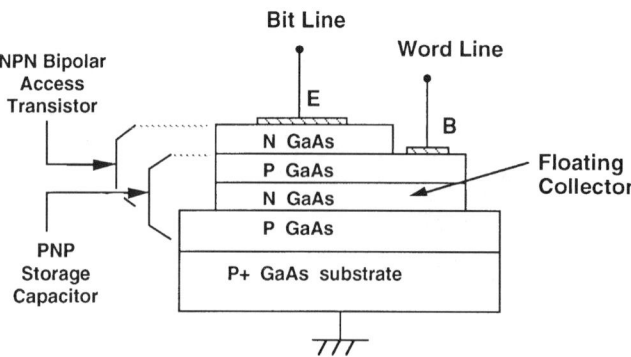

FIG. 3. Cross section of a GaAs bipolar memory cell. The top three n-p-n layers form the access transistor, which is merged with a p-n-p storage capacitor. Alternatively, the structure may be viewed as an n-p-n bipolar transistor with a floating collector, where the floating collector has a large capacitance to ground.

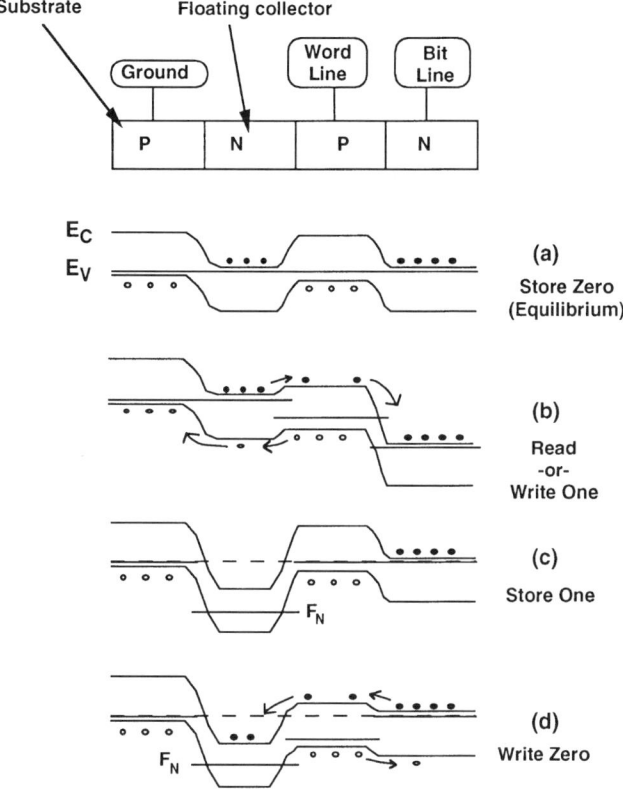

FIG. 4. Band diagrams of the bipolar memory cell in four phases of operation: (a) store "zero" (equilibrium) state, (b) read data or write "one" state, (c) store "one" state, and (d) write "zero" state.

generation in the depletion regions gradually restores the structure to equilibrium. Of course, in a wide bandgap semiconductor this "storage time" may be quite long. Finally, to write a "zero" into the cell, the emitter is held at ground and the base is taken slightly positive, Fig. 4(d). Electrons are injected from the emitter into the base and diffuse across the base into the collector, returning it to ground.

To observe the transient charge recovery of the bipolar RAM cell in the "store one" state, the potential of the floating collector can be monitored indirectly by observing the capacitance at the base of the bipolar access transistor (Stellwag, Cooper, and Melloch, 1992). Such a recovery transient is shown in Fig. 5 for the bipolar RAM cell of Fig. 3. Prior to $t = 10,000$ sec.,

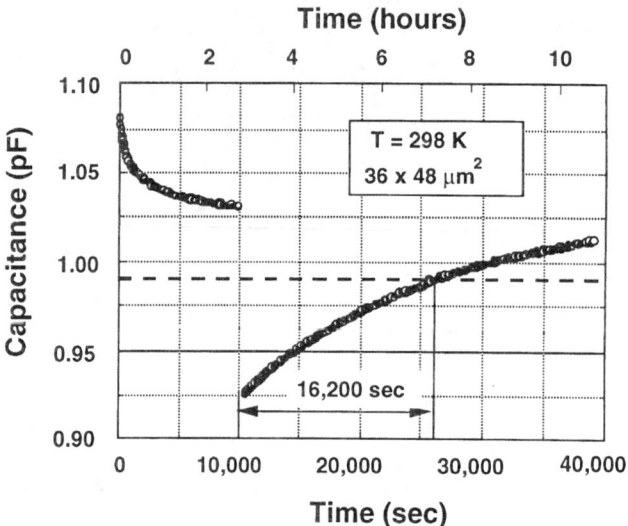

FIG. 5. Room temperature capacitance recovery transient of a GaAs bipolar memory cell. The $1/e$ time constant is 4.5 h.

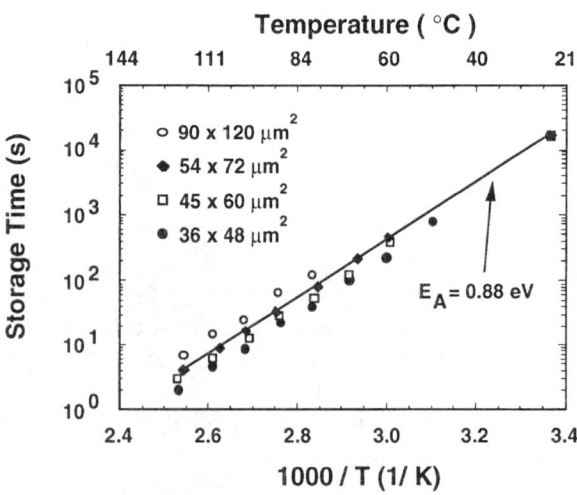

FIG. 6. Storage time as a function of temperature for several GaAs bipolar memory cells of different area. The recovery process is thermally activated with an energy of 0.88 eV, slightly greater than half the zero temperature bandgap.

the initial bias transient is allowed to die out, and the equilibrium condition is gradually established. At $t = 10{,}000$ sec., a short bias pulse is applied to remove electrons from the floating collector (the pulse is too short to be resolved on this time scale). This results in an almost instantaneous drop in capacitance, as the collector-base and collector-substrate depletion regions widen. For times after $t = 10{,}000$ sec., all terminals are held at ground potential. The capacitance transient is due to thermal generation acting to discharge the floating collector. The $1/e$ time constant of this cell is 16,200 sec., or about 4.5 h. Similar cells have exhibited storage times in excess of 10 hours at room temperature.

Figure 6 shows the dependence of storage time on temperature for several GaAs memory cells of different area (Stellwag, Cooper, and Melloch, 1992). As expected based on our earlier arguments, the recovery time is thermally activated. However, the activation energy is approximately 0.88 eV, slightly greater than half the zero-temperature bandgap. This suggests that the dominant generation center does not lie exactly at midgap, but rather at an energy 0.88 eV from one band.

III. Nonvolatile Memories in Silicon Carbide

1. Generation Mechanisms

Because the storage time increases exponentially with bandgap, a generation-limited memory will exhibit a very large increase in storage time if constructed in a semiconductor with even a slightly wider bandgap (every 0.12 eV increase in bandgap increases the storage time by an order of magnitude). Thus, if the bipolar memory cell of the last section were constructed in 6H-SiC (with a bandgap of 3.0 eV), the room temperature storage time should be about 13 orders of magnitude longer than GaAs and almost 16 orders of magnitude longer than silicon. This would translate into a room temperature storage time greater than 10^{10} years. Such a memory would certainly be considered nonvolatile!

To test these concepts, a series of 6H-SiC npn storage capacitors were fabricated at Purdue University (Wang, 1996; Wang, et al., 1996). Figure 7 shows storage time as a function of temperature for several of these memory cells. Note that these data are taken at very high temperatures, typically above 300°C, so that the recovery transient can go to completion in a reasonable time. The recovery time is thermally activated, as expected, with

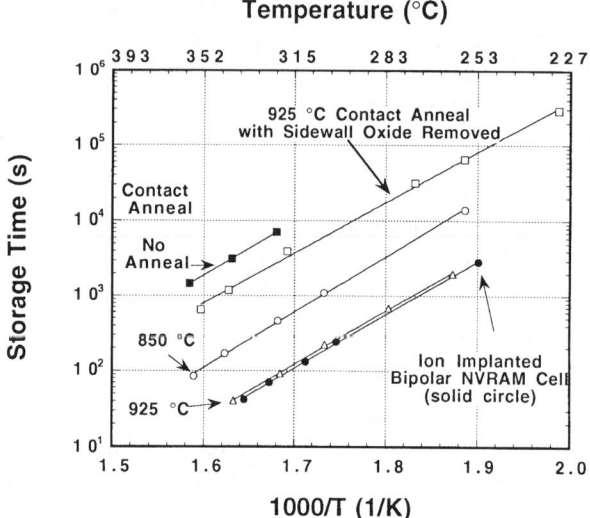

FIG. 7. Storage time as a function of temperature for several 6H-SiC n-p-n storage capacitors and an ion-implanted bipolar NVRAM cell. The capacitors differ in the nature of the ohmic contact anneal. Devices that have been annealed are dominated by surface generation and exhibit shorter storage times, while those without an anneal are dominated by bulk generation. All devices have an activation energy of 1.48 eV.

an activation energy of 1.48 eV, approximately half the bandgap of 6H-SiC. Extrapolation of these data to room temperature predicts a room temperature storage time greater than 10^6 years. (This is about four orders of magnitude lower than expected based on the ratio of bandgap energy to that of silicon or GaAs. The reduction may be due to the prefactor of the exponential term in the generation rate equation. This term, known as the *generation lifetime*, depends upon the density and capture coefficient of the dominant generation-recombination centers in the material.)

The data of Fig. 7 suggests a very long storage time at room temperature. However, other leakage mechanisms having activation energies lower than 1.48 eV may be present. If so, these may become dominant at lower temperatures, reducing the storage time below the extrapolated value. The only way to evaluate this possibility is to conduct long-term storage time measurements at room temperature. In a unique experiment (Wang, 1996; Wang, *et al.*, 1996), storage capacitors from several fabrication runs in both 4H and 6H-SiC have been continuously measured at room temperature for over two years. The devices are mounted in a light-tight package and maintained under short-circuit conditions in a shielded box. The charge recovery is monitored by measuring the capacitance at regular intervals—this measurement is nondestructive of the stored data. Figure 8 shows several capacitance transients after more than two years of nonvolatile storage. The two 4H storage capacitors exhibit a slow recovery, with time constants of 21 and 43 years, respectively. The 6H storage capacitors exhibit a slight recovery (about 2%) during the first 1500 h due to release of electrons from deep traps, but no measurable change after that. Based on the scatter in the data, we estimate that the true storage time is in excess of 100 years, and may be many orders of magnitude longer.

An investigation of charge recovery in SiC memory capacitors (Wang, 1996; Wang, *et al.*, 1996) has identified three primary generation mechanisms: (i) thermal generation in the p-n junction depletion regions, (ii) surface generation at the perimeter of the memory capacitors, and (iii) generation associated with defects in the bulk material. The first two mechanisms have activation energies of half bandgap, but defect generation is characterized by an activation energy much less than half bandgap. The exact mechanism of defect generation is not well understood, but the generation rate in defect-dominated samples is strongly field-dependent, leading to an easily-

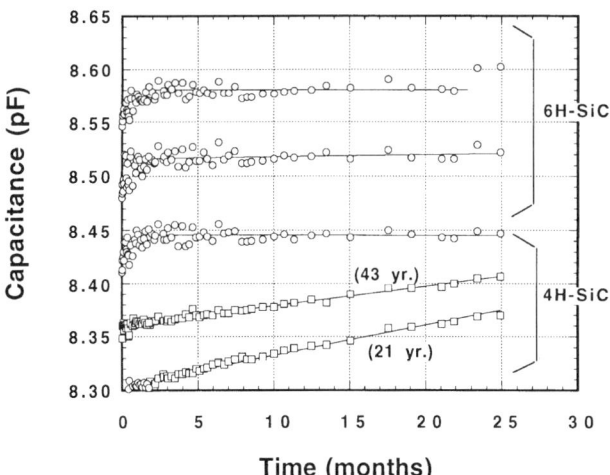

FIG. 8. Capacitance transients of five SiC storage capacitors at room temperature. The vertical scale has been magnified to show details—the full recovery would require a change of about 1 pF in capacitance. The two 4H samples show steady capacitance recovery with $1/e$ storage times estimated to be 21 and 43 years. The three 6H samples show a slight initial recovery during the first two months, but no measurable recovery thereafter. We estimate the $1/e$ storage time of these samples to be greater than 100 years.

recognized non-exponential capacitance transient. We speculate that this is due to field-induced barrier lowering at the generation sites. Defect generation is related to the quality of the material, and samples can easily be found which do not exhibit these effects. Surface generation can be eliminated by proper passivation of the etched sidewalls of the memory capacitor. When both surface and defect generation are eliminated, the recovery is dominated by generation in the junction depletion regions, and extremely long storage times are observed. The interested reader is referred to the literature for details on these experiments (Wang, 1996; and Wang, et al., 1996).

2. Memory Cell Design

As discussed earlier, to achieve true nonvolatile storage with no applied bias, it is necessary to use a *bipolar*-accessed memory cell (Cooper, 1993; Wang, 1996). The major design challenge with bipolar-accessed cells occurs during the readout process (Wang, 1996). Imagine the cell in the "store zero" state, with the floating collector at ground. The bit line is precharged to a positive potential and allowed to float. At time $t = 0$, the word line (connected to the base of the access transistor) is taken to a positive potential and electrons are injected from the collector into the base. Ideally, all these electrons diffuse across the base and are swept into the emitter. The electrons reaching the emitter reduce the positive potential of the bit line, allowing sense amplifiers to detect the charge state of the cell. However, in real cells two important factors limit the efficiency of the readout process. First, any electrons injected into the base *under the base contact* do not reach the emitter, but instead recombine at the base contact, producing base current. These electrons are lost to the readout process. Only those electrons injected into the base under the emitter junction have a chance of reaching the emitter. Unfortunately, the flow of base current laterally under the emitter junction produces a voltage drop across the base which reduces the injection. This effect, current crowding due to base spreading resistance, is exacerbated during the readout transient by the displacement current passing through the charging emitter-base junction. Because of the combined effects of recombination under the base contact and current crowding under the emitter junction, the readout efficiency of bipolar memory cells is quite low, typically in the range of 10–20% (Wang, 1996).

To improve readout efficiency, it is important to minimize the area of the base contact (to reduce recombination) and maximize the emitter periphery (since most of the injection from the collector occurs along the emitter periphery). To increase the voltage swing on the bit line during readout, the

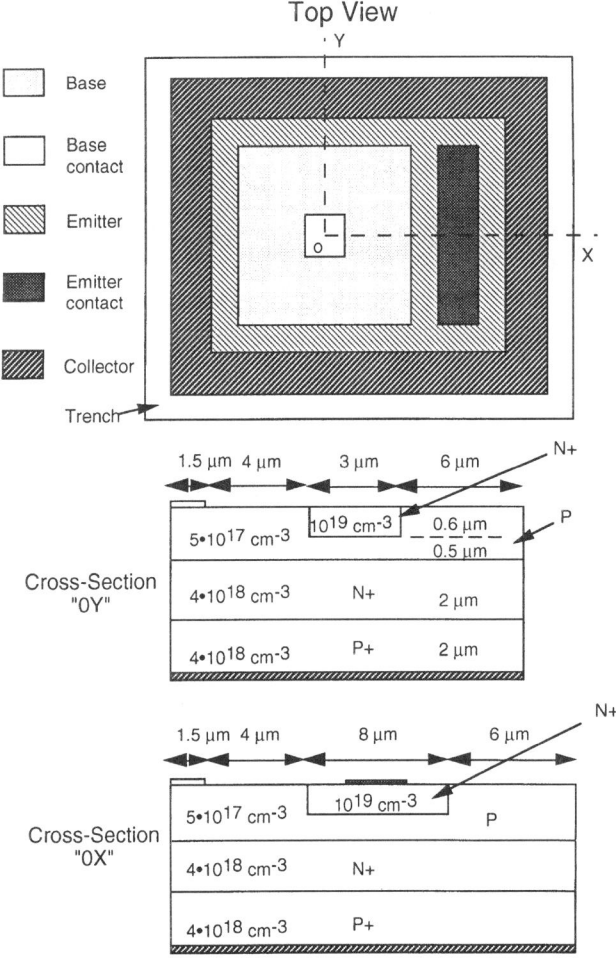

FIG. 9. Top view and cross sections of a prototype SiC NVRAM cell, including lateral dimensions and doping and thickness of each layer in the device. The lateral dimensions result from the use of rather conservative 3 μm design rules, and can be reduced with more aggressive processing.

emitter area should be minimized, since the emitter-base capacitance adds significantly to the bit line capacitance. Finally, to increase the charge available for readout, the area of the floating collector can be increased by extending the collector beyond the emitter. A SiC cell design employing all these features is shown in Fig. 9 (Wang, 1996).

3. Monolithic NVRAM Demonstration Chips

To make a practical memory product that can be sold commercially, it is necessary to include all peripheral logic and control circuitry on the same chip with the memory cell array. The peripheral circuitry includes sense amplifiers that detect the charge placed on the bit lines, row and column address circuits that select the desired cell location within the memory array, address and data latches that interface with the address and data busses, and a variety of control and timing circuits that provide sequencing and synchronizing signals. In the SiC NVRAM's currently under development, these functions are implemented in enhancement-mode NMOS logic fabricated on the same chip as the bipolar memory cell array using a compatible fabrication process. Monolithic NMOS integrated circuits were first demonstrated in SiC in 1994 (Xie, Cooper, and Melloch, 1994) and were implemented on the same chip with bipolar NVRAM cells later that same year (Xie, *et al.*, 1995). The first CMOS integrated circuits in SiC were demonstrated in 1995 (Slater, *et al.*, 1996). At the time of this writing, monolithic NVRAM demonstration chips containing 1024-bit cell arrays, complete with SiC NMOS peripheral circuitry, are under development (Cree Research, Inc.).

IV. Potential for Nonvolatile Memories in the AlGaN System

In principle, nonvolatile memories of the type described above can be implemented in the AlGaN material system with performance superior to SiC. The AlGaN system provides two significant advantages over SiC: wider bandgap and higher electron mobility. The wider bandgap should lead to even longer storage times than currently obtained in SiC. Since the storage time in SiC memory cells has been demonstrated to be longer than 100 years at room temperature (Wang, 1996; Wang, *et al.*, 1996), it would seem that little improvement is needed. However, this is not the case. Since storage times are thermally activated, as temperature is raised, the storage time decreases exponentially. For example, in SiC capacitors at 110°C, long-term measurements indicate that the storage time is on the order of 30 years. While this is still long enough to be considered nonvolatile, there are many systems applications where even higher temperatures are required. The use of memories constructed in the AlGaN system should allow nonvolatile opertion to be obtained at much higher temperatures than SiC.

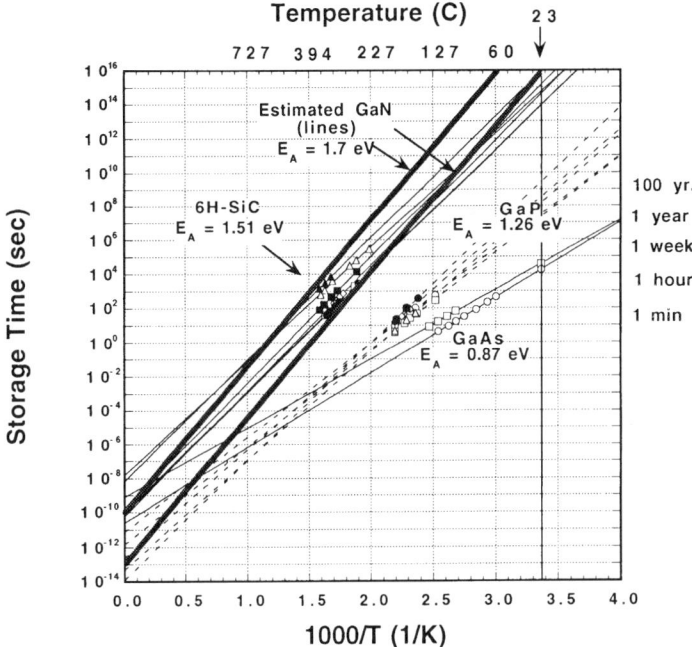

FIG. 10. Storage time as a function of temperature for p-n junction storage capacitors in GaAs, GaP, and 6H-SiC. The two heavy lines indicate the boundaries of the expected behavior of GaN storage capacitaors. GaN cells may have a significant advantage over SiC at temperatures below about 250°C.

Figure 10 shows a plot of storage time versus temperature for GaAs, GaP, SiC, and GaN. The curves for GaAs (Stellwag, *et al.*, 1992), GaP (Wang, *et al.*, 1993), and 6H-SiC (Wang, 1996; Wang, *et al.*, 1996) are derived from measurements taken at Purdue University. The $1/e$ storage time of a symmetrically doped pn junction storage capacitor is given by (Cooper, 1993)

$$\tau_S = \frac{N_B}{2\sqrt{N_C N_V}} \tau_p \exp\left(-\frac{\alpha}{2k}\right) \exp\left(\frac{E_{G0}/2 + \Delta E_T}{kT}\right) \quad (1)$$

where $N_B = N_A = N_D$ is the doping, N_C and N_V are the conduction and valence band density of states, τ_p is the hole lifetime (assumed equal to the electron lifetime), k is Boltzmann's constant, α is the linear coefficient of the temperature dependence of the bandgap, E_{G0} is the zero-temperature bandgap, ΔE_T is the difference between the dominant RG center energy and midgap, and T is absolute temperature. Equation (1) may be written

$$\tau_S = \tau_0 \exp\left(\frac{E_{G0}/2 + \Delta E_T}{kT}\right) = \tau_0 \exp\left(\frac{E_A}{kT}\right) \tag{2}$$

where E_A is the activation energy. The activation energies in Fig. 10, derived from the slope of the measured data, are slightly greater than half the zero-temperature bandgap for each semiconductor (E_{G0} = 1.52, 2.4, and 3.0 eV for GaAs, GaP, and 6H-SiC respectively). The prefactor τ_0 is given by the intercept of the linear extrapolation at $1/T = 0$. The prefactors for GaAs and 6H-SiC are comparable, approximately 10^{-10} sec., while the prefactor for GaP is in the range 10^{-11}–10^{-14} sec. (One notes that two SiC storage capacitors have prefactors around 10^{-8} sec. These are devices in which surface generation has been eliminated.)

To estimate the storage times obtainable from GaN, we consider two possible scenarios, indicated by the two dark lines in Fig. 10. These lines have activation energies of 1.7 eV, approximately half the 3.4 eV bandgap of GaN. Their intercepts are taken to be 10^{-10} and 10^{-13} sec., corresponding roughly to the values observed for SiC and GaP. If the upper line is the correct assumption, the GaN capacitor would exhibit 10 year (or longer) storage times at temperatures up to 190°C. By comparison, for 6H-SiC we would expect 10 year (or longer) storage times only up to about 125°C. The improved performance at high temperatures may be important in some applications. Of course, if the lower line turns out to be correct, then GaN would have no significant advantage over SiC at high temperatures. The true performance can only be determined by experimental measurements, which have not been conducted at the time of this writing.

GaN may offer other important advantages for NVRAM's. For example, the readout efficiency of the bipolar access transistor in SiC storage cells is typically in the range of 10–20%. This limitation arises because of the low hole mobility in SiC, leading to significant base spreading resistance and severe current crowding during the transient portion of the readout. As described earlier, the bipolar access transistor is operated in the inverse mode during readout, and current crowding under the emitter causes a large fraction of the stored electrons to be injected under the base contact, where they recombine and fail to reach the emitter. These electrons are therefore lost to the readout process. This limitation forces us to provide larger cells than would otherwise be necessary, and to either shorten the bit lines (to reduce bit line capacitance) or design very elaborate sense amplifiers to detect the small voltage swings on the bit line. These limitations could be avoided by using a heterojunction bipolar transistor (HBT) in the AlGaN material system. The emitter and collector would both be wider bandgap AlGaN, while the base would be heavily doped GaN. The heavy doping of the base and relatively high carrier mobility would minimize current crowding under the emitter, resulting in much higher readout efficiency. It is expected that read/write speeds would also be increased.

One disadvantage of the GaN system relative to SiC is the lack of a high-quality native oxide. SiC can be thermally oxidized to form SiO_2, and the MOS interface is of sufficient quality to make MOS transistors. This makes it possible to implement the necessary peripheral circuits in SiC using either NMOS or CMOS logic. To make a commercially viable product in GaN, the peripheral logic could be implemented using either MESFET's or HFET's (MODFET's). At this writing, no GaN logic circuits have yet been reported.

V. Conclusions

As we have seen, wide bandgap semiconductors exhibit very low thermal generation rates, making it possible to realize an entirely new class of memory device, the nonvolatile random-access memory (NVRAM). SiC NVRAM's can be read and written in a few hundred nanoseconds or less, and can store data almost indefinitely at room temperature with no power applied. Because of this unique combination of properties, NVRAM's can theoretically replace both high-speed RAM and low-speed magnetic storage (hard disks) in future computing systems, thereby saving power, reducing weight, and increasing reliability. The design of the NVRAM involves many considerations, including thermal generation in the memory capacitor, leakage current in the access transistor, readout efficiency of the cell, and design and implementation of peripheral logic. Several semiconductors have been explored to date, but so far SiC appears to be the most promising material to satisfy all the above requirements. This is especially true of the peripheral logic, where the availability of MOS integrated circuits in SiC make it possible to implement designs closely analogous to those of silicon dynamic RAM's.

GaN and materials in the AlGaN family promise longer storage times than SiC at elevated temperatures, and AlGaN HBT's could improve the readout efficiency of the NVRAM cell, eliminating one of the problems of SiC access transistors. However, a viable integrated circuit technology must be developed in the AlGaN system to implement the required peripheral logic before commercial NVRAM devices can be realized.

REFERENCES

Cooper, J. A., Jr., Qian, Q-D., and Melloch, M. R., (1986). "Evidence of Long-Term Storage of Minority Carriers in N^+-GaAs/AlGaAs/P-GaAs MIS Capacitors," *IEEE Electron Device Letters*, **7**, 374.

Cooper, J. A., Jr. (1993). "Recent Advances in GaAs Dynamic Memories," In *Advances in Electronics and Electron Physics*, (Hawkes, P. W., ed.), Vol. 86, pp. 1–79. Academic Press.

Dungan, T. E., Cooper, J. A., Jr., and Melloch, M. R. (1987). "A Thermal-Generation-Limited Buried-Well Structure for Room-Temperature GaAs Dynamic RAM's," *IEEE Electron Device Letters*, **8**, 243.

Dungan, T. E., Neudeck, P. G., Melloch, M. R., and Cooper, J. A., Jr., (1990). "One-Transistor GaAs MESFET- and JFET-Accessed Dynamic RAM Cells for High-Speed Medium Density Applications," *IEEE Transactions on Electron Devices*, **37**, 1599.

Kleine, J. S., Qian, Q-D., Cooper, J. A., Jr., and Melloch, M. R. (1989). "Electron Emission from Direct Bandgap Heterojunction Capacitors," *IEEE Transactions on Electron Devices*, **36**, 289.

Kleine, J. S., Cooper, J. A., Jr., and Melloch, M. R. (1992). "Characterization of a GaAs/AlGaAs Modulation-Doped Dynamic Random Access Memory Cell," *Applied Physics Letters*, **61**, 834.

Melloch, M. R., Qian, Q-D., and Cooper, J. A., Jr., (1986). "Long Term Storage of Inversion Holes at a Superlattice/GaAs Interface," *Applied Physics Letters*, **49**, 1471.

Neudeck, P. G., Dungan, T. E., Melloch, M. R., and Cooper, J. A., Jr., (1989). "Electrical Characterization of a JFET-Accessed GaAs Dynamic RAM Cell," *IEEE Electron Device Letters*, **10**, 477.

Qian, Q-D., Melloch, M. R., Cooper, J. A., Jr. (1986). "Multi-Day Storage of Holes at the AlAs/GaAs Interface," *IEEE Electron Device Letters*, **7**, 607.

Qian, Q. D., Melloch, M. R., and Cooper, J. A., Jr., (1989). "Electrical Behavior of a Static Hole Inversion Layer at the i-AlAs/N-GaAs Heterojunction," *Journal of Applied Physics*, **65**, 3118.

Slater, D. B., Jr., Johnson, G. M., Liplin, L. A., Suvorov, A. V., and Palmour, J. W. (1996). "Demonstration of 6H-SiC CMOS Technology," *IEEE Device Research Conference*, Santa Barbara, CA.

Stellwag, T. B., Cooper, J. A., Jr., and Melloch, M. R. (1992). "A Vertically Integrated GaAs Bipolar Dynamic RAM Cell with Storage Times of 4.5 h at Room Temperature," *IEEE Electron Device Letters*, **13**, 129.

Stellwag, T. B., Melloch, M. R., Cooper, J. A., Jr., Sheppard, S. T., and Nolte, D. D. (1992). "Increased Thermal Generation Rate in GaAs Due to Electron-Beam Metallization," *Journal of Applied Physics*, **71**, 4509.

Wang, Y., Ramdani, J., He, Y., Bedair, S. M., Cooper, J. A., Jr., and Melloch, M. R. (1993). "Long-Term Storage in GaP PN Junction Capacitors," *Electronics Letters*, **29**, 1154.

Wang, Y. (1996). "Analysis and Optimization of Bipolar Nonvolatile Random Access Memory Cells in 6H Silicon Carbide," Ph.D. Thesis, Purdue University.

Wang, Y., Cooper, J. A., Jr., Melloch, M. R., Sheppard, S. T., Palmour, J. W., and Lipkin, L. A. (1996). "Experimental Characterization of Electron-Hole Generation in Silicon Carbide," *Journal of Electronic Materials*, **25**, 899.

Xie, W., Cooper, J. A., Jr., and Melloch, M. R. (1994). "Monolithic NMOS Digital Integrated Circuits in 6H-SiC," *IEEE Electron Device Letters*, **15**, 455.

Xie, W., Johnson, G. M., Wang, Y., Cooper, J. A., Jr., Palmour, J. W., Lipkin, L. A., Melloch, M. R., and Carter, C. H., Jr., (1995). "Cell Design and Peripheral Logic for Nonvolatile Random Access Memories in 6H-SiC," *Inst. Phys. Conf. Ser.*, No. 141, Ch. 4, pp. 395–398.

Index

A

Ab initio methods, 207, 410, 412
Absorption coefficient, near bandgap, 312–315
Acceptor bound exciton (ABE)
 bound excitons in GaN, 349, 353–355
 related optical spectra, 355–361
Aixtron GmbH, 20
AlGaInN
 applications of, 432
 methods used to grow, problems with, 148–149
AlGaN
 applications of, 433
 doping of, 271
 InGaN/AlGaN double-heterostructure LEDs, 434–439
 luminescence in, 295–299
 nonvolatile random access memory (NVRAM) in, 486–489
AlGaN/GaN
 growth of, 29
 triangular quantum wells, 230–241
 two-dimensional electron gas, 231–233
AlInGaP, 432, 438, 440
AlInN, problems with methods used to grow, 146–147
Aluminum nitride (AlN)
 band structure calculations, 372
 crystal structure, 173–174, 176–177
 lattice constants, 173–174
 mechanical and thermal properties of, 173–174
 sputtered thin films, 44–46
 thermal stability of, 79–84
 wet etching of, 104, 105–106

AlN/InN, band discontinuity determination and, 218–221
Arsenide compounds, surface segregation, 129–131
Atomic force microscopy (AFM), 114, 452
Atomic layer epitaxy (ALE)
 growth of InGaN and, 142–143, 145
 growth of InN and, 140–141
 temperature affects on growth and, 132–138
Atomic sphere approximation (ASA), 375
Atoms
 divalent, 265–269
 rare earth, 274
 tetravalent, 269
Auger electron spectroscopy (AES), 113, 122
Augmentation, 375

B

Band diagrams, for MODFETs, 237–239
Band discontinuity/band offsets
 determination, 215
 in Aln/InN, 218–221
 in GaN/AlGaN, 216–217
 in GaN/InN, 218–221
Bandgap
 optical properties of above, 306–311
 refractive index and optical properties of below, 342–345
Bandgap region, exciton effects and near, 311
 absorption coefficient, measuring, 312–315
 intrinsic excitonic structure, 315–325

Bandgap region (*continued*)
  strain effects on excitons, 325–334
Band structure
  confined states, calculating, 199–214
  future outlook for, 404–405
  of GaN, 195–199
  methods, 374–375
  near the edges, 399–403
  photoemission and, 391–394
  relationship between Brillouin zones of wurtzite and zincblende, 379–385
  trends in, in terms of ionicity and cation, 385–391
  ultraviolet photoelectron spectroscopy and, 394–397
  x-ray absorption and (known as NEXAFS or XANES), 397–398
Band structure, calculating
  confined states and, 199–214
  density functional theory (DFT), 372–374
  early empirical studies, 370–372
  GW approximation, 374, 378, 379
  Hartree-Fock approximation, 374, 378
  local density approximation (LDA), 372–374, 378–379
  local density functional, 372–377
  for nitrides, 375–377
Berylium (Be), as a dopant, 269
Binary reactions, free energy of, 67–73
Bipolar-accessed memory cells, 478–481, 484
Bond-orbital approximation, 73
Boron nitride (BN), 168
Bound excitons (BE) in GaN, 349–355
Brillouin zone (BZ), 281, 282, 307–308
  relationship between Brillouin zones of wurtzite and zincblende, 379–385

C

Capacitance-voltage technique, 267
Cathodoluminescence, band discontinuity determination and, 216
Cation, band structure trends and, 385–388
Chemically assisted ion bean etching (CAIBE), 107, 118
Chemical vapor deposition (CVD)
  analysis of GaN thin film growth, 92–98

chlorine transport CVD (CTCVD), 92–95
organometallic (OMCVD), 267
Coincidence site lattice (CSL), 184
Conduction band minimum (CBM), 281, 282, 292
Confined states, calculating, 199–214
Confinement energies, 201, 205, 231
Continuum elasticity theory, 184
  epitaxy of GaN and, 185–190
c-plane sapphire, growth of GaN on, 23–25
Crystal field splitting, 401–402
Crystal structures
  AlN, 173–174, 176–177
  defects, 179–184
  epitaxial growth, 184–190
  GaN, 174–175, 177
  InN, 165–176, 177–178
  phase stability and transitions, 178
  polarity of, 172
  polytypism, 178
  rock-salt, 167, 169
  wurtzite, 167, 168–171
  zinc-blende, 167, 169–171, 176–178
Cubic GaN, optical properties of, 345–348

D

Debye theory, 62
Decomposition reaction, 79–84
Defect-related optical properties, 348
  bound excitons in GaN, 349–355
  donor- or acceptor-related optical spectra, 355–361
  transition metal, 361–362
Defects
  micro-twins, 181, 183, 188
  in nitrides, 181–184
  planar, 181–182
  stacking faults, 181, 182–183, 188
  strain relief and, 182–183
  threading dislocations, 181
Defects and impurities in GaN, high pressure techniques and
  background information, 279–280
  n-type conductivity in undoped GaN, 283–289
  oxygen and silicon impurities, 289–290
  pressure dependence of the electronic states, 281–283

INDEX

yellow luminescence, 280, 291–295
Density functional theory (DFT), 308, 372–374
Disilane, 28
Donor-acceptor pair (DAP), 357–359
Donor bound exciton (DBE), 336
  bound excitons in GaN, 349–355
  related optical spectra, 355–361
Dopants, common
  acceptors, 265–269
  berylium, 269
  divalent atoms, 265–269
  donors, 262–265
  germanium, 265
  magnesium, 267–269
  nitrogen vacancies, 262, 283–284, 285
  oxygen, 262, 280, 284, 289–290
  silicon, 263–265, 280, 284, 289–290
  tetravalent atoms, 269
  zinc, 266–267
Doping
  during growth, 271–272
  of indium-based nitride compounds, 155, 269–271
  p- and n-type GaN, 28
  post-growth (ion implantation), 272–274
  undoped GaN, 260–261
Dry etching, 8
  boiling points, 110, 111
  chemically assisted ion bean etching (CAIBE), 107, 118
  damage levels and selectivity for the various methods, 118–122
  electron cyclotron resonance (ECR), 43, 49–51, 107–111, 114, 118
  inductively coupled plasma (ICP), 107, 109–111, 118
  magnetron enhanced RIE (MERIE), 107, 109, 110
  plasma chemistries for, 110
  rates, 109–111
  reactive ion etching (RIE), 107, 108–111, 118
Dynamic random access memory (DRAM) in GaAs
  basic storage capacitor designs, 475–477
  bipolar-accessed memory cells, 478–481
  FET-accessed memory cells, 477–478
Dyson equation, 373

E

Einstein theory, 61
Electrically-alterable read-only memory (EAROM), 473, 474
Electroluminescence (EL)
  in AlGaN, 295–299
  in GaN, 295–299
  in InGaN, 35, 295–299
  in InGaN/AlGaN, 435, 436
  in InGaN single quantum well, 442, 443
Electron cyclotron resonance (ECR), 43, 49–51, 107–111, 114, 118, 194
Electron-hole exchange interaction, 315
Electron-hole plasma (EHP), 454–455
Electronic defect states, 281–283
Emcore Corp., 20
Energy despersive spectroscopy (EDS), 136, 137, 149
Enthalpy
  of condensed phases, 62–63
  of gases, 63–67
  of ternary nitrides, 73–74
Entropy
  of condensed phases, 62–63
  of gases, 63–67
  of ternary nitrides, 74, 76
Envelope function approximation, 198, 199, 200, 207
Epitaxial growth, 184–190
Etching
  chemically assisted ion bean etching (CAIBE), 107, 118
  damage levels and selectivity for the various methods, 118–122
  dry, 8, 107–117
  electron cyclotron resonance (ECR), 43, 49–51, 107–111, 114, 118
  inductively coupled plasma (ICP), 107, 109–111, 118
  low energy electron enhanced etching (LE4), 8, 118
  magnetron enhanced RIE (MERIE), 107, 109, 110
  reactive ion etching (RIE), 107, 108–111, 118
  wet, 8, 104–106
Evaporation coefficient, 80–81
Exciton densities, high, 339–342
Exciton effects, near bandgap region and, 311

Excitation effects (*continued*)
  absorption coefficient, measuring, 312–315
  intrinsic excitonic structure, 315–325
  strain effects on excitons, 325–334
Exciton recombination rates, 334–339

F

Fabry-Perrot {Perot} cavity, 6, 452
Fick's first law of diffusion, 15
Field-effect transistors (FETs)
  accessed memory cells, 477–478
  triangular quantum well and modulation doped, 239–241
Fourier transformation (FT), 179
Frank-van der Merwe growth mode, 40
Free energy
  of binary reactions, 67–73
  of condensed phases, 62–63
  of gases, 63–67
  of ternary nitrides, 76–79

G

Gallium arsenide (GaAs), dynamic random access memory in
  basic storage capacitor designs, 475–477
  bipolar-accessed memory cells, 478–481
  FET-accessed memory cells, 477–478
Gallium nitride (GaN)
  See also Defects and impurities in GaN, high pressure techniques and; Optical properties of GaN
  band structure of, 195–199
  chemical vapor deposition analysis of thin film growth, 92–98
  coincidence site lattice (CSL) and epitaxy of, 185–190
  compressibility of, 413–415
  crystal structure, 174–175, 177
  defects, 181
  dry etching of, 107–117
  history of research on, 1–9
  hydrogen and, 83
  impurities, 7
  injection laser for, 6
  lattice constants, 175
  luminescence in, 295–299
  mechanical and thermal properties of, 174–175
  molecular beam epitaxy analysis of thin film growth, 84–93
  phonons and phase transitions, 409–426
  sputtered thin films, 46–48
  thermal stability of, 79–84
  undoped, 260–261
  wet etching of, 104–105, 106
GaN/AlGaN
  band discontinuity determination and, 216–217
  calculating confined states, 199–214
  optical properties of, 222–230
GaN/InN, band discontinuity determination and, 218–221
GaP, 432
Generation lifetime, 482
Generation-limited cells, 475–476
Germanium (Ge), as a dopant, 265
Grüneisen parameters, 421
GW approximation, 374, 378, 379, 393

H

Hall measurements, 260, 261, 264, 267
Hall mobility, 50
Hartree-Fock approximation, 374, 378
Heat capacity
  of condensed phases, 62–63
  of gases, 63–67
Hertz-Langmuir equation, 80, 83
Heterobipolar transistor (HBT), advantages of, 7
Heterojunction emitter, gallium nitride and, 7
High pressure techniques. *See* Defects and impurities in GaN, high pressure techniques and
Hollow anode source, 91–92
Hydride vapor phase epitaxy (HVPE), 271–272, 289, 339, 351–352
Hydrogen
  decomposition and, 83
  effects of, on InGaN films, 150–155

I

Indium-based nitride compounds
  doping of, 155

effects of hydrogen on indium incorporation, 150–155
problems with the growth of, 127–129
reaction pathways for, 131–132
surface segregation, 129–131
temperature affects on growth, 132–138
Indium gallium nitride (InGaN)
applications of, 433
based heterostructures, 156–163
doping of, 269–271
effects of hydrogen on films, 150–155
growth of, 29–35
luminescence in, 295–299
methods used to grow, problems with, 141–146
multi-quantum well laser diodes, 448–455
phase separation issue, 153–155
single quantum wells, 439–444
InGaN/AlGaN double-heterostructure LEDs, 434–439
InGaN/GaN
calculating confined states, 199–214
quantum wells, 241–248
InGaN/InGaN
calculating confined states, 201–214
quantum wells, 248–253
Indium nitride (InN)
band structure calculations, 372
crystal structure, 175–176, 177–178
dry etching of, 121
mechanical and thermal properties of, 175–176
methods used to grow, problems with, 138, 140–141
sputtered thin films, 48–51, 140
thermal stability of, 79–84
wet etching of, 105
Inductively coupled plasma (ICP), 107, 109–111, 118
Inert gas diode sputtering, 41–42
Ion beam implantation, 272–274
Ionicity, band structure trends and, 388–391

K

Kinetic barriers, 56–57, 80, 86, 96
Kohn-Luttinger Hamiltonian, 399, 403

Kohn-Sham equation, 373
Kramers-Kronig relations, 306, 308, 344

L

Laser diodes (LDs)
background of research on, 431–434, 459–462
basic structure, 462
carrier and optical confinement, 466–469
conductivity control, 464–466
continuous wave (CW), 461
critical layer thickness, 463–464
current injection, 459–460
InGaN/InGaN quantum wells and, 250–253
InGaN multi quantum well, 448–455
low threshold structure, 470
transparent carrier density, 469–470
Lattice constants
AlN, 173–174
GaN, 175
Lattice stability, GaN, 410–415
Leakage-limited storage cells, 475
Light emitting diodes (LEDs)
antistokes, 2
background of research on, 431–434
displays using, 438–439, 444
emission mechanism of single quantum well, 445–448
grown on sapphire, 265
InGaN/AlGaN double-heterostructure, 434–439
InGaN/GaN quantum wells and, 246–248
InGaN single quantum well, 439–444
M-i-n type, 1–2
MOCVD and, 12
nichia bright, 5
traffic lights, 438, 444
Linear combinations of atomic orbitals (LCAO) approximations, 195–196, 207
band discontinuity determination and, 216
Linearized augmented plane wave (LAPW), 375
Linear muffin tin orbital (LMTO), 371–372, 375, 412

Linear muffin tin orbital (*continued*)
  band discontinuity determination and, 216, 217
  full-potential, 74
Liquid phase epitaxy (LPE), 17, 93
Local density approximation (LDA), 308, 412
  band structure calculations, 372–374, 378, 392–394
Local vibrational modes, 424–426
Low energy electron beam irradiation (LEEBI), 3–4, 28, 267, 465
Low energy electron enhanced etching (LE4), benefits of, 8, 118

M

Magnesium (Mg)
  as a dopant, 267–269
MgO, as substrate material, 22, 23
MgAlO, 432–433
Magnetron enhanced RIE (MERIE), 107, 109, 110
Magnetron sputtering devices, 43
Memory
  devices, 9
  dynamic random access memory (DRAM) in GaAs, 475–481
  nonvolatile random access memory (NVRAM), 474–475, 481–489
  types of, 473–474
Metal organic chemical vapor deposition (MOCVD), 7–8, 40, 85, 94, 194
  applications of, 11–12
  band discontinuity determination and, 215
  doping, 28, 271–272
  growth of AlGaInN and, 149
  growth of AlGaN and AlGaN/GaN and, 29
  growth of GaAlN/GaN and, 214
  growth of GaN and, 26–27
  growth of InGaN and, 29–35, 141–146
  growth of InGaN and InGaN/GaN and, 29–35
  growth of InN and, 140
  growth on c-plane sapphire, 23–25
  growth rate calculations, 16
  reaction chemistry, 13–17
  reactor designs, comparison of, 20
  substrates, 20, 22–25
  system and reactor design issues, 17–19
  techniques, 12–13
  temperature affects on growth and, 132–138
Metal organic molecular beam epitaxy (MOMBE)
  growth of InGaN and, 146
  growth of InN and, 140
Metal organic vapor phase epitaxy (MOVPE), 147, 214, 289
Metastable DX centers, 280
Modulation doped field effect transistors (MODFET), 194
  AlGaN/GaN triangular quantum wells, 230–241
  band diagrams for normally on and off, 237–239
Molecular beam epitaxy (MBE), 40
  analysis of GaN thin film growth, 84–92
  band discontinuity determination and, 215
  criterion for metastable growth, 90
  doping during growth, 271–272
  forward reaction, 86–89
  growth of InN, 140
  kinetic energy from plasma, role of, 90–92
  low energy species, impact on growth, 92
  modified, 194
  reactive ammonia, 194
  reverse reaction, 89
Mole fraction, 29
Multi-quantum wells (MQWs), 6, 33–34, 159
  InGaN laser diodes, 448–455
Murnaghan equation, 413, 414

N

n doping, 28
Nichia Chemistries Inc., 19
Nippon Sanso, 20
Nitrogen vacancies, 262, 283–284, 285
Nonvolatile random access memory (NVRAM)
  in AlGaN, 486–489

development of, 474–475
generation mechanisms, 481–484
memory cell design, 484–485
monolithic demonstration chips, 486
in SiC, 481–486
Nucleation layers, implementation of, 22

O

Optical microscopy, 136
Optical properties, GaN
  above bandgap energy, 306–311
  cubic GaN and, 345–348
  defect-related, 348–363
  exciton effects and the near bandgap region, 311–334
  exciton recombination dynamics, 334–339
  high exciton densities, 339–342
  intrinsic versus extrinsic, 305
  refractive index and below bandgap, 306, 342–345
Optical properties, of quantum well structures, 221–230
Optical Spectrum Analyzer, 450
Optical transitions, in bulk GaN, 195–199
Organometallic vapor phase epitaxy (OMVPE), 460, 465
Orthogonalized plane wave (OPW) calculations, 372
Oxygen, as a dopant, 262, 280, 284, 289–290

P

p doping, 28–33
Perkin-Elmer Fourier spectrometer, 286
Phase diagrams, use of, 59–61
Phase equilibria of ternary nitrides, 73–79
Phase separation issue, InGaN and, 153–155
Phase stability and transitions, in crystal structures, 178
Phonons and phase transitions, 409
  internal stress and, 423
  lattice stability, 410–415
  local vibrational modes, 424–426
  pressure dependence and, 420–421
  temperature dependence and, 421–423
  two-phonon Raman spectra, 424
  zone-center modes at zero pressure, 415–420
Phosphide compounds, surface segregation, 129–131
Photoconductivity, in GaN, 5–6
Photoemission, band structure and, 391–394
Photoluminescence (PL)
  in AlGaN, 295–299
  band discontinuity determination and, 216
  cubic GaN and, 346–348
  exciton recombination rates and use of, 335–339
  exciton structure and use of, 319–321
  in GaN, 295–299
  in InGaN, 31–35, 295–299
  in InGaN single quantum well, 445–448
  optical properties and, 221–230
  time resolved, 33
  of undoped GaN, 261
Photoluminescence excitation (PLE) spectra, 324–325
Piezoelectric effect, 214, 218–221
Plasmon-phonon coupling, 288
Poisson's equation, 231, 233
Polariton-phonon scattering, 322–324
Polarity, of crystal structures, 172
Polytypism, 178
Power-law equation
  for gases, 66
  for solids and liquids, 63
Pressure dependence, phonons and, 420–421

Q

Quantum wells (QWs)
  band discontinuity determination, 215–221
  confined states, calculating, 199–214
  doping and, 6
  emission mechanism of single quantum well LEDs, 445–448
  InGaN/GaN, 241–248
  InGaN/InGaN, 248–253
  InGaN multi, 448–455
  InGaN single, 439–444

Quantum wells (*continued*)
multi, 6, 33–34, 159, 448–455
optical properties of, 221–230
piezoelectric effect, 214, 218–221
single, 31–33, 35, 159, 439–444
triangular, 230–241
Quasi-cubic model, 318

R

Radiative lifetimes, excitons and, 337–339
Raman scattering, 47, 286, 415, 417, 423
two-phonon, 424
Random access memory (RAM), 473
dynamic, in GaAs, 475–481
nonvolatile (NVRAM), 474–475, 481–489
static versus dynamic, 474
Raoultian behavior, 76
Rare earth (RE) atom, 274
Rashba-Sheka-Pikus (RSP) Hamiltonian, 399, 403
Reactive diode sputtering, 42–43
See also Sputtering, reactive
Reactive evaporation, 138, 140
Reactive ion etching (RIE), 107, 108–111, 118, 462
Reactor designs
atmospheric and low pressure, 18–19
closed space rotating disc, 20
high speed rotating disc, 20
horizontal, 18
problems when designing, 19
two-flow horizontal planetary rotation, 20
vertical, 17–18
Read-only memory (ROM), 473, 474
Reflectance measurements, intrinsic excitons and, 315–317
Refractive index
below bandgap optical properties and, 342–345
calculating, 306
laser diode optical confinement and, 468–469
Regular solution model, 73, 79
Rocking curve widths, 47–48, 261
Rock-salt crystal structures, 167, 169, 411–412

S

Scanning electron microscope (SEM), 3, 113, 136
Schrödinger equation, 200, 231
Secondary ion mass spectroscopy (SIMS), 27, 159, 262, 284
Shockley partial dislocations, 183
Silane, 28
Silicon (Si), as a dopant, 263–265, 280, 284, 289–290
Silicon carbide (SiC), 261, 330, 431, 432–433
generation mechanisms, 481–484
memory cell design, 484–485
monolithic demonstration chips, 486
nonvolatile random access memory (NVRAM) in, 481–486
3C-SiC, as substrate material, 22, 23
Single quantum wells (SQWs), 31–33, 35, 159
emission mechanism of, in LEDs, 445–448
InGaN, 439–444
Spectroscopic ellipsometry, 311, 313
Spin-orbit coupling, 195–196, 208–209, 212, 311, 400
Sputtering, reactive
AlN thin films and, 44–46
description of, 40
GaN thin films and, 46–48
inert gas diode, 41–42
InN thin films and, 48–51, 140
magnetron devices, 43
reactive diode, 42–43
Stark effect, quantum confined, 467–468
Stirling's approximation, 79–84
Strain effects, excitons and, 325–334
Stranski-Krastanov growth mode, 40
Stress, phonons and internal, 423
Substrates, 20, 22–25
Subthreshold leakage, 477–478
Superlattice structures (SLSs), 156, 158
optical properties of, 221–230
piezoelectric effect, 214, 218–221
Surface acoustic wave (SAW) sensors, 44–45

T

Temperature dependence, phonons and, 421–423

INDEX

Ternary nitrides, phase equilibria of, 73–79
Thermodynamics
　chemical vapor deposition analysis of GaN thin film growth, 92–98
　decomposition reaction, 79–84
　differences in free energy and, 56, 58
　formalism and, 58–59
　free energy of binary reactions, 67–73
　free energy of condensed phases, 62–63
　free energy of gases, 63–67
　hydrogen, role of, 83
　molecular beam epitaxy analysis of GaN thin film growth, 84–92
　phase diagrams, use of, 59–61
　phase equilibria of ternary nitrides, 73–79
　regular solution model, 73–79
　role of, 56
　thermal stability of GaN, InN, and AlN, 79–84
Thomas Swan Ltd., 20
Three dimensional island growth mode, 40
Transition metal defects, 361–362
Transmission electron microscopy (TEM), 24, 25, 45, 155, 172
　cross section (XTEM), 160
　high resolution (HRTEM), 179, 186–187, 188
　of InGaN multi quantum well, 455
Transparent carrier density, 469–470
Triangular quantum wells, 230–241
Trimethylaluminum (TMAl), 12, 17, 29
Trimethylgallium (TMGa), 12, 14, 29
Trimethylindium (TMIn), 12
Two-dimensional electron gas, 231–237
Two dimensional layer-by-layer growth mode, 40

U

UV injection laser, 5
Ultraviolet photoelectron spectroscopy (UPS), 393
　band discontinuity determination and, 216
　band structure and, 394–397

V

Valence band maximum (VBM), 281, 282, 292

Valence-band splittings, 400
Valence force field (VFF) approach, 73–74, 153
Van Hove singularities, 308
Volmer-Weber growth mode, 40

W

Wet etching, 8, 104–106
Wurtzite (Wz) crystal structures, 167, 168–171
　relationship between Brillouin zones of zincblende and, 379–385
　stability of, 410–413

X

X-ray absorption near edge fine structure (known as NEXAFS or XANES), 397–398
X-ray absorption spectroscopy (XAS), 411
X-ray diffraction (XRD), 45, 47, 136, 137, 411
　compressibility of GaN and, 413–415
　growth of AlGaInN and, 149
　growth of InGaN and, 141, 142
　growth of InN and, 141
X-ray photoelectron spectroscopy (XPS), 199, 393
　band discontinuity determination and, 216, 218

Y

Yellow luminescence, 280, 291–295, 361

Z

Zinc (Zn), as a dopant, 1, 266–267
Zinc-blende (Zb) crystal structures, 167, 169–171, 176–178
　relationship between Brillouin zones of wurtzite and zinc-blende, 379–385
ZnO, as substrate material, 22, 23
Zone-center modes at zero pressure, 415–420

Contents of Volumes in This Series

Volume 1 Physics of III–V Compounds

C. Hilsum, Some Key Features of III–V Compounds
Franco Bassani, Methods of Band Calculations Applicable to III–V Compounds
E. O. Kane, The k-p Method
V. L. Bonch-Bruevich, Effect of Heavy Doping on the Semiconductor Band Structure
Donald Long, Energy Band Structures of Mixed Crystals of III–V Compounds
Laura M. Roth and Petros N. Argyres, Magnetic Quantum Effects
S. M. Puri and T. H. Geballe, Thermomagnetic Effects in the Quantum Region
W. M. Becker, Band Characteristics near Principal Minima from Magnetoresistance
E. H. Putley, Freeze-Out Effects, Hot Electron Effects, and Submillimeter Photoconductivity in InSb
H. Weiss, Magnetoresistance
Betsy Ancker-Johnson, Plasma in Semiconductors and Semimetals

Volume 2 Physics of III–V Compounds

M. G. Holland, Thermal Conductivity
S. I. Novkova, Thermal Expansion
U. Piesbergen, Heat Capacity and Debye Temperatures
G. Giesecke, Lattice Constants
J. R. Drabble, Elastic Properties
A. U. Mac Rae and G. W. Gobeli, Low Energy Electron Diffraction Studies
Robert Lee Mieher, Nuclear Magnetic Resonance
Bernard Goldstein, Electron Paramagnetic Resonance
T. S. Moss, Photoconduction in III–V Compounds
E. Antoncik ad J. Tauc, Quantum Efficiency of the Internal Photoelectric Effect in InSb
G. W. Gobeli and I. G. Allen, Photoelectric Threshold and Work Function
P. S. Pershan, Nonlinear Optics in III–V Compounds
M. Gershenzon, Radiative Recombination in the III–V Compounds
Frank Stern, Stimulated Emission in Semiconductors

Volume 3 Optical of Properties III–V Compounds

Marvin Hass, Lattice Reflection
William G. Spitzer, Multiphonon Lattice Absorption
D. L. Stierwalt and R. F. Potter, Emittance Studies
H. R. Philipp and H. Ehrenveich, Ultraviolet Optical Properties
Manuel Cardona, Optical Absorption above the Fundamental Edge
Earnest J. Johnson, Absorption near the Fundamental Edge
John O. Dimmock, Introduction to the Theory of Exciton States in Semiconductors
B. Lax and J. G. Mavroides, Interband Magnetooptical Effects
H. Y. Fan, Effects of Free Carries on Optical Properties
Edward D. Palik and George B. Wright, Free-Carrier Magnetooptical Effects
Richard H. Bube, Photoelectronic Analysis
B. O. Seraphin and H. E. Bennett, Optical Constants

Volume 4 Physics of III–V Compounds

N. A. Goryunova, A. S. Borschevskii, and D. N. Tretiakov, Hardness
N. N. Sirota, Heats of Formation and Temperatures and Heats of Fusion of Compounds $A^{III}B^{V}$
Don L. Kendall, Diffusion
A. G. Chynoweth, Charge Multiplication Phenomena
Robert W. Keyes, The Effects of Hydrostatic Pressure on the Properties of III–V Semiconductors
L. W. Aukerman, Radiation Effects
N. A. Goryunova, F. P. Kesamanly, and D. N. Nasledov, Phenomena in Solid Solutions
R. T. Bate, Electrical Properties of Nonuniform Crystals

Volume 5 Infrared Detectors

Henry Levinstein, Characterization of Infrared Detectors
Paul W. Kruse, Indium Antimonide Photoconductive and Photoelectromagnetic Detectors
M. B. Prince, Narrowband Self-Filtering Detectors
Ivars Melngalis and T. C. Harman, Single-Crystal Lead-Tin Chalcogenides
Donald Long and Joseph L. Schmidt, Mercury-Cadmium Telluride and Closely Related Alloys
E. H. Putley, The Pyroelectric Detector
Norman B. Stevens, Radiation Thermopiles
R. J. Keyes and T. M. Quist, Low Level Coherent and Incoherent Detection in the Infrared
M. C. Teich, Coherent Detection in the Infrared
F. R. Arams, E. W. Sard, B. J. Peyton, and F. P. Pace, Infrared Heterodyne Detection with Gigahertz IF Response
H. S. Sommers, Jr., Macrowave-Based Photoconductive Detector
Robert Sehr and Rainer Zuleeg, Imaging and Display

Volume 6 Injection Phenomena

Murray A. Lampert and Ronald B. Schilling, Current Injection in Solids: The Regional Approximation Method
Richard Williams, Injection by Internal Photoemission
Allen M. Barnett, Current Filament Formation

R. Baron and J. W. Mayer, Double Injection in Semiconductors
W. Ruppel, The Photoconductor-Metal Contact

Volume 7 Application and Devices
Part A

John A. Copeland and Stephen Knight, Applications Utilizing Bulk Negative Resistance
F. A. Padovani, The Voltage-Current Characteristics of Metal-Semiconductor Contacts
P. L. Hower, W. W. Hooper, B. R. Cairns, R. D. Fairman, and D. A. Tremere, The GaAs Field-Effect Transistor
Marvin H. White, MOS Transistors
G. R. Antell, Gallium Arsenide Transistors
T. L. Tansley, Heterojunction Properties

Part B

T. Misawa, IMPATT Diodes
H. C. Okean, Tunnel Diodes
Robert B. Campbell and Hung-Chi Chang, Silicon Junction Carbide Devices
R. E. Enstrom, H. Kressel, and L. Krassner, High-Temperature Power Rectifiers of $GaAs_{1-x}P_x$

Volume 8 Transport and Optical Phenomena

Richard J. Stirn, Band Structure and Galvanomagnetic Effects in III–V Compounds with Indirect Band Gaps
Roland W. Ure, Jr., Thermoelectric Effects in III–V Compounds
Herbert Piller, Faraday Rotation
H. Barry Bebb and E. W. Williams, Photoluminescence I: Theory
E. W. Williams and H. Barry Bebb, Photoluminescence II: Gallium Arsenide

Volume 9 Modulation Techniques

B. O. Seraphin, Electroreflectance
R. L. Aggarwal, Modulated Interband Magnetooptics
Daniel F. Blossey and Paul Handler, Electroabsorption
Bruno Batz, Thermal and Wavelength Modulation Spectroscopy
Ivar Balslev, Piezopptical Effects
D. E. Aspnes and N. Bottka, Electric-Field Effects on the Dielectric Function of Semiconductors and Insulators

Volume 10 Transport Phenomena

R. L. Rhode, Low-Field Electron Transport
J. D. Wiley, Mobility of Holes in III–V Compounds
C. M. Wolfe and G. E. Stillman, Apparent Mobility Enhancement in Inhomogeneous Crystals
Robert L. Petersen, The Magnetophonon Effect

Volume 11 Solar Cells

Harold J. Hovel, Introduction; Carrier Collection, Spectral Response, and Photocurrent; Solar Cell Electrical Characteristics; Efficiency; Thickness; Other Solar Cell Devices; Radiation Effects; Temperature and Intensity; Solar Cell Technology

Volume 12 Infrared Detectors (II)

W. L. Eiseman, J. D. Merriam, and R. F. Potter, Operational Characteristics of Infrared Photodetectors
Peter R. Bratt, Impurity Germanium and Silicon Infrared Detectors
E. H. Putley, InSb Submillimeter Photoconductive Detectors
G. E. Stillman, C. M. Wolfe, and J. O. Dimmock, Far-Infrared Photoconductivity in High Purity GaAs
G. E. Stillman and C. M. Wolfe, Avalanche Photodiodes
P. L. Richards, The Josephson Junction as a Detector of Microwave and Far-Infrared Radiation
E. H. Putley, The Pyroelectric Detector—An Update

Volume 13 Cadmium Telluride

Kenneth Zanio, Materials Preparations; Physics; Defects; Applications

Volume 14 Lasers, Junctions, Transport

N. Holonyak, Jr. and M. H. Lee, Photopumped III–V Semiconductor Lasers
Henry Kressel and Jerome K. Butler, Heterojunction Laser Diodes
A Van der Ziel, Space-Charge-Limited Solid-State Diodes
Peter J. Price, Monte Carlo Calculation of Electron Transport in Solids

Volume 15 Contacts, Junctions, Emitters

B. L. Sharma, Ohmic Contacts to III–V Compounds Semiconductors
Allen Nussbaum, The Theory of Semiconducting Junctions
John S. Escher, NEA Semiconductor Photoemitters

Volume 16 Defects, (HgCd)Se, (HgCd)Te

Henry Kressel, The Effect of Crystal Defects on Optoelectronic Devices
C. R. Whitsett, J. G. Broerman, and C. J. Summers, Crystal Growth and Properties of $Hg_{1-x}Cd_xSe$ alloys
M. H. Weiler, Magnetooptical Properties of $Hg_{1-x}Cd_xTe$ Alloys
Paul W. Kruse and John G. Ready, Nonlinear Optical Effects in $Hg_{1-x}Cd_xTe$

Volume 17 CW Processing of Silicon and Other Semiconductors

James F. Gibbons, Beam Processing of Silicon
Arto Lietoila, Richard B. Gold, James F. Gibbons, and Lee A. Christel, Temperature Distribu-

tions and Solid Phase Reaction Rates Produced by Scanning CW Beams
Arto Leitoila and James F. Gibbons, Applications of CW Beam Processing to Ion Implanted Crystalline Silicon
N. M. Johnson, Electronic Defects in CW Transient Thermal Processed Silicon
K. F. Lee, T. J. Stultz, and James F. Gibbons, Beam Recrystallized Polycrystalline Silicon: Properties, Applications, and Techniques
T. Shibata, A. Wakita, T. W. Sigmon, and James F. Gibbons, Metal-Silicon Reactions and Silicide
Yves I. Nissim and James F. Gibbons, CW Beam Processing of Gallium Arsenide

Volume 18 Mercury Cadmium Telluride

Paul W. Kruse, The Emergence of $(Hg_{1-x}Cd_x)Te$ as a Modern Infrared Sensitive Material
H. E. Hirsch, S. C. Liang, and A. G. White, Preparation of High-Purity Cadmium, Mercury, and Tellurium
W. F. H. Micklethwaite, The Crystal Growth of Cadmium Mercury Telluride
Paul E. Petersen, Auger Recombination in Mercury Cadmium Telluride
R. M. Broudy and V. J. Mazurczyck, (HgCd)Te Photoconductive Detectors
M. B. Reine, A. K. Soad, and T. J. Tredwell, Photovoltaic Infrared Detectors
M. A. Kinch, Metal-Insulator-Semiconductor Infrared Detectors

Volume 19 Deep Levels, GaAs, Alloys, Photochemistry

G. F. Neumark and K. Kosai, Deep Levels in Wide Band-Gap III–V Semiconductors
David C. Look, The Electrical and Photoelectronic Properties of Semi-Insulating GaAs
R. F. Brebrick, Ching-Hua Su, and Pok-Kai Liao, Associated Solution Model for Ga-In-Sb and Hg-Cd-Te
Yu.Ya. Gurevich and Yu. V. Pleskon, Photoelectrochemistry of Semiconductors

Volume 20 Semi-Insulating GaAs

R. N. Thomas, H. M. Hobgood, G. W. Eldridge, D. L. Barrett, T. T. Braggins, L. B. Ta, and S. K. Wang, High-Purity LEC Growth and Direct Implantation of GaAs for Monolithic Microwave Circuits
C. A. Stolte, Ion Implantation and Materials for GaAs Integrated Circuits
C. G. Kirkpatrick, R. T. Chen, D. E. Holmes, P. M. Asbeck, K. R. Elliott, R. D. Fairman, and J. R. Oliver, LEC GaAs for Integrated Circuit Applications
J. S. Blakemore and S. Rahimi, Models for Mid-Gap Centers in Gallium Arsenide

Volume 21 Hydrogenated Amorphous Silicon
Part A

Jacques I. Pankove, Introduction
Masataka Hirose, Glow Discharge; Chemical Vapor Deposition
Yoshiyuki Uchida, di Glow Discharge
T. D. Moustakas, Sputtering
Isao Yamada, Ionized-Cluster Beam Deposition
Bruce A. Scott, Homogeneous Chemical Vapor Deposition

Frank J. Kampas, Chemical Reactions in Plasma Deposition
Paul A. Longeway, Plasma Kinetics
Herbert A. Weakliem, Diagnostics of Silane Glow Discharges Using Probes and Mass Spectroscopy
Lester Gluttman, Relation between the Atomic and the Electronic Structures
A. Chenevas-Paule, Experiment Determination of Structure
S. Minomura, Pressure Effects on the Local Atomic Structure
David Adler, Defects and Density of Localized States

Part B

Jacques I. Pankove, Introduction
G. D. Cody, The Optical Absorption Edge of a-Si:H
Nabil M. Amer and Warren B. Jackson, Optical Properties of Defect States in a-Si:H
P. J. Zanzucchi, The Vibrational Spectra of a-Si:H
Yoshihiro Hamakawa, Electroreflectance and Electroabsorption
Jeffrey S. Lannin, Raman Scattering of Amorphous Si, Ge, and Their Alloys
R. A. Street, Luminescence in a-Si:H
Richard S. Crandall, Photoconductivity
J. Tauc, Time-Resolved Spectroscopy of Electronic Relaxation Processes
P. E. Vanier, IR-Induced Quenching and Enhancement of Photoconductivity and Photoluminescence
H. Schade, Irradiation-Induced Metastable Effects
L. Ley, Photoelectron Emission Studies

Part C

Jacques I. Pankove, Introduction
J. David Cohen, Density of States from Junction Measurements in Hydrogenated Amorphous Silicon
P. C. Taylor, Magnetic Resonance Measurements in a-Si:H
K. Morigaki, Optically Detected Magnetic Resonance
J. Dresner, Carrier Mobility in a-Si:H
T. Tiedje, Information about band-Tail States from Time-of-Flight Experiments
Arnold R. Moore, Diffusion Length in Undoped a-Si:H
W. Beyer and J. Overhof, Doping Effects in a-Si:H
H. Fritzche, Electronic Properties of Surfaces in a-Si:H
C. R. Wronski, The Staebler-Wronski Effect
R. J. Nemanich, Schottky Barriers on a-Si:H
B. Abeles and T. Tiedje, Amorphous Semiconductor Superlattices

Part D

Jacques I. Pankove, Introduction
D. E. Carlson, Solar Cells
G. A. Swartz, Closed-Form Solution of I–V Characteristic for a a-Si:H Solar Cells
Isamu Shimizu, Electrophotography
Sachio Ishioka, Image Pickup Tubes

P. G. LeComber and W. E. Spear, The Development of the a-Si:H Field-Effect Transistor and Its Possible Applications
D. G. Ast, a-Si:H FET-Addressed LCD Panel
S. Kaneko, Solid-State Image Sensor
Masakiyo Matsumura, Charge-Coupled Devices
M. A. Bosch, Optical Recording
A. D'Amico and G. Fortunato, Ambient Sensors
Hiroshi Kukimoto, Amorphous Light-Emitting Devices
Robert J. Phelan, Jr., Fast Detectors and Modulators
Jacques I. Pankove, Hybrid Structures
P. G. LeComber, A. E. Owen, W. E. Spear, J. Hajto, and W. K. Choi, Electronic Switching in Amorphous Silicon Junction Devices

Volume 22 Lightwave Communications Technology
Part A

Kazuo Nakajima, The Liquid-Phase Epitaxial Growth of IngaAsp
W. T. Tsang, Molecular Beam Epitaxy for III–V Compound Semiconductors
G. B. Stringfellow, Organometallic Vapor-Phase Epitaxial Growth of III–V Semiconductors
G. Beuchet, Halide and Chloride Transport Vapor-Phase Deposition of InGaAsP and GaAs
Manijeh Razeghi, Low-Pressure Metallo-Organic Chemical Vapor Deposition of $Ga_x In_{1-x} As P_{1-y}$ Alloys
P. M. Petroff, Defects in III–V Compound Semiconductors

Part B

J. P. van der Ziel, Mode Locking of Semiconductor Lasers
Kam Y. Lau and Ammon Yariv, High-Frequency Current Modulation of Semiconductor Injection Lasers
Charles H. Henry, Special Properties of Semiconductor Lasers
Yasuharu Suematsu, Katsumi Kishino, Shigehisa Arai, and Fumio Koyama. Dynamic Single-Mode Semiconductor Lasers with a Distributed Reflector
W. T. Tsang, The Cleaved-Coupled-Cavity (C^3) Laser

Part C

R. J. Nelson and N. K. Dutta, Review of InGaAsP InP Laser Structures and Comparison of Their Performance
N. Chinone and M. Nakamura, Mode-Stabilized Semiconductor Lasers for 0.7–0.8- and 1.1–1.6-μm Regions
Yoshiji Horikoshi, Semiconductor Lasers with Wavelengths Exceeding 2 μm
B. A. Dean and M. Dixon, The Functional Reliability of Semiconductor Lasers as Optical Transmitters
R. H. Saul, T. P. Lee, and C. A. Burus, Light-Emitting Device Design
C. L. Zipfel, Light-Emitting Diode-Reliability
Tien Pei Lee and Tingye Li, LED-Based Multimode Lightwave Systems
Kinichiro Ogawa, Semiconductor Noise-Mode Partition Noise

Part D

Federico Capasso, The Physics of Avalanche Photodiodes
T. P. Pearsall and M. A. Pollack, Compound Semiconductor Photodiodes
Takao Kaneda, Silicon and Germanium Avalanche Photodiodes
S. R. Forrest, Sensitivity of Avalanche Photodetector Receivers for High-Bit-Rate Long-Wavelength Optical Communication Systems
J. C. Campbell, Phototransistors for Lightwave Communications

Part E

Shyh Wang, Principles and Characteristics of Integrable Active and Passive Optical Devices
Shlomo Margalit and Amnon Yariv, Integrated Electronic and Photonic Devices
Takaoki Mukai, Yoshihisa Yamamoto, and Tatsuya Kimura, Optical Amplification by Semiconductor Lasers

Volume 23 Pulsed Laser Processing of Semiconductors

R. F. Wood, C. W. White, and R. T. Young, Laser Processing of Semiconductors: An Overview
C. W. White, Segregation, Solute Trapping, and Supersaturated Alloys
G. E. Jellison, Jr., Optical and Electrical Properties of Pulsed Laser-Annealed Silicon
R. F. Wood and G. E. Jellison, Jr., Melting Model of Pulsed Laser Processing
R. F. Wood and F. W. Young, Jr., Nonequilibrium Solidification Following Pulsed Laser Melting
D. H. Lowndes and G. E. Jellison, Jr., Time-Resolved Measurement During Pulsed Laser Irradiation of Silicon
D. M. Zebner, Surface Studies of Pulsed Laser Irradiated Semiconductors
D. H. Lowndes, Pulsed Beam Processing of Gallium Arsenide
R. B. James, Pulsed CO_2 Laser Annealing of Semiconductors
R. T. Young and R. F. Wood, Applications of Pulsed Laser Processing

Volume 24 Applications of Multiquantum Wells, Selective Doping, and Superlattices

C. Weisbuch, Fundamental Properties of III–V Semiconductor Two-Dimensional Quantized Structures: The Basis for Optical and Electronic Device Applications
H. Morkoc and H. Unlu, Factors Affecting the Performance of (Al,Ga)As/GaAs and (Al,Ga)As/InGaAs Modulation-Doped Field-Effect Transistors: Microwave and Digital Applications
N. T. Linh, Two-Dimensional Electron Gas FETs: Microwave Applications
M. Abe et al., Ultra-High-Speed HEMT Integrated Circuits
D. S. Chemla, D. A. B. Miller, and P. W. Smith, Nonlinear Optical Properties of Multiple Quantum Well Structures for Optical Signal Processing
F. Capasso, Graded-Gap and Superlattice Devices by Band-Gap Engineering
W. T. Tsang, Quantum Confinement Heterostructure Semiconductor Lasers
G. C. Osbourn et al., Principles and Applications of Semiconductor Strained-Layer Superlattices

Volume 25 Diluted Magnetic Semiconductors

W. Giriat and J. K. Furdyna, Crystal Structure, Composition, and Materials Preparation of Diluted Magnetic Semiconductors

W. M. Becker, Band Structure and Optical Properties of Wide-Gap $A^{II}_{1-x}Mn_xB^{IV}$ Alloys at Zero Magnetic Field

Saul Oseroff and Pieter H. Keesom, Magnetic Properties: Macroscopic Studies

Giebultowicz and T. M. Holden, Neutron Scattering Studies of the Magnetic Structure and Dynamics of Diluted Magnetic Semiconductors

J. Kossut, Band Structure and Quantum Transport Phenomena in Narrow-Gap Diluted Magnetic Semiconductors

C. Riquaux, Magnetooptical Properties of Large-Gap Diluted Magnetic Semiconductors

J. A. Gaj, Magnetooptical Properties of Large-Gap Diluted Magnetic Semiconductors

J. Mycielski, Shallow Acceptors in Diluted Magnetic Semiconductors: Splitting, Boil-off, Giant Negative Magnetoresistance

A. K. Ramadas and R. Rodriquez, Raman Scattering in Diluted Magnetic Semiconductors

P. A. Wolff, Theory of Bound Magnetic Polarons in Semimagnetic Semiconductors

Volume 26 III–V Compound Semiconductors and Semiconductor Properties of Superionic Materials

Zou Yuanxi, III–V Compounds

H. V. Winston, A. T. Hunter, H. Kimura, and R. E. Lee, InAs-Alloyed GaAs Substrates for Direct Implantation

P. K. Bhattachary and S. Dhar, Deep Levels in III–V Compound Semiconductors Grown by MBE

Yu. Yu. Gurevich and A. K. Ivanov-Shits, Semiconductor Properties of Supersonic Materials

Volume 27 High Conducting Quasi-One-Dimensional Organic Crystals

E. M. Conwell, Introduction to Highly Conducting Quasi-One-Dimensional Organic Crystals

I. A. Howard, A Reference Guide to the Conducting Quasi-One-Dimensional Organic Molecular Crystals

J. P. Pouquet, Structural Instabilities

E. M. Conwell, Transport Properties

C. S. Jacobsen, Optical Properties

J. C. Scott, Magnetic Properties

L. Zuppiroli, Irradiation Effects: Perfect Crystals and Real Crystals

Volume 28 Measurement of High-Speed Signals in Solid State Devices

J. Frey and D. Ioannou, Materials and Devices for High-Speed and Optoelectronic Applications

H. Schumacher and E. Strid, Electronic Wafer Probing Techniques

D. H. Auston, Picosecond Photoconductivity: High-Speed Measurements of Devices and Materials

J. A. Valdmanis, Electro-Optic Measurement Techniques for Picosecond Materials, Devices, and Integrated Circuits

J. M. Wiesenfeld and R. K. Jain, Direct Optical Probing of Integrated Circuits and High-Speed Devices

G. Plows, Electron-Beam Probing

A. M. Weiner and R. B. Marcus, Photoemissive Probing

Volume 29 Very High Speed Integrated Circuits: Gallium Arsenide LSI

M. Kuzuhara and T. Nazaki, Active Layer Formation by Ion Implantation
H. Hasimoto, Focused Ion Beam Implantation Technology
T. Nozaki and A. Higashisaka, Device Fabrication Process Technology
M. Ino and T. Takada, GaAs LSI Circuit Design
M. Hirayama, M. Ohmori, and K. Yamasaki, GaAs LSI Fabrication and Performance

Volume 30 Very High Speed Integrated Circuits: Heterostructure

H. Watanabe, T. Mizutani, and A. Usui, Fundamentals of Epitaxial Growth and Atomic Layer Epitaxy
S. Hiyamizu, Characteristics of Two-Dimensional Electron Gas in III–V Compound Heterostructures Grown by MBE
T. Nakanisi, Metalorganic Vapor Phase Epitaxy for High-Quality Active Layers
T. Nimura, High Electron Mobility Transistor and LSI Applications
T. Sugeta and T. Ishibashi, Hetero-Bipolar Transistor and LSI Application
H. Matsueda, T. Tanaka, and M. Nakamura, Optoelectronic Integrated Circuits

Volume 31 Indium Phosphide: Crystal Growth and Characterization

J. P. Farges, Growth of Discoloration-free InP
M. J. McCollum and G. E. Stillman, High Purity InP Grown by Hydride Vapor Phase Epitaxy
T. Inada and T. Fukuda, Direct Synthesis and Growth of Indium Phosphide by the Liquid Phosphorous Encapsulated Czochralski Method
O. Oda, K. Katagiri, K. Shinohara, S. Katsura, Y. Takahashi, K. Kainosho, K. Kohiro, and R. Hirano, InP Crystal Growth, Substrate Preparation and Evaluation
K. Tada, M. Tatsumi, M. Morioka, T. Araki, and T. Kawase, InP Substrates: Production and Quality Control
M. Razeghi, LP-MOCVD Growth, Characterization, and Application of InP Material
T. A. Kennedy and P. J. Lin-Chung, Stoichiometric Defects in InP

Volme 32 Strained-Layer Superlattices: Physics

T. P. Pearsall, Strained-Layer Superlattices
Fred H. Pollack, Effects of Homogeneous Strain on the Electronic and Vibrational Levels in Semiconductors
J. Y. Marzin, J. M. Gerárd, P. Voisin, and J. A. Brum, Optical Studies of Strained III–V Heterolayers
R. People and S. A. Jackson, Structurally Induced States from Strain and Confinement
M. Jaros, Microscopic Phenomena in Ordered Suprlattices

Volume 33 Strained-Layer Superlattices: Materials Science and Technology

R. Hull and J. C. Bean, Principles and Concepts of Strained-Layer Epitaxy
William J. Schaff, Paul J. Tasker, Marc C. Foisy, and Lester F. Eastman, Device Applications of Strained-Layer Epitaxy

S. T. Picraux, B. L. Doyle, and J. Y. Tsao, Structure and Characterization of Strained-Layer Superlattices
E. Kasper and F. Schaffer, Group IV Compounds
Dale L. Martin, Molecular Beam Epitaxy of IV–VI Compounds Heterojunction
Robert L. Gunshor, Leslie A. Kolodziejski, Arto V. Nurmikko, and Nobuo Otsuka, Molecular Beam Epitaxy of II–VI Semiconductor Microstructures

Volume 34 Hydrogen in Semiconductors

J. I. Pankove and N. M. Johnson, Introduction to Hydrogen in Semiconductors
C. H. Seager, Hydrogenation Methods
J. I. Pankove, Hydrogenation of Defects in Crystalline Silicon
J. W. Corbett, P. Deák, U. V. Desnica, and S. J. Pearton, Hydrogen Passivation of Damage Centers in Semiconductors
S. J. Pearton, Neutralization of Deep Levels in Silicon
J. I. Pankove, Neutralization of Shallow Acceptors in Silicon
N. M. Johnson, Neutralization of Donor Dopants and Formation of Hydrogen-Induced Defects in *n*-Type Silicon
M. Stavola and S. J. Pearton, Vibrational Spectroscopy of Hydrogen-Related Defects in Silicon
A. D. Marwick, Hydrogen in Semiconductors: Ion Beam Techniques
C. Herring and N. M. Johnson, Hydrogen Migration and Solubility in Silicon
E. E. Haller, Hydrogen-Related Phenomena in Crystalline Germanium
J. Kakalios, Hydrogen Diffusion in Amorphous Silicon
J. Chevalier, B. Clerjaud, and B. Pajot, Neutralization of Defects and Dopants in III–V Semiconductors
G. G. DeLeo and W. B. Fowler, Computational Studies of Hydrogen-Containing Complexes in Semiconductors
R. F. Kiefl and T. L. Estle, Muonium in Semiconductors
C. G. Van de Walle, Theory of Isolated Interstitial Hydrogen and Muonium in Crystalline Semiconductors

Volume 35 Nanostructured Systems

Mark Reed, Introduction
H. van Houten, C. W. J. Beenakker, and B. J. van Wees, Quantum Point Contacts
G. Timp, When Does a Wire Become an Electron Waveguide?
M. Büttiker, The Quantum Hall Effects in Open Conductors
W. Hansen, J. P. Kotthaus, and U. Merkt, Electrons in Laterally Periodic Nanostructures

Volume 36 The Spectroscopy of Semiconductors

D. Heiman, Spectroscopy of Semiconductors at Low Temperatures and High Magnetic Fields
Arto V. Nurmikko, Transient Spectroscopy by Ultrashort Laser Pulse Techniques
A. K. Ramdas and S. Rodriguez, Piezospectroscopy of Semiconductors
Orest J. Glembocki and Benjamin V. Shanabrook, Photoreflectance Spectroscopy of Microstructures
David G. Seiler, Christopher L. Littler, and Margaret H. Wiler, One- and Two-Photon Magneto-Optical Spectroscopy of InSb and $Hg_{1-x}Cd_xTe$

Volume 37 The Mechanical Properties of Semiconductors

A.-B. Chen, Arden Sher and W. T. Yost, Elastic Constants and Related Properties of Semiconductor Compounds and Their Alloys
David R. Clarke, Fracture of Silicon and Other Semiconductors
Hans Siethoff, The Plasticity of Elemental and Compound Semiconductors
Sivaraman Guruswamy, Katherine T. Faber and John P. Hirth, Mechanical Behavior of Compound Semiconductors
Subhanh Mahajan, Deformation Behavior of Compound Semiconductors
John P. Hirth, Injection of Dislocations into Strained Multilayer Structures
Don Kendall, Charles B. Fleddermann, and Kevin J. Malloy, Critical Technologies for the Micromachining of Silicon
Ikuo Matsuba and Kinji Mokuya, Processing and Semiconductor Thermoelastic Behavior

Volume 38 Imperfections in III/V Materials

Udo Scherz and Matthias Scheffler, Density-Functional Theory of sp-Bonded Defects in III/V Semiconductors
Maria Kaminska and Eicke R. Weber, El2 Defect in GaAs
David C. Look, Defects Relevant for Compensation in Semi-Insulating GaAs
R. C. Newman, Local Vibrational Mode Spectroscopy of Defects in III/V Compounds
Andrzej M. Hennel, Transition Metals in III/V Compounds
Kevin J. Malloy and Ken Khachaturyan, DX and Related Defects in Semiconductors
V. Swaminathan and Andrew S. Jordan, Dislocations in III/V Compounds
Krzysztof W. Nauka, Deep Level Defects in the Epitaxial III/V Materials

Volume 39 Minority Carriers in III–V Semiconductors: Physics and Applications

Niloy K. Dutta, Radiative Transitions in GaAs and Other III–V Compounds
Richard K. Ahrenkiel, Minority-Carrier Lifetime in III–V Semiconductors
Tomofumi Furuta, High Field Minority Electron Transport in p-GaAs
Mark S. Lundstrom, Minority-Carrier Transport in III–V Semiconductors
Richard A. Abram, Effects of Heavy Doping and High Excitation on the Band Structure of GaAs
David Yevick and Witold Bardyszewski, An Introduction to Non-Equilibrium Many-Body Analyses of Optical Processes in III–V Semiconductors

Volume 40 Epitaxial Microstructures

E. F. Schubert, Delta-Doping of Semiconductors: Electronic, Optical, and Structural Properties of Materials and Devices
A. Gossard, M. Sundaram, and P. Hopkins, Wide Graded Potential Wells
P. Petroff, Direct Growth of Nanometer-Size Quantum Wire Superlattices
E. Kapon, Lateral Patterning of Quantum Well Heterostructures by Growth of Nonplanar Substrates
H. Temkin, D. Gershoni, and M. Panish, Optical Properties of Ga$_{1-x}$In$_x$As/InP Quantum Wells

Volume 41 High Speed Heterostructure Devices

F. Capasso, F. Beltram, S. Sen, A. Pahlevi, and A. Y. Cho, Quantum Electron Devices: Physics and Applications
P. Solomon, D. J. Frank, S. L. Wright, and F. Canora, GaAs-Gate Semiconductor–Insulator–Semiconductor FET
M. H. Hashemi and U. K. Mishra, Unipolar InP-Based Transistors
R. Kiehl, Complementary Heterostructure FET Integrated Circuits
T. Ishibashi, GaAs-Based and InP-Based Heterostructure Bipolar Transistors
H. C. Liu and T. C. L. G. Sollner, High-Frequency-Tunneling Devices
H. Ohnishi, T. More, M. Takatsu, K. Imamura, and N. Yokoyama, Resonant-Tunneling Hot-Electron Transistors and Circuits

Volume 42 Oxygen in Silicon

F. Shimura, Introduction to Oxygen in Silicon
W. Lin, The Incorporation of Oxygen into Silicon Crystals
T. J. Schaffner and D. K. Schroder, Characterization Techniques for Oxygen in Silicon
W. M. Bullis, Oxygen Concentration Measurement
S. M. Hu, Intrinsic Point Defects in Silicon
B. Pajot, Some Atomic Configurations of Oxygen
J. Michel and L. C. Kimerling, Electical Properties of Oxygen in Silicon
R. C. Newman and R. Jones, Diffusion of Oxygen in Silicon
T. Y. Tan and W. J. Taylor, Mechanisms of Oxygen Precipitation: Some Quantitative Aspects
M. Schrems, Simulation of Oxygen Precipitation
K. Simino and I. Yonenaga, Oxygen Effect on Mechanical Properties
W. Bergholz, Grown-in and Process-Induced Effects
F. Shimura, Intrinsic/Internal Gettering
H. Tsuya, Oxygen Effect on Electronic Device Performance

Volume 43 Semiconductors for Room Temperature Nuclear Detector Applications

R. B. James and T. E. Schlesinger, Introduction and Overview
L. S. Darken and C. E. Cox, High-Purity Germanium Detectors
A. Burger, D. Nason, L. Van den Berg, and M. Schieber, Growth of Mercuric Iodide
X. J. Bao, T. E. Schlesinger, and R. B. James, Electrical Properties of Mercuric Iodide
X. J. Bao, R. B. James, and T. E. Schlesinger, Optical Properties of Red Mercuric Iodide
M. Hage-Ali and P. Siffert, Growth Methods of CdTe Nuclear Detector Materials
M. Hage-Ali and P Siffert, Characterization of CdTe Nuclear Detector Materials
M. Hage-Ali and P. Siffert, CdTe Nuclear Detectors and Applications
R. B. James, T. E. Schlesinger, J. Lund, and M. Schieber, $Cd_{1-x}Zn_xTe$ Spectrometers for Gamma and X-Ray Applications
D. S. McGregor, J. E. Kammeraad, Gallium Arsenide Radiation Detectors and Spectrometers
J. C. Lund, F. Olschner, and A. Burger, Lead Iodide
M. R. Squillante, and K. S. Shah, Other Materials: Status and Prospects
V. M. Gerrish, Characterization and Quantification of Detector Performance
J. S. Iwanczyk and B. E. Patt, Electronics for X-ray and Gamma Ray Spectrometers
M. Schieber, R. B. James, and T. E. Schlesinger, Summary and Remaining Issues for Room Temperature Radiation Spectrometers

Volume 44 II–IV Blue/Green Light Emitters: Device Physics and Epitaxial Growth

J. Han and R. L. Gunshor, MBE Growth and Electrical Properties of Wide Bandgap ZnSe-based II–VI Semiconductors

Shizuo Fujita and Shigeo Fujita, Growth and Characterization of ZnSe-based II–VI Semiconductors by MOVPE

Easen Ho and Leslie A. Kolodziejski, Gaseous Source UHV Epitaxy Technologies for Wide Bandgap II–VI Semiconductors

Chris G. Van de Walle, Doping of Wide-Band-Gap II–VI Compounds — Theory

Roberto Cingolani, Optical Properties of Excitons in ZnSe-Based Quantum Well Heterostructures

A. Ishibashi and A. V. Nurmikko, II–VI Diode Lasers: A Current View of Device Performance and Issues

Supratik Guha and John Petruzello, Defects and Degradation in Wide-Gap II–VI-based Structures and Light Emitting Devices

Volume 45 Effect of Disorder and Defects in Ion-Implanted Semiconductors: Electrical and Physiochemical Characterization

Heiner Ryssel, Ion Implantation into Semiconductors: Historical Perspectives

You-Nian Wang and Teng-Cai Ma, Electronic Stopping Power for Energetic Ions in Solids

Sachiko T. Nakagawa, Solid Effect on the Electronic Stopping of Crystalline Target and Application to Range Estimation

G. Müller, S. Kalbitzer and G. N. Greaves, Ion Beams in Amorphous Semiconductor Research

Jumana Boussey-Said, Sheet and Spreading Resistance Analysis of Ion Implanted and Annealed Semiconductors

M. L. Polignano and G. Queirolo, Studies of the Stripping Hall Effect in Ion-Implanted Silicon

J. Stoemenos, Transmission Electron Microscopy Analyses

Roberta Nipoti and Marco Servidori, Rutherford Backscattering Studies of Ion Implanted Semiconductors

P. Zaumseil, X-ray Diffraction Techniques

Volume 46 Effect of Disorder and Defects in Ion-Implanted Semiconductors: Optical and Photothermal Characterization

M. Fried, T. Lohner and J. Gyulai, Ellipsometric Analysis

Antonios Seas and Constantinos Christofides, Transmission and Reflection Spectroscopy on Ion Implanted Semiconductors

Andreas Othonos and Constantinos Christofides, Photoluminescence and Raman Scattering of Ion Implanted Semiconductors. Influence of Annealing

Constantinos Christofides, Photomodulated Thermoreflectance Investigation of Implanted Wafers. Annealing Kinetics of Defects

U. Zammit, Photothermal Deflection Spectroscopy Characterization of Ion-Implanted and Annealed Silicon Films

Andreas Mandelis, Arief Budiman and Miguel Vargas, Photothermal Deep-Level Transient Spectroscopy of Impurities and Defects in Semiconductors

R. Kalish and S. Charbonneau, Ion Implantation into Quantum-Well Structures

Alexandre M. Myasnikov and Nikolay N. Gerasimenko, Ion Implantation and Thermal Annealing of III-V Compound Semiconducting Systems: Some Problems of III-V Narrow Gap Semiconductors

Volume 47 Uncooled Infrared Imaging Arrays and Systems

R. G. Buser and M. P. Tompsett, Historical Overview
P. W. Kruse, Principles of Uncooled Infrared Focal Plane Arrays
R. A. Wood, Monolithic Silicon Microbolometer Arrays
C. M. Hanson, Hybrid Pyroelectric-Ferroelectric Bolometer Arrays
D. L. Polla and J. R. Choi, Monolithic Pyroelectric Bolometer Arrays
N. Teranishi, Thermoelectric Uncooled Infrared Focal Plane Arrays
M. F. Tompsett, Pyroelectric Vidicon
T. W. Kenny, Tunneling Infrared Sensors
J. R. Vig, R. L. Filler and Y. Kim, Application of Quartz Microresonators to Uncooled Infrared Imaging Arrays
P. W. Kruse, Application of Uncooled Monolithic Thermoelectric Linear Arrays to Imaging Radiometers

Volume 48 High Brightness Light Emitting Diodes

G. B. Stringfellow, Materials Issues in High-Brightness Light-Emitting Diodes
M. G. Craford, Overview of Device issues in High-Brightness Light-Emitting Diodes
F. M. Steranka, AlGaAs Red Light Emitting Diodes
C. H. Chen, S. A. Stockman, M. J. Peanasky, and C. P. Kuo, OMVPE Growth of AlGaInP for High Efficiency Visible Light-Emitting Diodes
F. A. Kish and R. M. Fletcher, AlGaInP Light-Emitting Diodes
M. W. Hodapp, Applications for High Brightness Light-Emitting Diodes
I. Akasaki and H. Amano, Organometallic Vapor Epitaxy of GaN for High Brightness Blue Light Emitting Diodes
S. Nakamura, Group III-V Nitride Based Ultraviolet-Blue-Green-Yellow Light-Emitting Diodes and Laser Diodes

Volume 49 Light Emission in Silicon: from Physics to Devices

David J. Lockwood, Light Emission in Silicon
Gerhard Abstreiter, Band Gaps and Light Emission in Si/SiGe Atomic Layer Structures
Thomas G. Brown and Dennis G. Hall, Radiative Isoelectronic Impurities in Silicon and Silicon-Germanium Alloys and Superlattices
J. Michel, L. V. C. Assali, M. T. Morse, and L. C. Kimerling, Erbium in Silicon
Yoshihiko Kanemitsu, Silicon and Germanium Nanoparticles
Philippe M. Fauchet, Porous Silicon: Photoluminescence and Electroluminescent Devices
C. Delerue, G. Allan, and M. Lannoo, Theory of Radiative and Nonradiative Processes in Silicon Nanocrystallites
Louis Brus, Silicon Polymers and Nanocrystals

ISBN 0-12-752158-5